CALCIUM PHOSPHATES IN BIOLOGICAL AND INDUSTRIAL SYSTEMS

CALCIUM PHOSPHATES IN BIOLOGICAL AND INDUSTRIAL SYSTEMS

edited by

Zahid Amjad Ph.D.
The BFGoodrich Company

KLUWER ACADEMIC PUBLISHERS
Boston / Dordrecht / London

Distributors for North America:
Kluwer Academic Publishers
101 Philip Drive
Assinippi Park
Norwell, Massachusetts 02061 USA

Distributors for all other countries:
Kluwer Academic Publishers Group
Distribution Centre
Post Office Box 322
3300 AH Dordrecht, THE NETHERLANDS

Library of Congress Cataloging-in-Publication Data

A C.I.P. Catalogue record for this book is available
from the Library of Congress.

Printed on acid-free paper.

To my devoted family,

Rukhsana and Naureen,

with all my love

CONTENTS

CONTRIBUTORS

John Afflitto, Colgate-Palmolive Company, 909 River Road, P. O. Box 1343, Piscataway, New Jersey 08855-1343 USA

Zahid Amjad, The Advanced Technology Group, The BFGoodrich Company, 9921 Brecksville Road, Brecksville, Ohio 44141 USA

O.S.L. Bruinsma, Delft University of Technology, Laboratory for Process Equipment, Leeghwaterstraat 44, 2628, CA Delft, The Netherlands

Laurence C. Chow, American Dental Association Health Foundation, Paffenbarger Research Center, National Institute of Science and Technology, Gaithersburg, Maryland 20899 USA

Georges Daufin, National Institute for Agriculture, INRA, Dairy Technology Research Laboratory, LRTL, 65 rue de Saint-Brieuc, 35042 Rennes Cedex, France

Edward D. Eanes, Mineral Chemistry and Structure Section, National Institute of Dental Research, National Institutes of Health, Building 30, Room 106, MSC-4320, Bethesda, Maryland 20892 USA

Lynn M. Frostman, Baker Petrolite, 1600 Industrial Boulevard, Sugarland, Texas 77448 USA

Abdul Gaffar, Colgate-Palmolive Company, 909 River Road, P. O. Box 1343, Piscataway, New Jersey 08855-1343 USA

Jasbir S. Gill, Calgon Corporation, P. O. Box 1346, Pittsburgh, Pennsylvania 15230 USA

Koji Ioku, Department of Advanced Materials Science and Engineering, Faculty of Engineering, Yamaguchi University, 2557 Tokiwadai, Ube-Shi, Yamaguchi 755 Japan

Kazuaki Ito, Department of Environmental Science, Faculty of Engineering, Hiroshima University, 1-4-1 Kagamiyama, Higashi-Hiroshima 739, Japan

Amy T. Kan, Rice University, Environmental Science and Engineering MS 317, 600 S. Main Street, Houston, Texas 77005-1892 USA

Mansoor A. Khan, College of Pharmacy and Health Sciences, Northeast Louisiana University, Monroe, Louisiana 71209 USA

Saeed R. Khan, Department of Pathology, College of Medicine, P. O. Box J-275, JHMHC, University of Florida, Gainesville, Florida 32610-0275 USA

Petros G. Koutsoukos, Institute of Chemical Engineering and High Temperature Chemical Processes, P. O. Box 1414, Department of Chemical Engineering, University of Patras, GR-26500 Patras, Greece

Jean-Pierre Labbé, National School of Chemistry of Paris, 11 rue Pierre et Marie Curie, 75231 Paris Cedex 05, France

Berislav Markovic, Langmuir Center for Colloid and Interfaces, School of Engineering and Applied Sciences, Columbia University, New York, New York 10027 USA

Milenko Markovic, American Dental Association Health Foundation, Paffenbarger Research Center, National Institute of Science and Technology, Gaithersburg, Maryland 20899 USA

Mitsumasa Okada, Department of Environmental Science, Faculty of Engineering, Hiroshima University, 1-4-1 Kagamiyama, Higashi-Hiroshima 739, Japan

Libardo A. Perez, BetzDearborn Water Management Group, I Quality Way, Trevose, Pennsylvania 19053 USA

Kenneth P. H. Pritzker, Pathology and Laboratory Medicine, Mount Sinai Hospital, University of Toronto, 600 University Avenue, Toronto, Ontario, Canada M5G 1X5

Indra K. Reddy, College of Pharmacy and Health Sciences, Northeast University, Monroe, Louisiana 71209 USA

Christian Rey, Laboratoire des Matéreriaux, Physico-chimie des Phosphates, Ecole Nationale Superieure de Chimie, CNRS 5071, 38 rue des 36 Ponts - 31400 Toulouse, France

G.M. van Rosmalen, Delft university of Technology, Laboratory for Process Equipment, Leeghwaterstraat 44, 2628 CA Delft, The Netherlands

John D. Sallis, University of Tasmania, Department of Biochemistry, P. O. Box 252-58 GPO Hobart, Tasmania, Australia 7001

M.M. Seckler, IPT Institute for Technological Research of the State of Säo Paulo, Säo Paulo, S. P., Brazil

Saburo Shimabayashi, Faculty of Pharmaceutical Sciences, The University of Tokushima, Sho-machi 1-78-1, Tokushima 770, Japan

P. Somasundaran, Langmuir Center for Colloids and Interfaces, School of Engineering and Applied Science, Columbia University, New York, New York 10027 USA

Mason B. Tomson, Rice University, Department of Environmental Science and Engineering - MS 317, 6100 S. Main Street, Houston, Texas 77005-1892 USA

Ming S. Tung, American Dental Association Health Foundation, Paffenbarger Research Center, National Institute of Science and Technology, Gaithersburg, Maryland 20899 USA

Tadayuki Uno, Faculty of Pharmaceutical Sciences, The University of Tokushima, Tokushima, Sho-machi 1-78-1, Tokushima 770, Japan

James P. Yesinowski, Naval Research Laboratory, Washington DC 20375-5342 USA

ABOUT THE EDITOR

Zahid Amjad was born in Lahore, Pakistan; received his M.Sc. from Punjab University (Lahore) and Ph.D. from Glasgow University, Glasgow, Scotland. He is presently a Research Fellow in the Advanced Technology Group of The BFGoodrich Company in Brecksville, Ohio where he has served since 1982. Dr. Amjad was Lecturer at the Institute of Chemistry of Punjab University, and was Assistant Research Professor at the State University of New York at Buffalo (1977-79) before joining the Calgon Corporation. His areas of research include: the adsorption of polyelectrolytes at solid-liquid interface, crystal growth and inhibition, development and applications of water soluble/swellable polymers, and the control and removal of foulants from water treatment systems.

Dr. Amjad has authored or coauthored more than 90 technical publications and is a holder of 29 patents. He is also the editor of two books: Reverse Osmosis: Membrane Technology, Water Chemistry, and Industrial Applications (Van Nostrand Reinhold, New York, 1993) and Mineral Scale Formation and Inhibition (Plenum Press, New York, 1995). He has been inducted into the National Hall of Corporate Inventors and is listed in American Men and Women of Sciences, Who's Who in Technology, and Who's Who of American Inventors. Dr. Amjad is a recipient of 1997 EDI Innovation Award. He is a member of several professional organizations.

PREFACE

Research on calcium phosphates has attracted considerable attention in recent years. Calcium phosphates are the major inorganic constituents of biological hard tissues such as teeth and bones and other pathological minerals. Calcium phosphates have been widely produced in industry, in such forms as, ceramics, nutrient supplements, medicines, dentifrices, and stabilizers for plastics. They are utilized in solid state chemistry as fluorescent lamp phosphors and play a role in waste water treatment processes. Calcium phosphates are important in industrial water treatment (i.e., boiler and cooling) where deposition of these salts on heat exchanger surfaces can lead to loss of system efficiency, overheating, unscheduled shutdown time, and ultimately heat exchanger failures. Effective control of calcium phosphate deposits continue to challenge the industrial scientist. Calcium phosphate deposits have also been encountered during pasteurization of milk.

Calcium phosphates of interest include: dicalcium phosphate dihydrate (DCPD, $CaHPO_4{\cdot}2H_2O$), brushite; dicalcium phosphate anhydrous (DCP, $CaHPO_4$), monetite; octacalcium phosphate (OCP, $Ca_8H_2(PO_4)_6{\cdot}5H_2O$); tricalcium phosphate (TCP, $Ca_3(PO_4)_2$), whitlockite; and hydroxyapatite (HAP, $Ca_5(PO_4)_3OH$). Other calcium compounds of biological and industrial interest include: amorphous calcium phosphate (ACP); fluorapatite (FAP, $Ca_5(PO_4)_3F$); calcium pyrophosphate dihydrate (CPPD, $Ca_2P_2O_7{\cdot}2H_2O$); calcium fluoride; and calcium phosphonates.

This book is intended to provide a comprehensive discussion on calcium phosphates in the diverse areas of their investigations. The authors were selected because they are the specialists in their particular fields, possessing wide knowledge and experience and and able to analyze recent results and relate them to their respective areas of expertise. New information, as well as review of current concepts, generally highlights the individual contributions. Due to the broad scope of the subject covered and the large number of contributions, this book is divided into three parts. While each section contains a basic theme, there is a considerable overlapping of ideas and approaches. This reflects the considerable excitement and interdisciplinary nature of investigations by researchers interested in dissimilar aspects of calcium phosphates.

The book begins with an introduction to calcium phosphates in chapter 1 including a brief review of structural and solubility aspects of calcium phosphates. Chapter 2 deals with amorphous calcium phosphate, first addressing the thermodynamics and kinetics, followed by its role in the earlier stages of calcification. Chapter 3 covers the growth of calcium phosphates from epitaxial perspective. In chapter 4, physiochemical properties of fluorapatite is presented and compared with hydroxyapatite. The influence of various ions on the interfacial properties of synthetic and natural hydroxyapatite is discussed in chapter 5. In chapter 6, various NMR techniques for the studies of calcium phosphate surfaces are reviewed.

The focus of the second section is the mechanistic aspects of precipitation and inhibition of calcium phosphates and calcium pyrophosphate, especially in biological systems. Also included in this section is a discussion of the medical and industrial applications of calcium phosphates. Chapters 7-9 present discussion of the inhibition of calcium phosphates by metal ions, polymeric, and non-polymeric materials. Chapter 10 outlines the methods for the preparation of calcium phosphates and their unique properties as biomaterials for medical applications. Chapter 11 discusses the importance of calcium phosphates in urinary stones. In chapter 12, the crystal growth and dissolution of calcium pyrophosphate is reviewed. Chapter 13 and 14 cover the role of calcium phosphates in pharmaceutical and dentifrice applications. The final chapter of this section presents discussion on the medical and industrial applications of calcium phosphate fibers.

The final section concentrates on calcium phosphate scales commonly encountered in industrial systems and presents discussion on the removal of phosphate from waste water. Chapters 16-19 cover calcium phosphate scaling problem in cooling, boiler, and dairy industries. Included also in these chapters is the discussion of the important role played by polymeric inhibitors to control the deposition of calcium phosphates on heat exchanger and membrane surfaces. Phosphate removal methods for treating waste water as well as various phosphorous species present in sediments and their release control are reviewed in chapter 21. A discussion of the solution chemistry of various calcium phosphonates concludes the book.

Considering the general interest in calcium phosphates, this publication is directed to an audience of researchers in the fields of biology, chemistry, dentistry, geology, chemical engineering, environmental engineering, and medicine. It will also be useful to technology-focused researchers in industry whose investigations might be related directly or indirectly to calcium phosphates. It is my hope that this book will contribute to the better understanding of the role calcium phosphates play in a number of interesting fields including, but not limited to, biological, environmental, industrial, and medical applications.

I wish to thank all of the contributors and their institutions for making this book possible. I sincerely appreciate the support of The BFGoodrich Company, especially Dr. Victoria F. Haynes, for allowing me the opportunity to edit this book. Thanks are also extended to the editorial staff of the Kluwer Publishing Company for their patience and professionalism. Finally, a special thanks to my wife, Rukhsana, for her support and great patience during the completion of this book.

August, 1997 Zahid Amjad

1

CALCIUM PHOSPHATES: STRUCTURE, COMPOSITION, SOLUBILITY, AND STABILITY

Ming S. Tung

American Dental Association Health Foundation, Paffenbarger Research Center
National Institute of Standards and Technology, Gaithersburg, MD 20899 USA

INTRODUCTION

Calcium phosphates are important materials in the fields of biology, geology, industry, medicine, and dentistry. Their formation, functions, and applications depend on their structure, composition, solubility, and stability. These related basic properties have been extensively studied; the literature is vast and interdisciplinary.[1-3] An overview of these physicochemical properties is presented in this chapter with the compound of greatest interest and importance, hydroxyapatite (OHAp), discussed in more detail; a standard reference material of OHAp[4] is used as an example in discussion.

In the ternary system $Ca(OH)_2$-H_3PO_4-H_2O, there are eleven known calcium phosphates, with calcium-to-phosphate (Ca/P) molar ratios varying from 0.5 to 2: monocalcium phosphate monohydrate (MCPM), monocalcium phosphate anhydrous (MCPA), dicalcium phosphate dihydrate (DCPD), dicalcium phosphate anhydrous (DCPA), octacalcium phosphate (OCP), β-tricalcium phosphate (β-

TCP), two forms of α-tricalcium phosphate (α-TCP), amorphous calcium phosphate (ACP), hydroxyapatite (OHAp), and tetracalcium phosphate (TTCP). With the substitution of halide for the hydroxide ion, fluorapatite (FAp) or chlorapatite (ClAp) is obtained. Whitlockite can form in the presence of magnesium ions, and carbonate apatite (CO_3Ap) in the presence of carbonate. The calcium phosphates are listed in Table 1 showing their chemical formulas with their abbreviations, crystal space groups and lattice parameters, and solubility product.

Table 1. Abbreviations, formulas, crystal data and solubility products of calcium phosphates.

Abbreviations Names Formulas	Ca/P	Space Groups and lattice parameters (Å and degrees)	log K_{sp} at 25 °C
MCPM: monocalcium phosphate monohydrate $Ca(H_2PO_4)_2{\bullet}H_2O$	0.50	Triclinic[5,6,8]; P1 a=5.6261(5) =98.633(6) b=11.889(2) =118.262(6) c= 6.4731(8) =83.344(6)	highly soluble
MCPA: monocalcium phosphate anhydrous $Ca(H_2PO_4)_2$	0.50	Triclinic[7]; P1 a=7.5577(5) =109.87(1) b=8.2531(6) =93.68(1) c=5.5504(3) =109.15(1)	highly soluble
DCPD: dicalcium phosphate dihydrate $CaHPO_4.{\bullet}2H_2O$	1.00	Monoclinic[9,10]; Ia a=5.812(2) =116.42(2) b=15.180(3) c=6.239(2)	6.59[56]
DCPA: dicalcium phosphate anhydrous $CaHPO_4$	1.00	Triclinic1[1,12]; P1 a=6.910(1) =96.34(2) b=6.627(2) =103.82(2) c=6.998(2) =88.32(2)	6.90[57]
OCP: octacalcium phosphate $Ca_8H_2(PO_4)_6.{\bullet}5H_2O$	1.33	Triclinic[16]; P1 a=19.692(4) =90.15(2) b=9.523(2) =92.54(2) c=6.835(2) =108.65(2)	96.6[58]
α-TCP: α-tricalcium phosphate α- $Ca_3(PO_4)_2$	1.50	Monoclinic[13]; P21/a a=12.887(2) =126.20(1) b=27.280(4) c=15.219(2)	25.5[68]
β-TCP: β-tricalcium phosphate β- $Ca_3(PO_4)_2$	1.50	Rhombohedral[14]; R3c (hexagonal setting) a=b=10.439(1) c=37.375(6)	28.9[59]
whitlockite (mineral) $Ca_{18}(Mg,Fe)_2H_2(PO_4)_{14}$ $Ca_{18}(Mg,Fe)_2(Ca,)(PO_4)_{14}$	1.29 1.36	Rhombohedral; R3c (hexagonal setting) a=b=10.350(5) c=37.085(11)	
OHAp: hydroxyapatite $Ca_5(PO_4)_3OH$	1.67	Hexagonal[4,18,19]; P63/m a=b=9.4206(10) c=6.8844(9)	58.4[62]

Table 1. (continued)

Abbreviations Names Formulas	Ca/P	Space Groups and lattice parameters (Å and degrees)	log K_{sp} at 25 °C
FAp: fluorapatite $Ca_5(PO_4)_3F$	1.67	Hexagonal[20]; P63/m a=b=9.367(1) c=6.884(1)	60.5[76]
ClAp: chlorapatite $Ca_5(PO_4)_3Cl$	1.67	Monoclinic[21]; P21/b a=9.628(5) r=120o b=2a c=6.764(5)	
CO_3Ap: carbonated apatite A-CO_3Ap $Ca_{10}(PO_4)_6CO_3$ B-CO_3Ap $Ca_{10}(PO_4)_{6-x}(CO_3)_{3x/2}(OH)_2$		For A-CO_3Ap Monoclinic[22]; P6 a=9.557(3) b=2a c=6.87 γ =120.36(4)	
TTCP: tetracalcium phosphate $Ca_4(PO_4)_2O$	2.00	Monoclinic[23, 24]; P21 a=7.023(1) =90.90(1) b=11.986(4) c=9.473(2)	38[68]

STRUCTURE

The structures of calcium phosphates are mainly determined by single crystal x-ray diffraction (XRD).[5-24] Some structural studies also have been done with neutron diffraction[8,10,12] to determine the positions of hydrogen ions in the lattice. For crystals that are impossible to prepare as large single-crystals (e.g., synthetic hydroxyapatite), the structural parameters have been determined by the Rieveld method.[26] Infrared,[27-30] Raman,[31] and NMR spectroscopy[33,34] also have been used to study the local chemical and structural environments within the lattice of calcium phosphates.[19,20] The radial distribution function, calculated from XRD data,[35,36] and the extended x-ray absorption fine structure (EXAFS)[37,38] have been used to determine the atomic distances in ACP, which does not have long-range atomic order. Three important aspects of calcium phosphate structures are discussed below.

Acidic Calcium Phosphates: MCPM and DCPD

These two compounds are members of a group of compounds containing a "corrugated sheet" structure.[39] MCPM has two molecules in one unit cell. The sheets in the MCPM structure are comprised of $-CaH_2PO_4^-$ chains in which Ca shares two edges of tetrahedral PO_4 in the form of (HO)OPO(OH) = Ca = (OH)OPO(OH) = Ca. As shown in Figure 1, these chains are shaded on the b plane, alternating above and below the midplane in parallel arrangement. Water and $H_2PO_4^-$ occupy the space between the corrugated sheets.[5-8] DCPD has four molecules in one unit cell. Their chains are $CaHPO_4$, and only water molecules occupy the space between the corrugated sheets.[9,10]

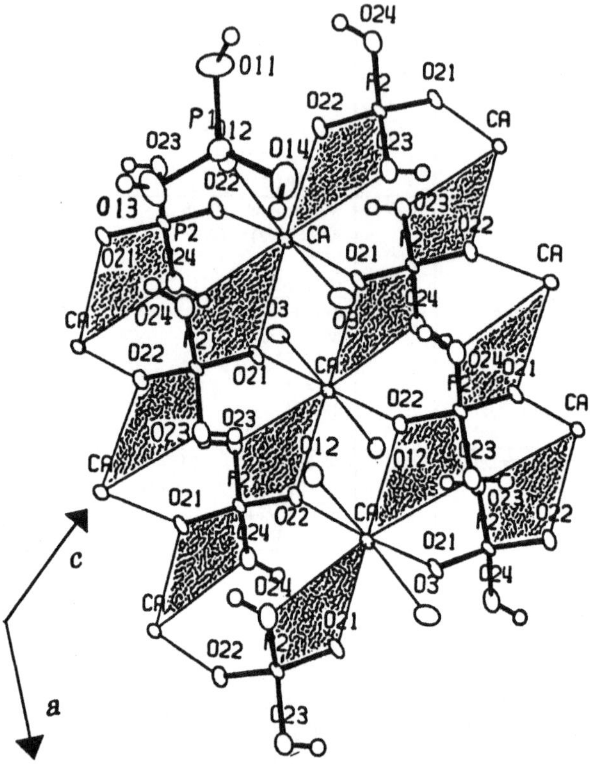

Figure 1. The structure of MCPM showing corrugated sheets of composition CaH_2PO_4 on the b plane. The chains are shaded and arranged in a parallel fashion to (100). The H_2PO_4 ion, P1, and the H_2O molecule, O3, occupy the space between the corrugated sheets.[6]

Whitlockite and Anhydrous Tricalcium Phosphates

Whitlockite occurs widely in biological systems and pathological calcifications, in lunar materials, and in meteoric samples. It can be prepared in aqueous systems[40] in the presence of Mg^{2+}, Fe^{2+}, and Mn^{2+}. Since the x-ray powder patterns of whitlockite and β-TCP are not easily distinguished, the two names have been used interchangeably and synonymously. Since the subtle structural differences between these two compounds have now been elucidated,[17] attention should be paid to the proper use of these two names. It was proposed[2] that the term "whitlockite" be used for the mineral or similar synthetic materials in which Mg^{2+} and HPO_4^{2-} ions play a structural role, and β-TCP for one of the polymorphs of $Ca_3(PO_4)_2$ in which Mg^{2+} ions are not present in the structure.

Around 1125 °C, β-TCP converts to α-TCP.[41] At higher temperatures (above 1430 °C), a new phase, super-α-TCP (α-TCP), forms; this phase is unable to withstand quenching to room temperature.[42] A high temperature phase diagram has been constructed.[43]

Apatitic Structures

OHAp and OCP

As is the case for whitlockite and β-TCP, OHAp and OCP also have similar x-ray powder diffraction patterns and crystal structures.[15,16] Their crystal structures are shown in Figure 2 projected down the c axes; Figure 2a shows two unit cells of OHAp[18,19] and Figure 2b shows one unit cell of OCP from +0.5 to -0.5 along the a axis.[16] The cell corners are marked. The OCP unit cell has a d100 spacing of 18.68Å, as compared to 8.16Å for OHAp. Consequently, a unit cell of OCP is approximately equivalent to two unit cells of OHAp. The hexagon of calcium ions, Ca2, shown in Figure 2a is actually made up of two triangles of calcium ions, one located at the plane up 1/4 the length of the unit cell along the c axis and the other 3/4 up the length. Two of the three calcium ions in each triangle are retained in OCP. Column calcium ions are shown as Ca1. Similarly, there are two triangles of phosphate ions in OHAp, and only two phosphate ions from the triangle are retained in OCP. The portion of the structure of OHAp shown with shaded atoms (Figure 2a) corresponds very closely to the shaded atoms of OCP (Figure 2b).

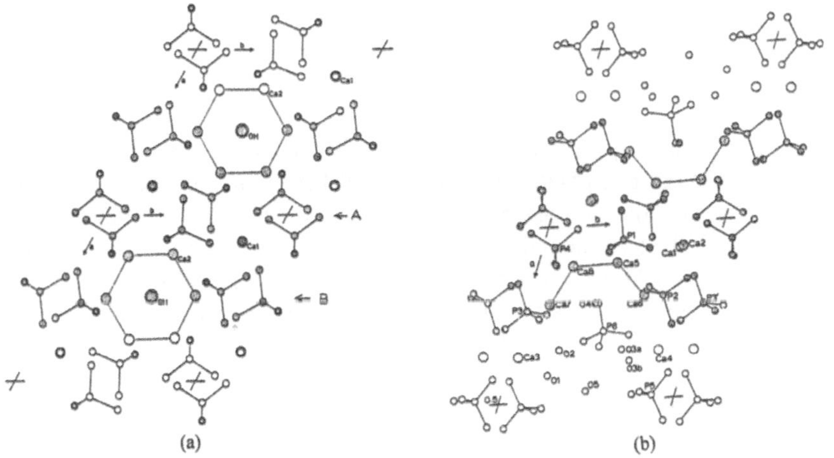

Figure 2. The structures of (a) OHAp and (b) OCP. The arrows indicate the cell edges.

A layer in TTCP also contains much of this portion of the structure. This shaded portion is called the apatitic layer (AL) and is believed to be relatively stable. It plays a role in the stability of OCP and in the formation of OHAp as will be discussed in the STABILITY section. An interlaying structure of OCP and OHAp with a unit cell ratio of 1 to 2 and a Ca/P ratio of 1.5 has been proposed for the often encountered tricalcium phosphate hydrate with an apatitic structure; the apatite with this structure is called "sesquiapatite."[15]

OHAp can be either hexagonal or monoclinic in structure.[18,19] The hexagonal and monoclinic structures are very similar; the main difference is an ordered arrangement of the OH ions above and below the calcium triangle in the monoclinic form and a disordered arrangement in the hexagonal form. This small difference is sufficient to affect some properties quite strongly, e.g., deuteration kinetics, and dissolution kinetics.[26]

Substituted Apatites

All three ion species (Ca^{2+}, PO_4^{3+}, OH^-) in the structure of apatite can be substituted by various other ions in varying amounts;[1] the replacement of Ca^{2+} by monovalent ions, Na^+, K^+, Li^+, divanlent ions, Sr^{2+}, Ba^{2+}, Pb^{2+}, Mn^{2+}, Sn^{2+}, Zn^{2+}, and polyvalent ions, Al^{3+}; the replacement of PO_4^{3-} by HPO_4^{2-}, CO_3^{2-}, SO_4^{2-}, MnO_4^-, VO_4^{2-}, and BO_3^{3-}; the replacement of OH^- by F^-, Cl^- and CO_3^{2-}. Only three important substituted apatites are discussed here: FAp,[20] ClAp,[21] and CO_3Ap.[22] In the case of CO_3Ap, there are two types of carbonate apatites, A and B.[1] The designations of A and B types are based on IR absorption spectra of carbonate ions and their lattice parameters.[1,2,25] In FAp, ClAp, and A-CO_3Ap, the OH ions are replaced; the substitution ranges from none to complete replacement. F- is located in the middle of the Ca-triangle plane, whereas as Cl^- and CO_3^{2-} are located between the Ca-triangles. Fully substituted A-CO_3Ap and ClAp both undergo monoclinic to hexagonal thermal phase transition at around 200 °C.[44] (There is evidence that this transition in ClAp occurs at ~320 °C).[45]

Type B carbonate apatites are prepared by either direct precipitation or hydrolysis of other calcium phosphates such as ACP, DCPD, DCPA, or OCP in carbonate-containing solutions and are believed to have the phosphate substituted with the carbonate.[25] The maximum incorporation of CO_3 in B-CO_3Ap is about 3 molecules of CO_3/cell.[1] A-CO_3Ap can be prepared by solid state reactions at high temperature (1000 °C). The bioapatites contain both A and B types carbonate apatites.[1]

COMPOSITION

The determination of the composition of calcium orthophosphate salts primarily involves chemical analysis of their Ca and P content and, in the case of hydrated compounds, water content. Calcium is usually determined by atomic absorption spectroscopy; phosphorus is usually determined colorimetrically as the phosphovanadomolybdate complex.[46,47] For the nonstoichiometric apatite, further determination of hydroxide content is needed.[49]

Stoichiometry of OHAp

Apatites and ACP are notorious for their occurrence in nonstoichiometric forms (i.e., various Ca/P ratios) and for the incorporation of other ions. The other calcium phosphates are much less likely to be so. The nonstoichiometric apatites, usually with Ca/P ratios lower than 1.67, are compounds with large variations in Ca/P ratios that otherwise appear to be OHAp in structure, as determined by x-ray powder diffraction. Proposals made to account for the nonstoichimetry of apatite have been reviewed[48,49] and include lattice substitutions, presence of calcium vacancies,[49] interlayering of OHAp and OCP,[50,51] undetected phases, surface adsorption, and surface layers.[52,53] It is unlikely that only one mechanism is involved in the nonstoichiometry of these apatites. Also, their properties vary with their Ca/P molar ratios. Therefore, it is important to determine the exact Ca/P molar ratios of these apatites. A thermal product analysis[71,78] offers a more accurate determination of Ca/P ratios than does the chemical method. The former method is illustrated for Ca-deficient OHAp below. This method can also be used in determining the Ca/P ratios of other calcium phosphates.

The thermal reaction of calcium-deficient OHAp with the general formula $Ca_{10-x}\ H_{2x-y}\ (PO_4)_6\ (OH)_{2-y}$, which decomposes into OHAp, TCP and H_2O, can be described by the equation:[71]

$$Ca_{10-x}\ H_{2x-y}\ (PO_4)_6\ (OH)_{2-y} \rightarrow (1-x)Ca_{10}(PO_4)_6\ (OH)_2 + 3xCa_3\ (PO_4)_2 + (2x-y)H_2O \qquad (1)$$

By using this equation, the Ca/P molar ratio can be calculated from the value of the experimentally determined mass fraction of TCP, wTCP, in the thermal product mixture:

$$\text{Ca/P molar ratio} = (10 - x)/6 \qquad (2)$$

where

$$x = 1.0796\ w_{TCP} / (1 + 0.0796\ w_{TCP}) \qquad (3)$$

A high-purity OHAp powder has been synthesized at the National Institute of Standards & Technology and will be issued as Standard Reference Material (SRM 2910) for OHAp.[4] This OHAp is stoichiometric within experimental error and is composed of a mass fraction of the hexagonal form (space groups P63/m)[19] of about 0.7 and a mass fraction of the monoclinic form (space group P21/b)[18] of about 0.3. The Ca/P molar ratio of this OHAp was analyzed by both chemical and thermal methods. The ratio obtained by the thermal method, calculated from the average value of 0.01 ± 0.003* mass fraction of -TCP (and 0.99 ± 0.003* mass fraction of OHAp) and equations (2) and (3), is 1.6649 ± 0.0005*. This agrees well with and has better precision than the chemically determined value of

* ± value is the expanded uncertainty that is equal to two times the standard uncertainty.

$1.664 \pm 0.006^{*}$. The lattice parameters were determined by the x-ray powder diffraction and are listed in Table 1.

SOLUBILITY

Solubility is the thermodynamic property indicating the equilibrium between the solid and the dissolved components in a solution. Gravimetric solubility expresses the mass of a solute that can be dissolved in a unit volume of solution. This expression is not suitable for describing the solubilities of various calcium phosphate salts since their solubilities can change dramatically with other solution parameters such as pH. Therefore, the thermodynamic solubility product constant, Ksp, is generally used instead. This fundamental parameter is related to the Gibbs free energy of the solid and is derived in the following paragraph with OHAp as an example.[54] The equilibrium reaction for calculating OHAp solubility is

$$Ca_5(PO_4)_3OH\,(solid) \rightarrow 5Ca^{2+}\,(aq) + 3PO_4^{3-}\,(aq) + OH^-\,(aq) \qquad (4)$$

The Gibbs free energy of dissolution, ΔG_d, for OHAp is

$$\Delta G_d = 5\mu_{Ca2+} + 3\mu_{(PO4)3-} + \mu^\circ_{OH-} - \mu^\circ_{OHAp} \qquad (5)$$

$$= 5\mu_{Ca2+} + 3\mu_{(PO4)3-} + \mu_{OH-} - \mu^\circ_{OHAp} + RT\ln(Ca^{2+})^5(PO_4^{3-})^3(OH^-) \qquad (6)$$

$$= \Delta G^\circ + RT\ln(Ca^{2+})^5(PO_4^{3-})^3(OH)^- \qquad (7)$$

where

$$\Delta G^\circ = 5\mu^\circ_{Ca2+} + 3\mu^\circ_{PO43-} + \mu^\circ_{OH-} - \mu^\circ_{OHAp} \qquad (8)$$

and is the standard free energy of dissolution.

μ_i, the chemical potential of the ion, i, is equal to $\{\mu^\circ_i + RT\ln(C_i)\}$, where μ°_i is the standard chemical potential of ion i, in the particular solvent and (C_i) is the activity of ion i.

μ°_{OHAp} is the chemical potential of the solid and is a constant independent of the solvent.

At equilibrium, $DG_d = 0$ and therefore

$$0 = \Delta G^\circ + RT\ln(Ca^{2+})^5(PO_4^{3-})^3(OH^-) \qquad (9)$$

$$\Delta G^\circ = -RT\ln K_{sp}(OHAp) \qquad (10)$$

where K_{sp} is the thermodynamic solubility product,

$$K_{sp} = (Ca^{2+})^5(PO_4^{3-})^3(OH^-) \text{ and } pK_{sp} = -\log (Ca^{2+})^5(PO_4^{3-})^3(OH^-) \quad (11)$$

Solution compositions that satisfy equation 11 are in equilibrium with OHAp and are defined by the solubility isotherm of OHAp as is shown in Figure 3.[55] In these Figures, the OHAp curves show the log [Ca] and log [P] of the solutions in equilibrium with this phase as a function of the pH. Since the isotherms define solution in equilibrium with respect to OHAp, they also demarcate the regions of undersaturation (lower region) and supersaturation (upper region). The solubility product is calculated from the data obtained in the solubility measurements using various dissociation constants of phosphoric acid and ion-pair formation constants.[58]

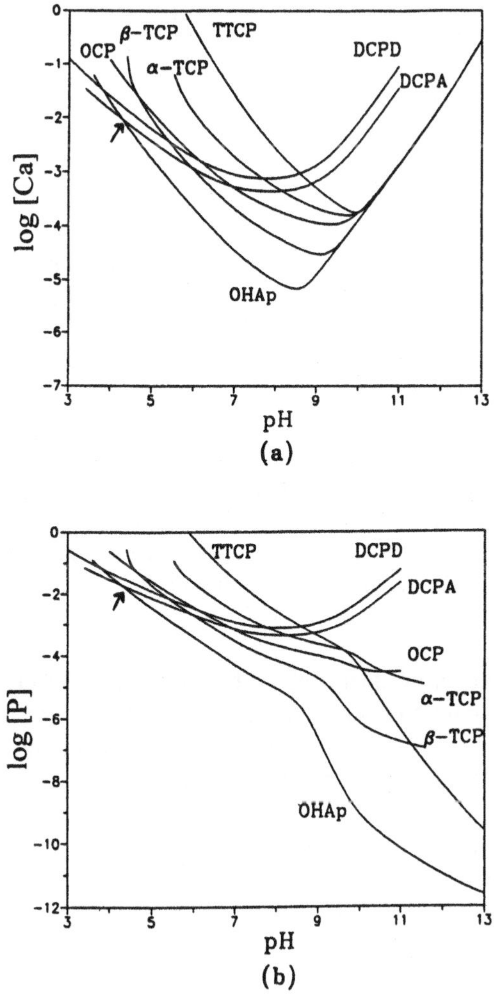

Figure 3. Solubility phase diagrams for the ternary system, $Ca(OH)_2$-H_3PO_4-H_2O, 37 °C showing two logarithms of the concentrations of (a) calcium and (b) phosphate as a function of the pH in solution saturated with various salt in the solubility isotherm.[55]

The isotherms of other calcium phosphates, indicating their respective solubilities,[55-64] also are shown in Figure 3. They show that the amount dissolved at equilibrium depends on the thermodynamic solubility product of the compound and the pH of the solution. At pH = 7.0, the solubility decreases in the order of TTCP > α -TCP > DCPD > DCPA ~ OCP > β -TCP > OHAp. The pH dependency is due to the fact that orthophosphoric acid is both a weak acid and a polybasic acid; the concentrations of HPO_4^-, $H_2PO_4^{2-}$, and PO_4^{3-} change dramatically with the pH of the solution.

When the isotherms of two solids intersect, the composition at that point is known as the "singular point." This point indicates that two solids are in equilibrium with each other and with the solution. The singular point for DCPA and OHAp has a pH = 4.2 at 37 °C as indicated by the arrow in Figure 3.

The solubility of OHAp has been extensively studied[60-64] and is a somewhat controversial property. Much of the confusion surrounding attempts to quantify this property has been resolved by Brown and coworkers.[60-62] Their studies indicate that the thermodynamic solubility product depends on the purity of the compound which, in turn, depends on the method of preparation. The choice of dissociation constants and ion-pair formation constants used to calculate ion activities further affect the value of the calculated Ksp.[57,58] Deviations from constant solubility can also be caused by a solid surface with high free energy and/or high solubility formed during thermal preparation. A constant solubility is obtained after this surface layer[64] is removed by repeated washing.

Fluoride,[20] carbonate[25] and magnesium[65] influence the structure of the calcium phosphates and, therefore, have specific effects on their solubilities.[1,25,76] In OHAp, fluoride decreases,[76] and carbonate and magnesium increase the solubility.[1] On the other hand, whitlockite has a significantly lower solubility than that of β-TCP. This indicates the increased stability in the β-TCP structure when certain Ca^{2+} ions are replaced by the smaller Mg^{2+} ions.[65]

The nonspecific effect of other ions is the increase in the ionic strength, which enhances the gravimetric solubility due to decreases in the ionic activity coefficients up to a certain ionic strength. Also, the equilibrium distribution among the various phosphate ion species is altered because the activity coefficient of a multivalent ion such as PO_4^{3-} is more affected by changes in ionic strength than that of $H_2PO_4^-$. Therefore, the effect of ionic strength on an acid calcium phosphate, e.g., $CaHPO_4$ is expected to be different from that of a more basic calcium phosphate, e.g., $Ca_5(PO_4)_3OH$.

Effect of Solvent

The effect of ethanol on the solubility of DCPD and OHAp has been reported.[66-67] The results indicate that the effect is mainly due to electrostatic interactions in solvents with different dielectric constants as expressed by the Born equation[77]

$$pK_{sp} - pK^{w}_{sp} = a(1/\varepsilon - 1/\varepsilon_w) \quad (12)$$

where pK_{sp} and pK^{w}_{sp} are the negative logarithms of the solubility products in a particular mixed solvent with dielectric constant ε and in water with dielectric constant ε_w. The constant, a, depends on the compound and temperature.

The relations between pK_{sp} and ε for two calcium phosphates obtained from experimental results[66,67] are

$$pK_{sp} - 6.69 = 575.2(1/\varepsilon - 0.0135) \text{ at 37 °C for DCPD} \quad (13)$$

and

$$pK_{sp} - 57.48 = 4136(1/\varepsilon - 0.0128) \text{ at 25 °C} \quad (14)$$

and

$$pK_{sp} - 57.44 = 2987(1/\varepsilon - 0.0133) \text{ at 37 °C for OHAp} \quad (15)$$

The equations for DCPD and OHAp indicate that as the dielectric constant decreases in the mixed ethanol/ water solution, the pKsp values increase, i.e., the solubilities of DCPD and OHAp decrease.

STABILITY

The stability of calcium phosphates in aqueous solutions is closely related to their structures, compositions, and solubilities. It determines the conversions, functions and applications of calcium phosphates. One classic application of the stability concept is the development of self-setting calcium phosphate cements.[55] The solubility isotherms in Figure 3 demonstrate the relative thermodynamic stability of various calcium phosphates in solutions. When the most soluble of the calcium phosphates, TTCP, is mixed with other calcium phosphates, the mixture becomes an apatite-forming cement in an aqueous milieu.

The kinetic stability of calcium phosphates has been investigated by precipitation[69,70] and hydrolysis studies.[72-74] Based on these studies, the formation and stability of OHAp, OCP, ACP, and a gel-like calcium phosphate described below at ~ neutral pH as the function of phosphate concentration (Ca/P molar ratio around 1.6) are schematically shown in Figure 4. The first solid phases that precipitate from the calcium phosphate solutions depend on the compositions of the solutions and are not necessarily the most stable phase. These precursors are not stable and will hydrolyze to apatite at neutral pH. In the solution of lowest supersaturation ([P] = 0.24 mM), indicated by the arrow 1, the OHAp forms directly.[69] At the higher concentrations ([P] = 1.2 mM), indicated by the arrow 2, OCP precipitates first and then converts to apatite.[69] At even higher concentrations ([P] = 4.2 mM), indicated by the arrow 3, a noncrystalline ACP is the first solid to form, and it then hydrolyzes to apatite through an OCP-like intermediary.[70] At highest concentrations ([P] > 100 mM), indicated by the arrow 4, the gel-like calcium phosphate forms first; this gel-like phase is a colloidal calcium phosphate with very high viscosity. This phase then solidifies to ACP, hydrolyzes to OCP, and finally to apatite.[75]

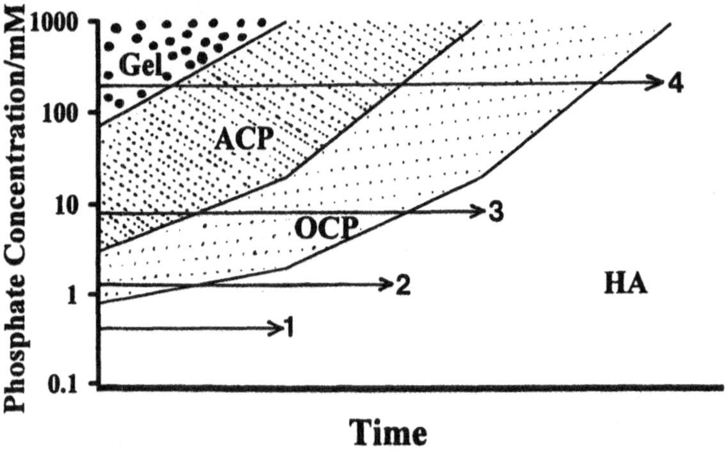

Figure 4. Formation, stability, and hydrolysis of calcium phosphates as a function of phosphate cocncentration (log (P)) in solutions at neutral pH.

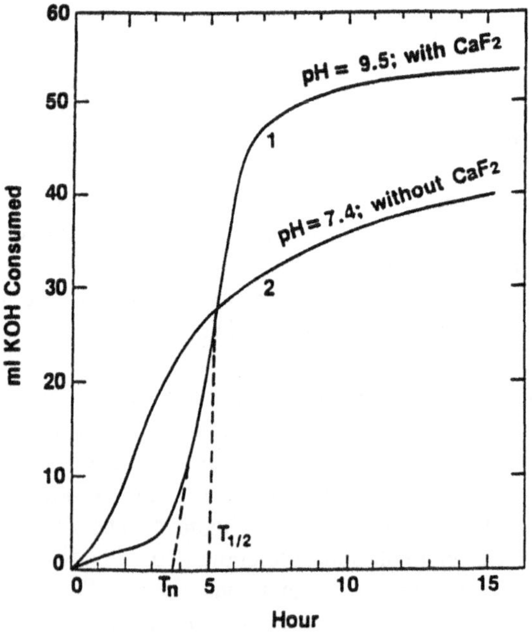

Figure 5. The rate of DCPD hydrolysis as indicated by the consumption of KOH as a functioin of time at 37 °C. Curve 2: pH = 7.4; Curve 1: pH = 9.5 and in the presence of CaF_2.[72]

The hydrolysis kinetics of DCPD, OCP, and ACP to apatite in solutions kept at constant pH in a pH-stat shows quantitatively their stability (or instability) and the effects of various ions on that stability. Hydrolysis of DCPD at pH = 7.4 and at pH = 9.5 in the presence of CaF_2 (as the powder suspended in the solution) is shown in Figure 5.[72] These sigmoidal curves are typical of a reaction involving nucleation in which the nucleation rate is slow compared to the crystal growth rate. The curves can be divided into three regions: (1) a "nucleation period" during which the DCPD precursor dissolves and nucleation of new crystals occurs; (2) a rapid "growth period" during which the nuclei of the crystals grow rapidly; (3) a gradual "tapering off" period during which the reaction approaches completion. The results also demonstrate that higher pH and the presence of fluoride increase hydrolysis or decrease stability. The effects of magnesium and temperature on the hydrolysis of OCP[73] are shown in Figure 6. Higher temperature increases the hydrolysis and magnesium decreases the hydrolysis. The hydrolysis of ACP, shown in Figure 7,[74] involves two processes. The first, conversion of ACP to an OCP-like phase, consumes acid, and the second, conversion of this OCP-like phase to apatite, consumes base.

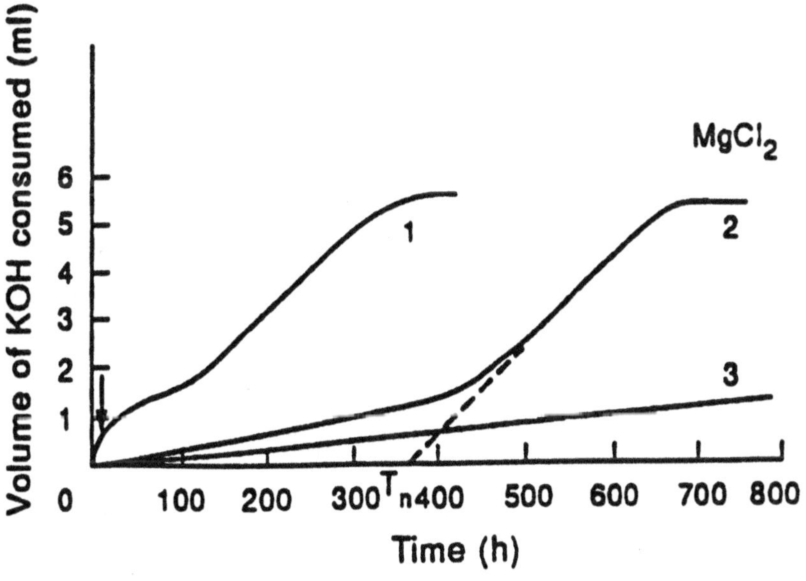

Figure 6. The rate of OCP hydrolysis as indicated by the consumption of KOH as a function of time at pH 7.5. Curve 1: OCP at 37 °C with arrow indicating the end of the initial surface reaction. Curve 2: OCP in the presence of 1 mM $MgCl_2$ at 37 °C. Curve 3: OCP at 25 °C. [73]

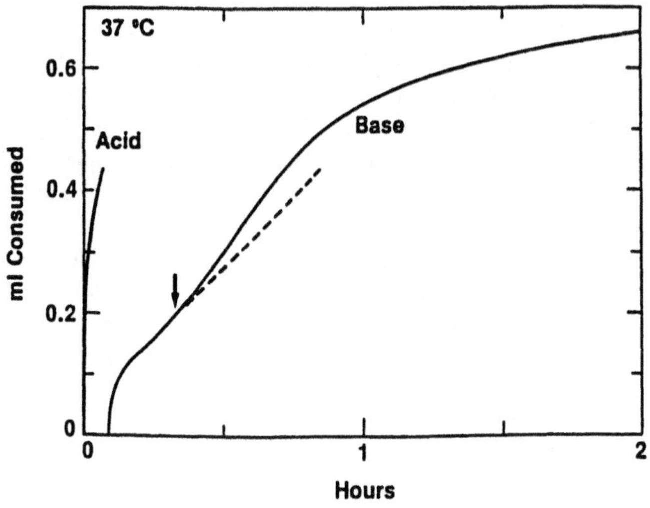

Figure 7. The rate of ACP hydrolysis in a solution that was equilibriated with ACP at pH 7.4 and 37 °C. The consumption of acid was due to the conversion of ACP to OCP, and the consumption of base was due to the cionversion of OCP to apatite.[74]

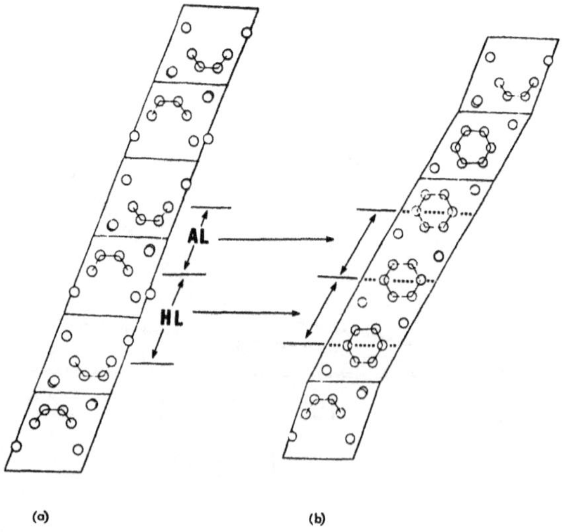

Figure 8 Proposed hydrolysis mechanism of two OCP layers to four unit cells of OHAp showing where reorganization takes place. The HL (hydrated layer) of OCP becomes the AL (apatitic layer) in OHAp. Only the Ca ions are shown: (a) three unit cells of OCP and (b) four unit cells of OHAp and two ½ unit cells of OCP.

The formation of apatite through OCP is important both theoretically[15,53] and practically.[79,80] The structural process of a layer-by-layer growth mechanism of OHAp through OCP is depicted in Figure 8 as projected down the c axes and only the Ca ions are shown. An OCP nucleus with few layers is formed first, as shown in Figurc 8a (thrcc laycrs arc shown). As calcium ions diffuse into the lattices and hydrogen ions and water diffuse out, the inner two unit cells of OCP convert to four unit cells of apatite by transforming the hydrated layer (HL) in OCP into an apatitic layer (AL), as shown in Figure 8b. The original AL is the same in OCP and OHAp and is retained. The outer two ½ unit cells of the OCP are retained in the process and are joined by epitaxy to the formed OHAp along the 100 plane. As layers of OCP are added on the outer ½ unit cell, the inner layers of OCP will further convert to apatite. This process is one of the mechanisms proposed for the incorporation of the impurities and structural defects. Some of impurites, water, strucural defects and calcium vacancies can be incorporated during the HL to AL conversion. Some of the HLs in the OCP cells can be only partially tranformed and/or are buried inside the crystals. This mechanism can also account for the cryptocrystalline and crystalline structures, nonstoichiometric compositions, and the varying solubilities and stabilities of apatites.[15,48]

ACKNOWLEDGMENT

The valuable discussions and assistance of Dr. Mathai Mathew, Dr. Drago Skrtic, and Mrs. M. W. Chalkley are highly appreciated.

REFERENCES

1. LeGeros RZ. *Calcium Phosphates in Oral Biology*. Basel: Karger, 1991.

2. Elliot JC. *Structure and Chemistry of the Apatites and Other Calcium Orthophosphates*. Amsterdam: Elsevier, 1994.

3. Driessens FCM. *Mineral Aspects of Dentistry*. Basel: Karger, 1982.

4. Standard Reference Material 2910: Calcium Hydroxyapatite, to be issued by National Institute of Standards and Technology, Standard Reference Materials Program, Gaithersburg, MD.

5. Brown WE, Smith JP, Lehr JR, Frazier AW. Crystallography of hydrated monocalcium phosphates containing potossium or ammonium. J PhysChem 1958;62:625-627

6. Dickens B, Bowen JS. Refinement of the crystal structure of $Ca(H_2PO_4)\cdot H_2O$. Acta Cryst1971;B27:2247-2255

7. Dickens B, Prince E, Schroeder LW, Brown WE. $Ca(H_2PO_4)_2$, a crystal structure containing unusual hydrogen bonding. Acta Cryst 1973;B29(Pt.10):2057-2070

8. Schroeder LW, Prince E, Dickens B. Hydrogen bonding in $Ca(H_2PO_4)_2\cdot H_2O$ as determined by neutron diffraction. Acta Cryst 1975;B31 (Pt. 1):9-12

9. Jones DW, Smith JAS. The structure of brushite, $CaHPO_4.2H_2O$. J Chem Soc 1962;1414-1420

10. Curry NA, Jones DW. Crystal structure of brushite, calcium hydrogen orthophosphate dihydrate: a neutron-diffraction investigation. J Chem Soc A 1971;3725-3729

11. Dickens B, Bowen JS, Brown WE. A refinement of the crystal structure of $CaHPO_4$ (synthetic monetite). Acta Cryst 1972;B28:797-806

12. Catti M, Ferraris G, Mason SA. Low-temperature ordering of hydrogen atoms in $CaHPO_4$ (Monetite): x-ray and neutron diffraction study at 145 K. Acta Cryst 1980;B36:254-259

13. Mathew M, Schroeder LW. Dickens B, Brown WE. The Crystal Structure of α-$Ca_3(PO_4)_2$. Acta Cryst 1977;B33:1325-1333

14. Schroeder LW, Dickens B, and Brown, W.E. Crystallographic studies of the role of Mg as a stabilizing impurity in β-$Ca_3(PO_4)_2$ II. Refinement of Mg-containing β-$Ca_3(PO_4)_2$. J Solid State Chem 1977;22:253-262

15. Brown WE, Mathew M, Tung MS. Crystal chemistry of octacalcium phosphate. Prog Crystal Growth Charact 1981;4:59-87 (Oxford, England: Pergamon Press Ltd.)

16. Mathew M, Brown WE, Schroeder LW, Dickens B. Crystal structure of octacalcium bis(hydrogenphosphate) tetrakis(phosphate) pentahydrate, $Ca_8(HPO_4)_2(PO_4)_4\cdot 5H_2O$. J Cryst and Spectr Res 1988;18(3):235-250

17. Gopal R, Calvo C, Ito J, Sabine WK. Crystal structure of synthetic Mg-whitlockite, $Ca_8Mg_2H_2(PO_4)_{14}$. Can J Chem 1974;52:1155-1164

18. Elliot JC, Mackie PE, Young RA. Monoclinic hydroxapatite. Science 1973;180:1055-1057

19. Posner AS, Perloff A, Diorio AF. Refinement of the hydroxyapatite structure. Acta Cryst 1958;308-309

20. Sudarsanan K, Markie PE, Young RA. Comparison of synthetic and mineral fluorapatite, $Ca_5(PO_4)_3F$, in Crystallographic Detail. Mat Res Bull 1972;7:1331-1338

21. Makie PE, Elliot JC, Young RA. Monoclinic structure of synthetic $Ca_5(PO_4)_3Cl$, chlorapatite. Acta Cryst 1972;B28:1840-1848

22. Elliott JC, Bonel G, Trombe JC. Space group and lattice constants of $Ca_5(PO_4)_3CO_3$. J Appl Cryst 1980;13:618-621

23. Brown WE, Epstein EF. Crystallography of tetracalcium phosphate. J Res NBS (Phys and Chem) 1965;47

24. Dickens B, Brown WE, Kruger GJ, Stewart JM. $Ca_4(PO_4)_2O$, tetracalcium diphosphate monoxide, crystal structure and relationships to $Ca_5(PO_4)_3OH$ and $K_3Na(SO_4)_2$. Acta Cryst 1973;10:2046-2056

25. LeGeros RZ, Trautz OR, LeGeros JP, Shirra WP. Apatite crystallites: Effects of carbonate on morphology. Science 1967;155:1409-1411

26. Young RA, Holcomb DW. Variability of hydroxyapatite preparations. Calcif Tissue Int 1982;34:S17-S32

27. Fowler BO. Infrared studies of apatites. I. Vibrational assignments for calcium, strontium, and barium hydroxyapatite utilizing isotopic substitution. Inorg Chem 1974;13:194-207

28. Fowler BO. Infrared studies of apatites. II. Preparation of normal and isotopically substituted calcium, strontium, and barium hydroxyapatites and spectra-structure-composition correlations. Inorg Chem 1974;13:207-214

29. Casciani F, Condrate RA. The vibrational spectra of brushite, $CaHPO_4.2H_2O$. Spectrosc Lett 1979;12:699-713

30. Fowler BO, Moreno EC, Brown WE. Infared spectra of hydroxyapatite, octacalcium phosphate and pyrolyzed octacalcium phosphate. Arch Oral Biol 1966;11:477-492

31. Casciani F, Condrate RA. The raman spectrum of monetite, $CaHPO_4$. J Solid State Chem 1980;34:385-388

32. Tropp J, Blumenthal NC, Waugh JS. Phosphous NMR study of solid amorphous calcium phosphate. J Am Chem Soc 1983;105:22-26

33. Yesinowski JP, Eckert H. Hydrogen environments in calcium phosphates: ^{1}H MAS NMR at high spinning speeds. J Am Chem Soc 1987;109:6274-6282

34. Rothwell WP, Waugh JS, Yesinowski JP. High-resolution variable-temperature ^{31}P NMR of solid calcium phosphates. J Am Chem Soc 1980;102:2637-2643

35. Betts F, Posner AS. An xray radial distribution study of amorphous calcium phosphate. Mater Res Bull 1974;9:353-360

36. Grynpas MD, Bonar LC, Glimcher MJ. X-ray diffraction radial distribution function studies on bone mineral and synthetic calcium phosphates. J Mater Sci 1984;19:723-736

37. Harries JE, Hukins DWL, Holt C, Hasnain SS. Conversion of amorphous calcium phosphate into hydroxyapatite investigated by EXAFS spectroscopy. J Crystal Growth 1987;84:563-570

38. Eanes ED, Powers L, Costa JL. Extended x-ray absorption fine structure (EXAFS) studies on calcium in crystalline and amorphous solids of biological interest. Cell Calcium 1981;2:251-262

39. Buerger MJ. *Elementary Crystallography*. New York: John Wiley, 1963.

40. Rowles SL. The precipitation of whitlokite from aqueous solutions. Bull Chem 1968:1798-1802

41. Welch JH, Gutt W. High-temperature studies of the system calcium oxide-phosphous pentoxide. J Chem Soc 1961;4442-4444

42. Nurse RW, Welch JH, Gutt W. High-temperature equilibria in the system dicalcium silicate-tircalcium phosphate. J Chem Soc 1959;1077-1083

43. Kreidler ER, Hummel FA. Phase relationships in the system SrO-P_2O_5 and the inflluence of water vapor on the formation of $Sr_4P_2O_9$. Inorg Chem 1967;6:884-891

44. Prener JS. The growth and crystallographic properties of calcium fluor- and chlorapatite crystals. J Electrochem Soc 1967;114:77-83

45. Bauer M. Rontgenographische und dielektrische untersuchungen an apatiten. Dissertatin, Fakultat fur Physik, Universitat Karlsruhe, 1991.

46. Moreno EC, Gregory TM, Brown WE. Solubility of $CaHPO_4 \cdot 2H_2O$ and formation of ion pairs in the system $Ca(OH)_2$-H_3PO_4-H_2O at 37.5 °C. J Res NBS (Phys and Chem) 1966;70A:545

47. Gee A, Deitz VR. Determination of phosphate by differential spectrophotometry. Anal Chem 1953;25:1320-1324

48. Brown WE. Crystal growth of bone mineral. Clinical Orthopaedics 1966;44:205

49. Meyer JL, Fowler BO. Lattice defects in nonstoichiometric calcium hydroxyapatite. A chemical approach. Inorg Chem 1982;21:3029-3035

50. Brown WE, Schroeder LW, Ferris JS. Interlayering of crystalline octacalcium phosphate and hydroxylapatite. J Phys Chem 1979;83:1385-1388

51. Dickens B, Schroeder LW. Investigation of epitaxy relationships between $Ca_5(PO_4)_3OH$ and other calcium orthophosphates. J Res Res 1980;85(5):347-362

52. Brown WE, Chow LC. Surface equilibria of sparingly soluble crystals. Colloids and Surfaces 1983;7:67-80

53. Brown WE, Mathew, M, Chow LC. "Roles of Octacalcium Phosphate in Surface Chemistry of Apatites." In *Adsorption on and Surface Chemistry of Hydroxyapatite*, D.N. Misra, ed. New York: Plenum Publishing Corp., 1984.

54. Brown WE, Chow LC. Chemical properties of bone mineral. Ann Rev Mat Sci 1976;6:213-236

55. Chow, LC. Development of self-setting calcium phosphate cements. J Ceram Soc Japan (The Centennial Memorial Issue) 1991;99(10):954- 964

56. Gregory TM, Moreno EC, Brown WE. Solubility of $CaHPO_4 \cdot 2H_2O$ in the system $Ca(OH)_2$-H_3PO_4-H_2O at 5, 15, 25 and 37.5 °C. J Res NBS (Phys and Chem) 1970;74A:461-475

57. McDowell H, Brown WE, Sutter J.R. Solubility study of calcium hydrogen phosphate: ion pair formation. Inorganic Chem 1971;10:1638-1643

58. Tung MS, Eidelman N, Sieck B, Brown WE. Octacalcium phosphate solubility product from 4 to 37 °C. J Res NBS 1988;93(5):613-624

59. Gregory TM, Moreno EC, Patel JM. Brown WE. Solubility of β-$Ca_3(PO_4)_2$ in the System $Ca(OH)_2$-H_3PO_4-H_2O at 5, 15, 25, and 37 °C. J Res NBS 1974;78A(6):667-674

60. Moreno EC, Gregory TM, Brown WE. Preparation and solubility of hydroxyapatite. J Res NBS (Phys and Chem) 196872A:773-782

61. Avnimelech Y, Moreno EC, Brown WE. Solubility and surface properties of finely divided HA. J Res NBS 1973;77A:149-155

62. McDowell H, Gregory TM, Brown WE. Solubility $Ca_5(PO_4)_3OH$ in the system $Ca(OH)_2$-H_3PO_4-H_2O at 5, 15, 25 and 37 °C. Res Natl Bur Stand 1977;81A:273-281

63. Verbeek RMH, Steyaer H, Thun HP, Verbeek F. Solubility of synthetic calcium hydroxyapatites. J C S Faraday I 1980;76:209-219

64. Brown WE. "Solubilities of Phosphates and Other Sparingly Soluble Compounds." In *Environmental Phosphorus Handbook.* New York: John Wiley & Sons, 1973.

65. Dickens B, Brown WE. The crystal structure of $Ca_7Mg_9(Ca,Mg)_2(PO_4)_{12}$. TMPM 1971;16:79-104

66. Tung MS, O'Farrell TJ. The effect of ethanol on the solubility of dicalcium phosphate dihydrate in the system $Ca(OH)_2$-HPO_4-H_2O at 37 °C. J Mol Liquids 1993;56:237-243

67. Tung MS, Lin C, Chow TH, Sung P. "The Effect of Ethanol on the Solubility of Hydroxyapatite in the System $Ca(OH)_2$-H_3PO_4-H_2O at 25 and 33 °C." In *Hydroxyapatite and Related Compounds.* P.W. Brown and B. Constantz, eds. Boca Raton: CRC Press, 1994.

68. Fowler BO, Kuroda S. Changes in heated and in laser-irradiated human teeth enamel and their probable effects on solubility. Calcif Tissue Int 1986;38-197-208

69. Nancollas GH, Tomazic BB. Growth of calcium phosphate on hydroxyapatite crystals, effect of supersaturation and ionic medium. J Phys Chem 1974;78:2218-2225

70. Eanes ED, Meyer JL. The Maturation of crystalline calcium phosphates in aqueous suspensions at physiologic pH. Calif Tiss Res 1977; 23:259-269

71. Markovic M, Fowler BO, Tung MS, Lagergren ES. "Composition and Solubility Product of a Synthetic Calcium Hydroxyapatite." In *Mineral Scale Formation and Inhibition*, Z. Amjad, ed., 1994. New York: Plenum Press, 1995.

72. Tung MS, Chow LC, Brown WE. Hydrolysis of dicalcium phosphate dihydrate in the presence or absence of calcium fluoride. J Dent Res 1985;64(l):2-5

73. Tung MS, Tomazic B, Brown W E. The effects of magnesium and fluoride on the hydrolysis of octacalcium phosphate. Arch Oral Biol 1992;37(7):585-591

74. Tung MS, Brown WE. An intermediate state in hydrolysis of amorphous calcium phosphate. Calcif Tissue Int 1983;35:783-790

75. Tung MS. Gelation of 1. J Dent Res 1996;75:Abstract 11

76. Moreno EC, Kresak M, Zahradnik RT. Fluoridated hydroxyapatite solubility and caries formation. Nature 1974, 247:64-65

77. Bates RG. Determination of pH. Chapter 8, New York: Wiley, 1973.

78. LeGeros RZ. Variability of -TCP/HAP ratios in sintered 'apatites'. J Dent Res 1986;65:292,abstr 110

79. Brown WE, Tung MS, Chow LC. Role of octacalcium phosphate in the incorporation of impurities into apatites. Proceedings, 2nd International Congress on Phosphorus Compounds, 1981; 59-71

80. Tung MS, Brown WE. The role of octacalcium phosphate in subcutaneous heterotopic calcification. Calcif Tissue Int 1985;37:329-331

2

AMORPHOUS CALCIUM PHOSPHATE: THERMODYNAMIC AND KINETIC CONSIDERATIONS

Edward D. Eanes

National Institute of Dental Research, Bone Research Branch
Bethesda, Maryland 20892, USA

INTRODUCTION

Crystalline hydroxyapatite (HAP) is generally considered to be the final, stable product in the precipitation of calcium and phosphate ions from neutral to basic solutions. However, over the broad range of solution conditions in which precipitation occurs spontaneously, unstable amorphous products precede the appearance of the HAP phase. These amorphous calcium phosphates (ACPs) are unique among calcium phosphate salts in that they lack the long-range, periodic atomic scale order of crystalline materials. Although this uniqueness has been the subject of considerable interest, of greater relevance to understanding the dynamics of HAP formation by spontaneous precipitation is the instability of these solids in solution. Above pH 9, ACPs convert directly into HAP. However, in the 7-9 pH range, ACPs are the immediate precursors to octacalcium phosphate (OCP)-like phases that, in turn, convert to stable HAP. These transformation processes and the possible role they play in biological calcification, more than anything else, raises the importance of the ACPs above being merely laboratory curiosities and places them in the mainstream of calcium phosphate chemistry. This chapter will cover the preparation and properties of the ACPs, their thermodynamic and kinetic relationship to crystalline calcium phosphates, especially OCP and HAP, the controversy surrounding their true structural nature, and their putative presence and role *in vivo*.

PREPARATION

Watson and Robinson were the first to describe a calcium phosphate salt with amorphous-like features.[1] They found that upon mixing $CaCl_2$ and Na_2HPO_4 solutions in sufficiently high concentrations to produce immediate flocculation at room temperature and neutral pH, the first solids formed were extremely fine in texture with no recognizably crystalline features when examined by transmission electron microscopy, and with no resolvable electron diffraction pattern. Although their procedure has undergone several refinements over the years, especially in the development of techniques to minimize post-reaction changes in solids sampled for analysis,[2,3] the basic approach of effecting spontaneous precipitation by the mixing of concentrated reactants is still widely used for preparing ACP. Another commonly used method[4] is to prepare acidic (pH 4-5), subsaturated calcium phosphate solutions, then induce precipitation by increasing the solutions to the desired pH with the rapid addition of concentrated base. With the use of these techniques, ACPs have been prepared from solutions that encompass a wide range of pHs (7-13), solution Ca/PO_4 molar ratios (0.1 to 10), and solution Ca and PO_4 concentrations (0.002 to 1.0 mol/ℓ). As discussed in greater detail later, the compositions of precipitated ACPs are remarkably insensitive to these broadly varying solution conditions. Solid Ca/PO_4 molar ratios are generally confined to the rather narrow range of 1.4 to 1.5. Only at neutral pHs and very low solution Ca/PO_4 ratios do solid Ca/PO_4 ratios drop to less than 1.3.[5] The stability of ACPs in solution, on the other hand, is very sensitive to preparative conditions. Generally, the most stable solids are formed around pH 10.[6] In contrast, especially rapid transformations can occur when solutions approach neutrality. Instability also increases rapidly with temperature which effectively limits preparations to 37 °C.

STRUCTURAL PROPERTIES

As already alluded to above, the most striking feature that distinguishes ACPs from other calcium phosphate salts is the apparent lack of crystalline order. The very diffuse electron diffraction profile first observed by Watson and Robinson,[1] as well as that observed later by x-ray diffraction,[3] is typical for substances that lack long-range periodic regularity (Figure 1). Radial distribution analysis of the x-ray diffraction profile[7,8] showed further that if orderly atomic arrangements exist within the body of the ACP, they are highly localized, inscribed within domains that do not exceed 0.9 nm in diameter. Regularity in the local environment about individual Ca^{2+} ions in ACP appears to be even more circumscribed. Extended x-ray absorption fine structure (EXAFS) spectroscopy[9,10] indicates that orderly Ca-Ca and Ca-P interactions do not extend beyond a distance of 0.3 nm. Infrared analysis of ACP shows a similar lack of crystalline order about the PO_4^{3-} anions in the structure.[11] Nevertheless, as discussed in greater detail below, other aspects of the structural data together with chemical, thermodynamic, and kinetic data suggest that, despite their apparent non-crystallinity, ACPs are a well-defined class of calcium phosphates with no known crystalline counterparts.

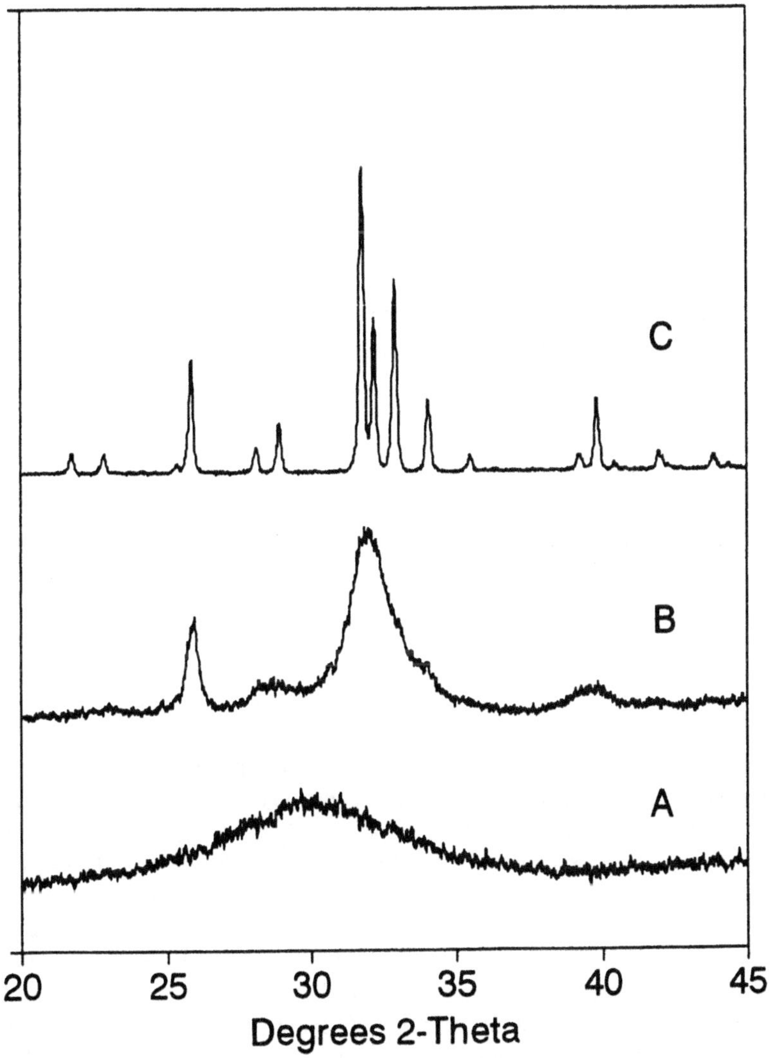

Figure 1. X-ray diffraction patterns of (A) amorphous calcium phosphate, (B) a poorly crystalline apatite, and (C) a well-crystallized hydroxyapatite. Patterns were recorded with CuK α radiation (= 0.154nm).

MORPHOLOGICAL FEATURES

As with their internal structure, particle morphology also distinguishes the ACPs from crystalline calcium phosphates. Transmission electron microscopic (TEM) observations[12,13] made of drops of ACP suspensions placed on Formvar-carbon coated grids, removed of excess solution, then air dried, show that the particles do not exhibit the angular shape, nor the flat, faceted surfaces characteristic of crystalline materials. Instead, dried ACPs occur most frequently as irregularly shaped, anastomosing aggregates of either low-contrast disks or high-contrast spheroids (Figure 2). The disk-shaped forms are most prominent in specimens prepared from freshly precipitated suspensions. They vary widely in lateral dimensions (10 nm to several μm) and represent the collapse during specimen drying of highly hydrated flocculant masses of unknown solution morphology.[13] The spherical forms are first seen as high-contrast buds extending from the low contrast disks. Generally much smaller (20-200 nm) than the disks from which they appear to emerge, the spherical forms eventually supplant the disks as they evolve by fusion into irregularly-shaped, branching clusters. Although the chain-like clusters vary widely in length, the diameters of the spheroids within the clusters remain quite uniform. The progressive change from discoidal to spheroidal morphologies with solution maturation indicates that the highly hydrated, loosely organized flocculants that first appear spontaneously in ACP preparations subsequently desolvate *in situ*[14] into denser, more tightly-packed noncrystalline arrays of Ca^{2+} and PO_4^{3-} ions. This intra-particle consolidation of the constituent ions with solution maturation is another feature of the ACPs that separates them from crystalline calcium phosphates, in which particle growth occurs without change in lattice ion packing density.

CHEMICAL COMPOSITION

Although first described qualitatively in 1953[1], quantitative chemical studies on the ACPs were not reported until 1965 when methods were devised to isolate large amounts of the unstable solids for analysis. To minimize changes during sample drying, these methods utilized filtration or centrifugation to wash excess solution ions from the ACP slurry, then freezing the ACP under high vacuum to remove any remaining entrapped solvent by sublimation.[2,3,15] The earliest chemical study on solids isolated in this way[2,3] showed that the molar Ca/PO_4 ratio of ACP prepared at pH 10.5 is very close to 1.5. Later analyses[15,16,17] showed that for preparative pHs ranging from 7 to 10.5, ACPs washed of adsorbed impurities retain a rather constant composition with Ca/PO_4 molar ratios between 1.4 and 1.5, about 5-15% of the PO_4s protonated, and no OH^- ions. In the absence of long-range periodic order, this relative constancy in composition over such a wide pH range suggest that the ACPs have a well-defined local chemical unit. This inference is further strengthened with the discovery[18] that a nearly constant solution ion activity product (IAP) of 1.6×10^{-25} could be obtained for ACP slurries over the pH range 7.4 to 9.25 and starting Ca/PO_4 molar ratios of 1.31 to 1.48 when the solid phase composition was postulated to be $Ca_3(PO_4)_{1.87}(HPO_4)_{0.2}$, i.e., a solid with a

Ca/PO_4 molar ratio of 1.45, containing approximately 10% HPO_4^{2-}, values comparable to experimental results. Moreover, at pH 7.4 the above IAP value was also obtained for ACP preparations in which a widely varying range of starting Ca/PO_4 molar ratios was employed (0.7 to 7.1).

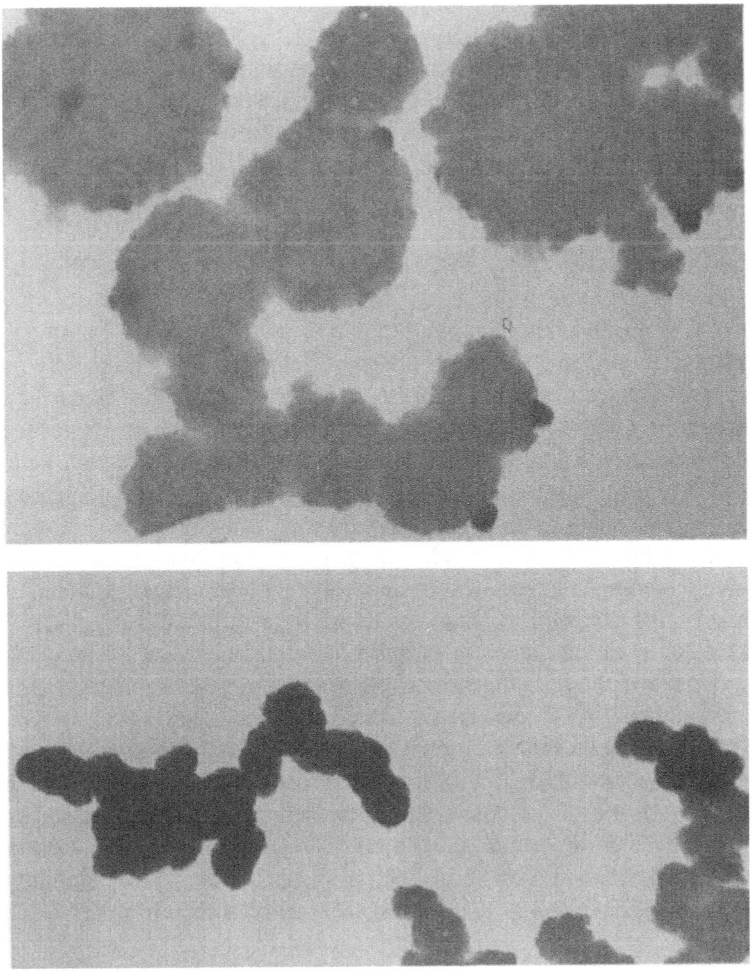

Figure 2. upper panel: Transmission electron micrograph of low-contrast disk-shaped particles formed during drying on the grid surface of a highly hydrated colloidal suspension of amorphous calcium phosphate collected immediately after precipitation at pH 7.4. Lower panel: Aggregates of high-contrast spherules formed in an older suspension by the spontaneous condensation in solution of the initially highly hydrated amorphous calcium phosphate into more compact, less hydrated particles. Scale: 0.5μm = 2.5 cm.

At pH s greater than 9.25, however, the ACPs do not appear to have a constant solution ion activity product for any single molecular formula like the one postulated for pHs 9.25 and below.[6,19] This breakdown in solution constancy suggest that the solubility-controlling structural units in ACP are subtly dependent compositionally on preparative solution conditions at these higher pH values. One possible explanation for this pH-dependent variability is that the HPO_4^{2-} content no longer remains constant but gradually decreases from 10% with increases in pH.

Lyophilized ACPs also contain approximately 10-15% water by weight.[20,21] A temperature programmed desorption analysis by Sedlak and Beebe[22] indicates further that about 75% of this retained water is tightly bound inside the ACP particle, the rest being more loosely held surface water. These data suggest that solution-matured ACPs are hydrated salts with 1-3 water molecules associated with each formula unit as defined above.

SOLUTION BEHAVIOR AND RELATION TO OCP AND HAP

Watson and Robinson, in their pioneering ACP study[1], were the first to observe the transient nature of this phase when kept in contact with its preparative medium. They found that the electron diffraction patterns of samples taken later in the precipitation reaction no longer were diffuse but resembled, instead, patterns typical of poorly crystalline HAP. Chemical and x-ray diffraction studies[2,3,4,18] showed further that this amorphous/crystalline transition is not gradual but occurs rather precipitously (Figure 3). After a period of relative stability, the transition is marked by the solid phase rapidly progressing from being barely crystalline to where the amorphous features completely disappear. Concomitantly, dramatic declines in the calcium and phosphate levels occur in the solution phase. The time it takes to reach this amorphous/crystalline boundary varies considerably with preparative conditions, being particularly sensitive to temperature and pH.[23] At pH 7.4, for example, ACP converts five times more rapidly at 37 °C than at 20 °C.[4] The conversion time shows a somewhat more complex bell-shaped dependency on pH.[6] At pH 7.4 and 25 °C, ACP in aqueous suspension has a lifetime of less than 0.3 hours. As the pH is raised, ACP becomes increasingly more stable until it reaches a maximum lifetime between pH 10.0 and 10.5 of over 9 hours, a greater than 30-fold increase in stability. At pH's above 10.5, ACP stability rapidly declines until at pH 12.8 its lifetime becomes nearly as short as at pH 7.4.[6,18]

The amorphous/crystalline transition appears to be initiated by the ACP itself. The first observable crystals of the product phase are intimately associated with the amorphous solids, most of them in contact with the surface of the spheroidal ACP clusters.[13] The direction of crystal growth, however, is away from the cluster surface, not inwardly into the body of the cluster (Figure 4). This pattern of outward growth from the surface indicates that the amorphous/crystalline transformation is not an *in situ* solid-state change, but one of heterogeneous nucleation on the ACP followed by a solution-mediated translocation of calcium and phosphate ions between the dissolving ACP and the growing crystals. The heterogeneous nucleating mechanism also accounts, in part, for the time that elapses before the first crystals appear. In addition to the time needed to consolidate the ACP into sufficiently dense particles with surfaces capable of supporting crystal nucleation, this induction period also includes the time required to form stablc nuclci on thcsc surfaccs.

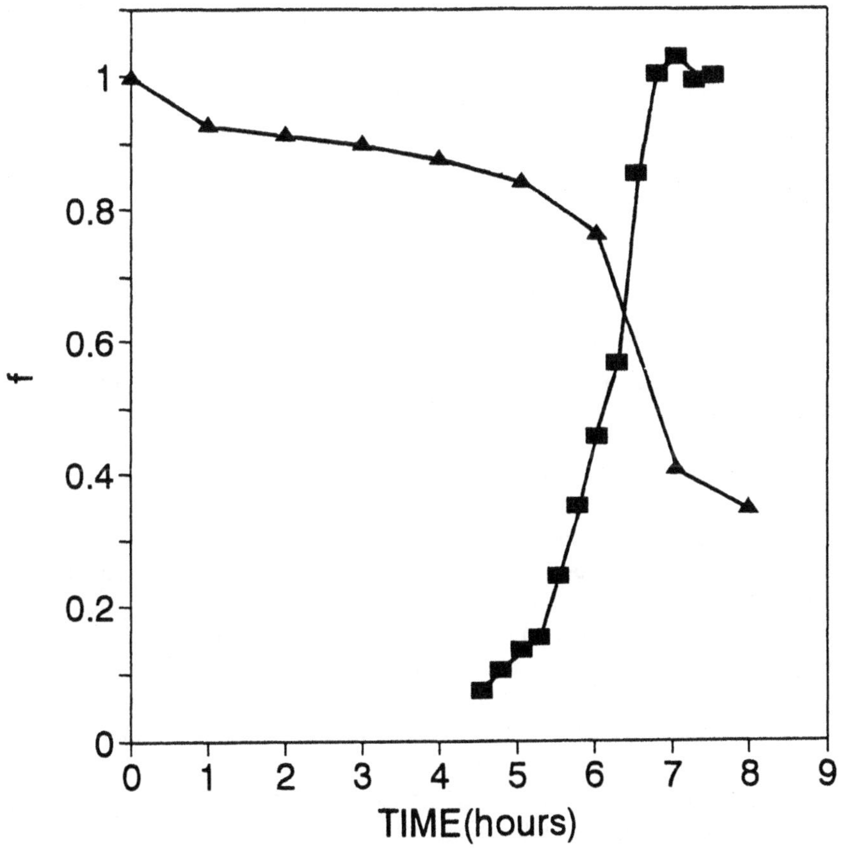

Figure 3. Curve connecting the closed squares shows the fractional conversion with time of amorphous calcium phosphate to crystalline hydroxyapatite. (1.0 = total conversion). Curve connecting the closed triangles shows the corresponding fractional drop in solution Ca concentration (1.0 = initial calcium concentration). Reaction was carried out at 25 °C and pH 9.8.[2,3]

The first crystals that nucleate on the ACP, however, do not grow very large (< 0.1 μm). The transformation advances, instead, by a multiplicative proliferation of small crystals about the amorphous particles until by the time the conversion is completed, the crystals have established a dense spherical envelope about the space previously occupied by the dissolved ACP (Figure 4).[13] X-ray diffraction studies,[2,3] reveal further that this proliferation occurs at an exponential rate, i.e. in proportion to the number of crystals already formed. This finding indicates that most of the crystals surrounding the space vacated by the dissolved ACP form by self-replication, possibly through a dendritic-like mechanism in which new crystals emerge on the surface of ones already formed. As already indicated above, only the earliest crystals in this process appear to be established by direct nucleation on the surface of the ACP.

The large disparity in the solubilities between the ACP and crystalline phases, resulting in a relatively high level of supersaturation with respect to the latter,[19] is most likely responsible for this self-replication. Unlike at low supersaturations where area-minimizing, stable surface tension forces keep crystals growing evenly and uniformly, high supersaturations produce steep, unstable ion diffusion gradients about the crystals that allow certain regions to grow ahead of the rest and so gain additional advantages in picking up ions for further growth. In particular, regions near edges and corners become more accessible to solute ions than the center of crystal faces. The result is overgrowth at these locations leading to dendritic-like projections which eventually develop into new crystals.[24]

As already indicated, the concentrations of calcium and phosphate ions in ACP preparative solutions remain relatively constant until the rapid conversion stage is reached when they drop to much lower values. The sharp decline in these concentrations occurs when insufficient ACP remains to sustain initial solubility and the solution reestablishes equilibrium with the rapidly emerging crystalline product.[4] At pHs greater than 9.25, the decline continues without pause until the reactant ion concentrations are well below corresponding solubility levels for all crystalline calcium phosphate salts except HAP. However, below pH 9.25, concentrations sharply level off when the preparative solutions reach the same thermodynamic stability as systems in equilibrium with OCP (Figure 5). This pause in the transformation process as the precipitate approaches the solubility of OCP is probably the most compelling thermodynamic evidence for OCP as the first crystalline phase to form at these lower pHs. Other evidence suggestive of OCP is the platy morphology of the earliest crystals, in contrast to the needle-like shapes more commonly associated with HAP, and a x-ray diffraction profile that, although generally apatitic in appearance, has an exceptionally large shift in the a-axis reflection (from 0.82 to 0.91 nm) consistent with an 0CP-like structural interpretation.[25] The compositional evidence, however, is somewhat more ambiguous. Solids harvested at this stage in the conversion have Ca/PO_4 molar ratios around 1.40 (vs. 1.33 for OCP) and HPO_4^{2-} contents less than 50% of the OCP value (16% vs. 33%). On the other hand, Meyer[19] presented kinetic evidence that showed that over the pH range 7.4 to 9.25 the ACP is unstable to a phase with an OCP-like composition. He found that for a range of solution Ca/PO_4 molar ratios the conversion time correlates with the solution composition above freshly precipitated ACP only when this composition is expressed as an OCP ion activity product (IAP). The inverse nature of this correlation, i.e., the conversion time decreases linearly with increases in this IAP, is also empirical evidence that this OCP-like phase forms by heterogeneous nucleation, a finding that is consistent with the TEM observations. The higher Ca/PO_4 molar ratios and lower % HPO_4^{2-} content of the solids indicates possibly that the OCP already begins to convert

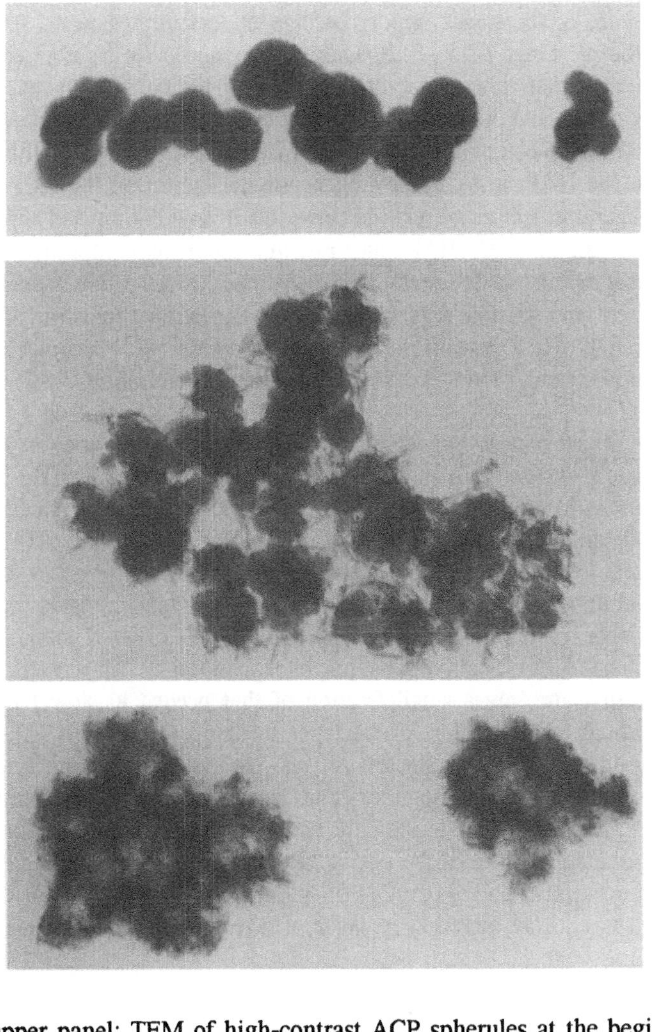

Figure 4. upper panel: TEM of high-contrast ACP spherules at the beginning of conversion to crystalline apatite. Middle panel: A partially crystallized preparation showing the amorphous spherules encrusted with crystals of apatite. Lower panel: The final stage of conversion showing crystals enveloping the space left by the dissolved ACP spherules. The needle-like particles are plate-like crystals seen edge-on. Reaction carried out at pH 7.4 and 25 °C. Scale: 0.5μm = 3.0 cm.

to HAP during this stage but in insufficient amounts to affect the solution equilibrium. This OCP/HAP transformation, unlike ACP/OCP, appears to be, at least in part, an *in situ* solid state rearrangement of the OCP structure, with concomitant water loss, into the anhydrous apatite structure.[4,26,27]

The hydrolysis of OCP to HAP is slowest at neutral pH, but steadily increases with pH until above 9.25 the lifetime of OCP is too short to obtain experimental evidence for its presence as an intermediate phase. At these higher pHs, ACP appears to convert directly to HAP without the formation of other transient intermediates such as -tricalcium phosphate (-TCP). As mentioned earlier, the solution phase at these pHs rapidly become undersaturated to all phases except HAP. In the pH range 9.25 -10.5, additional evidence for a direct conversion to HAP is the constant value of the product (IC) of the conversion time (I) and the magnitude of the HAP ionic activity product (C), i.e., degree of supersaturation, of the solution phase before conversion.[19] As is the case with OCP, this result also infers that the HAP forms by heterogeneous nucleation on the ACP. Above pH 10.5, however, the reciprocal relation between conversion time and supersaturation breaks down, and the ACP/HAP transformation occurs more rapidly than expected. This is most evident in the reversal in conversion time at these higher pH's as noted earlier. One possible explanation for the accelerated transformation at these highly alkaline pHs is that the surface structure of the ACP becomes increasingly like that of HAP and, therefore, a better substrate for nucleation.[6]

A variety of substances can delay or prevent the conversion of ACP. Some substances, e.g., magnesium and carbonate ions, coprecipitate with the calcium and phosphate ions to produce compositionally complex amorphous materials that are less soluble and, therefore, more stable in solution than pure ACP.[28-32] When incorporated in sufficiently high concentrations, these substances can actually reduce solubility to where the amorphous precipitates remain stable indefinitely in solution. Other inhibitory substances, especially highly charged polyanions such as pyrophosphate,[33] $P_2O_7^{4-}$, and organic polyphosphoesters,[32,34-36] are effective at very low levels that are only a small fraction of that needed to slow the conversion process through incorporation. These latter substances apparently act, instead, by selectively adsorbing to and blocking crystal nucleation at surface active sites on the ACP. Fluoride ions, on the other hand, slow somewhat the conversion of ACP to OCP, but greatly accelerate the OCP hydrolysis stage, with the overall effect of advancing the appearance of the HAP end product.[37] At sufficiently high solution fluoride concentrations (3% or more of initial solution phosphate at pH 7.4), the OCP stage is entirely bypassed in favor of direct HAP crystal formation on the surface of the ACP.[37]

STRUCTURAL NATURE OF ACP

The constancy in the core composition of ACP, especially below pH 9.25, indicates a reproducible local structural unit whose existence is independent of the broad

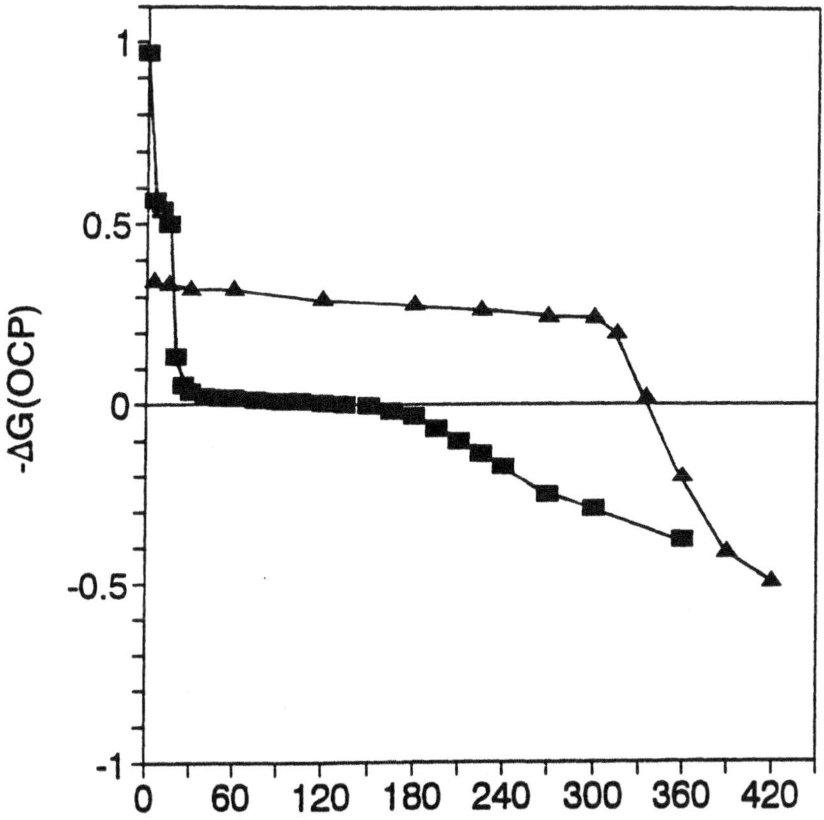

Figure 5. Changes with time in the degree of solution saturation (ΔG) with respect to octacalcium phosphate (OCP) for spontaneously precipitated reactions carried out at 25 °C and pH 7.4 (solid squares) and at pH 9.25 (solid triangles). The region above the solid line at $\Delta G = 0$ represents supersaturation; below this line, undersaturation. At pH 7.4, the abrupt decrease in ΔG at 15 minutes marks the end of the ACP stage, and the second decrease at 165 minutes denotes the end of the OCP stage. At pH 9.25, the ACP stage extends to 5 hours and there is no evidence of an intermediate OCP stage. The end product in both reactions is apatite.[18,38]

range of solution conditions under which ACP can be prepared. The apparent absence of observable crystalline features in ACP does not exclude *a priori* the possibility that this local unit may be a cryptostructural variant of one of the better known calcium phosphate salts such as dicalcium phosphate dihydrate (DCPD, $CaHPO_4 \cdot 2H_2O$), OCP, -TCP, or HAP. Several other lines of evidence suggest, however, that this is not the case. As already described, ACP is a slightly acidic salt whose composition lies between that of the acid calcium phosphates (DCPA, OCP) and of -TCP and HAP. The local structural unit is unlikely to be a phosphate-deficient form of DCPA or OCP as the structural integrity of these salts depends on a full complement of phosphate groups in the lattice. Also, the constancy in composition over a wide pH range appears to preclude the possibility that ACP is a cryptocrystalline mixture of these acidic salts with either -TCP or HAP. The overall Ca/PO_4 molar ratio in such mixtures would be expected to rise with pH as the component proportions would shift in favor of the basic salt. Phosphorus NMR spectra of ACP also precludes crystalline mixtures in favor of a unique, well-defined compound.[39]

The chance that ACP may be a Ca-deficient TCP or HAP cannot be as easily ruled out by such compositional arguments, especially since HAP with an ACP-like composition can be readily made at pH 7.4.[4] Bienenstock and Posner,[40] however, showed theoretically that the smallest structural unit of HAP would give rise to a x-ray diffraction pattern considerably different from that observed for ACP. Similarly, their NMR spectra are sufficiently different to suggest that ACP is not a poorly crystalline HAP.[39] In addition, as discussed above, ACP in solution does not convert directly into HAP below pH 9.25, or into TCP at any pH. Pure, Mg-free -TCP, in fact, cannot be precipitated at all from aqueous solutions.[41] If ACP was a cryptocrystalline form of either of these two salts, direct growth by crystal ripening would be expected to occur under these conditions. ACP, however, does convert preferentially into TCP upon ignition.[20] But the extremely high temperatures needed for this conversion (>550 C) together with the fact that Ca-deficient HAPs with similar Ca/PO_4 molar ratios (1.5) also convert into TCP upon ignition[20] indicates a compositionally driven thermochemical phase transition rather than one of crystal maturation.

Although the evidence appears to rule out ACP as being made up of highly disordered arrays of the unit cell contents of the common crystalline calcium phosphates, the possibility exits that a spatial subset of one of these cells may be the defining structural unit for this phase. Betts and Posner[7,8] were the first to explore this possibility in detail. They identified a sphere-like domain of the HAP unit cell with the composition of $Ca_9(PO_4)_6$ that, if expanded by 3%, produced a radial distribution function similar to that calculated from the x-ray diffraction profile of ACP. The domain consists of a central Ca^{2+} ion coordinated by the oxygens of 6 surrounding PO_4 anions which, in turn, are stabilized by another 8 Ca^{2+} ions spherically distributed along the domain's outer boundary. Betts and Posner[7,8] postulated that these domains are randomly clustered in ACP with water filling the intercluster space.

Their model, although attractive, has a number of weaknesses. First, EXAFS spectroscopy[9,10] indicates that the range of orderly Ca-Ca and Ca-P interactions in ACP is much shorter than expected from the regularity of the apatitic-like ionic arrangements in the model cluster. Second, no provision is made for the observation that approximately 10% of the phosphates in ACP are protonated. Lastly, the model is based on their assumption that water is not an essential part of the ACP structure, only filling the interstices between clusters, contrary to the experimental evidence of Sedlak and Beebe.[22] Possibly, a smaller spatial domain carved from the lattice of one of the acidic calcium phosphate salts, such as OCP, that contains a part of the hydration layer could provide a more satisfactory crystallographic model for ACP. It is also possible, however, that the defining structural unit of ACP is unique with no crystalline analogue.

ACP *IN VIVO*

Biogenic amorphous minerals are widely distributed in living organisms. They represent about 20% of the 60 or so different mineral types known to be formed by biologically controlled precipitation processes.[42] Among the biogenic amorphous minerals, ACPs are most prevalent in marine invertebrates. Although occasionally found in such intracellular locations as mitochondria,[43] these phases are most abundant in the teeth and exoskeletal structures of these organisms.[44,45] They are also exceptionally stable in these organisms, a result of having high magnesium and carbonate contents.

The widespread existence of ACPs in vertebrate organisms, on the other hand, is less clearly established. Except in some highly specialized locations such as inner ear structures of embryonic sharks and cartilage cell mitochondria,[42] sarcoplasmic reticulum,[46] and milk,[47] the experimental evidence for ACPs in these organisms is equivocal. Some electron microscopic studies[48-51] have reported the presence of a homogeneous haze or small spheroidal particles in bone tissue, concentrated primarily in actively metabolizing preosseous regions. Most hard tissue EM studies, however, do not report or even mention findings of such possibly amorphous structures. Early x-ray diffraction studies[52,53] reported the bone mineral of several animal species as exhibiting markedly less intense apatitic x-ray diffraction patterns than would be expected if the mineral was fully crystalline. The reduced intensity was attributed to the presence of 30% or more of non-crystalline mineral. Later studies,[54-56] however, questioned whether all of the x-ray amorphous bone mineral is truly noncrystalline. Some of this mineral fraction may include crystalline material too poorly developed to be discernible by the x-ray methods used. X-ray radial distribution analysis[54] of mature bone mineral places an upper limit of 10% on the amount that possibly is in an amorphous state. Further complicating a proper assessment of the presence of ACP in skeletal tissues is the fact that extracellular skeletal fluids are insufficiently supersaturated for spontaneous precipitation to occur. The minimum Ca x P solution product needed for ACP to form in physiological-like synthetic solutions at pH 7.4 is considerably greater than that found in serum.[57,58] Moreover, it is unlikely that Donnan equilibrium effects, release of macromolecular-bound reactants, and other passive, non-energy requiring processes can raise this product to homogeneous nucleation levels. Even if energy-driven, enzymatic processes were available to actively increase calcium and phosphate concentrations in skeletal fluids, heterogeneous nucleation on collagenous matrix components and secondary mineral nucleation would sufficiently accelerate to bring on substantial new crystal development in lieu of free ACP formation.[59]

Although subthreshold supersaturations and heterogeneous mineralization events would appear to prevent systemic *de novo* ACP formation in skeletal fluids, favorable local conditions may exist in skeletal tissues that would allow for ACP development. One such possible locale is in the aqueous core of matrix vesicles. Most commonly found in rapidly mineralizing tissues such as embryonic bone,[60] and growth plate cartilage,[61,62] matrix vesicles are submicron, membrane-enclosed spherical structures of cellular origin that appear to be the primary loci for the initial mineral in these tissues. Preceding the appearance of the first crystals in these structures is the accumulation of appreciable calcium and phosphate ions in their aqueous cores. These accumulations exceed by several fold the ion levels needed for homogeneous nucleation,[63,64] a possible indication that they exist in an amorphous state. Studies with synthetic liposomes[65] show, further, that ACP should be able to readily form under such vesicular-like conditions. The infrared spectroscopic finding[66] that the first crystalline material in matrix vesicles is possibly OCP is also consistent with the presence of ACP. As described earlier, OCP is an intermediate phase in the conversion of ACP to HAP at physiological pH. On the other hand, there was no direct infrared evidence of ACP preceding the OCP. Nor has there been any reports of discrete ACP-like particles in matrix vesicles. These latter findings suggest that a sizable fraction of the calcium and phosphate ions accumulated prior to the appearance of the OCP may be in a more diffuse amorphous state, possibly stabilized by complexation with proteins and other biomolecular species found in the vesicle interior or on the inner membrane surface.

To summarize, the physical and morphological evidence for ACP in skeletal tissues is ambiguous, but solution conditions in the extracellular matrix would appear to preclude the wide distribution of this phase throughout the mineralized regions of these tissues. The evidence is stronger for its presence in specialized organelles such as mitochondria and matrix vesicles. In the latter locale, ACP-organic complexes may be an important factor in the earliest stages of calcification by providing nucleating sites or reactant ions for the first mineral crystals that form in skeletal matrices.

CONCLUSION

In the introduction, ACP is described as a unique calcium phosphate in that it lacks crystalline order. This appellation is also exemplified, however, by the seemingly paradoxical nature of many of the other properties covered in this chapter. For example, the spontaneous precipitation of ACP from highly unstable solutions suggest kinetic factors control its formation, yet ACP is thermodynamically well-behaved over a wide pH range with a constant, solubility-determining composition. The chemical properties of ACP suggest further a well-defined local organizational unit despite an absence of the stabilizing long-range regularity found in crystalline materials. This structural order within disorder is undoubtedly ACP's most striking feature, but the resulting solution instability probably has a greater impact *in vivo*. As stated in the previous section, the transient appearance and subsequent transformation of ACP-like complexes in specialized extracellular mineral-forming *loci* such as matrix vesicles suggests a critical intermediary role for this phase in the initial deposition of HAP in many skeletal tissues.

REFERENCES

1. Watson ML, Robinson RA. Collagen-crystal relationships in bone. II. Electron microscope study of basic calcium phosphate crystals. Amer J Anat 93:25-60, 1953

2. Eanes ED, Gillessen IH, Posner AS. Intermediate states in the precipitation of hydroxyapatite. Nature 208:365-367, 1965

3. Eanes ED, Posner AS. Kinetics and mechanism of conversion of noncrystalline calcium phosphate to crystalline hydroxyapatite. Trans NY Acad Sci. II 28:233-241, 1965

4. Eanes ED, Meyer JL. The maturation of crystalline calcium phosphates in aqueous suspensions at physiologic pH. Calcif Tiss Res 23:259-269, 1977

5. Wuthier RE, Rice GS, Wallace JEB Jr, Weaver RL, LeGeros RZ, Eanes ED. In vitro precipitation of calcium phosphate under intracellular conditions: formation of brushite from an amorphous precursor in the absence of ATP. Calcif Tiss Int 37:401-410, 1985

6. Meyer JL, Weatherall CC. Amorphous to crystalline calcium phosphate phase transformation at elevated pH. J Colloid Interface Sci 89:257-267, 1982

7. Betts F, Posner AS. An x-ray radial distribution study of amorphous calcium phosphate. Mat Res Bull 9:353-360, 1974

8. Betts F, Posner AS. A Structural model for amorphous calcium phosphate. Trans Amer Cryst Assoc 10:73-84, 1974

9. Eanes ED, Powers L, Costa JL. Extended x-ray absorption fine structure (EXAFS) studies on calcium in crystalline and amorphous solids of biological interest. Cell Calcium 2:251-262, 1981

10. Harris JE, Hukins DWL, Holt C, Hasnain SS. Conversion of amorphous calcium phosphate into hydroxyapatite investigated by EXAFS spectroscopy. J Crystal Growth 84:563-570, 1987

11. Termine JD, Lundy DR. Vibrational spectra of some phosphate salts amorphous to x-ray diffraction. Calcif Tiss Res 15:55-70, 1974

12. Nylen MU, Eanes ED, Termine JD. Molecular and ultrastructural studies of non-crystalline calcium phosphates. Calcif Tiss Res 9:95-108, 1972

13. Eanes ED, Termine JD, Nylen MU. An electron microscopic study of the formation of amorphous calcium phosphate and its transformation to crystalline apatite. Calcif Tiss Res 12:143-158, 1973

14. Barton SS, Harrison BH. "Surface and Bulk Properties of Amorphous Calcium Phosphate" in *Colloid and Interface Science* V.III, Academic Press, Inc: New York, 1976

15. Termine JD, Eanes ED. Comparative chemistry of amorphous and apatitic calcium phosphate preparations. Calcif Tiss Res 10:171-197, 1972

16. Walton AG, Bodin WJ, Furedi H, Schwartz A. Nucleation of calcium phosphate from solution. Canadian J Chem 45:2695-2701, 1967

17. Meyer JL. Hydroxyl content of solution-precipitated calcium phosphates. Calcif Tiss Int 27:153-160, 1979

18. Meyer JL, Eanes ED. A thermodynamic analysis of the amorphous to crystalline calcium phosphate transformation. Calcif Tiss Res 25:59-68, 1978

19. Meyer JL. Phase transformations in the spontaneous precipitation of calcium phosphate. Croatica Chemica Acta. 56:753-767, 1983

20. Eanes ED. Thermochemical studies on amorphous calcium phosphate. Calcif Tiss Res 5:133-145, 1970

21. Holmes JM, Beebe RA. Surface areas by gas adsorption on amorphous calcium phosphate and crystalline hydroxyapatite. Calcif Tiss Res 7:163-174, 1971

22. Sedlak JM, Beebe RA. Temperature programmed dehydration of amorphous calcium phosphate. J Colloid Interface Sci 47:483-489, 1974

23. Boskey AL, Posner AS. Conversion of amorphous calcium phosphate to microcrystalline hydroxyapatite. A pH-dependent, solution-mediated, solid-solid conversion. J Phys Chem 77:2313-2317, 1973

24. Cahn JW. "On the Morphological Stability of Growing Crystals" In *Crystal Growth*, Peiser HS, ed. Pergamon:Oxford, 1967

25. Brown WE, Schroeder LW, Ferris JS. Interlayering of crystalline octacalcium phosphate and hydroxyapatite. J Phys Chem 83:1385-1388, 1979

26. Brown WE, Smith JP, Lehr JR, Frazier AW. Octacalcium phosphate and hydroxyapatite. Nature 196:1048-1055, 1962

27. Brown WE, Mathew M, Tung MS. Crystal chemistry of octacalcium phosphate. Prog Crystal Growth Charact 4:59-87, 1981

28. Bachra BN, Trautz OR, Simon SL. Precipitation of calcium carbonates and phosphates under physiological conditions. Arch Biochem Biophys 103:124-138, 1963

29. Bachra BN. Precipitation of calcium carbonates and phosphates from metastable solutions. Ann NY Acad Sci 109:251-255, 1963

30. Eanes ED, Posner AS. Intermediate phases in the basic solution preparation of alkaline earth phosphates. Calcif Tiss Res 2:38-48, 1968

31. Boskey AL, Posner AS. Magnesium stabilization of amorphous calcium phosphate: A kinetic study. Mat Res Bull 9:907-916, 1974

32. Blumenthal, NC, Betts F, Posner AS. Stabilization of amorphous calcium phosphate by Mg and ATP. Calcif Tiss Res 23:245-250, 1977

33. Meyer JL. Can biological calcification occur in the presence of pyrophosphate? Arch Biochem Biophys 231:1-8, 1984

34. Blumenthal NC, Betts F, Posner AS. Nucleotide stabilization of amorphous calcium phosphate. Mat Res Bull 10:1055-1060, 1975

35. Termine JD, Conn KM. Inhibition of apatite formation by phosphorylated metabolites and macromolecules. Calcif Tiss Res 22:149-157, 1976

36. Termine JD, Eanes ED, Conn KM. Phosphoprotein modulation of apatite crystallization. Calcif Tiss Int 31:247-251, 1980

37. Eanes ED, Meyer JL. The influence of fluoride on apatite formation from unstable supersaturated solutions at pH 7.4. J Dent Res 57:617-624, 1978

38. Eanes ED. Crystal growth of mineral phases in skeletal tissues. Prog Crystal Growth Charact. 3:3-15, 1980

39. Tropp J, Blumenthal NC, Waugh JS. Phosphorus NMR study of solid amorphous calcium phosphate. J Amer Chem Soc 105:22-26, 1983

40. Bienenstock A, Posner AS. Calculation of the x-ray intensities from arrays of small crystallites of hydroxyapatite. Arch Biochem Biophys 124:606-607, 1968

41. LeGeros RZ. *Calcium Phosphates in Oral Biology and Medicine.* Karger: Basel, 1991

42. Lowenstam HA, Weiner S. *On Biomineralization.* Oxford University Press:New York, 1989

43. Becker GL, Termine JD, Eanes ED. Comparative studies of intra- and extramitrochondrial calcium phosphates from the hepatopancreas of the blue crab (callinectes sapidus). Calcif Tiss Res 21:105-113, 1976

44. Lowenstam HA. Phosphatic hard tissues of marine invertebrates: their nature and mechanical function and some fossil implications. Chem Geol 9:153-166, 1972

45. Lowenstam HA, Weiner S. Transformation of amorphous calcium phosphate to crystalline dahllite in the radular teeth of chitons. Science 227:51-53, 1985

46. Raeymaekers L, Agostini B, Hasselbach W. The formation of intravesicular calcium phosphate deposits in microsomes of smooth muscle; a comparison with sarcoplasmic reticulum of skeletal muscle. Histochem 70:139-150, 1981

47. McGann TCA, Buchheim W, Kearney RD, Richardson T. Composition and ultrastructure of calcium phosphate-citrate complexes in bovine milk systems. Biochim Biophys Acta 760:415-420, 1983

48. Robinson RA, Watson ML. Crystal-collagen relationships in bone as observed in the electron microscope. III. Crystal and collagen morphology as a function of age. Ann NY Acad Sci 60:596-628, 1955

49. Thyberg J. Electron microscopic studies on the initial phases of calcification in guinea pig epiphyseal cartilage. J Ultrastruc Res 46:206-218, 1974

50. Schraer H, Gay CV. Matrix vesicles in newly synthesizing bone observed after ultracryotomy and ultramicroincineration. Calcif Tiss Res 23:185-188, 1977

51. Gay CV. The ultrastructure of the extracellular phase of bone as observed in frozen thin sections. Calcif Tiss Res 23:215-223, 1977

52. Harper RA, Posner AS. Measurement of non-crystalline calcium phosphate in bone mineral. Proc Soc Exp Biol Med 122:137-142, 1966

53. Termine JD, Posner AS. Amorphous/crystalline interrelationships in bone mineral. Calcif Tiss Res 1:8-23, 1967

54. Posner AS, Betts F. Synthetic amorphous calcium phosphate and its relation to bone mineral structure. Accounts of Chemical Research 8:273-281, 1975

55. Grynpas MD, Bonar LC, Glimcher MJ. Failure to detect an amorphous calcium phosphate solid phase in bone mineral : a radial distribution function study. Calcif Tiss Int 36:291-301, 1984

56. Grynpas MD, Bonar LC, Glimcher MJ. X-ray diffraction radial distribution function studies on bone mineral and synthetic calcium phosphates. J Mater Sci 19:723-736, 1984

57. Termine JD, Eanes ED. Calcium phosphate deposition from balanced salt solutions. Calcif Tiss Res 15:81-84, 1974

58. Boskey AL, Posner AS. Formation of hydroxyapatite at low supersaturation. J Phys Chem 80:40-45, 1976

59. Eanes ED. The interaction of supersaturated calcium phosphate solutions with apatitic substrates. Calcif Tiss Res 20:75-89, 1976

60. Bernard GW, Pease DC. An electron microscopic study of initial intramembranous osteogenesis. Amer J Anat 125:271-290, 1969

61. Bonucci E. Fine structure of early cartilage calcification. J Ultrastruct Res 20:33-50, 1967

62. Anderson HC. Vesicles associated with calcification in the matrix of epiphyseal cartilage. J Cell Biol 41:59-72, 1969

63. Wuthier RE. Electrolytes of isolated epiphyseal chondrocytes, matrix vesicles, and extracellular fluid. Calcif Tiss Res 23:125-133, 1977

64. Wuthier RE, Gore ST. Partition of inorganic ions and phospholipids in isolated cell, membrane and matrix vesicle fractions: evidence for Ca-P_i-acidic phospholipid complexes. Calcif Tiss Res 24:163-171, 1977

65. Eanes ED, Hailer AW, Costa JL. Calcium phosphate formation in aqueous suspensions of multilamellar liposomes. Calcif Tiss Int 36:421-430, 1984

66. Sauer GR, Wuthier RE. Fourier transform infrared characterization of mineral phases formed during induction of mineralization by collagenase-released matrix vesicles in vitro. J Biol Chem 263:13718-13724, 1988

3

GROWTH OF CALCIUM PHOSPHATES ON DIFFERENT SUBSTRATES: EPITAXIAL CONSIDERATIONS

Petros G. Koutsoukos

Institute of Chemical Engineering and High Temperature Chemical Processes and the Department of Chemical Engineering University of Patras GR 265 00 Patras, GREECE

INTRODUCTION

The formation of calcium phosphates takes place in supersaturated solutions either homogeneously or heterogeneously. In the former case, the supersaturated solutions must be absolutely free of any foreign surfaces which may reduce the free energy for the formation of new nuclei. Experimentally, such conditions are very difficult to achieve and most often the formation of crystals in solutions takes place heterogeneously on existing foreign surfaces. From a supersaturated solution of calcium phosphate a number of phases may form in the order of decreasing solubility: Amorphous calcium phosphate (ACP),[1] dicalcium phosphate dihydrate (DCPD) $CaHPO_4 \cdot 2H_2O$,[2,3] anhydrous calcium phosphate (DCPA) $CaHPO_4$[4] octacalcium phosphate $Ca_8H_2(PO_4)_6 \bullet 5H_2O$ (OCP),[5,6] β-Tricalcium phosphate (β-TCP), $Ca_3(PO_4)_2$,[7] and hydroxyapatite ($Ca_{10}(PO_4)_6(OH)_2$ HAP).[8,9] The various phases have markedly different solubilities and different crystallographic structures. The crystallographic characteristics of the various calcium phosphates are summarized in Table 1. The formation of the calcium phosphate phases in aqueous supersaturated solutions may take place either spontaneously or through the mediation of a substrate. In case the substrate consists of the calcium phosphate phase which shall subsequently grow, the process is termed as seeded crystal growth.

Spontaneous precipitation takes place at high supersaturations at conditions corresponding to the labile region, while seeded growth normally takes place at considerably lower supersaturations in which crystal growth proceeds only after the introduction of the seed crystals. The solutions in which seeded growth takes place are metastable and remain unchanged practically indefinitely.

Crystal growth of a crystalline phase may in principle take place on a crystalline substrate introduced in the corresponding metastable solution at conditions of equilibrium between the aqueous phase and the solid substrate. The particular case in which the formation of crystalline overgrowth on a substrate is oriented, the process is termed as epitaxial growth, from the composite Greek word επιταξιζ῀ (επι + ταξιζ) meaning "upon (= overgrowth) with orderness". Epitaxy involves the growth of crystalline phase over another provided that the atomic dimensions of one or more commonly occuring faces of each are similar. Epitaxy is behind artificial rain caused by seeding clouds with silver iodide, the lattice of which shows an excellent fit with the lattice of ice.

Epitaxy is a common crystallographic occurrence involving the growth of one crystal on another when atomic dimensions of one or more commonly occuring faces of each are similar. Although chemical similarity is not necessary it is helpful. Epitaxy is one reason why certain materials can function as seed nuclei for other materials. Some deposits require an epitaxial relationship before they will deposit and grow. Others, nucleate and grow on almost any substarte including gas bubbles.

The epitaxial growth of one calcium phosphate phase upon another or upon foreign (both organic and inorganic) substrates may be important steps during the crystallization of apatites and for biological mineralization in general. Both thermodynamics of phase formation the kinetics involved and the surface phenomena are important in the epitaxial formation of a mineral salt on a substrate. The present work aims at reviewing the considerable work done on the epitaxial growth of calcium phosphates on calcium phosphates and other substrates both inorganic and organic. Thermodynamic and kinetics results will be presented and the importance of the substrate for the epitaxial growth will be investigated through the published results.

Table 1. Crystallographic Characteristics of the Calcium Phosphate Phases[4]

Phase abbrev.	Formula	Crystallographic System	Space Group	a_0	b_0	c_0	α	β	γ
					Å			°	
DCPD	$CaHPO_4.2H_2O$	Monoclinic	2/m	5.812	15.180	6.239		116,41	
DCPA	$CaHPO_4$	Triclinic	$P\bar{1}$	6.906	8.577	6.634	93.90	91.50	127.63
OCP	$Ca_8(PO_4)_6.5H_2O$	Triclinic	$P\bar{1}$	19.715	9.534	6.839	90.14	92.52	108.67
β-TCP	β-$Ca_3(PO_4)_2$	Trigonal	R3C	9.460		6.874			
HAP	$Ca_{10}(PO_4)_6(OH)_2$	Hexagonal	$P6_3/m$	9.418	6.881	6.881	90	90	120

Fundamental thermodynamic considerations

Epitaxial crystallisation of a crystalline phase on a substrate takes place in solutions, provided that there is a chemical potential difference between the non-equilibrium state, $\mu_s = \left(\sum_i \mu_i\right)_s$ and the equilibrium, $\mu_\infty = \left(\sum_i \mu_i\right)_\infty$:

$$\Delta\mu = \mu_s - \mu_\infty \qquad (1)$$

The nonequilibrium state is created through the preparation of supersaturated solutions from the salt components (in the case of calcium phosphates from stock solutions of Ca^{2+}, PO_4^{3-} and OH^- ions) in such concentrations that the solutions are stable i.e., no spontaneous precipitation occurs for practically infinitely long time periods. Regardless whether the process of epitaxial growth is considered macroscopically or at an atomic level three zones may be distinguished: The substrate, normally crystalline material, the bulk solution and an intermediate zone, adjacent to the crystalline substrate which transforms into the overgrown phase by the gradual increase of the orderness of the growing phase. The order of the molecular or ionic units in this intermediate zone is less than the corresponding in the crystalline substrate and definitely higher than the corresponding to the fluid phase. All processes which determine epitaxy take place precisely at this layer which may be controlled from the conditions chosen.

Most often, in systems in which epitaxy is investigated, single crystals are selected as crystalline substrates. In the case of calcium phosphates however the majority of the reported investigations is concerned with polycrystalline materials in which it is difficult to distinguish whether there is true epitaxy i.e., oriented overgrowth of a calcium phosphate phase over a substrate. The term epitaxy is however often used in crystallography and the use of this term may be justified in view of the selectivity of the substrates for the overgrown crystalline phase.

The role of the substrate is very decisive for epitaxial growth, provided that the thermodynamic driving force in the solution is present. The key criterion for the relationship between the substrate and the overgrowth is the compatibility of their crystal lattices in geometrical terms. The lattice compatibility consists in the favorable arrangement of the atomic units in the crystal lattices of the substrate and the overgrowth which may be attained either through the accommodation of structural defects in the layer or by strain associated with a relevant interfacial potential energy. A quantitative measure for the lattice compatibility between substrates and overgrown phases has been termed as lattice misfit by van der Merwe.[10] Assuming a rectangular symmetry for both the substrate and the overgrowth, the lattice misfit, $\%\sigma_i$, may be defined as:

$$\delta_i = \frac{\alpha_{oi} - \alpha_{si}}{\alpha_{si}} \tag{2}$$

where α_{si}, α_{oi} are the lattice constants of the substrate and of the overgrowth respectively. Since the substrate surface however is not perfect the lattice misfit is expressed in relation with the fixed interatomic spacing.

The kinetics and energetics of solid/solid interfaces depend strongly on the spacing and structure of the lattices involved as well as on their relative orientations. There are two extreme cases for the relationship between the substrate and the overgrowth: A. complete coherence and B. complete incoherence. In the first case both the substrate and the overgrowth have the same or very similar lattice parameters (Figure 1a). In this case, the crystal lattice shows a continuity across the interface and it is only the chemical composition change which differentiates the overgrowth from the substrate. In the case of different materials the change in chemical composition may be sharp but in cases in which there is overgrowth of one calcium phosphate phase over another (e.g., HAP on DCPD) the composition change zone may be diffuse and spread over the surface of the substrate.[11] The crystal lattice coherence may be limited to one orientation only as is the case for the close packed planes in adjoining hexagonal and face centred cubic crystals. It is usually impossible to achieve coherence between a substrate and an overgrowth when their crystal structures are entirely different. In most cases however, partial coherence is encountered (Figure 1b). The overgrowth of a crystalline layer on a substrate occurs in the direction of energy minimization. Incoherent interfaces in which there is little or no bonding, the energy is relatively higher in comparison with coherent interfaces.

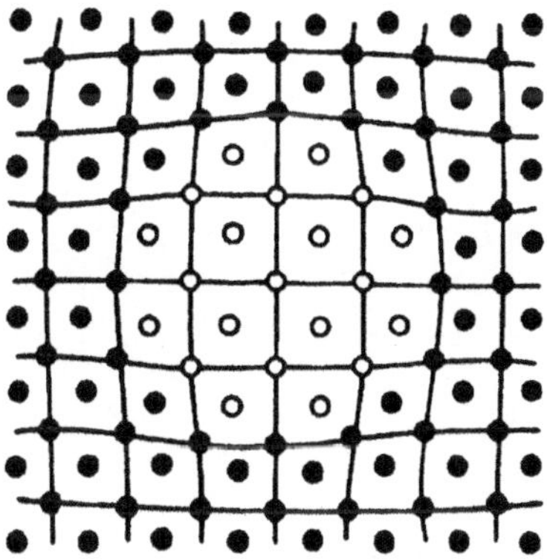

a

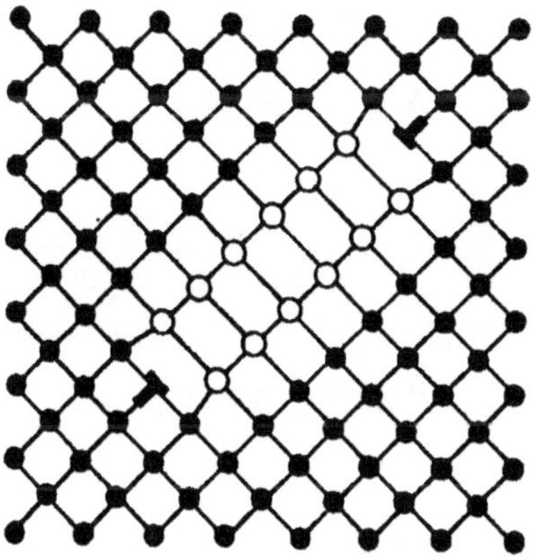

b

Figure 1. Interface structure for the nucleation of a crystalline solid on a crystalline substrate. a: completely coherent interface; b: semi-coherent interface.

In the majority of the cases however, interfaces are partially coherent and there are regions of high coherence separated by disordered regions. In the cases of formation of thin epilayers (1 or 2 layers thick) the interatomic distances at equilibrium may differ by 5% and equation (2) is used to express misfit. For thick epilayers however the definition of lattice misfit is modified so that both the substrate and the overgrowth are treated equivalently.[12]

$$\delta_i = \frac{2(\alpha_{si} - \alpha_{oi})}{\alpha_{si} + \alpha_{oi}} \tag{3}$$

When the lattice misfit between the substrate and the overgrowth is sufficiently small the strain between the two lattices is low and as a result a coherent epilayer is formed on the substrate, a few atomic layers thick. Upon increase of the epilayer thickness dislocations develop in order to reduce the increasing strain energy. It should be noted that the criteria of lattice misfit posed by equations (2) and (3) are mainly valid for substrate-overgrowth systems of the same crystallographic symmetry. If the heteroepitaxial systems differ crystallographically from the substrates the interfacial translational symmetry should be taken into consideration in the comparisons with respect to the characteristics of the bulk phases.[13] Thus, in such cases, favorable lattice matching between two crystalline systems exist provided that the interface translational symmetry is comparable with the symmetry on both sides of the interface.

The overgrowth of an epilayer over a substrate is preceded by the nucleation process in which the heteronuclei form on the substrate bound to the crystallographic limitations presented above. In case of favorable lattice matching the substrate may act as the catalyst of a series of stepwise bimolecular reactions which may be presented as:

$$\begin{aligned} &A + A \Leftrightarrow A_2 \\ &A + A_2 \Leftrightarrow A_3 \\ &\cdot \quad \cdot \quad \cdot \\ &\cdot \quad \cdot \quad \cdot \\ &\cdot \quad \cdot \quad \cdot \\ &A_{n-1} + A \Leftrightarrow A_n^* \end{aligned} \tag{4}$$

where A, A_2, ... A_n^* are clusters of growth units (ions or molecules) of the overgrowth. The substrate on which the growth units are attached may be thought of as a spherical particle of increasing thickness because of the continuous arrival of growth units. The growth in size continues until eventually a macroscopic nucleus of the overgrowth is formed. The mechanism is the same as in homogeneous nucleation except for the fact that in the presence of a substrate the supersaturations needed to observe the phenomenon are quite lower. The relationship between the driving force for homogeneous, ΔG_h, nucleation with the corresponding for heterogeneous, ΔG_{het} is:

$$\Delta G_{het} = \Phi \Delta G_{hom} \qquad (5)$$

where Φ is a factor ($0< \Phi <1$) related to the contact angles, φ, between the substrate and the heteroembryo forming on the substrate.[14] As may be seen in Figures 2a and 2b for the formation of coherent heteronuclei c on a phase s it is necessary to have a favorable balance of the surface energies[15] (Figure 2b).

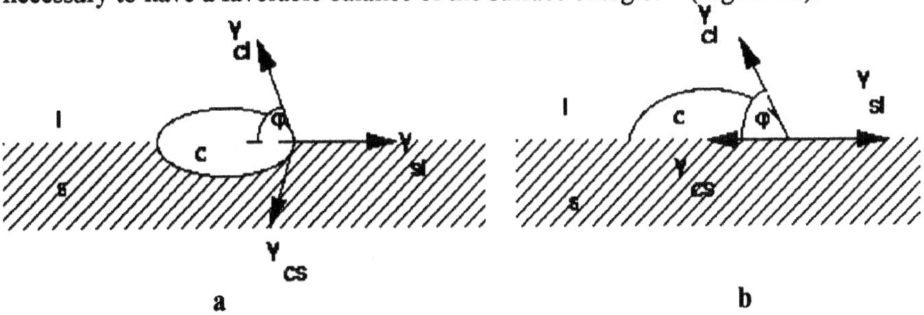

Figure 2. Nucleation of a crystalline solid, c, on a substrate, s: a: incoherent nucleus; b: fully coherent nucleus. l :liquid phase

The contact angle between the overgrowth and the substrate in terms of the interfacial energies is given by equation (6).[16]

$$\cos\varphi = \frac{\gamma_{sl} - \gamma_{cs}}{\gamma_{cl}} \qquad (6)$$

and the relationship between the contact angle and the factor φ is:

$$\Phi = \frac{(2+\cos\varphi)(1-\cos\varphi)^2}{4} \qquad (7)$$

Both in homogeneous and in heterogeneous nucleation, it is obvious that the supersaturation, is also very important as it is related with the driving force. The ratio:

$$S = \left(\frac{(\Pi\alpha_i)_s}{K_s^0} \right)^{1/\nu} \qquad (8)$$

of the activity product of the ions constituting the calcium phosphate phase forming containing ν ions over the respective thermodynamic solubility product is the measure of departure from equilibrium i.e., supersaturation. For the various calcium phosphates considered here the expressions for the activity products are:

DCPD : $(Ca^{2+})(HPO_4^{2-})$ $\qquad \nu = 2$

$OCP : (Ca^{2+})^4 (PO_4^{3-})^3$ $\quad$ $\nu = 8$
$\beta\text{-}TCP : (Ca^{2+})^3 (PO_4^{3-})^2$ $\quad$ $\nu = 5$
$HAP : (Ca^{2+})^5 (PO_4^{3-})^3 (OH^-)$ $\quad$ $\nu = 9$

The driving force for the formation of new calcium phosphate nuclei on a substrate is:

$$\Delta G = RT\ln S \tag{9}$$

in equation (9), R is the gas constant and T the absolute temperature. As may be seen from equations (8) and (9) it is very important to calculate precisely the supersaturation in an aqueous calcium phosphate solution. This may be done by taking into account the appropriate equilibria, mass and charge balance conditions and using the appropriate expressions for the activity coefficients of the various ionic species.[9,17]

Kinetics of formation of calcium phosphates on foreign substrates

The rate, J, of formation of new nuclei on favorable substrates is given by equation (10):

$$J = A\exp\left(-\frac{\Phi\Delta G}{kT}\right) \tag{10}$$

where A is a pre-exponential factor related with the mechanism of incorporation of the growth units in the embryo. Taking into consideration the Volmer-Weber-Becker theory, the rate of precipitation as a function of the solution supersaturation S may be expressed as.[18]

$$J = A\exp\left(-\frac{\Phi\beta\sigma^3\upsilon^2}{k^3T^3}\frac{1}{(\ln S)^2}\right) \tag{11}$$

where β, is a shape factor (=16π/3 for spherical nuclei), ν the molecular volume of the growing phase and σ its surface energy. In most cases, the formation of heteronuclei is preceded by induction times, τ, inversely proportional to the solution supersaturation. The induction times lapsed from the establishment of supersaturation in the presence of the substrate until the observation of the first changes in the solution properties, corresponding to the formation of calcium phosphates, are related with the time needed for the formation of critical nuclei which may grow further forming macroscopic crystalline overgrowth layers. Since the induction times are inversely proportional to the rates of growth of the new phase, a relationship similar to equation (11) may be derived for the dependence of the induction time on supersaturation:

$$\tau = \exp\left(\frac{\Phi\beta\sigma^3\upsilon^2}{(kT)^3}\frac{1}{(\ln S)^2}\right) \tag{12}$$

Experimentally, the investigations of the overgrowth of a calcium phosphate phase on a substrate suspended in supersaturated solutions may be done by monitoring the change of solution variables associated with the formation the respective salt. Such solution variables for calcium phosphates are the pH, changes in total calcium, ΔCa_t, and total phosphate, ΔP_t.[19] The main problems associated with this approach include the relatively large errors involved in the chemical analysis which render the indentification of calcium phosphates based on calcium: phosphate molar ratio questionable, and the possible initial surges observed in heterogeneous precipitation processes. The surges intoduce large uncertainties in the measurements of the rates of precipitation. These problems are more pronounced in investigations of epitaxial growth in which the overgrowth may proceed very slowly, past relatively long induction times. In some cases, the supersaturations involved in investigations of overgrowth on foreign substrates are so high that the possibility of formation of phases other than the investigated may be formed[20-22]. The best method available for the precise evaluation of the kinetics parameters is the method in which the chemical potential of the solution species is kept constant.[8,23] Following suspension of the substrate investigated for its ability to induce the formation of calcium phosphate in metastable supersaturated solutions, the process starts provided the critical supersaturation is reached, accompanied by proton release. The proton variation is monitored with a glass-saturated calomel pair of electrodes. Through the use of an automatic titrator, the lattice ions activities are held constant by the simultaneous addition of titrant solutions of calcium, phosphate and potassium hydroxide with the stoichiometry of the precipitating salt. It is thus possible to monitor overgrowth processes which proceed at very low rates and at the same time to measure accurately the precipitation rates from the traces of titrant solutions added as a function of time. The major advantage of this method is that through the maintenance of constant supersaturation and of pseudo-steady state conditions it is possible to extend the crystallization process growing quantities sufficient for the characterization of the growing calcium phosphate phases, even at very low supersaturations.[8] The invariance of the solution species activities throughout the crystal growth processes which depend strongly and directly on the titrant solutions composition is taken as a proof of the stoichiometry of the precipitating phase.

The growth of calcium phosphates on inorganic salts

DCPD is the calcium phosphate phase which has been suggested to form at the initial stages of bone mineralization, being the precursor of HAP.[24-27] As may be seen from the solubility isotherms of the calcium phosphates shown in Figure 3. Although HAP is the most stable phase, DCPD may become more stable if the solution is sufficiently acid. Upon increasing acidity, a more acid calcium phosphate may be expected to form on a calcium phosphate substrate.

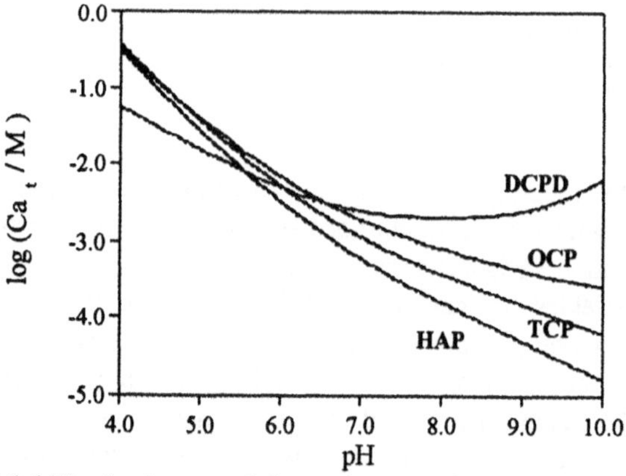

Figure 3. Solubility isotherms of the various calcium phosphates at 37°C, 0.15M NaCl.

Seeded growth studies have shown that even at pH 7.40 the formation of DCPD is possible in aqueous solutions supersaturated with respect to all calcium phosphate phases.[28] Moreover DCPD together with HAP have been identified in embryonic chick bone.[29] Experiments at lower pH (5.0 - 5.6) have confirmed that DCPD may be selectively grown on HAP seed crystals following the same kinetics as the growth of DCPD on DCPD seed crystals at similar conditions.[30] The experimental data available corroborate the suggestion of epitaxial growth of HAP on DCPD.[31] Comparison of the (010) and (110) planes of the unit cells at HAP and DCPD showed that in these planes lattice matching is favorable. The excellent comatibility between the unit cell lattice parameters for HAP and DCPD (misfit below 15%) may also be seen in Figure 4. Although the lattice misfit is better than

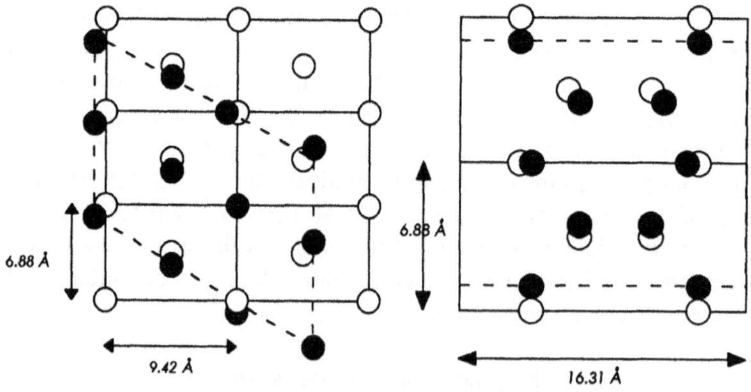

Figure 4. Calcium atoms in the unit cell at HAP (O) and DCPD (●)
a: (010) plane; b:(110) plane

15%, this does not consitute a necessary condition for epitaxial growth. In experiments however in which calcium phosphate solutions, supersaturated with respect to TCP and HAP only and saturated with respect to DCPD, were seeded with DCPD, HAP was grown selectively on the seed crystals past the lapse of induction times inversely proportional to the solution supersaturation. Moreover, the rates of growth of HAP on the DCPD seed crystals increased with increasing supersaturation and were independent of the amount of the inoculating seed crystals. These two indications suggested that the growth of HAP on DCPD is selective and it is not secondary nucleation and growth.[31] The dependence of the induction times and of the subsequent crystal growth rates on the solution supersaturation is shown in Figure 5.

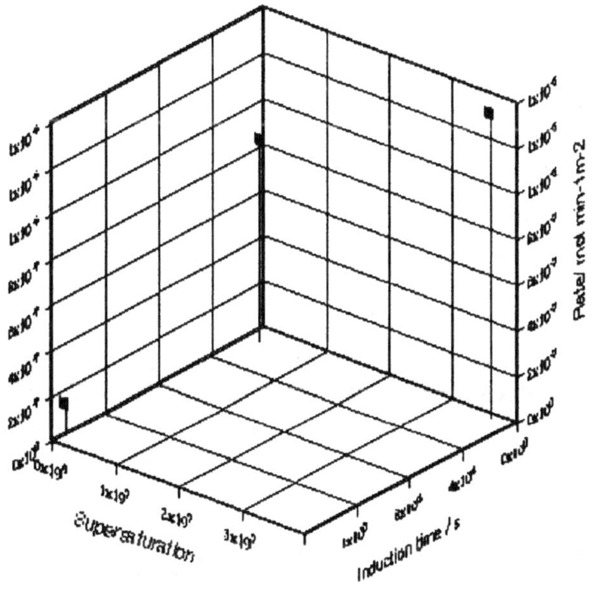

Figure 5. Overgrowth of HAP on DCPD seed crystals at conditions of constant supersaturation. Induction times and crystal growth rates dependence on the solution supersaturation.[33]

The measurements were done using the constant supersaturation method. Assuming a value of 100 mJm^{-2} for the surface energy of the HAP overgrowth on HAP a contact angle of 35° was calculated for the new nucleus. This angle should not be taken in a strict thermodynamic sense since it was obtained from kinetics data and not at equilibrium conditions. It may be used however, in addition to the lattice misfit as a measure for the capability of overgrowth of a salt on a substrate. The relationship between the contact angle of the new nucleus of the crystalline substance which nucleates on the substrate is schematically shown in Figure 6.

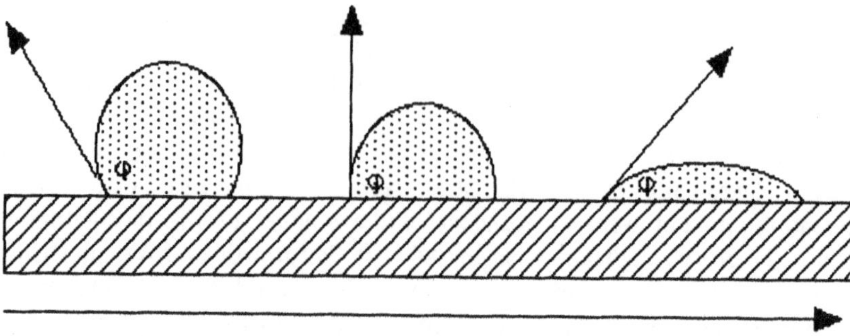

Figure 6. The importance of contact angle for the growth of a nucleus on a solid, crystalline substrate.

Analysis of the published data,[31] showed that the dependence of the rates, R_g, of HAP overgrowth on the solution supersaturation is of high order according to equation (13) which is widely used for the fit of crystal growth kinetics data:

$$R_g = k_g \sigma^n \tag{13}$$

here k_g is a constant depending on the active growth sites on the seed crystals, $\sigma =$ S-1 is the relative supersaturation and n the order of the process which is indicative of the mechanism. The value n=3 found suggested a polynuclear type of mechanism, which may be supported from the electron micrographs presented[31] in which HAP grown on DCPD appeared to form clusters on isolated locations of the DCPD seed crystals instead of being evenly distributed on the DCPD platelets. However, mechanistic conclusions cannot be safely reached because of the unavailability of large number of data, yielding statistical validity. It should be pointed however that these experiments are rather difficult to perform and they are limited within narrow concentration limits in order to avoid complications from the presence of precursor phases.

The growth of HAP on CaF_2 crystals has allso been reported.[31] In CaF_2, the calcium ions are found on a face centered cubic lattice. Each fluorine atom is at the center of one of the smaller cubes obtained by dividing the unit cube into eight parts. It is surrounded by four calcium atoms and each calcium atom is surrounded by eight fluorine atoms (8,4 coordination). The lattice matching for the (111) and (110) faces of CaF_2 with the corresponding faces for HAP is better than 8%[32] and the favorable lattice matching may be seen in Figure 7.

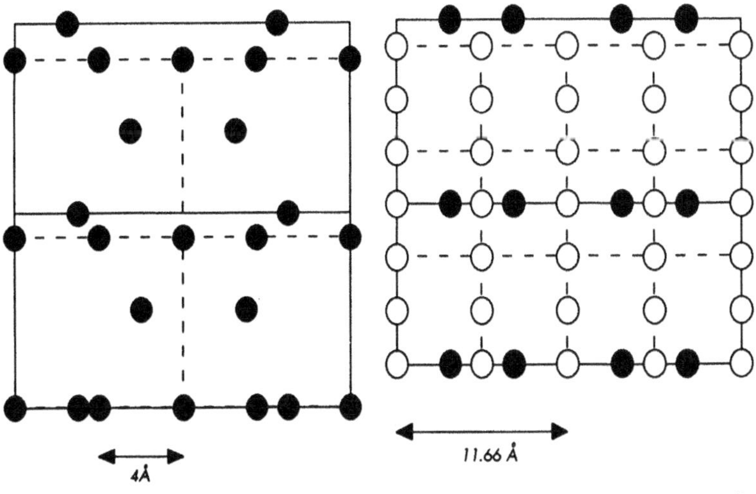

Figure 7. (110) (a) and (111) (b) faces for HAP (●) and CaF_2 (O) unit cells respectively.

Experiments at constant supersaturation were done in calcium phosphate solutions supersaturated with respect to HAP and TCP only. The solutions, were also saturated with respect to CaF_2 to avoid dissolution of the seed crystals. The kinetics parameters measured showed inverse dependence the induction times on the solution supersaturation and proportionality to the rates of crystal growth. The dependence of induction times preceding the overgrowth of HAP on CaF_2 and of the subsequent rates of crystal growth on the solution supersaturation are shown in Figure 8. The morphology of the grown crystals showed elongated prismatic crystals emanating from the microcrystalline (~0.5 μm) CaF_2 seed crystals.[31,32] The apparent habit modification of the HAP overgrowth was due to the presence of F^- present in he solution due to the saturation conditions with respect to the CaF_2 seed crystals. Examination of the solids (seeds + overgrowth) by powder x-ray diffraction did not give evidence for fluoroapatite formation.

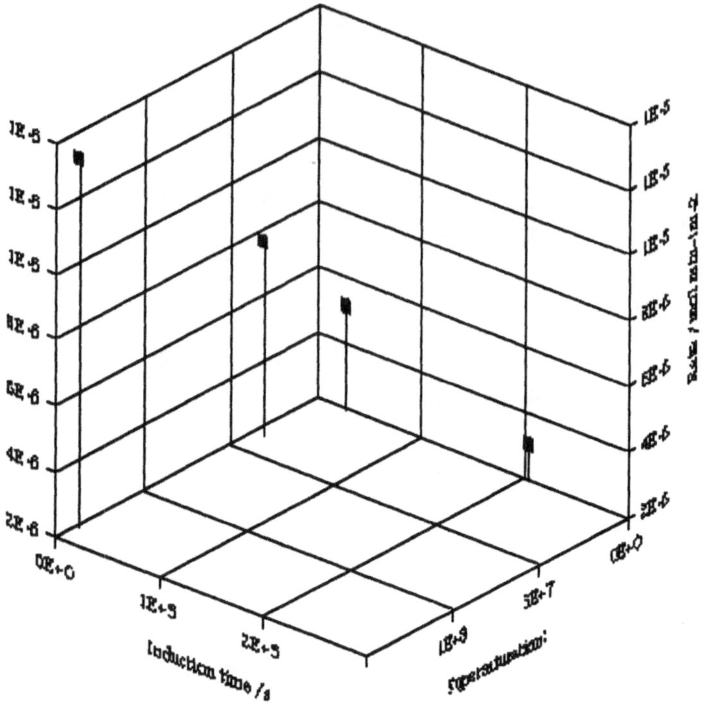

Figure 8. Overgrowth of HAP on CaF_2 seed crystals. Dependence of induction times and growth rates on solution supersaturation.[31,32]

The crystallization of HAP on calcium carbonate is also of particular interest not only in the oceans, where because of the increased phosphate input due to domestic use of detergents in several cases seawater becomes supersaturated with respect to calcite and HAP,[33] but also in industrial scale formation in cases where the water used may be supersaturated with respect to calcium phosphates and carbonates.[34] It is interesting to note that HAP formation is inhibited in the presence of carbonate ions[35,36] yet calcium carbonate polymorphs may be a substrate with crystallographic characteristics favorable for the overgrowth of HAP. The reports in the literature are concerned with the interrelationship between the thermodynamically most stable calcium carbonate calcite and HAP.[31,34] The comparison between the projected orthogonal dimensions of the (110) plane of HAP with the (1010) of calcite, shown in Figure 9, suggested a lattice misfit of about 5%, which is favorable for the overgrowth of one phase on thether. The kinetics of calcite formation on well characterized HAP seed crystals have been measured

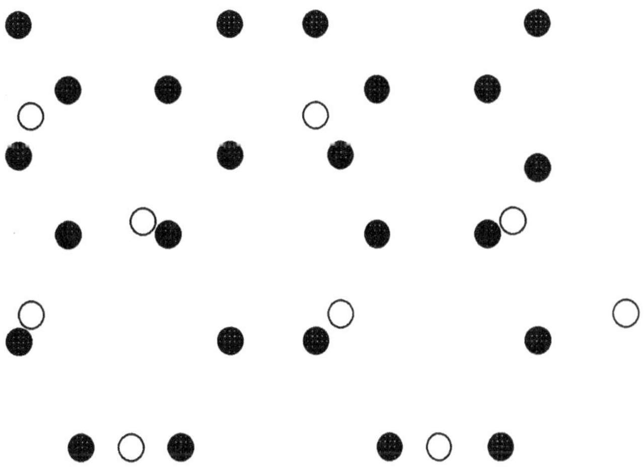

Figure 9. Calcium atoms in the (110) plane of HAP (O) and in the (1010)) of calcite (●)

at conditions of constant supersaturation[34]. The kinetics results are shown in Figure 10.

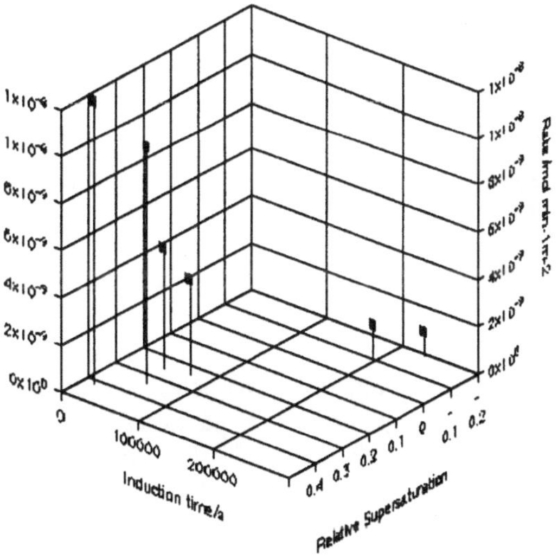

Figure 10. Crystallization of calcite on HAP seed crystals. Dependence of induction times ans crystal growth rates on the solution supersaturation with respect to calcite, pH 8.50, 25°C.[34]

Moreover, the scanning electron micrograph (SEM) shown in Figure 11, demonstrates clearly that the rhombohedral calcite crystals outgrow the microcrystalline HAP substrate.

Figure 11. Scanning electron micrograph of calcite crystals grown on HAP seed crystals; pH 8.50, 25° C.

Besides this type of overgrowth in which there is a favorable lattice matching between the substrate and the overgrowth an additional type of selective overgrowth of DCPD on titania surfaces has been reported.[37] In this case the growth of DCPD on the titanium oxide particles is preceded by the selective adsorption of calcium and phosphate ions. At the experimental conditions reported the titania surface was negatively charged[38,39] favoring adsorption of Ca^{2+} ions. This adsorption may be followed by surface binding of the phosphate species present in the solution. The binding, mainly of electrostatic nature, implies strong interactions between adsorbate-adsorbent, possibly yielding an ordered structure of the adsorbed calcium phosphate layer which provides the reduced surface energy needed for the nucleation of DCPD.

Overgrowth of calcium phosphate phases on calcium phosphate substrates has been reported over a range of temperatures up to 80 °C.[4] More specifically the nucleation of OCP on DCPA seed crystals was reported. The dependence of the rate and induction times on the solution supersaturation is shown in Figure 12. As may be seen, the rates were proportional and the induction times inversely proportional to the relative solution supersaturation with respect to OCP.

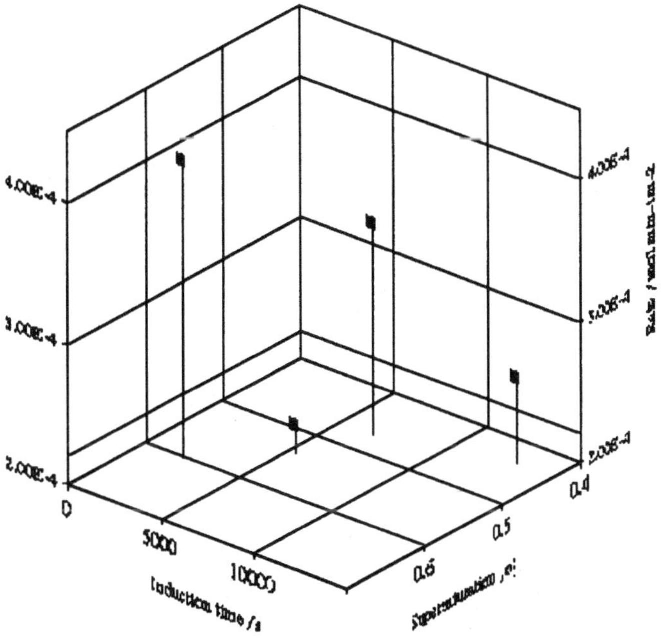

Figure 12. Growth of OCP on DCPA seed crystals. Dependence of the rates of crystallization and of the induction times measured on the solution supersaturation.[4]

The mechanism of overgrowth was explained fitting the kinetics data to the polynuclear model in agreement with earlier reports concerning OCP crystal growth kinetics studies done by a free-drift technique.[40,41] The dependence of the logarithm of the induction time on the inverse of the logarithm of the supersaturation ratio S yielded a single straight line suggesting that the overgrowth is heterogeneous.[42]

It is interesting to note that there is evidence that the size of crystals may influence the kinetics of heterogeneous nucleation of calcium phosphates. Thus, the overgrowth of OCP on DCPD seed crystals was characterized by decreasing induction times with increasing crystal size.[41] This finding was interpreted as an indication that the nucleation of OCP growth over the DCPD seed crystals was favoured by the larger number of crystal defects present in the larger crystals.

The growth of DCPD on DCPA seed crystals has also been demonstrated at 60 °C.[4] The rates obtained were higher as compared to the rates of the growth of DCPD on DCPD at 37 °C although in the lattter case the supersatuartion was higher[43] as may be seen in Figure 13. The higher rates for the DCPD overgrowth on DCPA suggested a high activation energy. The linear dependence of the rates on

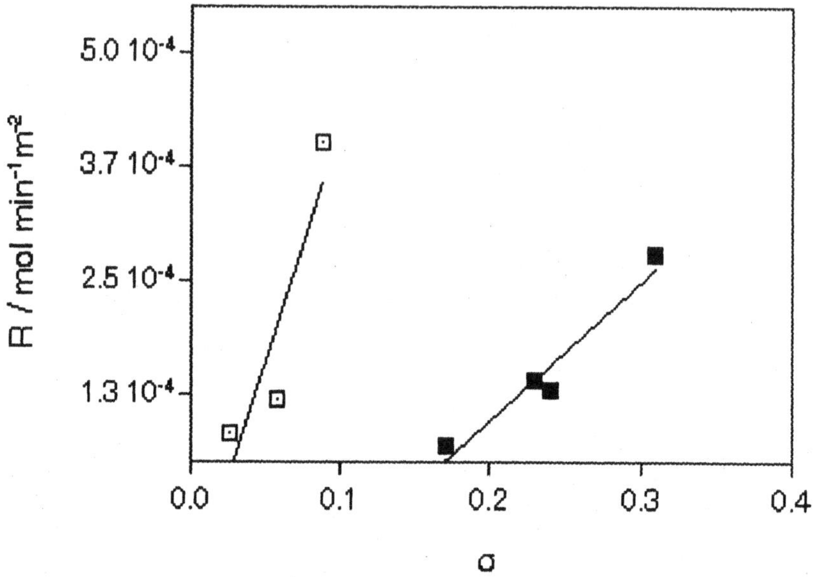

Figure 13. Crystal growth kinetics of (□) DCPD on DCPA seed crystals at 60 °C[4]; (■) DCPD on DCPD at 37 °C.[46]

the relative supersaturation suggested a surface controlled process. Moreover, it should be noted that using the constant composition it was possible to demonstrate that at pH 5.0 it was possible to induce the heterogeneous nucleation of DCPD on OCP seed crystals past induction times inversely proportional to the solution supersaturation[44]. Increase of pH to the domain where DCPD and OCP are not stable (pH 7.4-8.0) resulted in the hydrolysis of the substrate to non stoichiometric apatitic phases.

For applications of the various calcium phosphates for the preparation of ceramic implants, one of the interesting aspects is whether they have the capability of nucleation of apatite or apatitic minerals after the sintering and other related processes in which they are subjected, where phase changes may also occur.[45,46] HAP powders sintered at 900 °C were converted into mixtures of TCP an HAP and when indroduced in solution supersaturates with respect to HAP and TCP only, α-TCP was deposited on the calcined surfaces showing the direct formation of this, otherwise high temperature phase[46] in this study however there is no adequate data for the differentiation between selective overgrowth of DCPD on OCP and secondary nucleation of this phase due to the presence of OCP.

Overgrowth of calcium phosphates on organic substrates

The fact that bone and teeth consist mainly of a fibrous protein which is a matrix on which calcium phosphate mineral phases are depositied has attracted the interest of the researchers working in biological mineralization into the investigation of the relationship between the organic matrix and the inorganic salts found in hard tissues mainly calcium phosphates.[47-50] Although the mediation of biochemical processes is a very important part of the biological mineralization, in the present paper we shall view the mineralization process as heterogeneous nucleation. The substrate is provided by the biopolymers and the supersaturated biological fluids provide the supersaturation conditions which are prerequisite for the formation of the mineral phase. The formation of enamel e.g., is considered as an example of matrix directed mineralization where the reactions in the mineralization front, where HAP is formed by heterogeneous crystallization, are medaited by phosphophoryns. In bone, x-ray diffraction studies have revealed crystallographic relationship between small HAP crystallites (ca 5x100 nm x nm) and the collagenous matrix, formed from a simple protein containing polypeptide chains of 1000 amino acid residues.[50]

The investigation of heterogeneous nucleation of calcium phosphates on polymeric substrates has been used as model for the investigation of the significance of the structure of the polymers in inducing nucleation of calcium phosphates.[51] Polymers modified to contain phosphonic groups (>(P=0)-O-R) with various degrees of phosphorylation showed that increase of phosphorus content in the polymers resulted in higher rates of HAP deposition. Moreover, a favorable polymer conformation was found to be necessary for the enhancement of HAP deposition. It is interesting to note that in all polymers tested in which phosphorus content varied between 0.5 - 2.0% w/w the formation of HAP took place without any appreciable induction time.

Experiments using collagen type I showed also that HAP was formed at low supersaturation without the lapse of induction times.[52] This result is in agreement with the results obtained from the phosphorylated polymers, since exposure of collagen to phosphate containing solutions resulted in the uptake of considerable amounts of phosphate on the collagenous matrix.[53] It has been postulated that phosphorylation of the collageneous matrix and more specifically of the serine aminoacid is a prerequisite for the formation of nucleation sites on collagen.[54,55] Nucleation and growth of HAP on collagen type I from bovine Achilles tendon and on collagen obtained from demineralized bone[37] occured following the same mechanism, yet the rates of crystallization at constant supersaturation conditions were lower for the collagen type I substrate as may be seen in Figure 14.

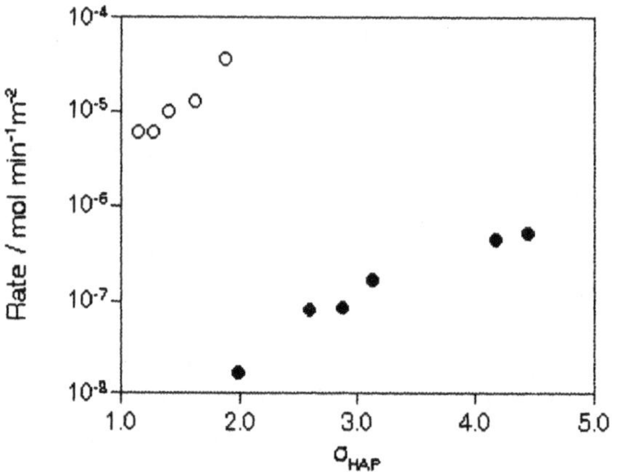

Figure 14. Crystallization of HAP on collagen substrates at conditions of constant supersaturation pH 7.40, 37 °C (●) Collagen type I from Achilles tendon;[52] (O) Collagen from demineralized bone.[37]

At higher supersaturations OCP could nucleate and grow past induction times both in collagen type I[52] and collagen from demineralized bone.[37] Experiments on glutaraldehyde treated porcine heart valves showed also the posibility of these surfaces to nucleate OCP without induction times.[56] The kinetics plots, including the kinetics of overgrowth of OCP on OCP seed crystals[57] for comparison reasons are shown in Figure 15.

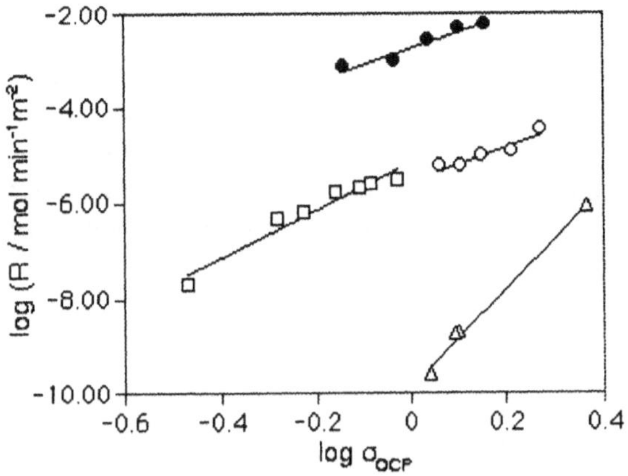

Figure 15. Logarithm of the crystal growth rates of OCP as a function of the logarithm of the relative supersaturation with respect to this phase at conditions of constant supersaturation 37 °C.; (●) OCP on heart valves;[56] (□) OCP on OCP seed crystals;[60] (O) OCP on collagen from demineralized collagen;[39] (Δ) OCP on Achilles tendon collagen type I.[55]

As may be seen the growth of OCP on collagen type I obtained from demineralized bone was closer to the growth of OCP on OCP seed crystals. Type I collagen from Achilles tendon yielded lower rates, fact which may be ascribed both to the different type of collagen but also from the lower purity of this product which was obtained commercially and may contain low molecular weight collagenous fractions which inhibit the formation of HAP. Heart valves yielded faster rates. It should be noted however in this case that the reported rates do not distinguish between deposition on the collagenous matrix and on the elastin rich matrix present in one of the two sides of the valve membranes. Elastin is a biopolymer highly cross-linked, found in arterial walls and ligaments. Experiments in which calcium phosphate supersaturated solutions were seeded with hydroxyapatite showed that elastin could induce the formation of HAP, past induction times.[58] As anticipated for selective overgrowth, the induction times were inversely proportional and the rates of HAP crystallization mesasured at constant supersaturation were found to increase with supersaturation. As in the case of collagen, significant amounts of orthophosphate were adsorbed from the phosphate containing solutions in contact with elastin, suggesting a similar mechanism for the initiation of the mineral depostion. A model for the deposition of HAP on the polymeric and biopolymers surfaces is shown in Figure 16.

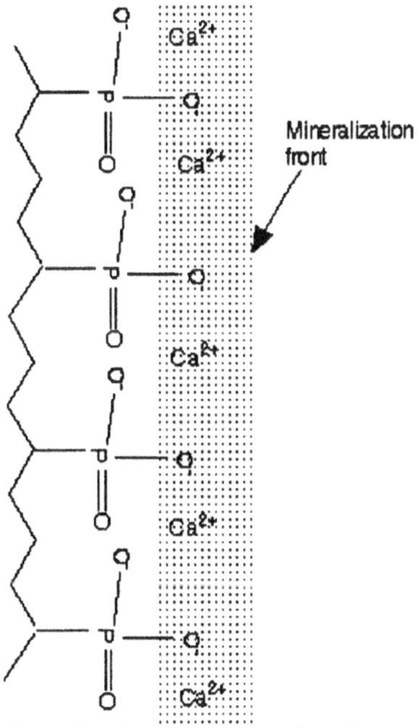

Figure 16. Model of mineralization of polymeric substrates containing phosphate groups (bound or adsorbed).

Although both in collagen[53] and on elastin[58] adsorption of inorganic orthophosphate is predominantly electrostatic in nature, no reversibility was reported suggesting strong interaction between substrate and the adsorbate. The kinetics results obtained for HAP overgrowth on a phosphorus containing copolymer (1.7% w/w),[51] collagen type I from Achilles tendon[52] and with elastin[58] are summarized in Figure 17.

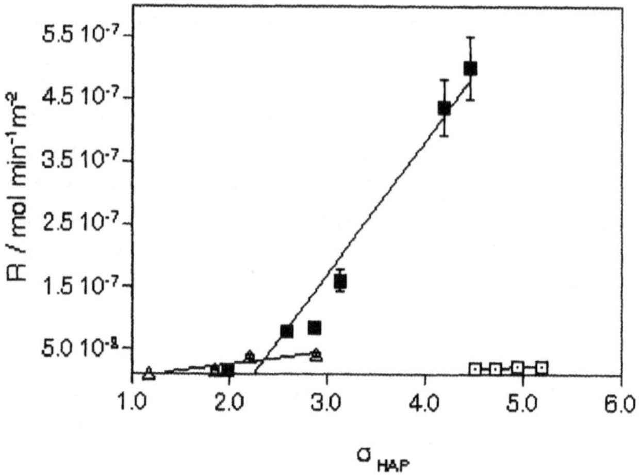

Figure 17. Plots of the rates of HAP formation at xonstant supersaturation as a function of the relative supersaturation, pH 7.40; (■) phosphorylated copolymer;[54] (□) collagen;[55] (Δ) elastin.[61]

As may be seen, within the range of supersaturation examined, a linear R-σ relationship was found suggesting a surface controlled process. Bulk diffusion controlled mechanism was precluded because of the higher rates calculated for this case (by several orders of magnitude) in the case e.g., of growth of HAP on HAP seed crystals in gels, where growth is controlled by diffusion a rate of $4x10^{-5}$ mol min^{-1} m^{-2} was calculated from the reported data.[59] It should be noted however that even in this case, the presence of type I collagen fibers facilitated the growth of HAP. The difference observed in the kinetics parameters measured may be ascribed to the nature of the substrate. Measurements by atomic force microscopy have revealed that the surface structure of the deposits depended on the nature of the substrate.[60] There is evidence that the organization of the organic matrix at the substrate / electrolyte interface is responsible for the recognition and subsequent oriented crystal nucleation and growth.[61,62] The influence of the organic matrix on the control of nucleation and growth of inorganic minerals in general may be ascribed to stereochemical, chemical affinity and electrostatic interaction factors. It is necessary that emphasis is placed on the molecular level for the discription of heterogeneous nucleation and epitaxial growth.[63,64] Alternatively, it has been suggested that mineralization may be initiated by clusterring of calcium and phosphate ions in the nanospace (5-6 nm) existing in collagen or elastin which is capable of providing the critical space needed for the growth of HAP nanoparticles (3 nm).[65] The dimensions of the nanospace ensure exclusion of the dissolved macromolecules which may inhibit HAP nucleation and crystal growth.

REFERENCES

1. Lundager Madsen HE, Lopez-Valero I, Lopez-Acevedo V, Boistelle R. The formation product of amorphous tricalcium phosphate at 37 °C, J Crystal Growth 1986;75:429-434

2. Bohner M, Lemaitre J. Hydraulic properties of tricalcium phosphate-phosphoric acid-water mixtures in *3rd Euro-Ceramics*. P. Duran, J.F. Fernandez Eds, Faenza Editrice Ib eria S.L. Castellon de la Plana Spain 1993.

3. Bohner M. Propriétés Physico-Chimiques et Ostéogéniques d' un Biociment Hydraulique a Basse de Phosphates de Calcium. Ph.D. Thesis No 1171, EPFL, Lausanne, Switzerland 1993.

4. Frèche M. Contribution à l; étude des phosphates de calcium, Thèse, No 162, INP Toulouse, Toulouse France, 1989.

5. Bermudez O, Boltong MG, Driessens FCM, Planell JA. Developement of an octacalcium phosphate cement. J Mater Sci Mater Me. 1994;5:144-1461

6. W.E. Brown in *Environmental Phosphorus Handbook*, E.J. Griffith, A. Becton, J.M. Spencer, D.T. Mitchell (eds) Wiley & Sons, New York 1973.

7. Hamad M. Contribution d' un mechanisme reactionnel d'une synthése de la fluorapatite, Thèse, No 81 INP Toulouse, Toulousse, France, 1980.

8. Koutsoukos PG, Amjad Z, Tomson MB, Nancollas GH. Crystallization of calcium phsophates. A constant composition study. J A. Chem Soc 1980;102:1553-1557

9. Nancollas GH. In vitro studies of calcium phosphate crystallization in *Biomineralization*, S. Mann, J. Webb, R.J. Williams Eds VCH Weinheim 1989.

10. van der Merwe JH. Misfitting monolayers and oriented overgrowth CRC Crit Rev. Solid Stat Mater Sci 1978;7:209-225

11. Russel KC. Nucleation in solids: The induction and steady state effects, Adv Coll Interface Sci 1980;13:205-318

12. Herman MA. Crystallization of epitaxial layers: Fundamentals and basic problems. In: *Elementary Crystal Growth*. K. Saugwall ed. Saan Publ. Lublin, 1994.

13. Zur A, McGill TC., Lattice match: An application to heteroepitaxy J Appl. Phys 1984;55:378-386

14. Nyvlt J. *Industrial Crystallization from Solutions* Butterworths, London 1971.

15. Kern R. Croissance épitaxique (aspects topologiques et structuraux). Bull Mineral 1978;101:202-233

16. Venables JA, Price GL. Nucleation of thin films In: *Epitaxial Growth*, John Wauchope Mathews, Ed. Acad. Press. New York 1975.

17. Papelis C, Hayes KF, Leckie JO. HYDRAQL: A program for the computation of chemical equilibrium composition of aqueous batch systems including surface complexation modeling of ion adsorption at the oxide/solution interface, Technical Report No 306 (Stanford University, Stanford CA 1988).

18. Nyvlt J, Sohnel O, Matuchova M, Broul M. *The Kinetics of Industrial Crystallization*; Elsevier, Amsterdam 1985.

19. Moreno EC, Zahradnik RT, Glazman A, Hwu R. Precipitaion of hydroxyapatite from dilute solutions upon seeding, Calcif Tissue Res 1977;24:47-57

20. Moreno EC, Varughese K, Hay DI. Effect of human salivary proteins on the precipitation kinetics of calcium phosphate. 1979;28:7-16

21. Meyer JL, Eanes ED. A. thermodynamic analysis of the amorphous to crystalline calcium phosphate. Calcified Tissue. 1978;25:59-68

22. Pak CYC, Diller EC, Smith GW, Howe ES. Renal stones of calcium phosphate: Physicochemical basis for their formation. Proc Soc Exp Biol Med 1969;130:753-757

23. Tomson MB, Nancollas GH. Mineralization kinetics: A constant composition approach. Science 1978;200:1059-1060

24. Francis MD, Webb NC. Hydroxyapatite formation from a hydrated calcium monohydrogen phosphate precursor. Calcif Tissue Res 19716:335-342

25. Neuman WF, Neuman MW. The Nature of the mineral phase of bone. Chem. Rev 1953;53:1-45

26. Neuman WF, Bareham BJ. evidence for the presence of secondary calcium phosphate in bone and its stabilization by acid production. Calcif. Tissue Res. 1975;18:161-172

27. Kani T, Kani M, Moriwaki Y, Doi Y. X-ray diffraction analysis of dental calculus J Dent Res 1983;62:92-95

28. Koutsoukos P, Amjad Z, Nancollas GH. The precipitation of hydroxyapatite at constant supersaturation. J Dent Res 1979;58A:167

29. Roufosse AH, Landis WJ, Sabine WK, Glimcher MJ. Identification of brushite in newly deposited bone mineral from embryonic chick. J Ultrastruct Res 1979;68:235-255

30. Barone JP, Nancollas GH. The seeded growth of calcium phosphates. The effect of solid/ solution ratio in controlling the nature of the grown phase. J Coll Interface Sci 1977;62:421-429

31. Koutsoukos PG, Nancollas GH. Crystal growth of calcium phosphates - Epitaxial considerations. J Crystal Growth 1981;53:10-19

32. Koutsoukos PG. *Kinetics of Precipitation of Hydroxyapatite from Aqueous solutions*, SUNYAB, Buffalo, NY 1980.

33. Brown WE. Solubilities of Phosphates and other sparingly soluble compounds, In *Environmetal Phsophorus Handbook*, E.J. Griffith, A. Burta, J.M. Spencer, D.T. Michael Eds. J. Wiley, New York (1973).

34. Klepetsanis PG, Drakia P, Koutsoukos PG. The calcite-hydroxyapatite system. Crystal growth studies in aqueous solutions In. *Mineral Scale Formation and Inhibition*, Z. Amjad Ed., Plenum Press New York, 1994.

35. Bachra BN, Trautz OR, Simon SL. Precipitation of calcium carbonates and phosphates. III. The effect of magnesium and fluoride ions on the spontaneous precipitation of calcium carbonates and phosphates. Arch Oral Biol 1965:731-738

36. Blumenthal NC, Betts F, Posner AS. Effect of carbonate and biological macromolecules on formation and properties of hydroxyapatite. Calcif Tiss Res. 1975;18:81-90

37. Combes C. Croissance cristalline de phosphates de calcium sur des substrats d'interêt biologique: Le titane et le collagene, Thèse doctorale, INPT, Toulouse, France, 1996. No 1189.

38. Koutsoukos PG, Lycourghiotis A. Regulation of the sorptive capacity of oxides used as catalyst carriers, Colloids and Surfaces 1991;55:297-308

39. Fokking LGJ, de Keizer A, Lyklema J. Temperature dependence of the electrical double layer on oxides: Rutile and hematite. J Coll Interface Sci 1989;127:116-131

40. Madsen HEL. Calcium phosphate crystallization. III. Overall growth kinetics of tetracalcium monohydrogen phosphate. Acta Chem Scand 1982;A36:239-249

41. Madsen HEL. Calcium phosphate crystallization. IV. Kinetics of heterogeneous nucleation of tetracalcium monohydrogen phosphate on brushite crystals. Acta Chem Scand 1983;A37:25-29

42. Mullin JW. *Crystallization*, 3rd Ed., Butterworth-Heinemann, Oxford 1993.

43. Salimi MH. Ph. D. Thesis, SUNY at Buffalo, Buffalo NY 1985.

44. De Rooij JF, Heughebaert JC, Nancollas GH. A pH study of calcium phosphate seeded precipitation, J Coll Interface Sci 1984;100:350-358

45. Osborn JF, Newesely H. The material science of calcium phosphate ceramics, Biomaterials 1980;1:108-111

46. Dalas E, Koutsoukos PG. The growth of calcium phosphate on ceramic surfaces. J Mater Sc. 1989;24:999-1004

47. Glimcher MJ. Composition, structure and organization of bone and other mineralized tissues and the mechanism of calcification. In: *Handbook of Physiology* vol. 7 G.D. Aurbach ed. Amer. Physiol. Soc. Washingtom DC 1976.

48. Jenkins GN. The physiology and biochemistry of the mouth, *Blackwell Sci*, 4th Edition, Oxford 1978.

49. Veis A, Sabsay B. Bone and tooth formation. Insights into mineralization strategies. In *Biomineralization and Biological Metal Accumulation.* P.Westbroek, E.W. de Jong Eds, D. Reidel Publ. Co., Dordrecht 1983.

50. Bills PM, Lewis D, Wheeler EJ. Mineral-Collagen orientation relationships in bone. J Crystallographic and Spectroscopic Res 1982;12:51-63

51. Dalas E, Kallitsis JK, Koutsoukos PG. Crystallization of hydroxyapatite on polymers. Langmuir 1991;7:1822-1826

52. Koutsoukos PG, Nancollas GH. The mineralization of collagen in vitro, Colloids Surfaces 1987;28:95-108

53. Koutsoukos PG, Nancollas GH. The adsorption of inorganic phosphate by collagen. Colloids & Surfaces 1986;17:81-90

54. Glimcher MJ. Molecular biology of mineralized tissues with particular reference to bone. Rev Mod Phys 1959;31:359-303

55. Glimcher MJ. Specificity of the molecular structure of organic matrices in mineralization. In *Calcification in Biological Systems*, R.F. Sognnaes (Ed) Am. Assoc. Adv. Sci., Washington DC 1960.

56. Kapolos I, Mavrilas D, Missirlis Y, Koutsoukos PG. A model experimental system for the investigation of heart valve calcification in vitro. Applied Biomaterials, In Press.

57. Heughebaert JC, Nancollas GH. Kinetics of crystallization of octacalcium phosphate. J Phys Chem 1984;88:2478-2481

58. Koutsopoulos S, Paschalakis PC, Dalas E. The calcification of elastin in vitro, Langmuir 1994;10:2423-2428

59. Boskey AL. Hydroxyapatite formation in a dynamic collagen gel system: Effects of Type I collagen, lipids and proteoglycans. J Phys Chem 1989:93:1628-1633

60. Siperko LM, Landis WJ. Atomic force microscopic determination of substrate effects on the structure of deposited biomineral phosphates. Analyst 1994;119:1935-1938

61. Addadi L, Moradian-Oldak J, Füredi-Milhofor H, Weiner S, Veis A. Stereochemical aspects of crystal regulation in calcium phosphate-associated mineralized tissues, In *Chemistry and biology of mineralized tissues.* H. Slavkin, P. Price Eds., Elsevier, Amsterdam, 1992.

62. Füredi-Milhofer H, Moradian-Oldak J, Weiner S, Veis A, Mintz KP, Addadi L. Interactions of matrix proteins from mineralized tissues with octacalcium phosphate. Conn. Tissue Res 1994;30:251-263

63. Mann S, Archibald DP, Didymus JM, Douglas T, Heyhood BR, Meldrum FC, Reeves NJ. Crystallization at inorganic-organic interfaces: Biominerals and biomimetic synthesis. Science 1993;261:1286-1292

64. Mann S. Molecular tectonics in biomineralization and biomimetic materials chemistry. Nature 1993;365:499-505

65. Katsura N. Nanospace theory ofr biominerlization Dentistry in Japan. 1990;27:57-63

4

PHYSICOCHEMICAL PROPERTIES OF FLUORAPATITE

Laurence C. Chow and Milenko Markovic

American Dental Association Health Foundation, Paffenbarger Research Center, National Institute of Standards and Technology, Gaithersburg, MD 20899 USA

INTRODUCTION

Fluoride (F) is a subject of great interest in dental caries research because F, applied in various ways, is the most effective factor in reducing caries. The cariostatic mechanisms of F are complex and have been studied extensively for several decades, and the large body of information in the literature on this subject was reviewed recently.[1,2]

The inorganic components of enamel, dentin, and bone are calcium hydroxyapatite (OHAp) minerals that contain significant amounts of various "impurities" such as carbonate, sodium, chloride, F, H_2O, etc.[3,4] The presence of these foreign ions in the apatite structure is known to produce significant effects on the solubility and other physicochemical properties of OHAp. Fluoride, which can substitute the hydroxyl ions in the apatite lattice, reduces the solubility of the apatitic mineral under a wide range of conditions including those found in the mouth.[5] Thus, a purported cariostatic mechanism of F is that apatitically incorporated F makes the tooth more resistant to demineralization caused by bacterial acids present in dental plaque. Two crystallographic forms of OHAp are known, hexagonal[6] and monoclinic[7] OHAp. Fluorapatite (FAp), an apatite in which all of the OH^- ions are replaced by F^-, has a hexagonal crystallographic form.[8,9] FAp has a lower thermodynamic solubility product, $K_{sp}(FAp) = 2.5 \times 10^{-60}$ at 25°C,[10] than OHAp, $K_{sp}(OHAp) = 3.04 \times 10^{-59}$ at 25°C.[11] Because F^- and OH^- ions have the same charge and similar ionic sizes, substitution of OH^- by F^- in the OHAp structure can occur to a continuously variable extent, forming a series of fluorhydroxyapatite solid solutions, represented by the general formula of $Ca_5(PO_4)_3(OH)_{1-x}F_x$ in which $0 < x < 1$. The solubility and other physicochemical properties of the solid solutions series, $OH_{1-x}F_xAp$, with OHAp and FAp as the end members, are of particular interest because of their roles in the caries process.

SOLUBILITY PROPERTIES OF FAp

The solubility product constant of FAp is about one order of magnitude (12 times) smaller than that of OHAp.[12] The K_{sp}'s of $OH_{1-x}F_xAp$ solid solutions are also lower than that of OHAp.[13] In addition, having a lower thermodynamic solubility, a very important factor that differentiates the dissolution characteristics of FAp from that of OHAp is the absence of OH^- ions in the FAp structure. There are two important consequences:

A. For any solution that is saturated with respect to both OHAp and FAp, equations (1) and (2) hold, and equation (3) was obtained by dividing equation (2) by equation (1). It may be inferred from equation (3) that FAp is the more stable (less soluble) phase whenever $(F^-) > 0.082 \times (OH^-)$. Thus, for neutral and acidic solutions, e.g., solutions with pH of 7 and 4, a solution saturated with respect to OHAp would be supersaturated with respect to FAp whenever (F^-) is greater than 8.2×10^{-9} and 8.2×10^{-12} mol/ℓ, respectively. Since a (F^-) activity of this magnitude can be easily exceeded after dissolution of a small amount of FAp, FAp is almost always less soluble than OHAp under neutral and acidic pH conditions. Conversely, FAp may become more soluble than OHAp in highly alkaline solutions. By taking logarithm of equation (3) and substituting the relationship, $(OH^-) = K_w/(H^+)$ where $K_w = 10^{-14}$, the equation (4) is obtained. The straight line $\log(F^-)$ *vs.* pH depicted in Figure 1 divides stability phase diagram in two regions. At the conditions in solution represented with straight line, OHAp and FAp are equally stable. For those solutions with (F^-) and pH that fall above the straight line (Figure 1), FAp is more stable than OHAp, i.e., if the solution is saturated with respect to FAp it is undersaturated with respect to OHAp, and if the solution is saturated with respect to OHAp it is supersaturated with respect to FAp. Since the relative stability of FAp and OHAp depends only on the F^- activity and pH and does not depend on the Ca^{2+} and PO_4^{3-} ion activities, a stability phase diagram such as in Figure 1 can be constructed without concerning the calcium and phosphate concentrations of the solutions.

$$K_{sp}(OHAp) = (Ca^{2+})^5(PO_4^{3-})^3(OH^-) = 3.04 \times 10^{-59} \quad (1)$$

$$K_{sp}(FAp) = (Ca^{2+})^5(PO_4^{3-})^3(F^-) = 2.5 \times 10^{-60} \quad (2)$$

$$(F^-) = 0.082 \times (OH^-) \quad (3)$$

$$\log(F^-) = -15.08 + pH \quad (4)$$

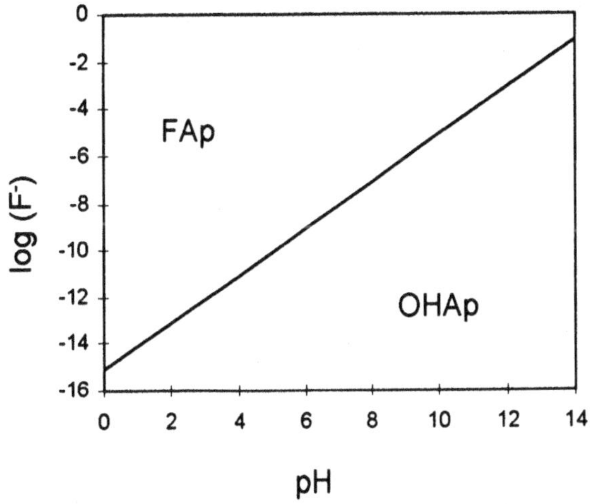

Figure 1. Stability phase diagram of FAp and OHAp.

B. When the same molar amount of mineral is dissolved, FAp does not neutralize as much acid and, consequently, it is not as effective in raising the pH of the solution as is OHAp. Equations (5) and (6) show, respectively, that 7 and 6 moles of protons are consumed when a mole of OHAp and FAp are dissolved.

$$Ca_5(PO_4)_3OH + 7H^+ \rightarrow 5Ca^{2+} + 3H_2PO_4^- + H_2O \quad (5)$$

$$Ca_5(PO_4)_3F + 6H^+ \rightarrow 5Ca^{2+} + 3H_2PO_4^- + F^- \quad (6)$$

As a result, when dissolving into a given acid solution, the FAp solution will have a lower pH, leading to a lower PO_4^{3-} activity for the same total phosphate concentration. This aspect of the FAp dissolution characteristics has the effect of increasing the solubility of FAp compared to OHAp.

In practice, the differences in the dissolution characteristics of OHAp and FAp described above are manifested in several ways. Figures 2 to 4 show the solubilities of FAp and OHAp under different conditions, computed with a commercially obtainable software (EQUIL, MicroMath Inc., 1996). When equilibrated with a strong acid such as HCl, FAp dissolves slightly less than does OHAp for a given acid concentration[14] (Figure 2a). However, the pH of the FAp-saturated solutions is always significantly lower than that of the OHAp-saturated solution obtained with the same acid concentration (Figure 2b). As a result, when viewing the solubility at the given pH, FAp is approximately 5 to 10 times less soluble than OHAp (Figure 2c). When equilibrated with a weak acid solution, e.g., acetic acid, which acts as a buffer at pH near its $pK_a = 4.8$ (where K_a is the acid dissociation constant), FAp is less soluble (Figure 3a) than OHAp at a given acid concentration. For a given acid concentration, the pH of the FAp-saturated solution is lower than that of the OHAp-saturated solution (Figure 3b). For a given pH, FAp is again much less soluble than OHAp (Figure 3c). Resting dental plaque (plaque that has not been recently exposed to sucrose) contains about 20 to 50 mmol/ℓ acetate and has a neutral pH. Soon after a sucrose exposure, lactic acid is produced by bacteria and tooth mineral is challenged in the same way as by a strong acid (because lactic acid is nearly totally dissociated at pH above 5) in the presence of a weak acid buffer. Accordingly, Figure 4a shows that when FAp and OHAp dissolve in strong acid solutions (HCl) that also contain 50 mmolℓ acetate buffer (at an initial pH of 7.0 before addition of HCl), FAp is significantly less soluble than OHAp. The presence of acetate buffer causes a difference in FAp and OHAp solubilities which are very similar in HCl without the buffer (Figure 2a). As before, the pH of the FAp-saturated solution in the buffer-containing system is lower than that of the OHAp-saturated solution at the same HCl concentration (Figure 4b), and at a given pH, FAp is much less soluble than OHAp (Figure 4c).

It may thus be concluded that, when equilibrated with the same acid solution, the FAp-saturated solution will have a lower calcium and phosphate concentrations and/or a lower pH. Both of these conditions may lead to a lower rate of demineralization for F-rich tooth enamel in the caries process.[3]

SOLUBILITY PROPERTIES OF $OH_{1-x}F_xAp$ SOLID SOLUTIONS

An apatite mineral that contains less F than does stoichiometric FAp may be considered as a solid solution of FAp and OHAp with the general formula $Ca_5(PO_4)_3(OH)_{1-x}F_x$, where $0 < x < 1$. The solubility properties of these solid solutions are more complex because as the solid-liquid equilibrium is approached, a change in composition (change in the x value) may occur on the surface of the crystalline particles such that the surface phase has the most stable (least soluble) composition for the particular aqueous solution composition. Thus, the solid phase that is in true equilibrium with the aqueous solution can have an x value that is different from that of the bulk solid.[15] Based on the solubility data for $OH_{1-x}F_xAp$ solid solutions reported in the literature,[13,16,17] Driessens[12] concluded that the presence of F ions in the solid reduces the solubility considerably, especially at low pH: a reduction of solubility of 80% can be expected at pH 4 for a $OH_{1-x}F_xAp$ solid solution with an x value of 0.1.

In the following section, we present an analysis that addresses rigorously the stability of the various $Ca_5(PO_4)_3(OH)_{1-x}F_x$ solid solutions as a function of the aqueous solution composition. It will be shown that the stability strongly depends on the solubility products of the solid solutions, $K_{sp}(OH_{1-x}F_xAp)$, at different x values. The analysis is given for several hypothetical situations.

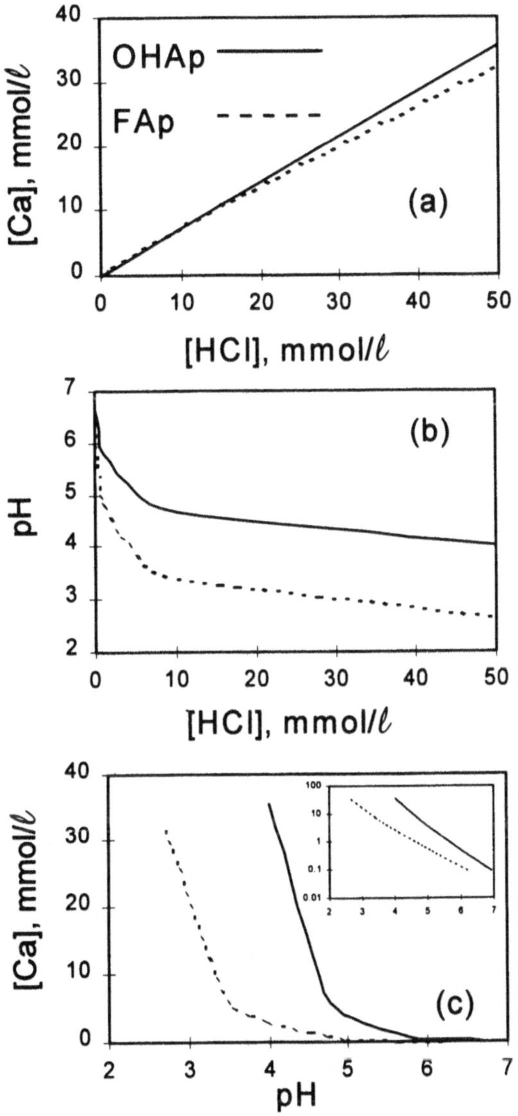

Figure 2. Solubility of OHAp and FAp in a strong acid: (a) Ca concentration, [Ca], of the saturated solution as a function of hydrochloric acid concentration, [HCl], (b) pH of the saturated solution as a function of [HCl], and (c) [Ca] of the saturated solution as a function of pH; inset showing the same plot with [Ca] in log scale.

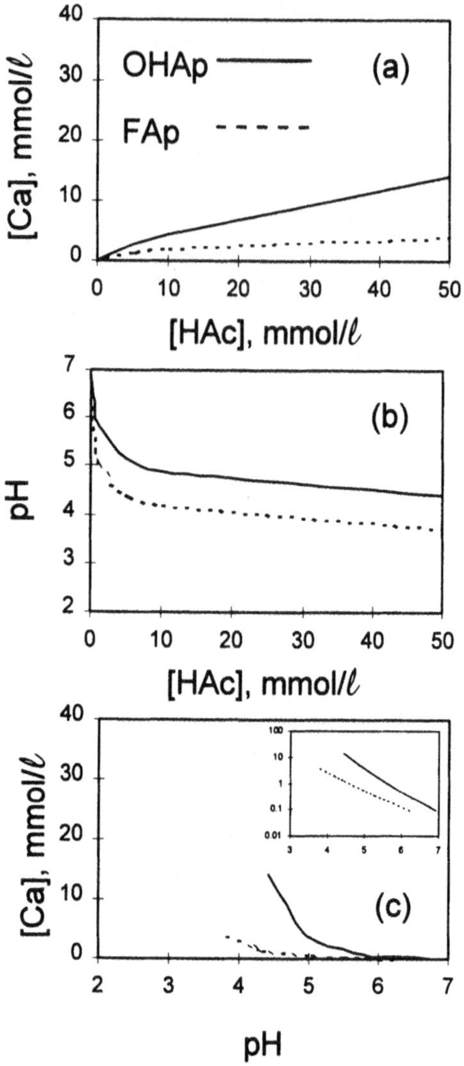

Figure 3. Solubility of OHAp and FAp in a weak acid: (a) [Ca] of the saturated solution as a function of acetic acid concentration, [HAc], (b) pH of the saturated solution as a function of [HAc], and (c) [Ca] of the saturated solution as a function of pH; inset showing the same plot with [Ca] in log scale.

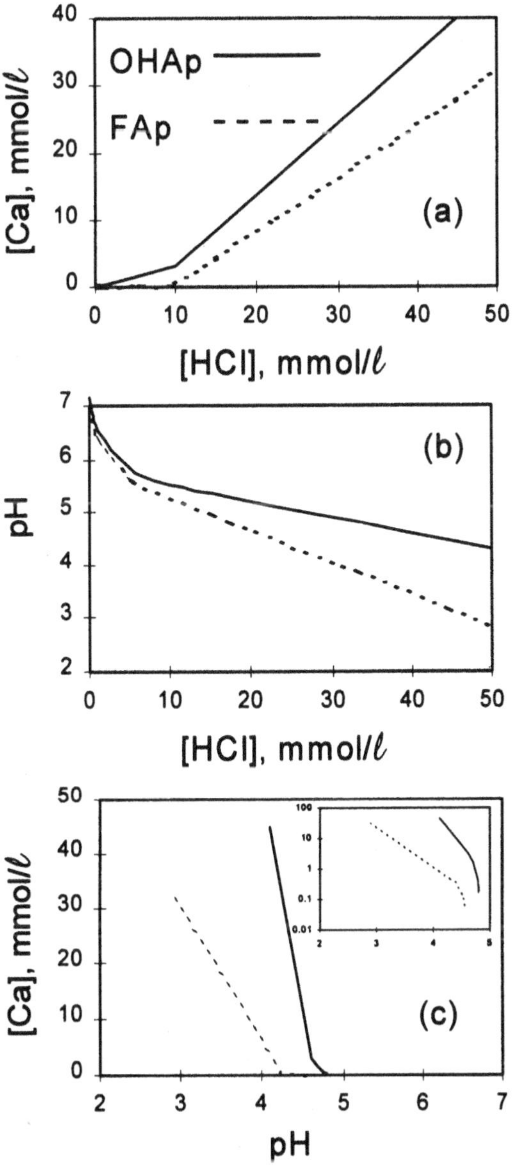

Figure 4. Solubility of OHAp and FAp in a strong acid in the presence of a acetate buffer: (a) [Ca] of the saturated solution as a function of [HCl], (b) pH of the saturated solution as a function of [HCl], and (c) [Ca] of saturated solution as a function of pH; inset showing the same plot with [Ca] in log scale.

A. In the simplest hypothetical case, the solubility products, log $K_{sp}(OH_{1-x}F_xAp)$, of the all solid solutions are assumed to be linear functions of the x value so that they lie on the straight line that connects log $K_{sp}(OHAp)$ = -58.52 and log $K_{sp}(FAp)$ = -59.60 values, as shown in Figure 5a. In this case the solubility product, expressed as log $K_{sp}(OH_{1-x}F_xAp)$, is defined by equation (7). For solutions that are saturated with respect to both OHAp and $OH_{1-x}F_{x}Ap$, equations (8) and (9) are valid.

$$\log K_{sp}(OH_{1-x}F_xAp) = (1-x)\log K_{sp}(OHAp) + x \log K_{sp}(FAp) = = -58.52 - 1.08x \quad (7)$$

$$(Ca^{2+})^5(PO_4^{3-})^3(OH^-) / (Ca^{2+})^5(PO_4^{3-})^3(OH^-)^{1-x}(F^-)^x = = K_{sp}(OHAp) / K_{sp}(OH_{1-x}F_xAp) \quad (8)$$

or

$$(OH^-)^x / (F^-)^x = 10^{1.08x}$$

and after taking logarithm and substituting pOH = pK_w - pH, the equation is

$$\log(F^-) = -15.08 + pH \quad (9)$$

For solutions that are saturated with respect to both FAp and $OH_{1-x}F_xAp$ equations (10) and (11) are valid.

$$(Ca^{2+})^5(PO_4^{3-})^3(F^-) / (Ca^{2+})^5(PO_4^{3-})^3(OH^-)^{1-x}(F-)^x = = K_{sp}(FAp) / K_{sp}(OH_{1-x}F_xAp) \quad (10)$$

or

$$(F^-)^{1-x} / (OH^-)^{1-x} = 10^{-1.08(1-x)}$$

giving

$$\log(F) = -15.08 - pH \quad (11)$$

It is seen that the condition for simultaneous equilibrium with OHAp and $OH_{1-x}F_xAp$ as represented by equation (9) and the condition for simultaneous equilibrium with FAp and $OH_{1-x}F_xAp$ as represented by equation (11) are identical and are independent of x. In fact, equations (9) and (11) are identical to equation (3) which describes the relative stability of FAp and OHAp only (solid solutions were not taken into consideration in equation (3)). The stability phase diagram (Figure 5b) for the solid solution in this example is essentially the same as that in the case where only FAp and OHAp are considered (Figure 1) except that in Figure 5b those aqueous solutions that are represented by the straight line are simultaneously saturated with respect not only to OHAp and FAp but also to every single solid solution (represented with every value of x in the series $OH_{1-x}F_xAp$). Thus, it can be concluded that in this hypothetical case the solid solutions play no significant role in the solubility behavior because they can exist only in the narrow range of solution compositions in which the solution is also saturated with respect to both OHAp and FAp.

B. In the next examples, the solubility product of a solid solution falls below the straight line shown in Figure 5a and described above. For simplicity, only one solid solution, $OH_{1-x}F_xAp$ with $x = 0.5$, will be considered. Figures 6a and 7a show the $K_{sp}(OH_{0.5}F_{0.5}Ap)$ in the two possible cases where (i) $K_{sp}(OHAp) > K_{sp}(OH_{0.5}F_{0.5}Ap) = 10^{-59.3} > K_{sp}(FAp)$ and (ii) $K_{sp}(OHAp) < K_{sp}(OH_{0.5}F_{0.5}Ap) = 10^{-60.5} < K_{sp}(FAp)$, respectively. The K_{sp} values for all other $OH_{1-x}F_xAp$ lie on the straight lines in Figures 6a and 7a.

(i) For the case where $K_{sp}(OH_{0.5}F_{0.5}Ap) = 10^{-59.3}$, the condition for simultaneous saturation with respect to OHAp and $OH_{0.5}F_{0.5}Ap$, derived as before, is represented by equation (12). The condition for simultaneous saturation with respect to FAp and $OH_{0.5}F_{0.5}Ap$ is given by equation (13).

$$\log(F^-) = -15.56 + pH \qquad (12)$$

$$\log(F^-) = -14.60 + pH \qquad (13)$$

The stability phase diagram (Figure 6b) shows that there is a region in which the solid solution is the most stable phase. The straight line defining the upper boundary of this region describes the compositions in which the solutions are simultaneously saturated with respect to the all solid solutions with $0.5 < x < 1$ and to FAp. This is a consequence of the fact that we assumed $K_{sp}(OH_{1-x}F_xAp)$ to be a linear function between $K_{sp}(FAp)$ and $K_{sp}(OH_{0.5}F_{0.5}Ap)$ (Figure 6a). Similarly, the straight line defining the lower boundary of the region describes those solutions that are simultaneously saturated with respect to the all solid solution with $0 < x < 0.5$ and to OHAp. Unlike in the case A. above, there are no solution compositions in which the solution is saturated with respect to both OHAp and FAp.

(ii) For the case where $K_{sp}(OH_{0.5}F_{0.5}Ap) = 10^{-60.5}$, the conditions for simultaneous saturation with respect to OHAp and $OH_{0.5}F_{0.5}Ap$ and to FAp and $OH_{0.5}F_{0.5}Ap$ are given by equations (14) and (15), respectively.

$$\log(F^-) = -17.96 + pH \qquad (14)$$

$$\log(F^-) = -12.20 + pH \qquad (15)$$

The stability phase diagram, shown in Figure 7b, is similar to that in Figure 6b except that the solid solution is the most stable phase over a wider region of the phase diagram; the lower the $K_{sp}(OH_{1-x}F_xAp)$ value means the wider composition region in which the solid solution is the most stable phase.

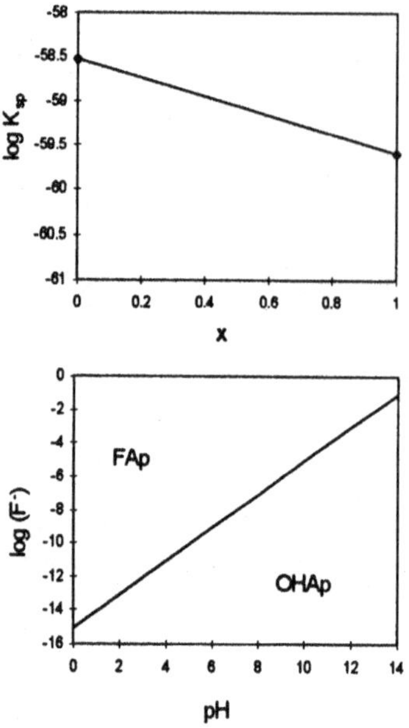

Figure 5 - (a) $K_{sp}(OH_{1-x}F_xAp)$ for $0<x<1$ is a linear function of x and (b) the corresponding stability phase diagram.

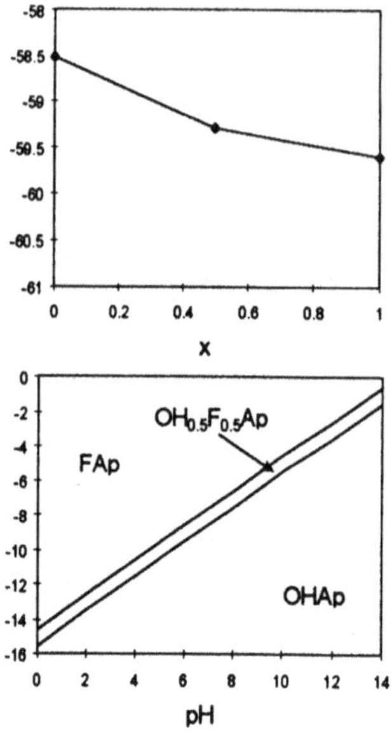

Figure 6 - (a) $K_{sp}(OH_{0.5}F_{0.5}Ap)$ value is between $K_{sp}(OHAp)$ and $K_{sp}(FAp)$ (b) the corresponding stability phase diagram.

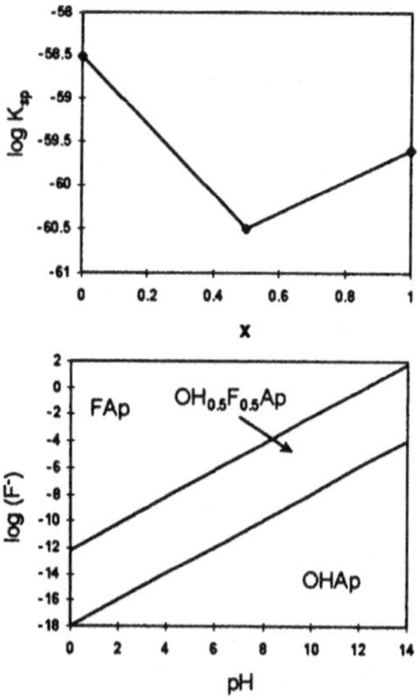

Figure 7 - (a) $K_{sp}(OH_{0.5}F_{0.5}Ap)$ is smaller than $K_{sp}(OHAp)$ and $K_{sp}(FAp)$ and (b) the corresponding stability phase diagram.

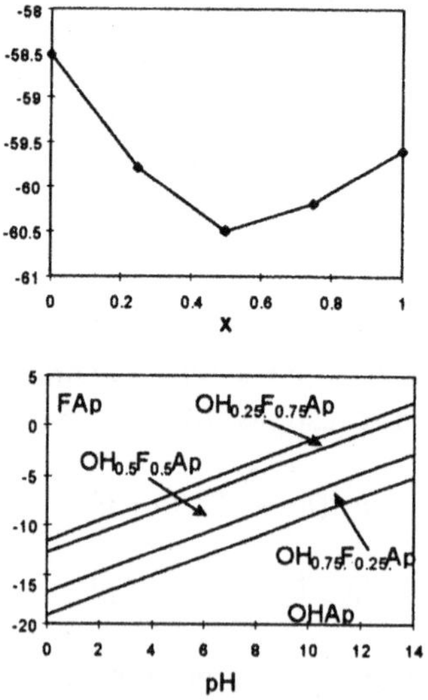

Figure 8 - (a) $K_{sp}(OH_{1-x}F_xAp)$ for x = 0.25, 0.5, and 0.75 are shown and (b) the corresponding stability phase diagram.

C. The example given here is similar to that in B. but here we consider three solid solutions with x = 0.25, 0.5, 0.75 for which the $K_{sp}(OH_{1-x}F_xAp)$ values are $10^{-59.8}$, $10^{-60.5}$, and $10^{-60.2}$, respectively, as shown in Figure 8a. There are four equations that describe conditions in solutions saturated with respect to the various pairs of solids as follows, $OHAp/OH_{0.75}F_{0.25}Ap$, equation (16); $OH_{0.75}F_{0.25}Ap/OH_{0.5}F_{0.5}Ap$, equation (17); $OH_{0.5}F_{0.5}Ap/OH_{0.25}F_{0.75}Ap$, equation (18); and $OH_{0.25}F_{0.75}Ap/FAp$, equation (19). The stability phase diagram (Figure 8b) shows that there are separate regions in which each of the three solid solutions is the most stable phase and the boundary between two regions define solution compositions at which the solutions are simultaneously saturated with respect to the two stable phases and all of the solid solutions with x values between those of the two phases.

$$\log(F^-) = -19.12 + pH \qquad (16)$$

$$\log(F^-) = -16.80 + pH \qquad (17)$$

$$\log(F^-) = -12.80 + pH \qquad (18)$$

$$\log(F^-) = -11.60 + pH \qquad (19)$$

RELATIVE STABILITY OF FAp AND CALCIUM FLUORIDE

The original purpose of topical F treatments was to form apatitically bound F in tooth mineral.[18] However, subsequent studies showed that calcium fluoride (CaF_2), not FAp, was the major product formed on the tooth surface.[19-21] Saliva is supersaturated with FAp but is undersaturated with respect to CaF_2. As a result, FAp is insoluble in the mouth under non-cariogenic conditions, whereas CaF_2 leaches away from the tooth surface gradually.[22,23] Analysis of the relative stability of FAp and CaF_2, addressed in this section, provides a means for understanding the conditions in which FAp or CaF_2 is the phase that would form or dissolve more readily. For solutions that are saturated with respect to both FAp and CaF_2, equations (20) and (21) hold and after rearrangement equation (22) is obtained.

$$(Ca^{2+})(F^-)^2 = K_{sp}(CaF_2) = 2.8 \times 10^{-11} \qquad (20)$$

$$(Ca^{2+})^5(PO_4^{3-})^3(F^-) = K_{sp}(FAp) = 2.5 \times 10^{-60} \qquad (21)$$

$$\log(F^-) = 0.33 \log(PO_4^{3-}) + 0.76 \qquad (22)$$

Eqn. (22) signifies that the relative stability of FAp and CaF_2 depends on the relative magnitudes of the F^- and PO_4^{3-} activities in the solution. A high (F^-) favors CaF_2 stability and a high (PO_4^{3-}) favors FAp stability. The (PO_4^{3-}), in turn, increases linearly with total phosphate concentration, $[PO_4]_{tot}$, but it increases more rapidly with increasing pH. Thus, FAp tends to be the more stable phase in neutral or alkaline solutions with a high $[PO_4]_{tot}$ and low $[F^-]$, whereas CaF_2 tends to be the more stable phase in acidic solutions with a high $[F^-]$. It is interesting to note that Ca^{2+} concentration plays no direct role in determining the relative stability of FAp and CaF_2. The linear function given by equation (22) defines the regions in which FAp or CaF_2 is the more stable phase in the stability phase diagram (Figure 9). Table 1 shows the (PO_4^{3-}) values in saliva and plaque fluid at pH 7 and 5, calculated (EQUIL) from the total phosphate concentrations, $[PO_4]_{tot}$, and ionic strengths, I, typical for these fluids.[24-28] Also calculated are the (F^-) and (PO_4^{3-}) values for acidulated phosphate fluoride (APF) solution, the most commonly used topical F agent administered in the dental clinic. Figure 9 shows that at pH 7, FAp is the more stable phase in saliva (and plaque) until the (F^-) reaches approximately 10^{-2} molℓ. At pH 5 that corresponds to pH of cariogenic plaque, FAp is the more stable phase until the (F^-) reaches approximately 10^{-3} molℓ. The (F^-) is typically 10^{-6} to 10^{-5} molℓ in saliva and plaque,[29,30] and CaF_2 is the less stable phase than FAp in these oral fluids. The (Ca^{2+}) is approximately 10^{-3} molℓ in saliva and (PO_4^{3-}) is approximately 10^{-8} molℓ at pH 7 and 10^{-12} molℓ at pH 5 (Table 1), and at these conditions saliva is supersaturated with respect to FAp and undersaturated with respect to CaF_2 at either pH. In contrast, the locus of APF is deeply in the region in which CaF_2 is the more stable phase (Table 2 and Figure 9) and, therefore, CaF_2 but not FAp is formed by the APF treatment.

Table 1. Calculated Values of (PO_4^{3-}) in Saliva and Dental Plaque, and Calculated Values of (PO_4^{3-}) and (F^-) in APF solution. All Concentrations are Expressed in molℓ.

System	$[PO_4]_{tot}$ [a]	pH [a]	I [a]	log (PO_4^{3-}) [b]	log (F^-)
Saliva	0.006	5	0.091	- 11.88	
Saliva	0.006	7	0.091	- 8.25	
Plaque	0.015	5	0.150	- 11.50	
Plaque	0.015	7	0.150	- 7.90	
APF	0.100	3.2	0.435	- 14.33	- 0.79 [c]

[a] Input parameters. [b] Calculated from input parameters. [c] Calculated from input parameters and $[F]_{tot}$ of 0.6 molℓ.

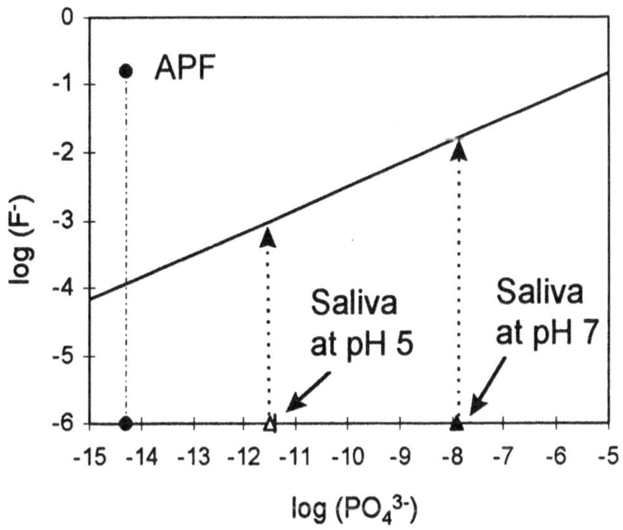

Figure 9. Stability phase diagram of FAp and CaF_2.

INCORPORATION OF TOOTH-BOUND F INTO ENAMEL AND DENTIN

Topical treatments intended to increase the tooth-bound F content in the apatitic mineral (enamel or root dentin) to a depth at least as large as that reached by the demineralizing acids. A shortened review[5] of treatment conditions which have been reported to incorporate tooth-bound F is given:

Acidulated phosphate fluoride (APF) treatments *in vitro* have been shown[22] to produce large amounts of CaF_2. The formed CaF_2 (labile F) dissolves slowly, probably due to formation of a surface coating.[31] Therefore, a very small amount of tooth-bound F was detected immediately after the treatment and significantly larger amount (although small in the absolute F level) were detected one day after the treatment.[23] It appears that after the treatment a small portion of the labile F reincorporated into tooth-bound F. Thus, CaF_2 can serve as a F reservoir for this reincorporation. *In vivo* results showed that one week[19] to three months[20,21] after the treatment the F level in the enamel decreased to approximately the same level as that before the treatment. Therefore, it seems that even with reduced rate of dissolution, CaF_2 did not persist over longer time period.

F Varnishes are believed to deposit CaF_2-like materials rather than apatitic F on the tooth surfaces. Significant F uptake has been detected for as long as 24 weeks after application.[32] No mechanisms which may allow for re-incorporation of some of the deposited F in the form of apatitically-bound F have been reported.

Calcium phosphate solutions containing low levels of F were used in several *in vitro* studies.[33,34] The results showed that when enamel specimens were immersed from several days to 1 month in a pH 7 artificial "saliva" solution containing 1 ppm (53 μmolℓ) F, the F content in the outer layers of the enamel increased significantly. Since the solutions were undersaturated with respect to CaF_2, most of the acquired F must be a result of fluoridated OHAp precipitation (firmly bound F) and of F adsorption in nonapatitic form[35] (labile F on the crystal surfaces[36]). These studies showed the firmly bound F content of teeth can be significantly increased by immersion in a F-containing calcium phosphate solution for several days. The extended immersion in a F containing calcium phosphate solution for several days may be difficult to achieve *in vivo.*

Dicalcium phosphate dihydrate (DCPD) was formed during topical F treatment as an intermediate.[37] This procedure involves a pretreatment with an acidic (pH 2 to 3) calcium phosphate solution that dissolves a small portion of the tooth mineral and re-precipitates it in the form of DCPD. The DCPD is subsequently converted to FAp through reactions with F from low or highly concentrated F agent. This treatment procedure thus reincorporates a portion of the labile F into tooth mineral in the form of tooth-bound F. DCPD-forming pretreatment solution is saturated with respect to FAp, and therefore: A. does not dissolve mineral at the tooth surface which is relatively F-rich, B. DCPD is formed under the tooth surface rather than on the tooth surface;[37] C. does not remove the existing F in the tooth; and D. penetrates into a deeper area of the tooth to react with the more soluble portions of the tooth mineral in each cycle of the multiple treatment. This multiple treatment provides a mechanism for building a reservoir of firmly bound F in the tooth through continuous conversion of DCPD into FAp[38] and it has been evaluated as a possible remineralization treatment.[39,40] When applied to sound enamel or dentin, the combination of DCPD-forming and F treatments should increase the tooth-bound F and reduce the porosity, both of which are desirable. A number of modifications in DCPD-forming treatment have been applied to make this treatment suitable for different clinical applications.[21,34,41-43]

ACKNOWLEDGMENT

This investigation was supported, in part, by USPHS Research Grant DE05354 to the American Dental Association Health Foundation from the National Institutes of Health - National Institute of Dental Research and is part of the dental research program conducted by the National Institute of Standards and Technology in cooperation with the American Dental Association Health Foundation.

Certain commercial materials and equipment are identified in this paper to specify the experimental procedure. In no instance does such identification imply recommendation or endorsement by the National Institute of Standards and Technology or the ADA Health Foundation or that the material or equipment identified is necessarily the best available for the purpose.

REFERENCES

1. Proceedings of a Joint IADR/ORCA International Symposium on Fluorides: Mechanisms of Action and Recommendations for Use; Dawes C, Cate JM ten, eds; 1989 March 21-24; Pine Mountain, Georgia, USA. J Dental Res 1990;69(Spec Iss)

2. *Fluoride in Dentistry*, 2nd ed. Fejerskov O, Ekstrand J, Burt BA, eds. Munksgaard: Copenhagen, 1996

3. LeGeros RZ. Calcium phosphates in oral biology and medicine. Karger: Basel, 1991

4. Elliott JC. *Structure and Chemistry of the Apatites and Other Calcium Orthophosphates.* Elsevier: Amsterdam, 1994

5. Chow LC. Tooth-bound fluoride and dental caries. J Dent Res 1990;69(Spec Iss):595-6007

6. Kay MI, Young RA, Posner AS. Crystal structure of hydroxyapatite. Nature 1964;204:1050-1052

7. Elliott JC, Mackie PE, Young RA. Monoclinic hydroxyapatite. Science 1973;180:1055-1057

8. Mehmel Z. Uber die struktur des apatits. Z Krist 1930;75:323-331

9. Naray-Szabo S. The structure of apatite $(CaF)Ca_4(PO_4)_3$. Z. Krist 1930;75:387-398

10. McCann HG. The solubility of fluorapatite and its relationship to that of calcium fluoride. Arch Oral Biol 1968;13:987-1001

11. McDowell H, Gregory TM, Brown WE. Solubility of $Ca_5(PO_4)_3OH$ in the system $Ca(OH)_2$ - H_3PO_4 - H_2O at 5, 15, 25 and 37 °C. J Res Natl Bur Stand 1977;81A:273-281

12. Driessens FCM. *Mineral Aspects of Dentistry*, Basel: Karger, 1982

13. Moreno EC, Kresak M, Zahradnik RT. Physicochemical aspects of fluoride-apatite systems relevant to the study of dental caries. Caries Res 1977;11(Suppl 1):142-171

14. Brown WE, Gregory TM, Chow LC. Effects of fluoride on enamel solubility and cariostasis. Caries Res 1977;11(Suppl 1):118-141

15. Driessens FCM. Thermodynamics of the solubility behavior of fluorhydroxyapatite solid solutions. Ber Bunsenges Phys Chem 1979;83:583-586

16. Narasaraju TSB, Rao VLN. A new interpretation of the solubility equilibria of hydroxyapatite. J Phys Chem 1974;255:655-660

17. Verbeeck RMH, Thun HP, Driessens FCM. Solubility behavior of defective fluorapatite and fluorhydroxyapatites. Ber Bunsenges Phys Chem 1980;84:159-163

18. Brudevold F, McCann HG, Nilson R, Richardson B, Coklica V. The chemistry of caries inhibition: Problems and challenges in topical treatments. J Dent Res 1967;46:37-45

19. Heifetz SB, Mellberg JR, Winter LJ, Doyle J. *In vivo* fluoride uptake by enamel of teeth of human adults from various topical fluoride procedures. Arch Oral Biol 1970;15:1171-1181

20. Wei SHY, Shulz E. *In vivo* microsampling of enamel fluoride concentrations after topical treatment. Caries Res 1975;9:50-58

21. Chow LC, Guo MK, Hsieh CC, Hong YC. Apatitic fluoride increase in enamel from a topical treatment involving intermediate $CaHPO_4{\bullet}x2H_2O$ formation: An *in vivo* study. Caries Res 1981;15:369-376

22. Caslavska V, Brudevold F, Verbic V, Moreno EC. Response of human enamel to topical application of ammonium fluoride. Arch Oral Biol 1971;16:1173-1180

23. Caslavska V, Moreno EC, Brudevold F. Determination of the calcium fluoride formed from *in vitro* exposure of human enamel to fluoride solutions. Arch Oral Biol 1975;20:333-339

24. Lagerlof F. Determination of ionized calcium in human parotid saliva. Clin Chim Acta 1980;102:127-135

25. Lagerlof F. Effects of flow rate and pH on calcium phosphate saturation in human parotid saliva. Caries Res 1983;17:403-411

26. Edgar WM. A method for studying the free aqueous phase of plaque. Caries Res 1972;6:80

27. Tatevossian A. "The Chemistry of Plaque Fluid." In *Frontiers of Oral Physiology*, Vol. 3, Ferguson DB, ed. Basel: Karger, 1981.

28. Margolis HC, Moreno EC. Composition and cariogenic potential of dental plaque fluid. Crit Rev Oral Biol Med 1994;5:1-25

29. Vogel GL, Mao Y, Carey CM, Chow LC, Takagi S. *In vivo* fluoride concentrations measured for two hours after a NaF or a novel two-solution rinse. J Dent Res 1992;63:1276-1278

30. Vogel GL, Mao Y, Carey CM, Chow LC. Increased overnight fluoride concentrations in saliva plaque and plaque fluid after a novel two-solution rinse. J Dent Res 1997; in press

31. Lagerlof F, Saxegaard E, Barkvoll P, Rolla G. Effects of inorganic orthophosphate and pyrophosphate on dissolution of calcium fluoride in water. J Dent Res 1988; 67:447-449

32. Dijkman AG, De Boer P, Arends J. *In vivo* investigation on the fluoride content in and on human enamel after topical applications. Caries Res 1983;17:392-402

33. Crall JJ, Bjerga JM. Enamel fluoride retention after DCPD and APF application and prolonged exposure to fluoride *in vitro.* J Dent Res 1986;65:387-389

34. Takagi S, Chow LC, Yamada EM, Brown WE. Enhanced enamel F uptake by monocalcium phosphate monohydrate gels. J Dent Res 1987;66:1523-1526

35. Takagi S, Chow LC, Schreiber CT. Enhanced root fluoride uptake by monocalcium phosphate monohydrate gels. Caries Res 1990;24:18-22

36. Laufer B, Mayer I., Gedalia D, Deutsch D, Kaufman HW, Tal H. Fluoride-uptake and fluoride-residual of fluoride-treated human root dentine *in vitro* determined by chemical, scanning electron microscopy and x-ray diffraction analyses. Arch Oral Biol 1981;26:159-163

37. Chow LC, Brown WE. Formation of $CAHPO_4$•x2H2O in tooth enamel as an intermediate product in topical fluoride treatments. J Dent Res 1975;54:65-76

38. Chow LC, Brown WE. An approach to remineralization *via* saliva. *Proceedings of the Third Annual Conference on Foods, Nutrition and Dental Health*, Vol. 2; 1979 October 10-12; Chicago, IL. Chicago: American Dental Association Health Foundation, 1982

39. Wefel JS, Harless JD. The use of saturated DCPD in remineralization of artificial caries lesions *in vitro.* J Dent Res 1987;66:1640-1643

40. Chow LC, Takagi S, Sugawara A. Remineralization of enamel caries lesion by a Na_2SIF_6 containing $CaHPO_4$•$x2H_2O$-forming solution. J Dent Res 1988;67(Spec Iss):257, Abst No 1158

41. Hong HC, Chow LC, Brown WE. Enhanced fluoride uptake from mouthrinses. J Dent Res 1985;64:82-84

42. Shern R.J, Mirth DB, Li SH, Bartkiewicz A, Monell-Torrens E, Schreiber CT, Chow LC. Effects of CPS rinses and the IFRD on caries and fluoride in rat molars. J Dent Res 1987;66(Spec Iss):302, Abst No 1561

43. Eidelman N, Takagi S, Chow LC. Hydrolysis of F complexes: Application in topical fluoride treatment. J Dent Res 1988;67(Spec Iss):290, Abst No 1416

44. Hoppenbrouwers PMM, Groenendijk E, Tewarie NR, Driessens FCM. Improvement of the caries resistance of human dental roots by a two-step conversion of the root mineral into fluoridated hydroxyapatite. J Dent Res 1988;67:1254-1256

5

INTERFACIAL PROPERTIES OF CALCIUM PHOSPHATES

P. Somasundaran and Berislav Markovic

Langmuir Center for Colloids and Interfaces
Columbia University, New York, NY 10027 USA

INTRODUCTION

Interfacial properties such as surface charge are important properties of solids since they control significant processes like adsorption, wettability, dissolution, chemical uptake and adhesion and thus the overall interfacial behavior of the solid. Adsorption of surfactants and macromolecules (polymers, proteins, polysaccharides) itself is widely used in order to modify and control wettability as well as suspension characteristics of solids (flotation, dispersion, flocculation, sedimentation). While considerable information is available on surface charge characteristics of oxides such as alumina, silica, and titania, much less is known about the behavior of sparingly soluble minerals such as apatite. Surface charge properties of these materials are affected by more variables and the charge generation mechanisms are much more involved than in the case of oxides.

It is known that the interfacial and physical properties of tooth apatite can be altered by chemical species such as fluoride,[1,2] chelating agents,[3,4] long-chain surfactants,[5,6] as well as some metals.[7-11] However, the role of the surface electrochemical properties in determining the uptake of these chemical species by apatite is not well established. Furthermore, earlier electrochemical studies on apatites have produced results that are often in conflict with each other. Systematic work by Saleeb and de Bruyn[12] yielded values for the point of zero charge of hydroxyapatite and fluorapatite that are in agreement with our earlier data for natural apatite.[13-16] Some results have been obtained using titration techniques but these are applicable strictly to insoluble materials.[17,18] Also, some of the prior work has been done with sodium salt solutions (as supporting electrolyte), and these results cannot be considered to truly represent those of apatite due to possible substitution of calcium in apatite lattice by sodium from solution.

Surface charge of a mineral is determined by the concentration of potential determining ions in solution. These ions in the case of apatite could be the lattice ions or their reaction products with water. Other inorganic as well as organic (surfactants, polymeric reagents) species can also affect the surface charge and that can inhibit or promote the transport of ions through the surface layers[19,20], but these secondary effects are again controlled primarily by potential determining ions. Thus it is important to identify these ions and the mechanisms involved since their concentrations control the overall surface behavior. In this work, the role of various ionic species of Ca, PO_4, OH^-, and F^- and adsorption properties of selected surfactants and polymers of different charge characteristics are reviewed.

MATERIALS

Apatite

Both natural apatite ($Ca_{10}(PO_4)_6(F,OH)_2$) and synthetic hydroxyapatite have been studied in the past. In our work particle zeta potential, as the most often used approximation for surface potential, was measured using streaming potential technique[21] in the case of apatite, and electrophoresis in the case of synthetic hydroxyapatite. Natural apatite was cleaned using dilute nitric acid[15] and streaming potential was measured using 35/65 mesh particles in potassium nitrate solutions at different preadjusted pH values. In addition, the fluoride content in solutions was determined using a fluoride electrode. Aqueous solutions containing various amounts of calcium nitrate, potassium dihydrogen phosphate and potassium fluoride were used to determine the role of calcium, potassium and fluoride ions.

Hydroxyapatite

Hydroxyapatite was synthesized using a precipitation technique by mixing appropriate amounts of alkaline (KOH) solution of K_2HPO_4 and aqueous solution of $Ca(NO_3)_2$ at boiling temperature. Precipitated crystals were freeze-dried and analyzed. X-ray diffraction pattern have shown hydroxyapatite structure and chemical analysis the stoichiometric ratio of calcium and phosphorous (Ca/P = 1.67). Zeta potential of the samples was measured using the Zeta-Meter System (Zeta-Meter, Inc., Stauton, VA) at constant ionic strength (KNO_3). Samples were aged overnight in electrolyte solution and equilibrated at the desired pH for 1 hour prior to zeta potential measurements. Calcium content in supernatant was determined using titration and atomic absorption spectroscopy, while total phosphates were measured using a colorimetric technique.

ELECTROKINETIC POTENTIAL

Zeta potential of synthetic hydroxyapatite under relevant physico-chemical conditions are given in Figure 1 and those for natural apatite in Figure 2.[14] It can be seen that the isoelectric point of synthetic hydroxyapatite is pH 7 while that of natural apatite is pH 5.6. These values are not affected by the concentration of KNO_3, which means that neither K^+ nor NO_3^- has any surface potential determining role and that this salt can be used as a background electrolyte in interfacial studies. It is clear that H^+ and OH^- are potential determining ions, and the examination of possible chemical reactions of apatite in water, listed in Table 1, reveals the development of surface potential under different pH conditions:

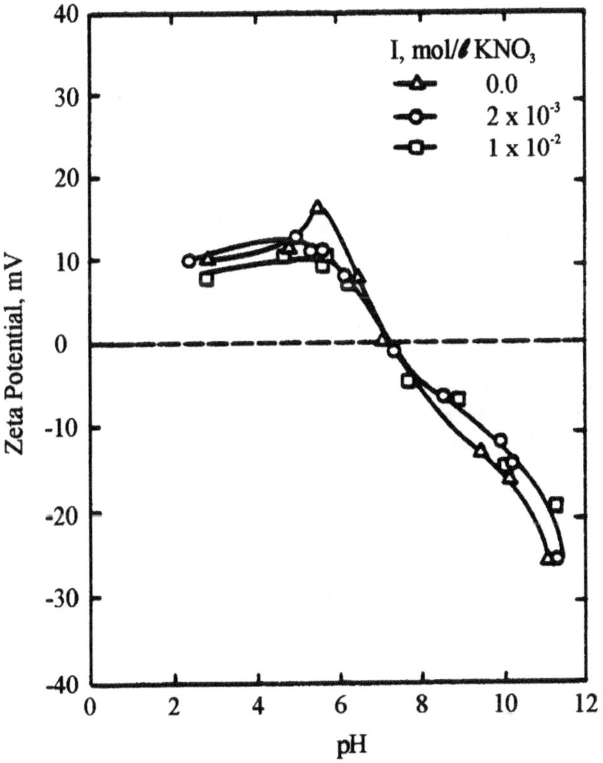

Figure 1. Zeta potential of synthetic hydroxyapatite as a function of pH at different ionic strengths.

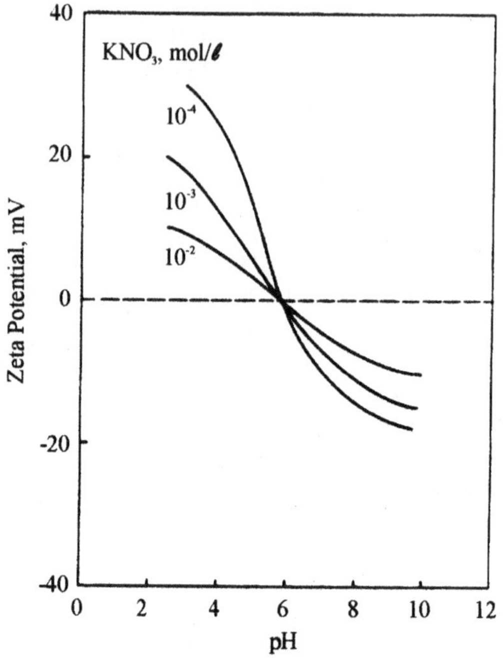

Figure 2. Zeta potential of natural apatite as a function of pH at different ionic strengths.

With an increase in pH of the system, the chemical equilibria in equations 1 through 6 will shift towards the right hand side, decreasing the activities of the positive species and increasing the activities of negative species in solution. The net result of such process will be an excess of negative potential determining ions in solution and hence a net negative charge on the surface. On the other hand, with a decrease in pH, the equilibria in the above reaction will move in the reverse direction and the surface becomes positively charged. If the above reactions are indeed responsible for the generation of the surface charge, then for apatite (represented by general formula $M_{10}(PO_4)_6Z_2$, where M can be calcium, barium, strontium etc., and Z stands for hydroxyl, fluoride, etc.) addition of any of the species should change surface potential in a manner dictated by the following equations:

$$M_{10}(PO_4)_6Z_2 \rightarrow 10M^{2+} + 6PO_4^{3-} + 2Z^- \qquad (10)$$

$$K_{sp} = (M^{2+})^{10} (PO_4^{3-})^6 (Z^-)^2 \quad \text{for } M^{2+} = Ca^{2+} \text{ and } Z^- = OH^- \qquad (11)$$

$$\psi_0 = - RT/F \times \ln (Ca^{2+})/(Ca^{2+})_{pzc} \qquad (12)$$

$$\psi_0 = - RT/F \times \ln (OH^-)/(OH^-)_{pzc} \qquad (13)$$

$$\psi_0 = - RT/3F \times \ln (PO_4^{3-})/(PO_4^{3-})_{pzc} \qquad (14)$$

Table 1. Chemical equilibria of apatite in water.

Chemical reaction	Equilibrium const.	Eqn.
$Ca^{2+} + OH^- \rightarrow CaOH^+$	$10^{1.4}$	(1)[22]
$CaOH^+ + OH^- \rightarrow Ca(OH)_2$	$10^{1.37}$	(2)[23]
$Ca(OH)_2(aq) \rightarrow Ca(OH)_2(s)$	$10^{2.45}$	(3)[23]
$H_3PO_4 \rightarrow H^+ + H_2PO_4^-$	$10^{-2.15}$	(4)[24]
$H_2PO_4^- \rightarrow H^+ + HPO_4^{2-}$	$10^{-7.2}$	(5)[24]
$HPO_4^{2-} \rightarrow H^+ + PO_4^{3-}$	$10^{-12.3}$	(6)[24]
$Ca^{2+} + HPO_4^{2-} \rightarrow CaHPO_4(aq)$	$10^{2.7}$	(7)[25]
$CaHPO_4(aq) \rightarrow CaHPO_4(s)$	$10^{4.3*}$	(8)
$Ca^{2+} + H_2PO_4^- \rightarrow CaH_2PO_4^+(aq)$	$10^{1.08}$	(9)[25]

Since more than two types of species (H^+ and OH^-; Ca and P species) can determine the surface potential of hydroxyapatite, there is not just one point of zero charge (as in the case of simple oxides like alumina or silica), but many points that will constitute a "line of zero charge" in the ternary phase diagram calcium-phosphate-hydroxyl. The requirement for constructing such a line is the condition of constant solubility product.

The role of calcium and phosphate as potential determining species for apatite is isolated by tests as a function of concentrations of these ions in the pH region below as well as above the original isoelectric point. If the cation is potential determining, then its addition will make the surface more positively charged above and below the isoelectric point. Similarly an anion, if potential determining, will make the surface more negatively charged both above and below the isoelectric point. Variation of surface potential with respect to the potential determining ions should follow the Nernst relationship. But in the system where mineral dissolves and the resultant species in supernatant undergo a complex step-wise reactions, it is difficult to estimate the activity of these dissolved species or the surface potential. In such systems, indirect tests on the effect of dissolved species under conditions when the mineral is similarly charged may be more appropriate. The effects of calcium, phosphate and fluoride ions on the zeta potential of hydroxyapatite at pH 10 (above the isoelectric point) under constant ionic strength conditions are illustrated in

* Calculated from equilibrium constants for reaction (7) and $CaHPO_4(s)$: Ca^{2+} + HPO_4^{2-}, $K_{sp} = 10^{-7}$

Figure 3. Calcium ions, as expected, make the mineral less negative and at concentrations higher than 10^{-4} mol/ℓ even reverse its charge. The role of phosphate as potential determining ion is also confirmed: the addition of phosphate makes the mineral more negatively charged even though the mineral is already negative. The addition of fluoride ions did not produce any significant effect under these conditions (pH 10). The effect of all the ions (calcium, phosphate and fluoride) on zeta potential of hydroxyapatite at the pH 5 (below the isoelectric point) is shown in Figure 4 for comparison. The addition of calcium again increases zeta potential, but now only at significantly higher concentration (above $4x10^{-3}$ mol/ℓ). This is because of the presence of significant amounts of dissolved calcium ($3x10^{-3}$ mol/ℓ) due to high mineral solubility at that pH. The phosphate ions have also expected the effect making the mineral more negative, while the fluoride, even though an anion, is observed to make the mineral more positive.

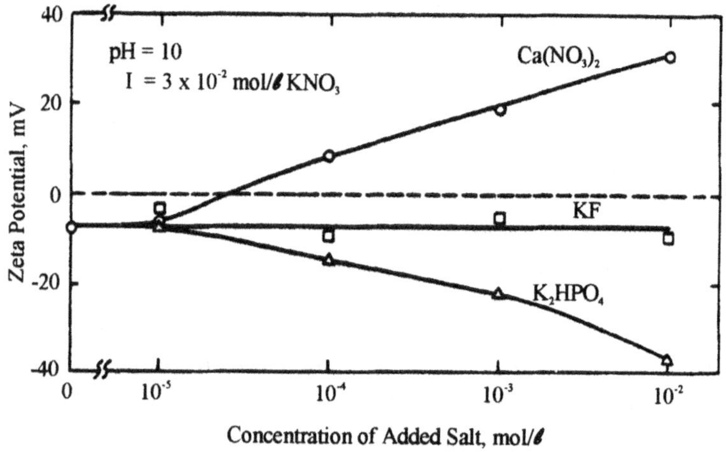

Figure 3. Zeta potential of synthetic hydroxyapatite as a function of concentration of Ca^{2+}, F^{-}, and HPO_4^{2-} ions at pH 10.

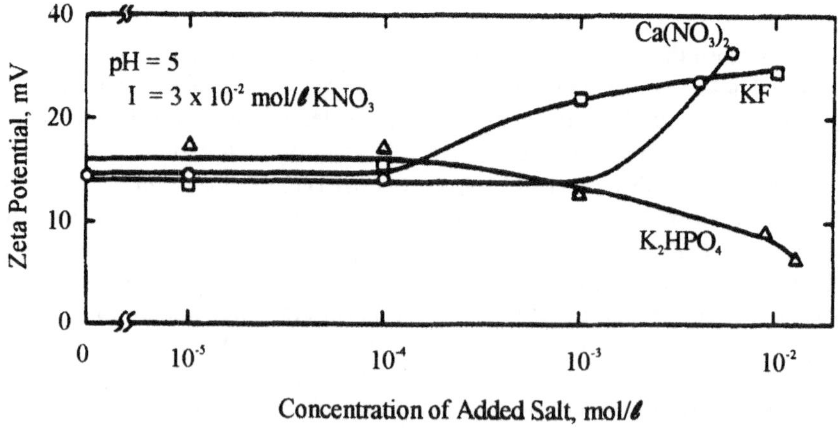

Figure 4. Zeta potential of synthetic hydroxyapatite as a function of concentration of Ca^{2+}, F^{-}, and HPO_4^{2-} ions at pH 5.

As shown earlier, the above ions can exist in hydroxyapatite supernatant in the form of various complexes with each other as well as with H^+ or OH^-. Therefore, their effect on surface potential can be either direct or due to such complexation. In order to better examine the role of different ions and complexes on zeta potential, the zeta potential in various electrolytes (at pH 5 and 10), is shown in Figures 5 to 8 as a function of the total concentration of relevant species (calcium and phosphate).

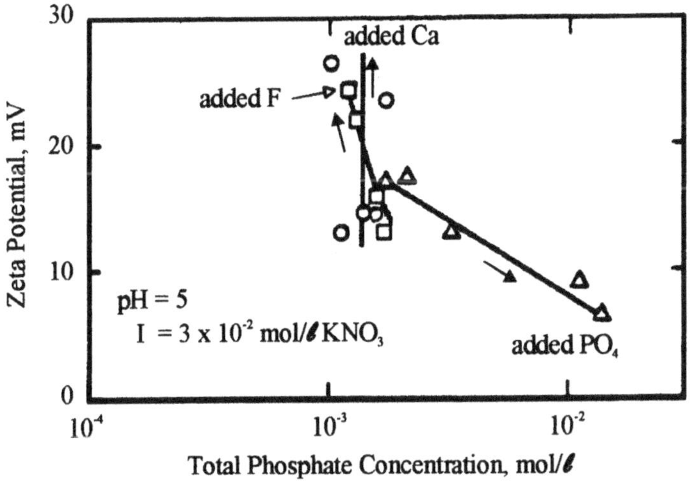

Figure 5. Zeta potential of synthetic hydroxyapatite in $Ca(NO_3)_2$, K_2HPO_4, and KF solutions at pH 5 as a function of total phosphate concentration.

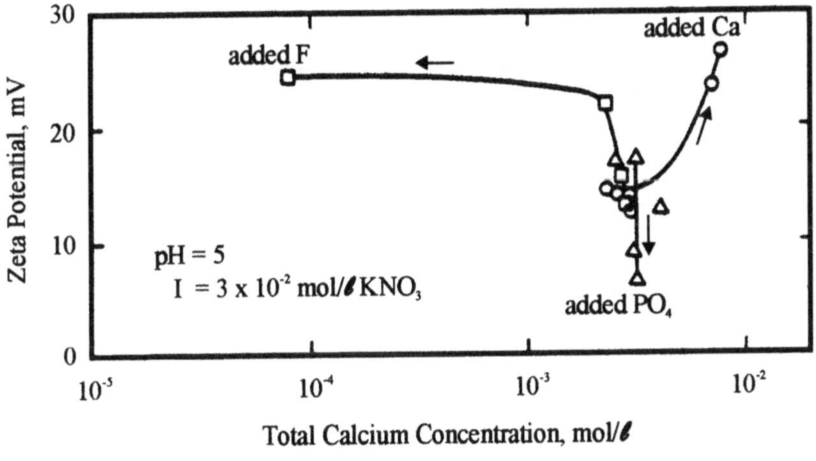

Figure 6. Zeta potential of synthetic hydroxyapatite in $Ca(NO_3)_2$, K_2HPO_4, and KF solutions at pH 5 as a function of total calcium concentration.

The addition of calcium changes zeta potential at pH 5 without altering the phosphate concentration significantly (Figure 5), but at pH 10 (Figure 7), the zeta potential changes upon the addition of calcium is accompanied by a reduction in the phosphate concentration. Stronger effects of calcium in the alkaline pH region than in the acidic (see Figure 9) can be attributed to this reduction in phosphate concentration and possible alterations of various calcium phosphate complexes present in the system. Also, more of the calcium will exist in the hydroxylated $CaOH^+$ form at pH 10 than at pH 5, and its potential determining role could be greater than that of Ca^{2+}. The presence of a larger amount of dissolved calcium in acidic solutions (Table 2) reduces the effect of added calcium and further complicates the interpretation of the results. The addition of phosphate also acts in a similar manner, directly at pH 5 (Figure 6) and with some reduction in calcium concentration at pH 10 (Figure 8).

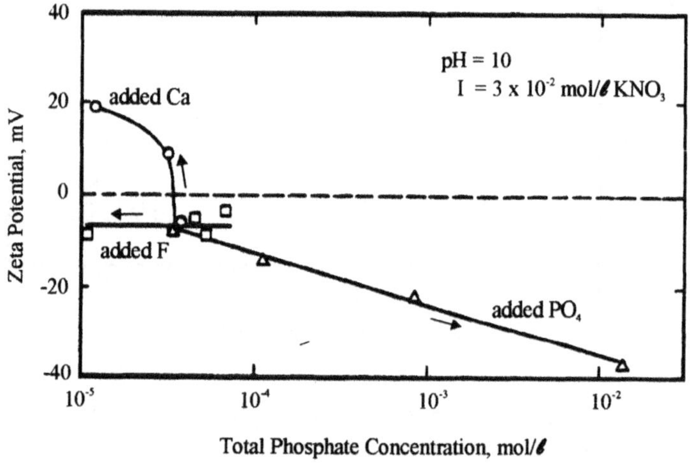

Figure 7. Zeta potential of synthetic hydroxyapatite in $Ca(NO_3)_2$, K_2HPO_4, and KF solutions at pH 10 as a function of total phosphate concentration.

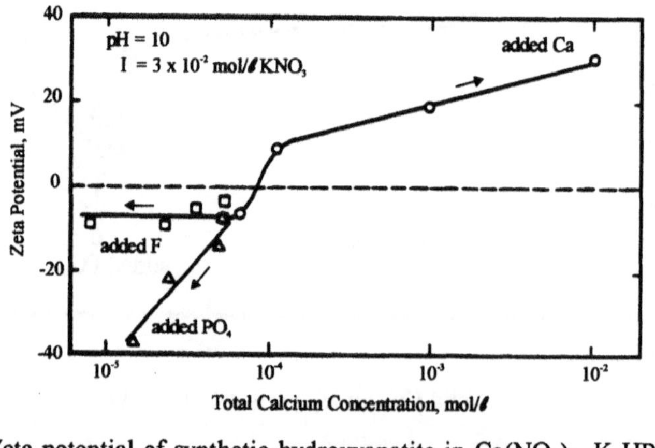

Figure 8. Zeta potential of synthetic hydroxyapatite in $Ca(NO_3)_2$, K_2HPO_4, and KF solutions at pH 10 as a function of total calcium concentration.

Table 2. Concentrations of Ca and P in the test solutions.

Solution	pH	Ca conc.(mol/ℓ)	P conc.(mol/ℓ)
Triple distilled water	11.2	3.75×10^{-6}	$< 1.6 \times 10^{-5}$
	10.3	$< 2.5 \times 10^{-5}$	$< 1.6 \times 10^{-5}$
	7.1	1.50×10^{-5}	$< 1.6 \times 10^{-5}$
	6.2	1.24×10^{-4}	1.13×10^{-4}
	5.6	1.17×10^{-3}	4.87×10^{-4}
	4.9	4.41×10^{-3}	2.70×10^{-3}
	4.7	5.12×10^{-3}	3.29×10^{-3}
	2.9	5.62×10^{-3}	3.35×10^{-3}
2×10^{-3} mol/ℓ KNO_3	11.4	$< 2.5 \times 10^{-5}$	$< 1.6 \times 10^{-5}$
	10.3	2.75×10^{-5}	$< 1.6 \times 10^{-5}$
	7.4	4.38×10^{-5}	$< 1.6 \times 10^{-5}$
	5.0	3.68×10^{-3}	2.73×10^{-3}
	2.5	5.51×10^{-3}	3.48×10^{-3}
1×10^{-2} mol/ℓ KNO_3	11.2	4.75×10^{-5}	$< 1.6 \times 10^{-5}$
	9.9	3.38×10^{-4}	$< 1.6 \times 10^{-5}$
	7.8	1.27×10^{-4}	8.06×10^{-5}
	5.8	4.20×10^{-3}	2.69×10^{-3}

The effect of fluoride addition on zeta potential is illustrated in Figure 9,[14] and it seems peculiar in that it makes the mineral more positive in the acidic region and more negative in the alkaline region. Additionally, the presence of fluoride ions in the supernatant produces a marked reduction in calcium concentration at all pH values (Figures 6 and 8). It has been suggested that fluoride in sufficient amounts can form fluorapatite at high pH values and fluorite at low pH values. The reported values for point of zero charge of fluorite are higher than that of hydroxyapatite,[26,27] while those for fluorapatite are lower.[12] Therefore, the zeta potential changes observed with the addition of fluoride, suggest the formation of fluorite at the surface of hydroxyapatite at low pH and fluorapatite at high pH. The concentrations of ions in the solution also dictate the above surface precipitation since the calcium concentration at pH 5 is 3×10^{-3} mol/ℓ and CaF_2 (solubility product of 3.95×10^{-11}) can be expected to form at about 10^{-4} mol/ℓ of fluoride.[28] The formation of CaF_2 is also indicated by the observed decrease in calcium concentration at this pH (Figure 6). At pH 10 with a total calcium concentration of 4×10^{-5} mol/ℓ, fluorite can precipitate above 10^{-3} mol/ℓ fluoride concentration. However, more of the calcium will be present in complex forms with hydroxyl and phosphate ions since the stability of $CaHPO_4$ is higher than that of $Ca_2HPO_4^+$ at high pH. Precipitation of fluorite will therefore require much larger concentrations of fluoride.

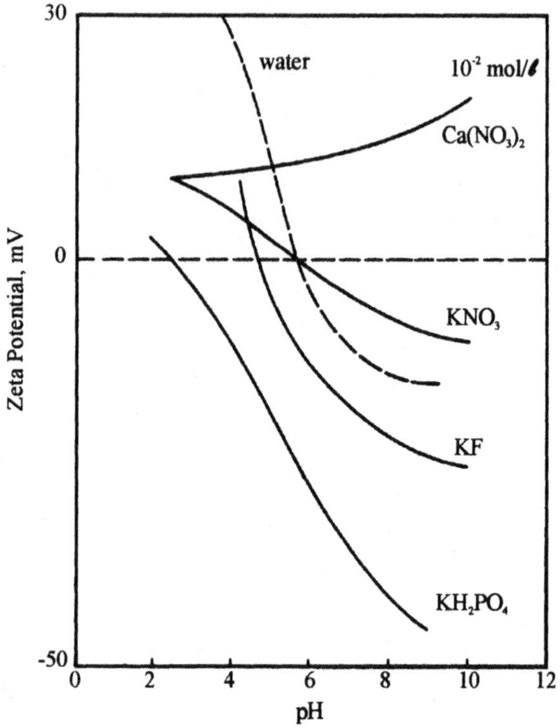

Figure 9. Effect of Ca, PO_4 and F on the zeta potential of natural apatite as a function of pH.

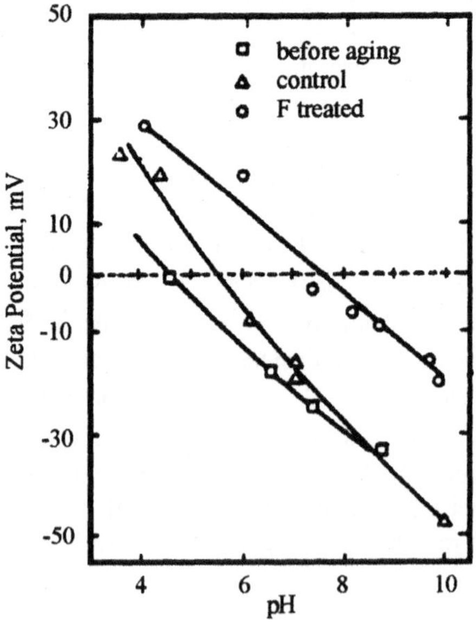

Figure 10. Effect of prolonged fluoride contact on the electrochemical properties of natural apatite.

SURFACE CONVERSIONS

When treated with concentrated fluoride solution (1 mol/ℓ) for prolonged time period (8 hours), zeta potential of natural apatite became more positive throughout the broad pH range (Figure 10). This again supports the possibility of fluorite precipitation on the apatite surface during contact with sufficiently concentrated fluoride solution. Such results have confirmed that the electrochemical properties of apatite are very much dependent upon the method of preparation of the mineral, cleaning, storing etc.[15] Also, the presence of other minerals in the system can alter the surface properties. For example, conditioning of synthetic hydroxyapatite in calcite supernatant (obtained by stirring calcite sample in water and removing all the solids by centrifugation) is found to shift the isoelectric point to that of calcite, suggesting precipitation of calcite on the hydroxyapatite surface. Zeta potential measurements conducted with apatite and calcite in water and in the supernatants of each other have shown that the surface conversion of mineral is possible (Figure 11).[29] It can be seen that the two minerals have almost interchanged their respective points of zero charge.

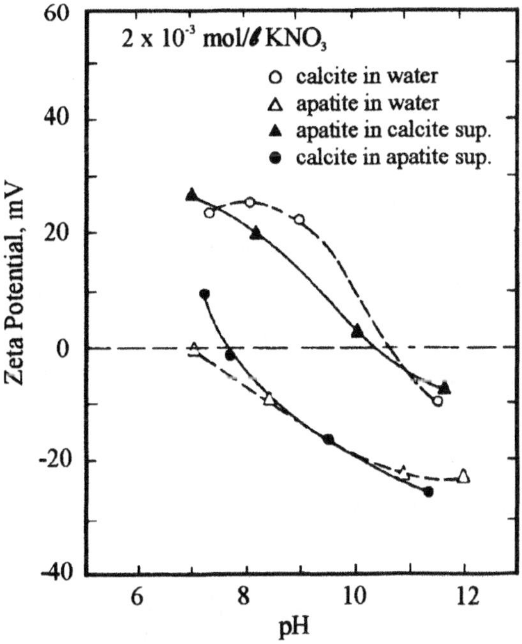

Figure 11. The zeta potential of apatite, calcite, apatite in calcite supernate, and calicte in apatite supernate as a function of solution pH.

Additional electrokinetic experiments in relevant organic solutions, as well as solubility[30] and spectroscopic (ESCA)[31] studies have also shown the surface conversion of apatite to calcite and vice versa under appropriate solution conditions. The ESCA results (Figure 12)[31] clearly show the surface conversion process. Apatite conditioned in the supernatant of calcite at pH 12 exhibits spectroscopic properties which are characteristic of both calcite and apatite. Solid-solution equilibria in mineral systems containing apatite can be extremely complex which leads to the possibility of precipitation of different mineral depending upon the solution conditions such as pH.[32-35] It is to be noted that these surface conversions of apatite are very much dependent upon whether the system is open to atmosphere (containing CO_2) or not.[29] The "real life" system always contain more than one mineral and these results suggest that the behavior of the minerals can be totally different from their behavior when present alone.

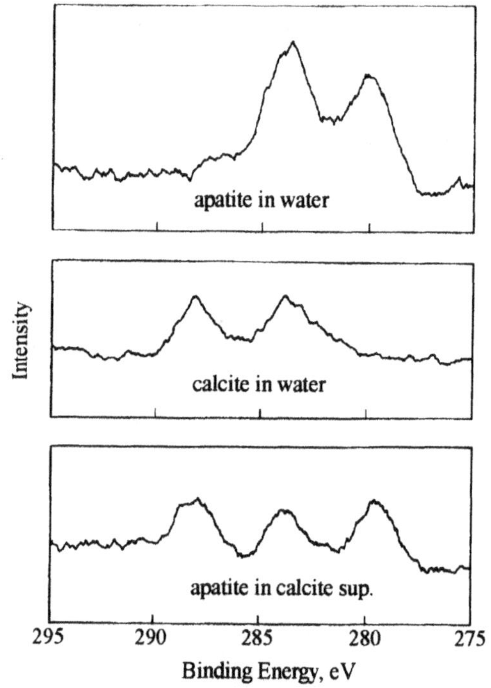

Figure 12. The ESCA spectra of apatite, calcite, and apatite in calcite supernatent.

SURFACTANT ADSORPTION

Mineral surface charge characteristics are extremely important in determining the adsorption of different species from solution. This is illustrated by the results for the adsorption of anionic surfactants such as dodecylsulfonate (NaDDS) and cationic surfactants such as dodecylamine (DDA HCl) on hydroxyapatite (Figures 13 and 14). While dodecylsulfonate adsorbs significantly at pH 6.7, there is no adsorption at pH 10.7. In contrast, cationic dodecylamine is found to adsorb at both pH values. Also, other zeta potential results[34] suggest adsorption of dodecyltrimethylammonium chloride and dodecylammonium chloride under all pH conditions at surfactant concentration of 10^{-3} mol/ℓ, while at lower concentrations a significant shift in zeta potential was observed only above the isoelectric point. Anionic dodecylsulfonate, on the other hand, lowers the zeta potential at all levels in the pH range from 5 to 10. It is known that oleate also lowers the zeta potential, shifting it towards that of Ca-oleate at high oleate to apatite ratios. Evidently, precipitation of calcium salts of given surfactant can also be a major phenomenon in these systems.

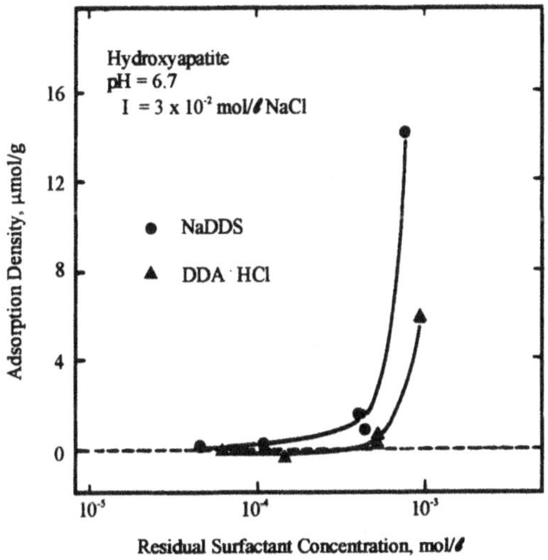

Figure 13. Adsorption of dodecylsulfonate (NaDDS) and dodecylamine (DDA HCl) on synthetic hydroxyapatite at pH 6.7.

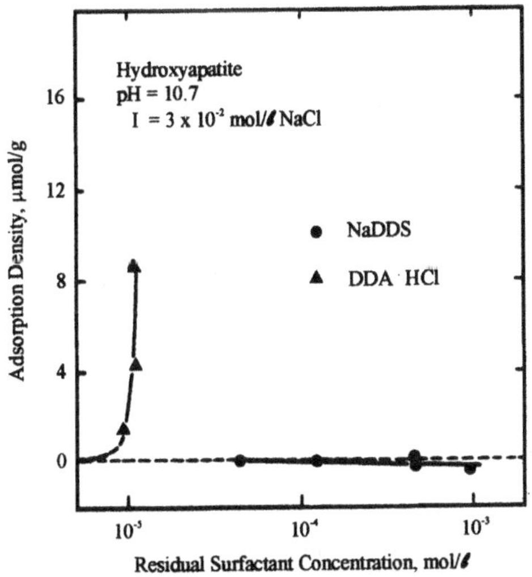

Figure 14. Adsorption of dodecylsulfonate (NaDDS) and dodecylamine (DDA HCl) on synthetic hydroxyapatite at pH 10.7.

POLYMER ADSORPTION

Polymers of different type and charge are often used for solids in processing. The adsorption of polymers on charged surfaces will be influenced by their charge characteristics. Figures 15 and 16 show the adsorption of selected polymers (cationic, anionic and nonionic polyacrylamide) on hydroxyapatite at pH 11.1 and 6.6. In the neutral pH range, where the mineral is positively charged, all the polymers strongly adsorb, while in the alkaline pH, where the mineral is negatively charged, only the cationic polymer is found to adsorb. Polymer adsorption on solids is considered to result mainly from hydrogen bonding, electrostatic interaction and covalent bonding depending on the mineral/polymer system.[36,37]

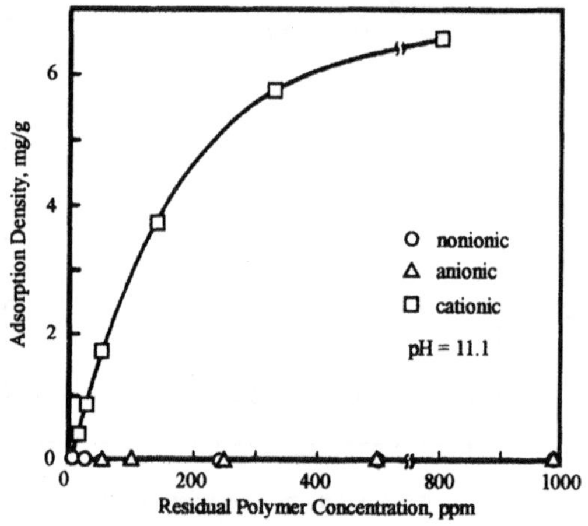

Figure 15. Adsortption of non-ionic, anionic and cationic polymers on synthetic hydroxyapatie at pH 11.1.

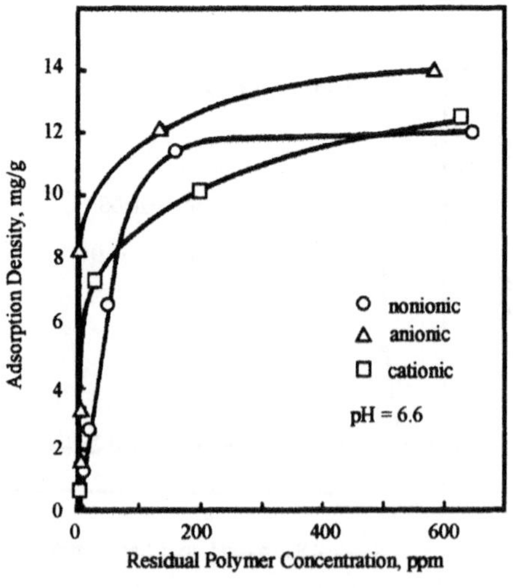

Figure 16. Adsortption of non-ionic, anionic and cationic polymers on synthetic hydroxyapatie at pH 6.6.

Hydrogen bonding might be the main mechanism in the case of adsorption of nonionic and cationic polymer in the neutral pH range. However, at high pH, even the nonionic polymer does not adsorb. This is attributed to the hydrolysis of the polymer to the anionic form in the alkaline solutions[38] and the electrostatic repulsion between the resultant functional groups on the polymer and similarly charged mineral surface. The adsorption of macromolecules on the solid surfaces will be evidently a complex function of not only properties of polymer and that of the solid, but also the solution properties and the possible alteration of both the solid and polymer in the solution. It is well known that polymer adsorption on the minerals can also depend on the presence of surfactant as they are often present together in many mineral processing operations.[39-43]

CONCLUSIONS

Electrochemical properties of apatite minerals are a complex function of solution conditions: pH, concentration of various ions, surfactants and polymers and the type of solid pretreatment. Hydrogen, hydroxyl, calcium and phosphate ions play a potential determining role either directly or by altering the concentration of other potential determining ions by shifting the chemical equilibria in solution. The addition of phosphate ion makes the mineral more negative and the addition of calcium more positively charged. However, fluoride is found to make the mineral more positively charged in acidic solutions and more negatively charged in alkaline solutions possibly due to the formation of fluorite and fluorapatite respectively. Addition of a large amount of fluoride in solution produces a marked increase in the isoelectric point of apatite suggesting fluorite precipitation on the surface. Similarly, the contact of hydroxyapatite with calcite supernatant, produces a shift of the zeta potential curve towards that of calcite. It is clear that surface alternations of apatite in systems with the other minerals are possible and depend on the mineral-solution equilibria involved. The resultant electrochemical nature of mineral surface plays an important role in the adsorption process of different surface active reagents and macromolecules of various charge characteristics and subsequent interfacial behavior

ACKNOWLEDGMENTS

The authors acknowledge the financial support from National Science Foundation (Grant No. CTS-9622781) and Industrial partners: Unilever and Nalco Chemical Co.

REFERENCES

1 Elliott JC. Recent progress in the chemistry, crystal chemistry, and structures of apatites. Calcif Tissue Res 1969;3:293-307

2 Gray JA. Acid dissolution rate of sound and white-spot enamel treated with tin (II) and fluoride compounds. J Dent Res 1965;44:493-501

3 Schatz A, and Martin JJ. Proteolysis-chelation theory of dental caries. J Am Dent Assoc

4 Johnson NW. Differences in the shape of human enamel crystallites after partial destruction by caries. Arch Oral Biol 1966;11:1421-1424

5 Manly RS, and Manly KF. Influence of cephalin on solution rate of tooth enamel. J Dent Res 1963;42:565-567

6 Malaowalla A, and Myers HM. Interaction of Sodium fluoride and synthetic apatite. J Dent Res 1963;41:413-419

7 Meckel AH, and Francis MD. Intraoral transfer of topically applied stanous fluoride. J Dent Res 1964;43:78-85

8 Healy WB, Ludwig TG, and Losee FL Soils and dental caries in Hawke's Bay, New Zeeland, Soil Sci 1961;92:359-366

9 Buttner W, and Muhler JC. Effects of salivary fluoride on enamel solubility. J Dent Res

10 Tank G, and Storvick CA. Effect of naturally occurring selenium and vanadium on dental caries. J Dent Res 1960;39:473-488

11 Geyer CF. Vanadium, a caries-inhibiting trace element in the Syrian hamster. J Dent Res

12 Saleeb FZ, and de Bruyn PL. Surface properties of alkaline earth apatites. Electrochemical and Interfacial Electrochemistry 1972;37:99-118

13 Somasundaran P. Zeta potential of apatite in aqueous solutions and its charge during equilibration. J Coll Interf Sci1968;27:659-666

14 Somasundaran P, and Agar GE. Further streaming potential studies of hydroxyapatite in inorganic solutions. Trans AIME 1972;252:348-352

15 Somasundaran P. Pretreatment of Mineral Surface and its Effect on their Properties, in *Clean Surfaces, Their Preparation and Characterization for Interfacial Studies*, G. Goldfinger ed., Marcel Dekker, New York, 1970.

16 Wang YH. *Zeta Potential Studies on Hydroxyapatite*, M.S. Thesis, Columbia University, New York, 1975

17 Bell LC, Posner AM, and Quirk JP. Surface charge characteristics of hydroxylapatite and fluorapatite. Nature 1972;239:515-517

18 Bell LC, Posner AM, and Quirk JP. Point of zero charge of hydroxyapatite and fluorapatite in aqueous solutions. J. Coll Interf Sci 1973;42:250-261

19 Sollner K. Outline of basic electrochemistry of porous membranes. J Dent Res

20 Somasundaran P, and Kulkarni RD. A new streaming potential cell and study of temperature effects using it. J Coll Interf Sci 1973;45:591-600

21 Bell RP, and George JHB. Incomplete dissociation of some thallous and calcium salts at different temperatures., Trans Faraday Soc 1953;49:619-627

22 Hedberg DD. *Sargent Chart of Equilibrium Constants of Inorganic Compounds*, E.H. Sargent and Co., Chicago, 1963.

23 Lide RD. *Handbook of Chemistry and Physics*, 77th edition, CRC Press, Boca Raton, Fl, 1996.

24 Bjerrum J, Schwarzenbach G, Sillen LG. *Stability Constants of Metal Ion Complexes, with Solubility Products of Inorganic Substances*, Part II, The Chemical Society, London, 1958.

25 Ney P. *Zeta-Potentiale und Flotierbarkeit Von Mineralen*, Springer-Verlag, New York, 1973.

26 Miller JD, and Hiskey JB. Electrokinetic behavior of fluorite as influenced by surface carbonation. J Coll Interf Sci 1972;41:567-573

27 Dean JA. *Lange's Handbook of Chemistry*, 13th edition, McGraw Hill, New York, 1972.

28 Nancollas GH, Amjad Z, and Koutsoukas P. Calcium Phosphates-Speciation, Solubility and Kinetic Considerations, in *Chemical Modelling in Aqueous Systems*, J.A. Jenne ed., ACS, Washington D.C., 1979.

29 Somasundaran P. Amankonah JO, and Ananthapadmanabhan KP. Mineral-solution equilibria in sparingly soluble mineral systems. Colloids and Surfaces 1985;15:309-333

30 Amankonah JO, Somasundaran P, and Ananthapadmanabhan KP. Effect of dissolved mineral species on the dissolution/precipitation characteristics of calcite and apatite. Colloids and Surfaces 1985;15:295-307

31 Amankonah JO, and Somasundaran P. Effect of dissolved mineral species on the electrokinetic behavior of calcite and apatite. Colloids and Surfaces 1985;15:335-353

32 Avnimelech Y, Moreno EC, and Brown WE. Solubility and surface properties of finely divided hydroxyapatite. J Res NBS 1973;77A(1):149

33 Mishra RK, Chander S, and Fuerstenau DW. Effect of ionic surfactants on the electrophoretic mobility of hydroxyapatite. Colloids and Surfaces 1980;1:105-120

34 Chander S, and Fuerstenau DW. On the dissolution and interfacial properties of hydroxyapatite, Colloids and Surfaces 1982;4:101-120

35 Sresty GC, Raja A, and Somasundaran P. Selective Flocculation of Mineral Slimes using Polymers, in *Recent Developments in Separation Science*, Vol 4., N.N. Li, S.A. Stern, R.B. Lond, and P. Somasundaran eds., CRC Press, West Palm Beach, FL, 1978

36 Somasundaran P, Markovic B, Krishnakumar S, and Yu X. Colloidal Systems and Interfaces: Stability of Dispersions through Polymer and Surfactant Adsorption, In *Handbook of Surface and Colloid Chemistry*, K.S. Birdi ed., CRC Press, West Palm Beach, FL, 1997

37 Hollander AF, Somasundaran P, and Gryte CC. Adsorption of Polyacrylamide and Sulfonated Polyacrylamide on Na-kaolinite, in *Adsorption from Aqueous Solution*, P. Tewari ed., Plenum Press, New York, 1981, 143-162

38 Moudgil BM, and Somasundaran P. Adsorption of charged and uncharged polyacrylamides on hematite. SME Preprint, 1982;160

39 Somasundaran P, and Lee LT. Polymer-surfactant interactions in flotation of quartz. Separation Sci. Techn 1981;16:1475-1490

40 Moudgil BM, and Somasundaran P. Effect of interactions between sulfonated polyacrylamide and sodium dodeclysulfonate-dodecylamine hydrochloride on adsorption. Colloids and Surfaces 1985;13:87-101

41 Maltesh C. *Doctoral Dissertation*, Columbia University, New York, 1993

42 Maltesh C, and Somasundaran P. Binding of sodium dodecylsulfate to polyethylene oxide at silica-water interface. J Coll Interf Sci 1992;153:298-301

6

NUCLEAR MAGNETIC RESONANCE SPECTROSCOPY OF CALCIUM PHOSPHATES

James P. Yesinowski

Naval Research Laboratory
Washington DC 20375-5342 USA

INTRODUCTION

The class of crystalline and amorphous compounds generically known as calcium phosphates has a widespread importance across a variety of disciplines, as can be seen from other chapters in this book and from other books.[1] In particular the apatite lattice of calcium phosphates, $Ca_5X(PO_4)_3$, with its wide variety of possible atomic substitutions, forms a structurally rich basis for many important materials, ranging from the bones and teeth of mammals to phosphors in fluorescent lamps to phosphate ores. Other calcium phosphates have such diverse applications as fertilizers, food additives, dentifrice abrasives, and bone implants, and they can occur in soils, suspended in milk, and as undesirable scaling products in pipes or in blood vessels.

Nuclear Magnetic Resonance (NMR) spectroscopy has been applied to investigate the structure of solid calcium phosphates as early as 1960, when the proton NMR spectrum of a single crystal of "dicalcium phosphate dihydrate" (brushite), $CaHPO_4 \cdot 2(H_2O)$, provided information about distances between hydrogen atoms in the crystal structure.[2] (Note that in this chapter calcium phosphate salts will be referred to by their well-established, if potentially misleading, common names.) However, detailed studies of a wide variety of calcium phosphates using more advanced NMR techniques have appeared only since about 1980. This chronology undoubtedly reflects the continual improvements in NMR instrumentation and concurrent development of NMR techniques that have taken place, especially since the early 1970's. Both factors resulted in improvements in the sensitivity of detection, in higher spectral resolution, and in new experimental approaches to obtaining detailed chemical and structural information. This chapter will provide scientists and technologists interested in calcium phosphates with an overview of how NMR spectroscopy can be profitably applied to problems in the area. Such problems range from the simple "fingerprint" identification of stoichiometric crystalline calcium phosphate phases to the detailed structural study of samples that may be amorphous, nonstoichiometric, or doped with various atoms. Problems involving the latter category appear often with both apatites and octacalcium phosphate, and present a continuing challenge to NMR spectroscopists. The surface chemistry of calcium phosphates can also be investigated with NMR techniques, especially for materials with relatively high specific surface areas. The identification and characterization of solid calcium phosphates by solid-state NMR techniques should now be considered to be a standard method. However, there is also great potential for non-standard extensions to various problem areas. Both aspects are discussed in this chapter.

In order to carry out this aim, a brief background will be provided on those aspects of NMR (including terminology) that are most important for calcium phosphate studies. Italics will be used to highlight key phrases and NMR terms whose definition should be useful to the calcium phosphate researcher trying to understand the NMR literature on calcium phosphates or seeking to effectively interact with an NMR spectroscopist in solving some problem. After a brief introduction to the basics of NMR spectroscopy, the important anisotropic interactions that result in much broader NMR spectra of solids (compared to those of liquids) will be discussed. A summary of some of the many NMR techniques that have been developed to obtain more information from solids, often by improving sensitivity and/or resolution will be presented. Due to space limitations, only techniques that have already been applied to calcium phosphates will be discussed, although undoubtedly new ones will continue to be developed and applied. References will be cited to provide an introduction to NMR techniques, but the primary literature references for the various techniques will not be given. These can generally be found in the cited review articles, encyclopedia, textbooks, and NMR studies of calcium phosphates. The application of NMR techniques to calcium phosphates is not a matter of simply obtaining "the" NMR spectrum of a given calcium phosphate. Appropriate experiments, best suited to answering the questions being posed, can be chosen from a large repertoire of NMR techniques.

Following the discussion of NMR basics and solid-state techniques, the next section will summarize the characterization by ^{1}H (or ^{19}F in the case of fluorapatite) and ^{31}P NMR of nearly all of the known stoichiometric crystalline calcium phosphates, emphasizing the primary studies. The results from these stoichiometric crystalline compounds are important, since they generally form the basis for interpretation of NMR experiments of less well-ordered systems.

The final section will summarize studies of calcium phosphates in less well-defined forms, beginning with a section on amorphous calcium phosphate. Two sections that follow will discuss a variety of ^{1}H and ^{19}F NMR studies of fluorohydroxapatite solid solutions, including the end-members hydroxyapatite and fluorapatite. These studies demonstrate new techniques for investigating "defects" in crystal structures. They also help in interpreting the fluoridated hydroxyapatite surface results discussed in the next section, which is followed by one on other surface NMR studies of calcium phosphates. Examples of how new ^{19}F NMR techniques can be applied in determining the location of a specific dopant, antimony (III), in the fluorapatite lattice are also presented. In addition, NMR investigations of nonstoichiometric and carbonated hydroxyapatites will be discussed. Experimental strategies for studying mineralized tissue and NMR results on bone and teeth will be discussed in the final section. The experiments and strategies referred to throughout this last chapter division should have widespread applicability to a variety of problems in calcium phosphate structural chemistry.

The standard disclaimer of personal familiarity is offered as the selection of examples were primarily drawn from work with which I have been personally involved. No attempt has been made to provide either a complete coverage or a critical review of all work in the field, and I apologize in advance to authors whose work has not been cited due to space limitations. A fairly comprehensive reference to NMR studies of calcium phosphates up to about 1994 may be found in Elliott's authoritative book.[1]

ASPECTS OF NMR RELEVANT TO CALCIUM PHOSPHATES

Basics of NMR

NMR is a type of spectroscopy that relies on the minute energy level splittings of nuclear spin states that occur when a suitable nuclear isotope is placed in an external magnetic field. The transitions between these levels take place in the radiofrequency region, typically 10-500 MHz, and the so-called Larmor precession, or Zeeman, frequencies are dictated by both an intrinsic parameter characterizing the particular stable nuclear isotope (the “magnetogyric ratio”, γ) and the external magnetic field strength B_0. The resonance frequency is linearly proportional to both γ and B_0. Higher magnetic fields are generally desirable for improved sensitivity and resolution, and this is true for the calcium phosphates as well. Thus, NMR peaks for the stable isotope ^{31}P (100% natural abundance) in a magnetic field of 11.7 Tesla (117 kGauss) occur in a narrow range near 202 MHz, whereas peaks for the stable isotope ^{1}H (99.985% natural abundance) in this same field occur in a narrow range near 500 MHz. The sample may be either liquid, solid, or gas. The modern NMR spectrometer usually has a superconducting magnet and the ability to operate at a variety of frequencies. Since the frequency region chosen determines which isotope is observed, NMR spectroscopy is inherently very selective in the sense that it focusses in on the properties of a particular isotope and hence element. Another important characteristic is that NMR is inherently quantitative: under the proper conditions the signal strength is directly proportional to the number of resonating nuclei. An example is provided below in the discussion of ^{1}H NMR results for hydroxyapatite.

Most current NMR spectrometers use radiofrequency (rf) pulses of several microsecond duration, have the ability to generate complex sequences of pulses (needed for all but the simplest experiments), and rely on Fourier transformation (FT) of repetitively-acquired signals in the time-domain (so-called free-induction decays or FID's) in order to improve sensitivity. The sensitivity, or signal-to-noise ratio, improves as the square root of the number of scans accumulated. The ability to provide rf at a number of different frequencies simultaneously (e.g., for double- and triple-resonance experiments) is common. The major factors determining whether a given spectrometer is suitable for solid-state NMR studies of calcium phosphates is the availability of higher-power rf amplifiers and suitable high-power probes and receiver sections.

Numerous introductory undergraduate-level textbooks on NMR spectroscopy exist,[3] as well as books describing basic experimental techniques.[4] Graduate level textbooks covering fundamental principles of solid-state NMR and containing many literature references to the experimental techniques discussed here can also be found.[5] Descriptions of specialized solid-state NMR techniques and applications can be found in books,[6] in recent reviews,[7] and in a multi-volume encyclopedia of NMR.[8]

The ^{1}H, ^{19}F, and ^{31}P isotopes are of greatest relevance in the present chapter, since elements containing these nuclei in 100% natural abundance occur in many important calcium phosphates. Furthermore, all of these isotopes have high resonance frequencies for a given magnetic field strength that result in favorable detection sensitivity, and have nuclear spin quantum numbers I=1/2 that in some ways simplify the NMR behavior. The ^{13}C isotope (natural abundance 1.1%) also has a nuclear spin I=1/2, and although less favorable from the standpoint of detection sensitivity, is of relevance to the study of carbonate-containing apatites and adsorbed organic molecules. This chapter will describe results only from these four spin-1/2 nuclei.

The remaining potentially-relevant isotopes generally have nuclear spin quantum numbers I>1/2, and are termed quadrupolar since their NMR spectra are influenced by an additional parameter: the nuclear quadrupole coupling constant. Two of these isotopes, ^{43}Ca (I= 7/2, natural abundance 0.045%) and ^{17}O (I= 5/2, natural abundance 0.037%), unfortunately have a very poor NMR detection sensitivity partly due to their low natural abundance, and it is very expensive to isotopically label samples with these isotopes. Labeling with the ^{2}H isotope (I=1, 0.015% natural abundance) is much less expensive, but although solid-state ^{2}H NMR techniques have been highly-developed in other applications, applications to calcium phosphates so far have been limited.

In order to describe the various features influencing NMR spectra, as well as the differences between solution state and solid-state NMR, experiments involving the ^{31}P nucleus are very illustrative, and will be described in this section. Perhaps the single chemically most useful parameter obtained from NMR spectra is the chemical shift. This is a dimensionless quantity representing the very small fractional change in the resonance frequency of a nucleus resulting from the fact that the external magnetic field is partially shielded or screened (or the opposite, deshielded) by the circulating electrons in the chemical bonds around a nucleus. Chemical shielding, or the chemical shift, is conventionally expressed in units of ppm (parts per million) with respect to the resonance of a single peak in some convenient but arbitrary reference compound (of course, these units should not be confused with concentration measurements!). Spectra are conventionally plotted such that peaks to the right in the plotted spectrum are at lower frequency and said to be upfield; peaks to the left are at higher frequency, or downfield. As an example, the chemical shift of ^{31}P in solutions of dilute phosphate varies over the pH range 0.5 to 13 by about 5 ppm, due to the formation of various protonated orthophosphate species.[9] Measured with respect to 85% H_3PO_4 as a reference, the chemical shifts of the dilute species H_3PO_4, $H_2PO_4^-$, HPO_4^{2-}, and PO_4^{3-} steadily increase (in the same order): 0.48 ppm, 0.67 ppm, 3.15 ppm, and 5.6 ppm. Although increasing success has been achieved in the calculation of chemical shifts from first-principles quantum mechanics, for most purposes the chemical shift can be regarded as a measurable parameter characteristic of a given local electronic environment around a nucleus (generally influenced to a measurable extent by nearest neighbors and next-nearest neighbors only).

One important aspect of the high-resolution solution state ^{31}P NMR spectra described above is that only a single resonance peak is observed at all pH values, even though mixtures of the different protonated species are present. The reason for this behavior is the NMR phenomenon of chemical exchange. In aqueous solution, protons form and break bonds with a given phosphate group very rapidly on the so-called NMR time scale (in this case, the inverse of the chemical shift difference between the various phosphate species, which is roughly $[100\ Hz]^{-1}$ = 10 ms). The result is that the observed chemical shift is a weighted-average of the chemical shifts of the various species present. Similar effects may also occur in the solid state, as discussed below for octacalcium phosphate and nonstoichiometric hydroxyapatite samples. In addition to averaging the chemical shifts, the exchange process also averages out the so-called J spin-spin coupling between the ^{31}P nuclei and ^{1}H nuclei. Since J-coupling does not play a significant role in the NMR of calcium phosphates, it will not be discussed further. It is interesting to note that solution complexes of phosphate with Ca^{2+} also undergo rapid chemical exchange, and that by observing the chemical shift changes in highly supersaturated metastable solutions of calcium phosphates the chemical shift of the $CaH_2PO_4^+$ complex could be calculated to be 0.13 ppm (the shift of the neutral solution complex $CaHPO_4$ was also estimated to be within 0.5 ppm of the uncomplexed HPO_4^- ion resonance at 3.13 ppm).[10]

Anisotropic Interactions and Solid State NMR

The ^{31}P NMR linewidths of phosphate in solution can be less than one Hz, making possible very precise frequency and chemical shift measurements and leading to the term "high-resolution solution-state NMR." In solid calcium phosphates, however, the ^{31}P NMR peaks are inherently 3 or 4 orders of magnitude broader. Protonated phosphate groups yield generally broader peaks than unprotonated groups, and such broad peaks may be essentially “invisible” to conventional solution-state high-resolution NMR spectrometers for a simple reason. The NMR signal is acquired in the "time-domain" after the intense radiofrequency pulse by measuring a weak radiofrequency voltage induced by the nuclear spins. In high-resolution NMR spectrometers, there is typically an interval of tens of microseconds after the pulse during which the signal cannot be observed (the "dead time"). Since the signal from spins in solids decays very rapidly, it often cannot be observed, or is very distorted, unless a spin-echo technique is used. An example of the "invisibility" of the NMR signal from a solid in a conventional high-resolution solution-state NMR spectrometer is an investigation using ^{31}P NMR to monitor the spontaneous precipitation of calcium phosphates from supersaturated solution. In this study, the formation of the solid dicalcium phosphate dihydrate from solution was indirectly observed as a sigmoidal decrease over time in the intensity of the solution phosphate resonance, since the NMR signal of the solid phase was too broad to be detected by the spectrometer used.

The fundamental reason for the difference between solution-state and solid-state NMR linewidths is the presence of anisotropic NMR interactions that cause the resonance frequency to depend upon the orientation of axes fixed on the molecule with respect to the external magnetic field. Except for situations involving a single crystal, all possible orientations of such axes are normally equally probable, giving rise to innumerable resonance frequencies across a broad powder pattern spectrum. In solution, molecules reorient very rapidly (on a nanosecond time scale or less), and these anisotropic interactions are averaged on the NMR time scale, very similar to the type of averaging described above that occurs in the case of chemical exchange. A direct and dramatic demonstration of this effect is provided by the ^{31}P NMR spectrum of a colloidal suspension of hydroxyapatite: the peak of the colloidal solid in suspension is some 30-times narrower than that of the same sample in a dry powdered form, reflecting the rapid tumbling motion of the $<0.1 \mu m$ diameter colloidal particles.[11,12]

What are the nature of these anisotropic interactions giving rise to the broadening in solids? Figure 1 depicts two major anisotropic interactions for a hypothetical

HPO_4^{2-} group in a solid. One is the chemical shift anisotropy (csa). It was mentioned above that the electrons in chemical bonds

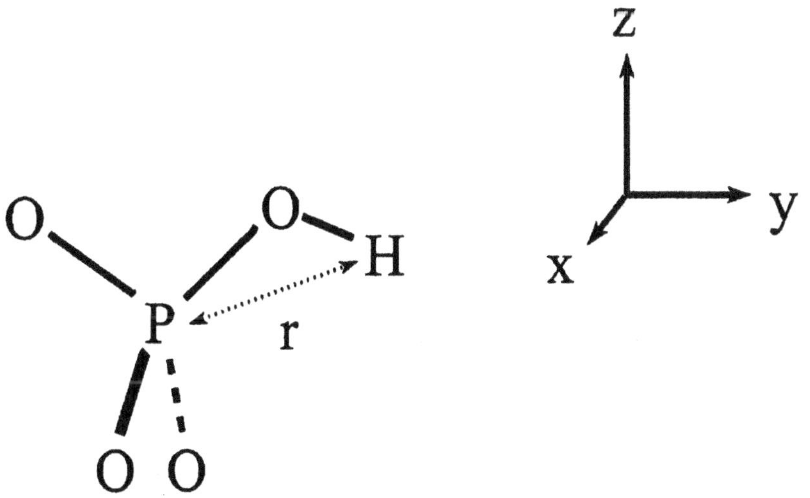

Figure 1. Schematic HPO_4^{2-} group in a solid illustrating two anisotropic ^{31}P NMR interactions, the chemical shift anisotropy and the dipolar coupling. The shielding of the ^{31}P nucleus by the surrounding electrons in the bonds depends upon the relative orientation of the external applied magnetic field (not represented). The resulting chemical shift anisotropy (csa) can be represented by the three principal values of a chemical shielding tensor, which correspond in this diagram to the chemical shielding experienced by the ^{31}P nucleus when the external magnetic field is along either the x, y, or z directions (the so-called "principal-axis system"). The second anisotropic interaction is the heteronuclear dipolar coupling between ^{31}P and a ^{1}H nucleus. This coupling is proportional to $1/r^3$, and hence falls off rapidly with distance. It also depends upon the angle (not shown) the vector r makes with respect to the magnetic field, as $(3\cos^2 -1)$.

serve to shield (or deshield) a nucleus from the applied magnetic field, giving rise to a chemical shift; it should therefore appear reasonable that the degree of shielding (and hence the resonance frequency) depends in general upon the orientation of the molecule or group with respect to the external magnetic field. Although the number of such possible orientations is essentially infinite, it can be shown that the frequency in any orientation can be predicted by knowing the three principal values of the chemical shielding tensor and the two angles that relate the shielding tensor principal-axis system to the crystallographic or molecule-fixed axes. In the isolated group shown in Figure 1, the principal values are directed along the x, y, and z axes, and represent the chemical shielding that would be experienced by the ^{31}P nucleus when the magnetic field is along either the x, y, or z axis. If two of these principal values are equal to each other, the tensor is called axially-symmetric. If all three values are equal, the csa is zero. As an example, the

HPO_4^{2-} group in dicalcium phosphate dihydrate has a sizeable ^{31}P chemical shift anisotropy, with the three (non-axially-symmetric) principal values having been measured[13] to be 70 ppm, -12 ppm, and -53 ppm. The isotropic chemical shift is the average of the three principal values, 1.7 ppm, and in the case of a molecule that can be dissolved would correspond to the chemical shift measured in solution NMR due to the isotropic rotational tumbling of molecules. In order to characterize chemical "perturbations" such as the protonation of a phosphate group, knowing the three principal values of the chemical shielding tensor is considerably more informative than simply knowing the isotropic chemical shift. For example, the above csa in an HPO_4^- group can be compared to the ideally zero csa expected for a PO_4^{3-} group having perfect tetrahedral site symmetry crystallographically, and to that measured for the slightly-distorted phosphate tetrahedron in calcium hydroxyapatite (19 ppm, 4 ppm, and -15 ppm).[14,15] There do not appear to be full quantum-mechanical calculations of the chemical shielding tensors in calcium phospates, but empirical correlations of ^{31}P isotropic chemical shifts and csa values in orthophosphate salts have been noted.[16]

The second important anisotropic interaction giving rise to broadening in solid-state NMR is the dipolar coupling between pairs of nuclear spins. In Figure 1, the dipolar coupling between ^{31}P and ^{1}H is represented. This interaction is proportional to $1/r^3$, where r represents the distance between the two nuclei, and the dipolar coupling thus falls off rapidly with distance. It also depends upon the angle between the vector connecting the two nuclei and the external magnetic field, the exact dependence being $(3\cos^2 - 1)$. In solution, the isotropic rotational tumbling results in this angular expression averaging to zero. In solids, atomic or molecular motions that are fast on the NMR time scale (the inverse of the dipolar coupling strength) can also result in partial averaging of dipolar couplings. Finally, the dipolar coupling is proportional to the product of the individual magnetogyric ratios of the nuclei involved. The strength of dipolar couplings can range from large (e.g., ca. 50 kHz for proton pairs in the water molecules of dicalcium phosphate dihydrate) to moderate (e.g., ca. 2 kHz for neighboring hydroxyl protons in hydroxyapatite) to vanishingly small (for any pair of nuclei separated by > 1 nm). It is important to realize that a given nucleus typically has many dipolar interactions with other nuclei, in a pairwise fashion. If the two nuclei are of the same type (e.g., two protons), then the interaction is said to be homonuclear, otherwise it is heteronuclear. The dipolar coupling plays a significant role in the solid state NMR of calcium phosphates in different ways. Because its functional form is precisely known, the measurement of dipolar couplings provides a measurement of internuclear (and hence interatomic) distances. Secondly, dipolar couplings are essential for the very useful cross-polarization NMR experiment discussed below.

In addition to these anisotropic parameters available for characterizing calcium phosphates, there are various relaxation time parameters that are relevant and potentially useful. The spin-spin, or transverse, relaxation time, T_2, is strictly-speaking most applicable to solution NMR studies, but it can be loosely defined as the decay time of the FID due to homonuclear dipolar interactions. For calcium phosphates T_2 values are typically on the order of milliseconds. A second transverse relaxation time, T_2^* , can be defined as the decay time of the FID due to all sources of broadening, and it is thus simply proportional to the inverse linewidth of the peak in the spectrum. Often the linewidths of solid-state spectra are characterized instead by the "moments" of spectral intensity about a central value, the second-moment being the most commonly-used measure. In cases where the linewidths are determined solely by dipolar couplings, it is possible to calculate rigorously the second moment for any arbitrary arrangement and spacing of the relevant nuclei and to compare this to the experimental value.

The spin-lattice, or longitudinal, relaxation time, T_1 , is the time constant governing the exponential return of the spin system to equilibrium after a radio frequency pulse is applied. Thus, the T_1 value limits the rate at which multiple acquisitions can be acquired to improve the signal-to-noise ratio. Pulsing too rapidly results in saturation, with an associated decrease in the signal strength from each acquisition. For calcium phosphates the T_1 of ^1H, ^{19}F or ^{31}P can typically be seconds or even minutes. Motions taking place in the solid at the appropriate frequency (the Larmor frequency) lead to shorter T_1 values, and thus T_1 varies with temperature. The presence of paramagnetic impurities, typically transition metal ions such as Mn^{2+} and Fe^{3+}, can markedly shorten T_1 values. Such impurities may provide the predominant T_1 relaxation mechanism of calcium phosphates in the solid state even in chemically "pure" samples, via the ability of dipolar couplings to relay such effects. This relay process is called spin-diffusion, although the nuclei themselves are stationary, and can be a source of valuable structural information as discussed in the section on antimony (III)-doped fluorapatite. Although some measurements of T_1 values in calcium phosphates have been reported, no detailed investigations and mechanistic interpretations of relaxation times appear to have been carried out.

Solid State NMR Experimental Techniques

The two major anisotropic interactions resulting in broad NMR spectra for solid calcium phosphates have been discussed, but it has not yet been indicated how these anisotropic parameters can be experimentally determined. If a large enough single crystal is available, the orientation dependence of the csa or dipolar couplings can be measured by changing the orientation of the crystal axes with respect to the external magnetic field; if necessary, heteronuclear dipolar interactions can be effectively removed by high-power decoupling (i.e. a double-resonance method involving irradiation at the NMR frequency of one nucleus while detecting at the frequency of the nucleus being observed). A number of ^1H and ^{19}F NMR studies have been reported using naturally-occurring single crystals of hydroxyapatite with fluorine substitutions and fluorapatite with hydroxyl substitutions, and internuclear distances have been determined for various configurations as summarized. However, since most calcium phosphate samples are usually available only as powders, this chapter will describe only results obtained from powdered samples. In such cases, the single most important technique for obtaining higher-resolution NMR spectra of solids is magic-angle spinning (MAS).

In the MAS-NMR method the powdered sample (typically 100-300 mg) is packed into a plastic or ceramic rotor that fits tightly inside a stator housing which provides jets of air to rapidly spin the rotor. Typical spinning speeds used in MAS-NMR studies of calcium phosphates have been 4-8 kHz, and even higher spinning speeds are currently obtainable with commercial instrumentation. The axis of spinning is designed to make an angle Ω of 54.7° with respect to the external magnetic field, the "magic angle" where $(3\cos^2\Omega - 1) = 0$. Under these conditions the chemical shift tensor is averaged, in this case by the rapid mechanical rotation rather than by molecular tumbling in solution, to its isotropic chemical shift value. However, the csa manifests itself in the presence of spinning sidebands that appear in the spectrum at multiples of the spinning frequency. Figure 2 shows as an example the ^{19}F MAS-NMR spectrum of fluorapatite at various

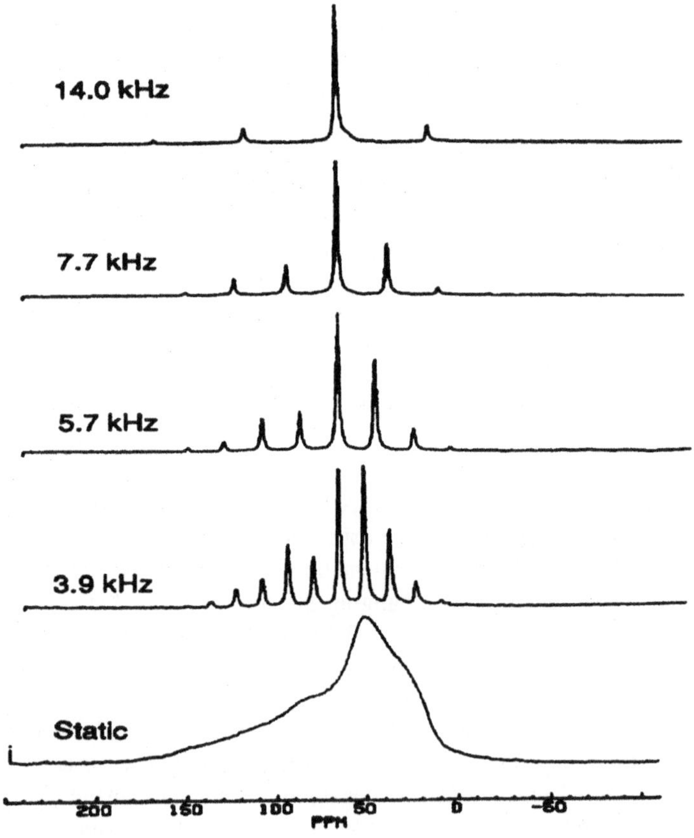

Figure 2. Comparison of ^{19}F MAS-NMR spectra at 282 MHz of calcium fluorapatite obtained at the different spinning speeds shown, and a spectrum of the static sample.

spinning speeds, and for comparison the "static" spectrum of a non-spinning sample. The centerband, not necessarily the most intense peak in the spectrum, can be readily identified by changing the spinning speed. Spinning more rapidly results in fewer and less intense sidebands, until finally only a center peak remains (such a limit is not always experimentally achievable). The intensity profile of the spinning sideband pattern directly reflects the anisotropic interactions. In cases where the csa is the predominant anisotropic interaction, the principal values of the chemical shielding tensor can be obtained by numerical or graphical procedures, or by direct simulation. Spinning sidebands due to the csa are more intense in higher magnetic fields. The reason is that the csa, although usually expressed in the field-independent units of ppm, in true frequency units of Hz actually depends linearly upon the magnetic field strength. Typical MAS linewidths (at peak half-height) for well-crystallized calcium phosphates are about 1 ppm, at best 0.5 ppm, and thus chemical shifts can be measured with a reproducibility of typically 0.2-0.4 ppm. However, chemical shift values for the same compound reported by different laboratories may exhibit greater differences, especially if hydroxyapatite or fluorapatite were not used as a secondary chemical shift reference (see below).

The effect of MAS upon dipolar couplings is similar. The dipolar coupling is averaged to zero (as it is in solution), but the anisotropy (which is directly related to the dipolar coupling strength) manifests itself in the sideband intensities. The difference between the two cases arises from a somewhat subtle property of homonuclear dipolar couplings, which results in a significant degree of linebroadening of the MAS-NMR spectra when the spinning speeds do not exceed the homonuclear dipolar coupling strength. In the calcium phosphates, this situation commonly arises only when there are water molecules of hydration having large ^{1}H-^{1}H dipolar interactions. In such cases, specialized multiple-pulse techniques can be used to eliminate in effect the homonuclear dipolar couplings while retaining chemical shift information. These "homonuclear dipolar decoupling" or "line-narrowing" techniques were originally developed for static samples. However, when they are combined with MAS to remove both the heteronuclear dipolar couplings and the csa, the resulting technique is known as CRAMPS (Combined Rotation And Multiple-Pulse Spectroscopy).

The other technique that has proven useful in solid-state NMR studies of calcium phosphates, usually in combination with MAS, is cross-polarization (CP). This is a double resonance scheme in which simultaneous irradiation at two frequencies is applied during a cross-polarization interval. This interval is referred to as the "mixing time" or the "contact time," and is typically milliseconds in duration. The term "polarization" refers to the population difference between the two nuclear energy levels, which is directly proportional to the observed NMR signal strength. The following discussion will assume that the two types of nuclei are ^{1}H and ^{31}P, although many combinations are possible (in fact, the first cross-polarization experiments were carried out on ^{19}F and ^{43}Ca). The two spin systems in the case of pairs such as ^{13}C and ^{1}H may be referred to as "rare" and "abundant" respectively, due to the widely different number of spins. For ^{1}H and ^{31}P nuclei in calcium phosphates, however, such terminology is often less appropriate. The amplitudes B_1 of both radiofrequency pulses during this interval must be adjusted such that $\gamma(^{31}P)B_1(^{31}P) = \gamma(^1H)B_1(^1H)$, the so-called "Hartmann-Hahn matching condition." In the CP experiment, the polarization of protons will initially be larger than that of ^{31}P because $\gamma(^1H) > \gamma(^{31}P)$, i.e., because the splitting of the proton energy levels is greater than that of ^{31}P nuclei in the same magnetic field. During the contact time, polarization is transferred from the protons to the ^{31}P nuclei at a rate proportional to the square of the ^1H-^{31}P dipolar coupling strength. In the ^1H-^{31}P CP experiment, the enhanced ^{31}P signal resulting from the cross-polarization by protons is subsequently obtained under conditions of ^1H high-power decoupling in order to remove the ^1H dipolar coupling to ^{31}P.

The advantages of the CP experiment are an enhanced sensitivity of detection and an improved selectivity. The sensitivity advantage arises not only from the larger initial polarization of the protons, but also because the protons usually relax back to equilibrium considerably more rapidly than the ^{31}P nuclei (i.e., have shorter T_1 values) and thus permit a more rapid acquisition of repetitive scans. The selectivity arises from the fact that the CP process depends upon the existence of heteronuclear dipolar couplings (e.g., those between ^1H and ^{31}P); therefore, ^{31}P signals can be preferentially obtained from phosphorus atoms that have nearby hydrogen atoms.

One important aspect of CP experiments is that the relative intensities of signals within a given sample (either arising from different compounds, or perhaps even from different types of groups in the same compound) can be altered by a variety of parameters under experimental control. For example, the cross-polarization mixing time can be varied, or a delay without proton decoupling may be inserted before the signal is acquired with high-power decoupling. The latter dipolar suppression or dipolar dephasing experiment can be very useful, since it results in a loss of signal intensity from ^{31}P nuclei strongly coupled to ^1H; the signal strength relative to a CP spectrum obtained without such a delay provides a quantitative index useful in investigating samples containing mixtures of strongly-coupled and weakly-coupled ^{31}P nuclei. Another recent experiment which has been used to suppress the signal from unprotonated orthophosphate groups and to observe selectively that from protonated phosphates is differential cross-polarization. It relies on the difference in the cross-polarization time constants T_{CP} of the two type of groups to effectively null the signal arising from one. This is accomplished by reversing the phase of the rf field after a forward transfer time (contact time), such that the ^{31}P polarization is transferred back to ^1H during the reverse transfer time t_r. This causes the ^{31}P spectrum to be successively reduced in amplitude, pass through a null, and finally become inverted. An example of ^{31}P differential CP-MAS NMR spectra of a bone sample is shown in Figure 3,

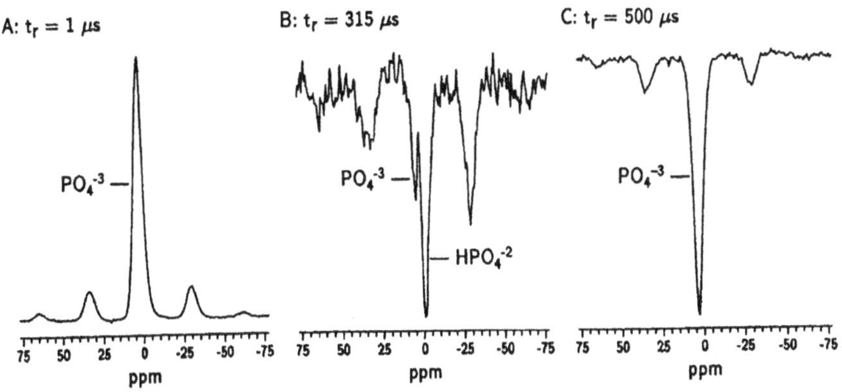

Figure 3. Differential cross-polarization ^{31}P MAS-NMR spectra phosphate spectra at 162 MHz of 19 day old chick embryo bone at different reverse CP times t_r (as shown). The spectrum (A) is essentially that of a conventional CP experiment, and shows only the apatitic PO_4^{3-} resonance. At the longer reverse CP time in spectrum (B) the PO_4^{3-} signal is nearly nulled and the (inverted) centerband and sidebands of the HPO_4^{2-} group are revealed. At the still longer reverse CP time in spectrum (C), the apatitic resonance is well past the null, and again obscures the HPO_4^{2-} signal.

and will be discussed in the last section.[17] By subtracting in the computer spectra obtained under different conditions from one another, it is also possible to obtain difference spectra that may correspond to a single component or resonance. Simple single-pulse, or Bloch-decay, spectra can be used to obtain quantitative intensities as long as the "recycle delay" between acquisitions is at least about 3 times greater than the longest T_1 value in the sample. The FID is acquired after a single short pulse of rf, usually in the presence of high-power decoupling.

There are several other NMR experiments that have only recently been applied to the study of calcium phosphates, but which have the potential of providing unique information. In multiple-quantum (MQ) NMR, groups of spins are collectively excited via the dipolar couplings existing between the spins in a given group, and information about the size of the "cluster" of spins can be obtained. So far, this technique has been applied only to the case of the 1-dimensional distribution of spins corresponding to fluorine nuclei or protons in fluorapatite or hydroxyapatite or solid solutions of both.[18-20] The technique of MQ NMR as it has been applied to apatites differs considerably from the other solid-state NMR techniques discussed above. First of all, it has been applied to static samples rather than to those undergoing MAS. Secondly, the end result of the MQ NMR experiment is not a conventional spectrum, but rather an intensity profile for various orders of coherence (e.g., double-quantum or 2-quantum coherence, 4-quantum coherence, etc. in pulse sequences that produce only even-orders of coherence). The practical meaning of these terms may become clearer in the discussion of experimental results in the separate section describing MQ NMR of fluorohydroxyapatite solid solutions and their end members. Although it is not possible to describe the experiment adequately in a short space, it involves applying a sequence of pulses to create MQ coherences in the sample for a period of time known as the preparation time. The preparation time may be increased up to the point where signal is no longer observable, due either to relaxation effects or to instrumental limitations. To obtain the intensity profile of the various coherence orders, it is necessary to increment the phases of the rf pulses in a prescribed manner.

A second new technique that is very promising for obtaining detailed structural information is Rotational Echo Double Resonance (REDOR). In this double-resonance method, the heteronuclear dipolar coupling between a relatively isolated pair of spins can be measured and used to obtain the internuclear distance. The REDOR experiment is carried out by observing either one or the other of the two nuclei. For instance, ^{19}F-observed ^{31}P-^{19}F REDOR is carried out by obtaining difference ^{19}F spectra in the presence or absence of the appropriate ^{31}P pulses, and it enables one to observe only those ^{19}F nuclei that are dipolar-coupled to ^{31}P nuclei.[21] By measuring the relative decrease in the ^{31}P FID intensity in the presence or absence of ^{19}F pulses, the reverse experiment of ^{31}P-observed ^{31}P-^{19}F REDOR enables one to measure a single P-F distance (about 3.6 Å in fluorapatite). Another experiment that is potentially quite promising is a specific version of a general class of two-dimensional NMR experiments called HetCor (heteronuclear correlation). In ^{1}H-^{31}P HetCor experiments the results may be presented on a contour plot like a topographical map, that has as one axis the ^{1}H chemical shift and as the other axis the ^{31}P chemical shift.[22] A peak in the contour plot then represents a proton at a given chemical shift value that is dipolar-coupled to a phosphorus atom with its own chemical shift value. There is a great practical advantage to this technique in that it can eliminate interfering proton NMR signals arising from components that are not dipolar-coupled to phosphate groups (e.g., surface water molecules or organic groups in bone). However, it is a technically-demanding experiment requiring careful adjustment of the radiofrequency pulses.

Although NMR spectroscopy is typically carried out by placing the powdered sample inside a radiofrequency coil, it is also possible to study biological calcium phosphates in bone and teeth in-vivo, relying upon the use of planar "surface coils" that can detect signals beneath the plane.[23] This is an instance of localized, or topical, NMR spectroscopy. In the work cited, a type of cross-polarization technique known as adiabatic demagnetization in the rotating frame (ADRF) CP was used and shown to be capable of distinguishing protonated from unprotonated phosphate groups, and to have the significant advantage for in-vivo samples of requiring much less rf power. It is distinct from NMR imaging, or Magnetic Resonance Imaging (MRI), in which a spatial map of the distribution of NMR nuclei is obtained. Such NMR images of hydroxyapatite[24] and bone mineral[25] have been obtained. Both techniques offer considerable promise for characterizing the nature of the mineral phase in a non-disruptive manner. The methods used generally offer lower resolution than traditional "high-resolution NMR of solids" techniques since sample spinning is not feasible, and may require new strategies to obtain information about the nature of the mineral phases.

NMR CHARACTERIZATION OF STOICHIOMETRIC CRYSTALLINE CALCIUM PHOSPHATES

Monocalcium Phosphate Monohydrate, $Ca(H_2PO_4)_2 \cdot H_2O$

The ^{31}P CP-MAS NMR spectrum (at 68 MHz) of this compound shows two distinct peaks of comparable intensity at chemical shift values of -0.1 and -4.6 ppm with associated sets of intense spinning sidebands. The two isotropic chemical shifts correspond to two inequivalent phosphate groups in the crystal structure, which are known to differ in their hydrogen bonding to the water molecule. The shifts are further upfield (to lower frequency) than those in the monoprotonated or unprotonated calcium phosphates, a trend reproducing that observed in solution. Analysis of the intense spinning sideband patterns observed yielded the chemical shielding tensor for both groups: 49, -1, and -48 ppm principal values for the downfield peak at -0.1 ppm, and 59, -7 and -66 ppm for the upfield peak at -4.6 ppm.

Anhydrous Monocalcium Phosphate, $Ca(H_2PO_4)_2$

The ^{31}P CP-MAS NMR spectrum (at 68 MHz) of this compound shows two only partially resolved resonances at = 0.5 and -0.5 ppm, corresponding to the two crystallographically inequivalent phosphates in the crystal structure. Individual chemical shift anisotropy values could not be obtained from the intense spinning sideband patterns because of the peak overlap.

Dicalcium Phosphate Dihydrate (Brushite), $CaHPO_4 \cdot 2(H_2O)$

Brushite has been used as a model for the protonated phosphate group that appears to be present in bone, and thus its ^{31}P CP-MAS NMR spectrum has been obtained by a number of groups. The isotropic chemical shift and chemical shielding tensor are reported above in the general NMR discussion on chemical shielding tensors.

The ^{1}H MAS-NMR spectrum (at 500 MHz) obtained at a spinning speed of 8 kHz reveals a peak at 10.4 ppm (all proton shifts reported relative to tetramethylsilane, TMS) with a pair of weaker spinning sidebands that could be assigned to the acidic proton.[26] The ^{1}H chemical shift value agrees with a linear correlation observed to exist between proton isotropic chemical shift values and hydrogen bonding strengths (whose strength is inversely related to the O—H—O distance). A second set of peaks with approximately four times the integrated intensity of the first has a broad centerband at 6.4 ppm and strong spinning sidebands extending over a 100 kHz range. These resonances can be assigned to the protons in the structural water molecules of hydration. The intense set of spinning sidebands arises from the strong homonuclear dipolar coupling between the two protons on a given water molecule, and was observed in other crystalline hydrates as well. By using the CRAMPS technique to eliminate the homonuclear dipolar coupling, it was possible to resolve two separate peaks at about 5 and 6.5 ppm corresponding to water protons with dissimilar hydrogen bonding strengths.

Anhydrous Dicalcium Phosphate (Monetite), $CaHPO_4$

The ^{31}P CP-MAS NMR spectrum (at 68 MHz) of this compound shows two only partially resolved resonances at 0.0 and -1.5 ppm, corresponding to the two crystallographically inequivalent phosphates in the crystal structure. Half of the phosphates are bonded to one hydrogen as HPO_4^{2-} groups, while the other half have one proton in a symetrically bridging hydrogen bond and one proton statistically distributed between two centrosymmetric positions. Static disorder involving the latter proton would result in further inequivalences among the phosphate groups. The intense spinning sideband pattern indicates that the csa is considerably larger than that of hydroxyapatite, but somewhat smaller than that in brushite, but the shielding tensor principal values could not be exactly determined because of the peak overlap. (The three values have been reported to be 59.5, 5.5, and -60.1 ppm in a later work, possibly by assuming both shielding tensors are equal.[27])

The ^{1}H MAS-NMR spectrum (at 500 MHz) obtained at a spinning speed of 8 kHz shows two peaks and an associated set of weaker sidebands at 16.2 ppm and 13.6 ppm in the intensity ratio 1:3. The low field peak could be assigned to the most strongly hydrogen bonded acidic proton in the crystal structure, with the upfield peak arising from three less-strongly hydrogen-bonded acidic protons. Complications due to disorder and two phases have been previously discussed.[13]

Octacalcium Phosphate, $Ca_8(HPO_4)_2(PO_4)_4{\cdot}5H_2O$

This compound has come to be recognized as an important possible transient intermediate in the precipitation of hydroxyapatite and biological apatites, and has consequently drawn much attention in recent decades. Its NMR properties, like its chemistry, are considerably more complex than those of stoichiometric hydroxyapatite, and bear some discussion. The ^{31}P CP-MAS NMR spectrum at a modest magnetic field strength (4.0 Tesla, ^{31}P frequency = 68 MHz) shows two center peaks corresponding to the two HPO_4^{2-} groups (upfield peak at -0.1 ppm) and four PO_4^{3-} groups (downfield group of unresolved peaks at 3.4 ppm). The assignments could be made on the basis of the observed cross-polarization behavior: upon increasing the cross-polarization mixing time from 1 ms to 5 ms, the downfield peak increased in intensity relative to the upfield peak, suggesting that the former arises from phosphorus nuclei having a smaller dipolar coupling to protons. This downfield peak appears to arise from the "apatitic layer" of phosphate groups, whose close structural similarity to the corresponding atoms in the lattice of hydroxyapatite has been reported by Brown and co-workers.[1] The assignments were also supported by Bloch-decay spectra that showed that the high-field peak arises from 1/3 of the total phosphates. The recycle delay of 5 minutes used to obtain the Bloch decay spectrum was judged to be sufficiently long to avoid saturation effects by comparison with a spectrum obtained at 1-minute intervals.

Significantly increased resolution was obtained in ^{31}P MAS spectra at considerably higher field strength (11.7 Tesla, 202 MHz for ^{31}P), such that the downfield peak could be partially resolved into three peaks at 3.5 ppm, 3.2 ppm, and 2.0 ppm, whereas a single upfield peak at -0.3 ppm was still observed.[28] A mathematical analysis technique called abstract factor analysis was applied to ^{31}P CP-MAS spectra obtained at 101 MHz with varying CP contact times; analysis of the results enabled three peaks to be identified at -0.7 ppm, 1.8 ppm and 3.2 ppm.[29] Similar shifts and assignments were made on the basis of ^{31}P CP-MAS NMR experiments involving a short (ms) interval after the cross-polarization mixing time in which the proton-decoupling was absent. Such a dipolar-suppression experiment suppresses the signal arising from ^{31}P nuclei strongly dipolar-coupled to ^{1}H, leaving only the downfield peak with its much weaker coupling to protons.

At room temperature the spinning sideband intensities of the upfield ^{31}P peak at -0.1 ppm are greater than those of the downfield peak, as would be expected for protonated phosphate groups, but they are weaker than those seen in other monoprotonated calcium phosphates. One of the potential problems that makes it difficult to use ^{31}P MAS-NMR sideband intensities (and the chemical shielding tensor values that can be obtained from them) as a unique signature of protonation of phosphate groups is that not only will they depend upon the specific structural details in the solid state, they will also be affected by any atomic or molecular motion that "averages" the chemical shielding tensor. Such an averaging process is related to the chemical exchange effect described in the introductory section. In the present case, it could be caused by a proton that hops onto each of the four oxygen atoms of the phosphate tetrahedron on a time-scale rapid compared to the inverse of the csa (ca. 1/[100 ppm] 0.1 ms). An indication that some such motion is taking place in octacalcium phosphate is obtained from ^{31}P CP-MAS NMR spectra obtained near liquid nitrogen temperature (at -165 C), which show considerably increased sideband intensities for a broadened peak (the high-field and low-field peaks are not resolved).

The ^{1}H MAS-NMR spectrum (at 500 MHz) of octacalcium phosphate presents some complexity. There is a major very broad peak centered at 5.5 ppm that has a strong set of associated spinning sidebands characteristic of structural hydrate water molecules as in brushite. However, the center peak height is relatively speaking somewhat greater than that observed for other hydrates, indicating the possibility of a more mobile water component, either as a structural water molecule or as surface-adsorbed water. The CRAMPS spectrum (at 200 MHz) revealed a narrowing of the water peak upon cooling the sample from room temperature to -20 °C,[22] indicative of the presence of slowed molecular/atomic motions. A well defined shoulder at 9.6 ppm in this spectrum was attributed to one of the (two) acid phosphate protons, the one weakly hydrogen bonding to a water molecule. The other acid phosphate proton, which is involved in a strong hydrogen bond between two phosphate groups, was assigned to a peak at 12 ppm (13 ppm in the MAS-NMR study). Two shoulders flanking the central water resonance at -20 C most likely arise from water molecules with different hydrogen-bonding strengths.

One of the most peculiar characteristics of the ^{1}H MAS-NMR spectrum of octacalcium phosphate is a set of sharp peaks at 1.5 ppm, 1.1 ppm, and 0.2 ppm. The intensity of these peaks relative to the structural water peaks at 5.5 ppm is quite variable from sample to sample, but they retain their characteristic sharpness even down to low spinning speeds, where sidebands begin to appear. The two peaks at 1.5 and 1.1 ppm have been assigned to a specific water molecule in the crystal structure that is isolated from other water molecules, may be undergoing large motions, and may be present in nonstoichiometric amounts due to a zeolitic nature of the hydrate water of octacalcium phosphate. The peak at 0.2 ppm is virtually indistinguishable from that arising from the hydroxyl group in pure hydroxyapatite (see below); however, it may well arise from a positionally-disordered hydroxyl group in the octacalcium phosphate lattice. The ^{1}H-^{31}P HetCor spectrum of octacalcium phosphate demonstrated that both the apatitic hydroxyl group (near 0.2 ppm) and the combined water/acid phosphate resonances downfield were dipolar-coupled to phosphate groups; however, more detailed correlations between the ^{1}H and the ^{31}P spectra were not made. Overall, the complexities of the ^{1}H NMR spectra reflect the structural complexities of octacalcium phosphate and the close structural and chemical relationship between it and hydroxyapatite.

The tendency of octacalcium phosphate to hydrolyze to hydroxyapatite is a further complicating factor. Indeed, the ^{1}H MAS-NMR spectrum of one sample that had been stored in air for several years at room temperature as a dry powder revealed clearly the presence of a solid-state disproportionation reaction to monetite and hydroxyapatite.

β - Tricalcium Phosphate, β-$Ca_3(PO_4)_2$

The ^{31}P MAS-NMR spectrum of this compound, which is closely related to the mineral whitlockite and which is used as a bone implant material, can be obtained without CP or ^{1}H-decoupling. The spectrum at a ^{31}P frequency of 122 MHz shows three partially-resolved peaks.[30] The chemical shifts are 0.6 ppm, 1.7 ppm, and 5.1 ppm, with the last peak having about one-half the total integrated intensity of the upfield pair. Similar results but with chemical shifts of 0.1, 1.2 and 4.2 that differ by roughly -0.5 ppm have been reported, along with a common T_1 value for the three peaks of 0.9 s.[31] It was noted in this work that the chemical shift values reported previously appeared to be offset by about 4 ppm downfield, possibly due to a chemical-shift referencing error. Two peaks with chemical shifts of 1.5 and 4.3 ppm have also been reported. Relatively weak spinning sidebands are indicative of a small csa for the unprotonated phosphate groups. There are three inequivalent phosphate groups in the crystal structure, but a statistically-disordered calcium site with half-occupancy in the crystal structure would lead to six different phosphate configurations.[30]

Tetracalcium Phosphate Monoxide, $Ca_4(PO_4)_2O$

The ^{31}P MAS-NMR spectra of this, the most basic of calcium phosphates, and one formed under anhydrous conditions, can like that of tricalcium phosphate above be obtained without CP or ^{1}H-decoupling. The spectrum at a ^{31}P frequency of 62 MHz shows a peak at 3.8 ppm and a weaker partly-resolved peak at 4.7 ppm, both with weak sideband intensities indicating a small csa for the unprotonated phosphate groups. Although the compound is structurally related to hydroxyapatite, there are four nonequivalent phosphate groups in the crystal structure.

Calcium Hydroxyapatite, $Ca_{10}(PO_4)_6(OH)_2$

The ^{31}P CP-MAS spectrum (at 68 MHz) of crystalline hydroxyapatite consists of a single peak at 2.8 ppm with a weak pair of spinning sidebands reflecting the small csa arising from the slight distortion of the phosphate group from being a perfect tetrahedron.' The chemical shift of this peak does not appear to depend greatly upon the sample being highly stoichiometric, and since the sample is commercially available (generally as a nonstoichiometric sample), it forms a convenient secondary chemical shift reference for both ^{31}P and ^{1}H MAS-NMR studies. A highly stoichiometric hydroxyapatite sample consisting of ca. 70% of the monoclinic form (the remainder being the hexagonal form typical of most synthetic and naturally-occurring hydroxyapatite samples) had an identical ^{31}P CP-MAS spectrum; apparently the slight inequivalences of the phosphate groups induced by the long-range ordering of the hydroxyl groups in the monoclinic form could not be resolved. A fairly long mixing time of the order of 5 ms is needed to obtain the maximum CP signal, due to the weak dipolar interactions between the hydroxyl protons and phosphate ^{31}P nuclei.

A detailed ^{1}H NMR investigation at both 200 MHz and 500 MHz has been reported for a pure synthetic precipitated sample of hydroxyapatite that served as a reference standard ("HA-N" or "HAP-N") for a large number of physical and chemical characterization techniques.[32] The static (non-spinning) spectrum shows two overlapping signals from adsorbed water molecules and the structural hydroxyl groups, making it difficult to extract useful information. The 200 MHz ^{1}H MAS-NMR spectrum shows a sharp peak at 0.18 ppm with an associated set of weak spinning sidebands arising from the crystallographically unique structural hydroxyl group protons, as well as a broader downfield peak at 5.6 ppm with much weaker sidebands arising from adsorbed water molecules. The lack of an intense spinning sideband pattern as seen for water molecules in hydrates such as brushite (see above) indicates that these water molecules are very mobile on the NMR time scale (ca. 10^{-5} s). The central peak and spinning sidebands of the hydroxyl group protons remain sharp even at low spinning speeds, due to the special behavior expected for a linear chain of spins as exist for ^{1}H and ^{19}F in the hydroxyapatite or fluorapatite lattice. This arrangement is depicted schematically in Figure 4.

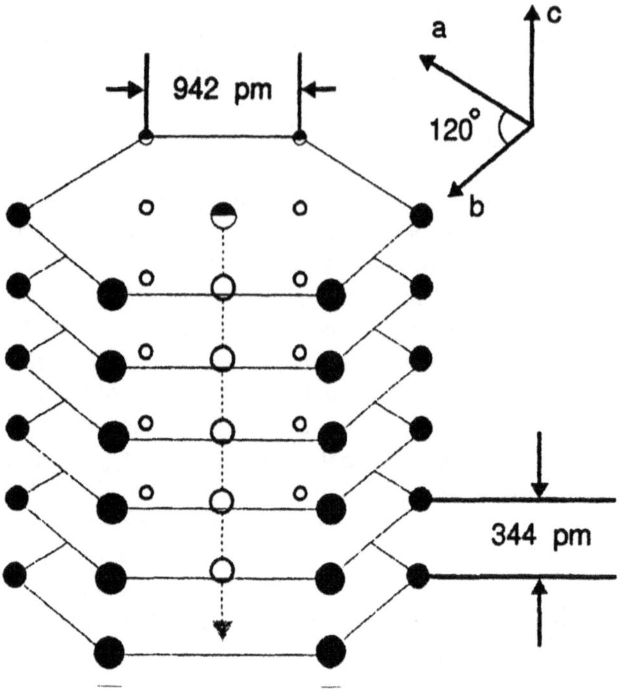

Figure 4. Schematic idealized arrangement of linear columns of hydroxyl-group protons in calcium hydroxyapatite, $Ca_5(PO_4)_3(OH)$. The fluorine atoms in fluorapatite have a similar arrangement.

The 200 MHz ^1H MAS-NMR results demonstrate how NMR spectroscopy can be used to quantitate the absolute amounts of various species in a sample. This is accomplished by measuring the integrated signal intensities from known weights of samples obtained under non-saturating conditions (i.e., obtained with a repetition interval or recycle delay sufficiently long to allow spin-lattice relaxation to restore the full signal strength). A sample containing a known concentration of protons is necessary for calibration (the mineral pyrophyllite was used in the hydroxyapatite quantitative study). The hydroxyl concentration for HAP-N was determined to be 3.13 wt. %, compared to a theoretical stoichiometric value of 3.39 wt. %. The ^1H multiple-quantum NMR results discussed below are sensitive to this slight degree of nonstoichiometry of the hydroxyl group content. The water content of the HAP-N sample was measured by NMR to be 2.82 wt. %, which corresponds to several monolayers of surface-adsorbed water in this moderately high specific surface area sample. The possibility of water in inclusions cannot be ruled out, but the chemical shift argues against the 5.6 ppm peak arising from any isolated water molecules in an apatite lattice that would resonate higher upfield due to the lack of hydrogen bonding.

Interestingly, the ^1H-^{31}P HetCor experiment on a sample of hydroxyapatite, whose CRAMPS spectrum showed a large water resonance, yielded a two-dimensional NMR spectrum whose projection onto the ^1H axis exhibited only the hydroxyl peak. This result is a clear indication that the surface-adsorbed water is not significantly dipolar-coupled to the ^{31}P nuclei of phosphate groups, and it suggests a means of eliminating interfering ^{1}H resonances in samples (e.g., bone) that are not dipolar coupled to ^{31}P groups of the mineral phase. At the present stage of development, the ^{1}H spectra obtained are not quantitative, however.

The 500 MHz ^{1}H MAS-NMR spectrum of HAP-N revealed the additional presence of two minor peaks. One at 1.5 ppm was suggested to arise possibly from an octacalcium phosphate central planar defect or surface unit cell, since a similar peak is observed for octacalcium phosphate (see above). The second peak at 8.7 ppm was assigned to the small amounts of HPO_4^{2-} groups known to be present in this sample.

Calcium Fluorapatite, $Ca_5(PO_4)_3F$

The ^{31}P MAS-NMR spectrum (at 68 MHz) of fluorapatite is virtually identical to that of hydroxyapatite, consisting of a single sharp peak at 2.8 ppm and a weak set of spinning sidebands. This reflects the close structural similarity of the two compounds, and the small ^{31}P csa value. It is also possible to obtain ^{19}F-^{31}P CP-MAS NMR spectra.

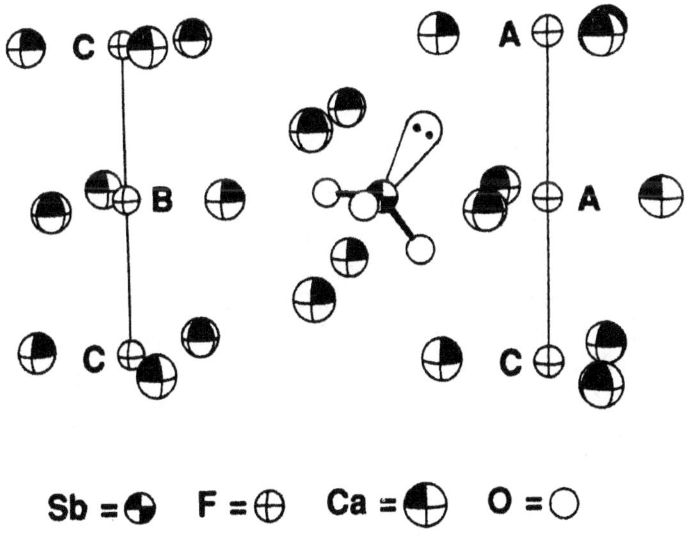

Figure 5. ORTEP drawing of the fluorapatite lattice, with antimony (III) and its lone pair electrons placed exactly at the phosphate site. The fluorine atoms form infinite linear chains along the crystallographic c-axis (vertical), as depicted in Figure 4. The assignment of fluoride ions to peaks A,B, and C in the ^{19}F MAS-NMR spectra is indicated (see text).

An ORTEP drawing of the crystal lattice of fluorapatite is shown in Figure 5 depicting an SbO_3^{3-} group replacing a phosphate group and producing inequivalent fluorine atoms (as discussed in a later section). Each fluorine atom sits at the center of an equilateral triangle of calcium ions. The ^{19}F MAS-NMR spectrum (at 282 MHz) of calcium fluorapatite consists of a centerband at 64.0 ppm from the hexafluorobenzene reference[33] and a set of strong spinning sidebands.[34] The half-height linewidth of ^{19}F (and of ^{31}P as well) was observed to be considerably narrower (only 0.5 ppm) in a high-temperature synthetic sample[35] compared to samples prepared by aqueous precipitation, presumably due to better crystallinity. The spinning sidebands arise principally from the axially-symmetric ^{19}F chemical shielding anisotropy of ca. 84 ppm, although ^{19}F-^{19}F dipolar couplings and ^{19}F-^{31}P dipolar couplings also contribute. For comparison, the ^{19}F isotropic chemical shift value in strontium fluorapatite is 97.2 ppm, and that in barium fluorapatite 184.8 ppm.[36] The corresponding axially-symmetric csa values are 107 ppm and 159 ppm. In all cases the two (equal) principal values of the chemical shielding tensors corresponding to the so-called “perpendicular edge” are located upfield, and the remaining principal value (“parallel edge”) is downfield. It is apparent that the ^{19}F chemical shift is very sensitive to the nature of the divalent cation.

NMR STUDIES OF OTHER CALCIUM PHOSPHATES: AMORPHOUS, SOLID-SOLUTION, SURFACE-TREATED, DOPED, NONSTOICHIOMETRIC AND BIOLOGICAL SAMPLES

Amorphous Calcium Phosphate

The first ^{31}P CP-MAS NMR investigation of the structure of amorphous calcium phosphate, one of several possible intermediates in the formation of hydroxyapatite by precipitation, was reported in 1983.[37] The Ca/P molar ratio of this preparation was 1.50±0.02, corresponding to the empirical formula $Ca_3(PO_4)_2 \bullet xH_2O$, and the sample contained about 20 wt. % tightly bound water. A single broad central peak plus associated sidebands was observed, at a chemical shift identical to that of hydroxyapatite (reported here as 1.7 ppm). The linewidth of 7 ppm is considerably broader than those observed for crystalline calcium phosphates (typically 1 ppm), and presumably reflects some heterogeneity of the structural environments. The spinning sidebands are about twice as intense as those of hydroxyapatite, but considerably weaker than those observed for protonated calcium phosphates. The fact that the ^{31}P CP-MAS NMR spectra did not change significantly with the CP contact time argued against amorphous calcium phosphate consisting of mixtures of phases such as hydroxyapatite, brushite, or octacalcium phosphate with very small crystal sizes or poor crystallinity. Furthermore, spectra obtained at temperatures as low as -120 °C failed to provide any evidence for protonated phosphate groups, although the possible presence of averaging of the ^{31}P csa by rapid motions of bound protons could not be excluded. Strong evidence that the ^{31}P signal arises from cross-polarization by protons in water molecules came from the observed disappearance of a ^{31}P CP-MAS signal when the water was removed by heating the sample in vacuo to 450 °C. The sideband intensities in this dried sample are undiminished in the Bloch-decay spectrum, providing strong evidence that the ^{31}P csa in amorphous calcium phosphate is due to structural distortion of an unprotonated PO_4^{3-} group.

Two other ^{31}P CP-MAS NMR studies of amorphous calcium phosphate and its hydrolysis to a poorly-crystalline hydroxyapatite have subsequently been reported.[38] Similar results and conclusions were obtained for the starting material, the reported linewidths being only slightly narrower (6 ppm). Both studies reported similar trends in the spectra as the hydrolysis times increased and the samples had undergone further conversion. Specifically, the linewidths of the CP spectra decreased, as did the Bloch-decay linewidths, which were somewhat sharper. Also, the CP spectra showed stronger sideband intensities than the Bloch decay spectra, indicating the presence of at least two types of phosphate environments in the partially hydrolyzed samples. The use of dipolar suppression experiments to eliminate the signal from ^{31}P nuclei strongly-coupled to protons led to an identification of one of these environments as an HPO_4^{2-} group, and the other as a PO_4^{3-} group in a poorly-crystalline apatite lattice where the waters of hydration present in amorphous calcium phosphate have been eliminated. These studies also reported comparisons to both bone mineral and carbonated amorphous calcium phosphates.

Fluorohydroxyapatite Solid Solutions, $Ca_5F_x(PO_4)_3(OH)_{(1-x)}$

Fluoride ions can replace hydroxyl groups in hydroxyapatite to form solid solutions over the entire compositional range. The substitution is depicted schematically on the left side of Figure 6,

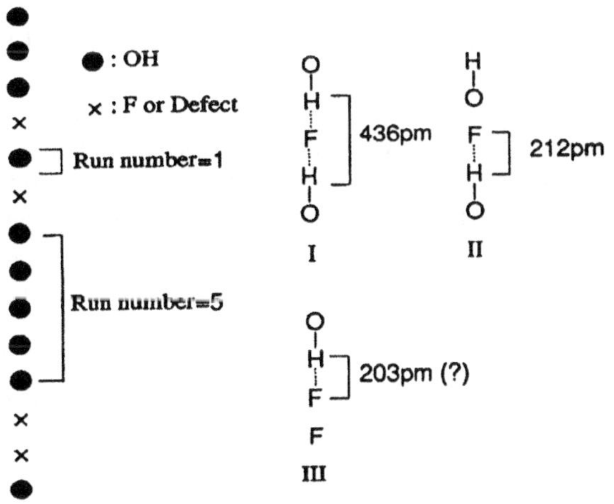

Figure 6. (left) Schematic arrangement of defects in hydroxyl chains of hydroxyapatite and fluorohydroxyapatite, illustrating definition of "run number." (right) Configurations along the crystallographic c-axis in fluorohydroxyapatites.

and some of the possible configurations are given in more structural detail on the right hand side. The ^{19}F MAS-NMR spectra at 282 MHz of fluorohydroxyapatite samples described by the formula above with x=1.00, 0.81, 0.41 and 0.24 have been reported. The single center peak observed showed a small monotonic change in the chemical shift from 64.0 ppm to 60.9 ppm. At least five distinct configurations about a given fluorine atom have been observed by single-crystal NMR as well as IR studies.[11] These include the configurations F-F-F (along the linear chain of fluoride ions in fluorapatite), (OH)-F-F, and (OH)-F-(OH). The latter two arrangements have different configurations depending upon whether or not an OH group hydrogen-bonds to the fluorine. The linewidths (ca. 4 ppm) are apparently too broad to permit resolution of the individual chain configurations. More dramatic than the change in isotropic chemical shifts was the marked reduction in spinning sideband intensities and a greatly increased relative intensity of the central peak as x decreases. Theoretical simulations of sideband intensities were carried out for the dilute fluoride environment (OH)-F-(OH) containing a single hydrogen bond to fluorine, with an assumed H-F distance of 2.12 Å as obtained from single-crystal NMR studies. This distance was used to calculate the ^{1}H-^{19}F dipolar coupling in an isolated pair of nuclei, which has a significant effect upon the calculated spectra. Reasonably good agreement of the simulations with the experimental results was obtained when an axially-symmetric ^{19}F csa of 30 ppm was assumed; similar reductions in the ^{19}F csa from its value of 84 ppm in pure fluorapatite were observed in the single-crystal work.[12] It is interesting to note that just as in the case of ^{31}P NMR, the ^{19}F isotropic chemical shifts are much less sensitive to structural perturbations than all three principal values of the shielding tensor themselves.

The ^{1}H MAS-NMR at 200 MHz of the same suite of fluorohydroxyapatite samples showed a rather broad peak at 5.8-6.3 ppm assigned to surface-adsorbed water, an upfield peak at 0.3-0.4 ppm assigned to the unperturbed structural hydroxyl group found in hydroxyapatite, and several partially-resolved resonances between 1.2 and 2.5 ppm arising from hydroxyl groups perturbed by fluorine. These peaks could be plausibly assigned to the various configurations discussed in detail in the paper, with the close H-F distance in a hydrogen-bonded OH group producing stronger sideband intensities from the large heteronuclear dipolar coupling. The only exception is a peak at 2.5 ppm, whose sharpness at low spinning speeds of 0.5 kHz and lack of sidebands suggested that it may arise from a highly mobile structural water group.

Multiple-Quantum NMR of Fluorohydroxyapatite Solid Solutions and their End-Members

Although up to this point the discussion has been organized based upon the sample rather than upon the NMR technique used, for multiple-quantum (MQ) NMR it is more useful to consider all of the experiments carried out on apatites in one section. The ^{1}H MQ NMR results obtained for hydroxyapatite and the same fluorohydroxyapatite samples referred to above will be discussed first. For the most highly stoichiometric hydroxyapatite studied, which contained a high percentage of the monoclinic form, the results are very interesting from a theoretical standpoint since they represent the first example of MQ growth in a quasi-one-dimensional distribution of spins. This is illustrated in Figure 4, which shows how the hydroxyl protons form an essentially infinite linear chain of uniformly-spaced spins along the c-axis of the crystal. Dipolar couplings between neighboring protons are over twenty times stronger than those between protons on different chains. Therefore, the involvement of ever-increasing numbers of spins in a MQ coherence as the preparation time is increased is expected to take place via dipolar couplings among spins within a given chain, rather than between chains. The number of spins effectively involved in a MQ coherence is observed to vary initially as the first power of the preparation time (i.e., linearly), as has been predicted by one model of MQ growth in one-dimensional distributions. In contrast, MQ growth in two- or three-dimensional distributions is characterized by larger exponents of the order of 2-3. Similar results, obtained using a somewhat different experimental approach, were also obtained for ^{19}F MQ growth in fluorapatite.

When the linear chains of uniformly-spaced spins are interrupted by defects as shown in Figure 6, which could arise from either ^{19}F nuclei or vacancies, the ^{1}H MQ growth is hindered. This behavior is understandable, since the ^{1}H-^{1}H dipolar coupling between protons separated by a single vacancy defect is some 8 times weaker than that between neighboring protons. By assuming that the defects are randomly distributed (i.e., ideal solid solutions), one can use probability arguments to calculate the discrete distribution of various “run numbers” that represent the number of contiguous spins terminated on both ends by defects. One can then assume that, e.g., for a run number equal to 5, it is possible to generate 2-quantum and 4-quantum coherences, but not 6-quantum coherences. In effect, one is postulating an isolated cluster of 5 spins dipolar coupled to one another, but not significantly coupled to other spins. For a truly isolated cluster of 5 spins, it is impossible by any means to create an MQ coherence of order >5, by the fundamental nature of MQ coherence. When no defects are present, it is assumed (and observed) that much higher orders of coherence can be created, especially as the preparation time is increased.

Although this simple model has limitations, it provides an intuitive grasp of the experimental ^{1}H MQ behavior at 400 MHz for hydroxyapatite and fluorohydroxyapatite samples plotted in Figure 7A. The ratio of the 4-quantum coherence order intensity to that of the 2-quantum coherence order is plotted in this graph. The number of spins involved in the formation of MQ coherences increases as the preparation time increases (linearly for the stoichiometric hydroxyapatite sample).

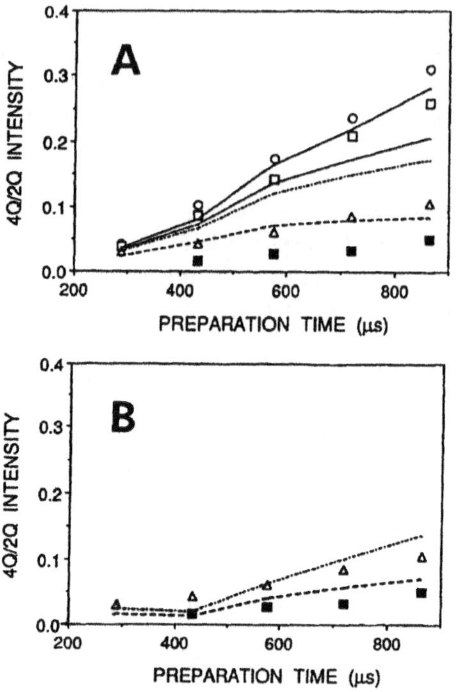

Figure 7. Effect of defects in apatites upon growth of multiple-quantum coherences in ^1H NMR. The experimental ratio of 4Q/2Q intensities is plotted versus the MQ preparation time for 4 samples: monoclinic (highly-stoichiometric) hydroxyapatite (open circles), slightly-nonstoichiometric hydroxyapatite (HAP-N, open squares), and two fluorohydroxyapatite samples with 24% substitution of hydroxyl sites by fluorine (open triangles) and 41% substitution of hydroxyl sites by fluorine (solid squares). (a) from top to bottom, the theoretical lines calculated using a 1-D cluster model (see text) correspond to 8%, 19%, 24% and 41% defect densities. (b) upper calculated line and lower calculated line correspond to 24% and 41% defect densities incorporating additional assumption that protons adjacent to fluoride ions are excluded from MQ growth.

What is immediately obvious from the experimental data points in Figure 7A is that the presence of defects, either vacancies as in the case of the precipitated HAP-N sample, or fluoride substitutions as in the case of the two fluorohydroxyapatite samples, results in a significant decrease in the growth rate of MQ coherences. The lines that are drawn on the graph represent the predictions of a simple 1-D cluster model assuming defect densities as described in the caption. By incorporating a more realistic assumption that hydroxyl protons adjacent to a fluoride ion are excluded from participating in MQ coherence formation because of the strong heteronuclear dipolar coupling, the agreement between experimental and predicted MQ growth is considerably improved (Figure 7B).

The MQ NMR experiment can thus provide valuable information of two kinds on apatites. First of all, from these and additional results one can construct calibration curves to estimate defect densities in unknown samples. A strategy of using highly-stoichiometric hydroxyapatite samples that are partially-deuterated at various levels to provide samples containing precisely-known levels of defects (deuterons) has been suggested. The approximately 8% hydroxyl deficiency in the HAP-N sample can be clearly distinguished from that of stoichiometric HAP-M in Figure 7A; the ability to detect even lower levels of defects by MQ NMR has not been investigated, but seems reasonable. In addition to providing an analytical approach to quantitating hydroxyl or fluoride nonstoichiometry in apatites, which is difficult to do accurately by other analytical methods, unlike most other analytical or spectroscopic methods the MQ NMR approach is intrinsically sensitive to any submicroscopic (atomic-level) heterogeneity in the distribution of hydroxyl groups or fluoride ions. Thus, if the sample contains regions of stoichiometric apatites and other regions of nonstoichiometric apatites, the MQ growth behavior should differ in detectable ways from that resulting from an apatite having the averaged value for the degree of nonstoichiometry. No applications of this approach have yet been reported.

The ^{1}H MQ NMR experiments described above used standard pulse sequences designed for spin systems where the homonuclear dipolar couplings are the dominant source of broadening. For ^{19}F in fluorapatite at high magnetic field strengths, however, the large ^{19}F csa produces sizeable shifts in the resonance frequency depending upon the crystallite orientation. For such cases, a chemical-shift-selective MQ NMR experiment dubbed CHESHIRE was devised to enable one to study MQ behavior at selected frequency positions on a powder pattern (which for an axially symmetric tensor such as that in fluorapatite correspond to unique crystallite orientations with respect to the magnetic field). This type of experiment broadens the classification of what constitutes a defect in the sense used above. For ^{1}H MQ NMR, a defect was anything other than a ^{1}H, i.e., a vacancy or a ^{19}F nucleus. For the ^{19}F CHESHIRE experiment, a defect can be anything other than a ^{19}F nucleus, or it may be a ^{19}F nucleus whose chemical shift is strongly perturbed by another atom. For instance, the substitution of Sr^{2+} for Ca^{2+} in fluorapatite is known to markedly affect the ^{19}F shielding tensor. Thus, the ^{19}F MQ growth in a 9/1 Ca/Sr fluorapatite solid solution was observed to be markedly slower than that for pure calcium fluorapatite. The results of such an experimental approach were described in terms of the correlation length, or persistence length in polymer terminology, of the shielding tensor, i.e. the length scale over which a shielding tensor retains its same orientation and principal values before being altered by some perturbation due to substitutions, etc..

NMR of Fluoridated Apatite Surfaces

Although NMR spectroscopy as generally practiced is not intrinsically a surface-selective technique, it has been used to obtain unique structural and dynamical information about many surfaces, including those of calcium phosphates. There are two main requirements for such studies: A. a specific surface area typically >> 10 m^2/g to provide a sufficient number of nuclear spins at the surface; B. the ability to distinguish spins at the surface from those of the bulk substrate, based on either their different spectral properties or the unique presence of an appropriate NMR-active isotope at the surface. The requirement of a high specific surface area can be readily met with precipitated hydroxyapatite samples. The ability to distinguish spins at the surface based upon their spectral properties is illustrated by the observation of a ^{1}H MAS-NMR signal from mobile surface-adsorbed water on hydroxyapatite, discussed above. Making such a distinction based upon the unique presence of an isotope is illustrated by the ^{19}F MAS NMR studies of fluoride adsorption onto hydroxyapatite that will be discussed below. By combining such studies with double-resonance approaches, specifically ^{19}F-^{31}P double-resonance in REDOR experiments, one can achieve surface selectivity for a nucleus such as ^{31}P that is also abundant in the underlying substrate. NMR spectroscopy is well suited for surface studies in the important respect that it is sensitive primarily to the local environment and does not require a crystalline environment. In many cases, however, the NMR results are interpreted by comparisons with the local environments in model crystalline compounds, such as fluorohydroxyapatites in the case of fluoride adsorption.

The reaction of fluoride ion with the hydroxyapatite surface has been extensively studied by many techniques due to the important anti-caries (prevention of cavities) action of fluoride. The first NMR study of this reaction using ^{19}F MAS NMR identified an initially-formed surface fluoride species shown in Figure 9a that was 1.4 ppm upfield from the resonance of fluorapatite and whose chemical shift and sideband intensity pattern both resembled the spectrum expected for a fluorohydroxyapatite solid-solution sample with $0.4 < x < 0.8$.[34] From the specific surface area of the sample (62 m^2/g) it was estimated that the 0.68 wt. % fluoride uptake corresponded to about 1.1 fluoride ions replacing each surface hydroxyl group. After 6 months storage of the nominally dry powdered sample, the spectrum changed to one characteristic of fluorapatite in a process suggested to occur via surface diffusion of ions in a hydration layer. It is important to realize that the comparison with the spectra of model compounds does not imply that bulk phases are present in the surface-treated samples; rather, it appears that the fluoride environment at the surface involves coordination by three calcium ions as in the bulk structure (Figure 4), and initially hydroxyl group neighbor(s) that is/are replaced by fluoride neighbor(s) over time. At higher concentrations, the resonance from the final species resembling fluorapatite is superimposed upon a broad peak from calcium fluoride, CaF_2. The calcium fluoride resonance is not appreciably narrowed at the 4 kHz MAS spinning speed used due to very strong homonuclear dipolar interactions, and its resultant short T_2 relaxation time enabled it to be selectively eliminated from the spectrum by using a Hahn spin echo sequence. High

MAS spinning speeds of 14 kHz resulted in a considerable narrowing of the calcium fluoride peak, permitting an accurate chemical shift to be measured (58.2 ppm).[39] The CRAMPS technique has also been used with ^{19}F at 282 MHz and spinning speeds of about 5kHz to narrow the line of CaF_2 in mixtures containing fluorapatite and to quantitate reletive amounts of the two componts.[40]

A more detailed study of the reaction of fluoride with the hydroxyapatite and enamel surface using solution chemical characterization methods as well as ^{19}F MAS NMR provided additional chemical insight.[41] An apatitic form of fluoride was seen under all conditions. Additionally, calcium fluoride was observed to form at high fluoride concentrations or low pH. An unusual peak at 44 ppm was observed in the fluoride-treated dental enamel samples, and it was argued that the presence of carbonate in enamel was not responsible for this peak since a fluoride-treated synthetic carbonate-hydroxyapatite failed to show a similar peak. A later study involving submonolayer coverages of fluoride on a synthetic hydroxyapatite surface also observed this 44 ppm peak as a prominent feature, and attributed it to nonspecifically-adsorbed F^- that is hydrogen-bonded to a protonated phosphate group at the surface.[39]

The ^{31}P-^{19}F REDOR experiment at a ^{31}P frequency of 121 MHz has recently been used for the first time to investigate the fluoridation of a high specific surface area hydroxyapatite sample (SSA=128 m^2/g). The ^{31}P observed REDOR intensities enabled one to infer a fluoridation level of 26-27% of the original hydroxyl sites in hydroxyapatite,[21] in apparent good agreement with the value from chemical analysis (25%) reported later.[42] A number of different models of the fluoride distribution were used to compute REDOR curves for comparison with the experimental REDOR behavior. For a solid-solution of fluorohydroxyapatite containing the same percentage of fluoride substitution as the surface-treated sample (25%) a random distribution was found to agree best with the REDOR curve. In contrast, the surface-treated sample results were best described by a model in which fluoride ions replace the OH ions in the first two or three layers of the unit cells along the (001) plane (i.e., the ab face of the crystal). Models involving fluoridation of other crystal faces did not agree with the data.[42] The ability to study the fluoridation process at this level of structural detail by the REDOR technique and modeling is very promising for the future.

The effect of fluoride ions upon the conversion of octacalcium phosophate to apatite has been studied using ^{31}P and ^{19}F MAS-NMR and ^{19}F-observed ^{31}P-^{19}F REDOR at 282 MHz.[43] A peak in the ^{19}F and MAS-NMR spectrum at 58-59 ppm was assigned to low levels of fluoride ion in a fluorohydroxyapatite environment. An intriguing small peak at 70.3 ppm in the spectra of octacalcium phosphate samples treated with very low levels of fluoride appeared to arise from a fluoride/octacalcium phosphate phase, and the REDOR spectrum indicated that the ^{19}F was dipolar coupled to ^{31}P nuclei.

Wide-line ^{19}F NMR Hahn spin-echo studies in a low magnetic field corresponding to a ^{19}F Larmor frequency of 27 MHz have been reported for fluoride in powdered rat bone mineral.[44,45] Low levels of fluoride could be quantitated by using large sample sizes (ca. 10 g), and the ^{19}F T_1 in the dry powder was measured to be 1.5 sec.[44] Two components could be identified on the basis of different Gaussian decay times, 90 μs for one and 300-400 μs for the other. These were assigned to different local fluoride environments based on calculations of the second moments due to both homonuclear and heteronuclear dipolar couplings.[45] The component with the longer decay time was assigned to fluoride ions having only hydroxyl nearest neighbors. The fluoride ions did not appear to be randomly dispersed, but tended to cluster. Higher field T_1 measurements at 80 MHz on two different types of bone samples showed a linear increase in the relaxation rate (T_1^{-1}) with fluoride concentration; the exact relaxation mechanisms are not known.[46]

Other Surface NMR Studies of Calcium Phosphates

The fluoridation of hydroxyapatite is not the only surface reaction that has been investigated by NMR. The strong chemisorption of diphosphonates (bisphosphonates) onto the hydroxyapatite surface has been studied using high-resolution ^{31}P NMR of colloidal suspensions.[11,12] Both ^{1}H and ^{31}P MAS and CP-MAS NMR have been used to investigate the sorption of inorganic phosphate onto the surface of calcium carbonate.[47] At the lowest concentrations the phosphate group appeared to be unprotonated, but the spectra did not resemble those of either hydroxyapatite or amorphous calcium phosphate. At higher phosphate concentrations a carbonated apatitic-like phase appeared. At concentrations exceeding monolayer coverage, both an apatitic-like phase and brushite were detected.[47]

There is also great interest in learning about the nature of adsorption of organic molecules, and particularly polypeptides and proteins, on the hydroxyapatite surface as a model for biological mineralization processes. One of the first such studies reported used natural-abundance ^{13}C CP-MAS NMR to study poly-L-lysine and poly-L-glutamic acid.[48] The sensitivity problem was alleviated somewhat by the use of a high-field spectrometer (400 MHz for ^{1}H) and extensive signal averaging (typically 40,000 scans). Information about the flat extended polypeptide conformation on the surface and specific interactions of side-chain functional groups with surface sites was obtained from ^{13}C chemical shift measurements and relaxation times and CP dynamics. Such studies promise to yield much valuable information, particularly if samples can be isotopically-labeled.

Dopants in Fluorapatite: NMR Spin Diffusion Approaches for Assignments

Detailed NMR studies of calcium fluorapatite doped with low levels (0.2-3 wt %) of antimony(III) have shown that the substitution in these important phosphor materials occurs via a SbO_3^{3-} group replacing a PO_4^{3-} group rather than a charge-coupled substitution of Sb^{3+} for Ca^{2+} as had been long-believed.[35,49] General aspects of strategies for investigating dopant substitution sites are revealed in this work. For instance, the ^{31}P MAS-NMR spectra showed no observable changes upon doping with antimony, whereas the ^{19}F MAS-NMR spectra showed additional peaks besides the main resonance at 64.2 ppm and its associated sidebands (called peak C for convenience). Two peaks consistently seen at 68.6 and 65.6 ppm are referred to as peaks A and B respectively, and were measured to have a 2:1 intensity ratio. They could be assigned using some of the spin-diffusion results discussed below to two types of fluoride ions in the fluorapatite lattice that are differently perturbed by the antimony substitution, as shown in detail in Figure 5. The intensity of peak A relative to that of the main peak C did not correlate with the total antimony content, since not all of the antimony appears to enter the apatite lattice but instead exists in other forms.

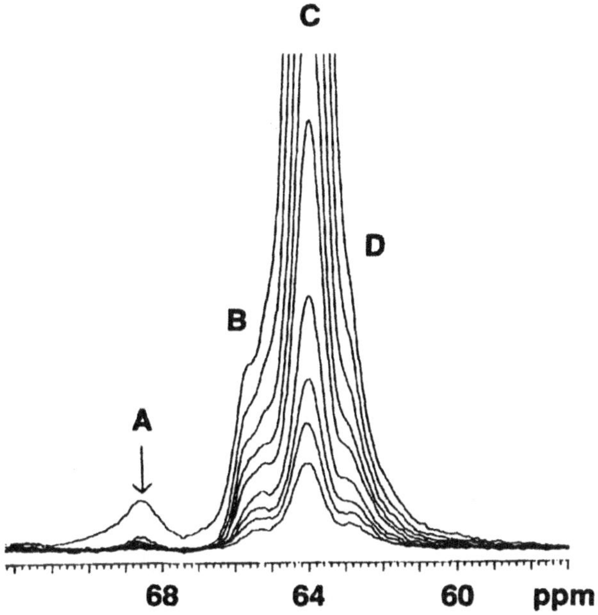

Figure 8. ^{19}F MAS-NMR (376 MHz) superimposed spectra of a calcium fluorapatite sample containing 1.7 wt. % Sb^{3+} dopant obtained using the SINK pulse sequence. The main peak C (off-scale in the first four spectra) corresponds to the unperturbed center resonance of pure calcium fluorapatite as seen in Figure 2 for a different sample. The other peaks are labeled to correspond to the assignments depicted in Figure 5. The arrow indicates the position of the radiofrequency transmitter used to saturate peak A. The spectra shown were obtained with an increasing number of saturating pulse trains from top to bottom, as described in the original reference. Note that peak A is immediately saturated, whereas cross-relaxation via spin-diffusion more gradually (and differentially) decreases the intensities of the other peaks.

The initial evidence for assigning peaks A and B to fluoride ions in the apatite lattice came from the observation that the ^{19}F T_1 relaxation times of all three peaks in a given sample were nearly equal to each other within experimental error, although they varied irregularly from sample to sample, from 129 to 378 s. The variation may be due to differing trace amounts of paramagnetic impurities in the different samples; the near-equality for a given sample is expected if spin-diffusion occurs between the different types of fluoride ions in the same lattice. In this case, the ^{19}F nuclei near a paramagnetic center will be rapidly relaxed and serve as relaxation "sinks" for other ^{19}F nuclei, resulting in a common T_1 relaxation time.

A more straightforward and informative experiment to demonstrate the existence of spin-diffusion between the different peaks is to selectively invert the signal from peak A and observe the cross-relaxation via dipolar coupling to other peaks as a function of a mixing interval in the so-called SPARTAN experiment.[35] Peak C was observed to be affected much more readily than peak B, leading to the structural assignments depicted in Figure 5.

In the SINK experiment shown in Figure 8[49] the minor peak A is continually saturated by selective pulses, which amplifies the cross-relaxation effects upon the other peaks. Thus, the main peak C is reduced in intensity by an order of magnitude, while peak B is relatively less affected. Quantitative spin-diffusion rates can be extracted from such experiments, but their exact interpretation is difficult. The main use of such experiments may be to assign minor peaks to dopant-related impurities in a lattice rather than to other unrelated species, and to better reveal subtle features such as the shoulder (peak D in Figure 8) on a much larger peak (Peak C) whose intensity can be greatly reduced. Extensions to ^{1}H and ^{31}P NMR may also prove useful.

NMR of Nonstoichiometric and Carbonated Hydroxyapatites

This section will discuss applications of NMR spectroscopy to the study of hydroxyapatite samples that are nonstoichiometric, including those containing carbonate. Such synthetic samples are models for the mineral phase in bone and teeth. The first ^{31}P CP-MAS NMR study of nonstoichiometric apatites investigated samples whose Ca/P molar ratios of 1.58, 1.46, 1.33 and 1.14 were considerably different from the stoichiometric value of 1.67.[13] Such samples are often referred to as "calcium-deficient." Except for the sample having the lowest ratio, the spectra were surprisingly similar in chemical shift and sideband intensities to that of stoichiometric hydroxyapatite, regardless of whether CP was used or Bloch decays were obtained. The sample with the lowest ratio (1.14) exhibited an additional peak that could be selectively observed and assigned to a monetite (anhydrous dicalcium phosphate) component. The monetite was crudely estimated to be less than 15% of the sample, indicating that the main peak arose from a highly nonstoichiometric hydroxyapatite as with the other samples. For the three other nonstoichiometric samples no evidence could be seen of definite phases of calcium phosphates more basic (i.e., having a lower Ca/P ratio) than hydroxyapatite, although the experiments should have been capable of revealing the presence of such phases. Since the degree of nonstoichiometry in these samples could not be accounted for by hydroxyl group deficiencies alone, but seemed to require the presence of protonated phosphate groups, spectra were obtained near liquid nitrogen temperatures in order to determine whether rapid proton motions at room temperature were producing a type of chemical-exchange effect that averaged the ^{31}P csa and reduced the spinning sideband intensities. Indeed spectra of samples with ratios of 1.46 and 1.33 at about -186 °C showed a marked increase in sideband intensities when CP was used, especially for the lower ratio sample, but not in Bloch decay spectra. These results indicated a heterogeneity of the phosphate groups at low temperature, with presumably the protonated ones yielding an increased sideband intensity but nearly identical chemical shift to that of the unprotonated orthophosphate. However, since based upon the Bloch decay spectra the amount of HPO_4^{2-} groups present appeared much too small to account for the nonstoichiometry, it was suggested that yet lower temperatures might be needed to freeze out all proton motion. Unfortunately, no such experiments have been subsequently reported, despite their feasibility (albeit with difficulty) and their potential for resolving the various proposals that have been advanced to account for hydroxyapatite nonstoichiometry.[13]

In the above study, the T_1 relaxation times of ^{31}P at a frequency of 68 MHz were observed to vary from 1 to 22 sec in the hydroxyapatite samples (stoichiometric and nonstoichiometric), and the value in human enamel was considerably longer (about 90 sec). The ^{31}P T_1 values measured at 40 MHz on related samples showed comparable variation: 2.0 sec for hydroxyapatite and about 30 sec for both a nonstoichiometric hydroxyapatite with a molar Ca/P=1.5 and rabbit bone. The reasons for these wide variations remain yet to be explained, although in a general sense they must relate to motions occurring in the solid lattice or at the surface. The ^{31}P chemical shifts of the two hydroxyapatite samples in this study were identical, but a somewhat greater linewidth was reported for the nonstoichiometric sample and ascribed to increased phosphate disorder.[31]

The ^1H MAS-NMR spectrum at 300 MHz of the above nonstoichiometric hydroxyapatite sample having a molar Ca/P=1.5 shows two peaks at 0.8 ppm and 6 ppm.[31,50] The former peak clearly belonged to structural hydroxyl groups, and the latter was assigned to acid phosphate groups. The presence of surface-adsorbed water signals as observed in other samples,[26] was ruled out on the basis of the large crystallite size. A broadening of the latter resonance was observed upon cooling from 22 to -23 °C, and attributed to decreased motion resulting in stronger dipolar coupling of the acid proton to the hydroxyl proton.[50] The possibility that the 6 ppm peak does not arise from acid phosphate groups should be considered. The hydrogen-bonding suggested to occur between these two protons would be expected to produce downfield chemical shifts[26] that are not observed. Also, the two peaks do not appear to have identical areas in the spectrum, the hydroxyl peak appearing distinctly less intense. It is possible that the 6 ppm signal may correspond to mobile water either in the apatite lattice or as inclusions, somewhat similar to that observed in dental enamel (see next section). The chemical shift is certainly appropriate for this assignment, and the behavior upon cooling is not inconsistent with the lack of a distinct freezing transition observed for water in confined spaces.

Carbonate-containing apatites have been studied by both ^1H and ^{13}C MAS and CP-MAS NMR at 400 MHz and 80 MHz respectively (on different spectrometers).[51]

Isotopic-labeling of synthetic samples was used to improve the signal-to-noise ratio of the spectra. It was possible to observe the natural-abundance ^{13}C signal from carbonate in dental enamel because of the low organic content of this mineralized tissue, which produces fewer problems from interfering resonances of organic groups. The ^{13}C spectra were able to reveal differences among the different types of carbonate substitution known to occur (so-called types A, AB, and B). The Bloch decay ^{13}C MAS NMR spectrum of type A carbonated apatite shows a single centerband at 166.5 ppm (from tetramethylsilane, TMS) and sidebands over a 10 kHz range due to the csa (not reported). Since type A carbonated apatite is known to involve replacement of a OH^- group with a CO_3^{2-} group, ^1H-^{13}C CP experiments were not expected to be possible because of the depletion of protons. The type B carbonated apatite gives an asymmetric signal with a maximum at 170.2 ppm, and a small signal at 166.5 ppm apparently due to trace type-A carbonate. In type B apatites the CO_3^{2-} group replaces a PO_4^{3-} group. The type AB carbonated apatite, which is believed to involve carbonate substitution at both hydroxyl and phosphate sites, shows a low intensity peak due to type A carbonate and an intense peak due to type B carbonate. In addition, a third peak is seen at 168.2 ppm, the same position as the broad peak observed for amorphous carbonate phosphate. Dipolar suppression experiments indicate that this 168.2 ppm peak is close to protons, and it was argued on several grounds that these protons are most likely associated with water molecules in the lattice. The ^1H MAS-NMR spectra appeared very similar to those of pure hydroxyapatite,[26] and no perturbations of the hydroxyl proton chemical shift by interactions with carbonate could be observed.

An example of the application of ^{31}P MAS-NMR in environmental science is a study of the form in which low levels of inorganic phosphate occurs in wastewater sludges and soils treated with such sludges.[52] One difficulty in applying NMR to such "dirty" systems is the presence of paramagnetic ions of Fe and Mn, which were observed in this study to result in spinning sideband intensities more intense than those of any protonated phosphate. This effect had been observed before (it was first observed in the ^{31}P MAS-NMR of mineral fluorapatite) and has been quantitatively explained.[53] A chemical reducing treatment was effective in decreasing the paramagnetic content and the sideband intensities. Carbonated apatite was the species identified in a number of the samples.

NMR Studies of Biological Mineralized Tissue and Possible Strategies

The first high-resolution NMR spectra of biological mineralized tissue, obtained using ^{31}P CP-MAS NMR, were reported in 1980 for bone[54] and for dental enamel.[13] The initial bone experiments provided evidence for the presence of a large proportion of brushite (dicalcium phosphate dihydrate) in the mineral phase of embryonic chick bone at the earliest stages of development; decreasing amounts of brushite and increasing amounts of a form of hydroxyapatite were observed in the bone obtained at later stages of development. The increased ^{31}P linewidth of bone relative to synthetic well-crystallized calcium phosphates was attributed to a distribution of chemical shifts due to differing local environments, and has been observed in other studies as well.

These results were extended in a very extensive characterization of the mineral in chick bone at various stages of development.[55] Three types of ^{31}P experiments, Bloch decay, CP, and dipolar suppression, were applied to bone samples and compared to results from a variety of crystalline and noncrystalline calcium phosphate standards. The NMR results ruled out the possibility that amorphous calcium phosphate was the major phase of newly-deposited bone mineral, a hypothesis which had been advanced on the basis of other evidence, since its NMR characteristics differed considerably from those of the bone samples. Instead, it was deduced that the youngest bone contains significant quantities of HPO_4^{2-} groups in a brushite-like environment, and that this fraction decreases with increasing age of the bone. The model advanced for the major mineral phase of bone mineral was a hydroxyapatite containing 5-10% CO_3^{2-} and 5-10% HPO_4^{2-} groups in a brushite-like configuration.[55] Further studies have supported the existence of a brushite-like environment for the HPO_4^{2-} groups most abundant in early bone, but not the presence of crystalline brushite.[56] A study of the mineral phase of calcified cartilage of young calves by ^{31}P MAS-NMR at 119 MHz has been reported.[57] The youngest (least dense) fractions contained only 2.8 wt. % mineral, illustrating the sensitivity of ^{31}P MAS-NMR. Spinning sidebands indicative of protonated phosphates in a presumed poorly-crystalline apatite lattice were observed, although the experimental conditions were not given.

The meaning of the term "brushite-like" as applied to HPO_4^{2-} groups in bone has received clarification from a later study of bone from three species at several developmental stages using the method of differential CP (DCP) ^{31}P MAS-NMR at 162 MHz to distinguish between protonated and unprotonated phosphate groups.[17] Spectra illustrating this technique are shown in Figure 3. The unprotonated orthophosphate PO_4^{3-} resonances of bone mineral, synthetic hydroxyapatite, and synthetic octacalcium phosphate share identical isotropic chemical shifts of 2.8 ppm and similar chemical shift anisotropies. The orthophosphate resonances in bone are somewhat broader, due presumably to a distribution of environments. The resonance in bone assigned to HPO_4^{2-} groups has an isotropic chemical shift of -0.4 ppm, which is very close to the value of -0.5 ppm for the HPO_4^{2-} group in octacalcium phosphate but different from the value of 1.4 ppm for the HPO_4^{2-} in brushite. However, the chemical shielding tensor of this protonated group in bone agrees closely with that observed in brushite, but not with that for the protonated phosphate in octacalcium phosphate. The latter has a much smaller csa. Thus, the HPO_4^{2-} peak in bone is "brushite-like" as far as the anisotropic part of its chemical shielding tensor is concerned (the part giving rise to spinning sideband intensities), but more like the HPO_4^{2-} group in octacalcium phosphate in terms of its isotropic chemical shift. An additional resemblance between bone and octacalcium phosphate is the similarity between their cross-polarization time constants T_{CP} for both PO_4^{3-} and HPO_4^{2-} groups . The corresponding time constants for PO_4^{3-} in hydroxyapatite and HPO_4^{2-} in brushite are quite different. It was suggested that the feature of octacalcium phosphate that may enable it to share some (but not all) of the same ^{31}P NMR characteristics as bone is the fact that it is a mixed crystal, containing both PO_4^{3-} and HPO_4^{2-} groups in the same lattice.[17] The development of

DCP and other techniques based upon spin-diffusion characteristics for determining whether these two groups are located in close spatial proximity, or widely separated, is an important area for future work.

The potential utility of ^{1}H MAS-NMR for investigating the mineral phase of bone was found to be limited by the presence of significant amounts of triglycerides (approximately 5 wt. %), which could be identified by ^{13}C MAS-NMR and two-dimensional ^{1}H MAS-NMR (COSY).[58] Interestingly, similar triglyceride signals were obtained from a number of Type I collagen samples, which is the primary constituent of the organic portion of bone. The degree to which triglyceride is "contamination" or intimately associated with the collagen/mineral matrix remains to be determined. Residual water present after lyophilization was also a complicating factor. Although in mineralized turkey tendon a peak assignable to the hydroxyl group in hydroxyapatite was observed, no such peak was observed in rabbit mandibular bone. The absence of structural hydroxyl groups in bone has been noted in earlier IR studies and briefly noted in one ^{1}H MAS-NMR result,[59] which may make observation of significant ^{1}H NMR signals from the mineral phase of bone problematic. Techniques to observe ^{1}H nuclei that are dipolar-coupled to ^{31}P nuclei such as HetCor or reverse CP may offer the most promise.

Mineralized tissue such as enamel containing less organic matter may be more suited to ^{1}H NMR investigations. Both ^{1}H MAS at 317 MHz and ^{31}P NMR have been used to study porcine enamel at various stages of maturation.[59] An example of how static ^{1}H NMR T_1 and T_2 relaxation times can be used to study enamel is a study of human enamel in which various components could be distinguished by widely differing T_2 values.[60] There is a solid-like water component with a short T_2 14 μs that could not be exchanged with the deuterium isotope, an apatitic component (mostly hydroxyl groups) with a T_2 61 μs, and semiliquid water with a T_2 10 ms, some of which could be exchanged by deuterium (the measured T_2^* was 200 μs due to the heterogeneous environments in enamel).

Despite the importance of these biological mineralized tissues, it is fair to say that the unambiguous determination of their detailed structure by NMR methods has presented problems which have not been entirely overcome. Some of these problems may originate in biological variability, some in the crucial effects of sample preparation and treatment, and some in the difficulty of using NMR techniques to learn about distributions of structural environments in poorly-crystalline samples having high surface areas. The continuing development of NMR techniques such as REDOR, HetCor and multiple-quantum NMR provides hope that the last-mentioned problem may be at least partially overcome. Sample preparation artifacts can in principle be overcome by the ability to carry out "low-resolution" NMR experiments in-vivo. Indeed, surface coils have recently been used to obtain in-vivo a ^{31}P single-pulse NMR spectrum of a human ankle and a CP spectrum of a rabbit skull.[61] The challenge is to obtain useful information in the absence of the isotropic chemical shift information that can be obtained from MAS NMR experiments. The measurement of T_1 or other relaxation times or of the spectral second moments offers one potential characterization tool for static in-vivo samples, as discussed above for ^{19}F in bone mineral. Double-resonance techniques such as differential CP and ADRF-CP offer another.

ACKNOWLEDGMENT

I am grateful to my former students (Dr. Gyunggoo Cho and Dr. Liam Moran) and many collaborators over the years for their hard work and for many interesting discussions. I am also grateful to the Office of Naval Research for support while writing this chapter.

REFERENCES

1. Elliott JC. *Structure and Chemistry of the Apatites and Other Calcium Orthophosphates.* Studies in Inorganic Chemistry Vol. 18, New York: Elsevier, 1994.

2. Jones DW, Smith JAS. Proton Magnetic Resonance in Brushite. Trans Faraday Soc 1960;56:638-647

3. see, e.g., Becker ED. *High Resolution NMR: Theory and Chemical Applications.* New York: Academic Press, 1980.

4. see, e.g., Derome AE. *Modern NMR Techniques for Chemistry Research.* New York: Pergamon Press, 1987.

5. see, e.g., Slichter CP. *Principles of Magnetic Resonance.* Third Edition. New York: Springer-Verlag, 1989.

6. Fyfe CA. *Solid State NMR for Chemists.* Guelph, Ontario, CRC Press, 1984.

7. Eckert H. "Structural Characterization of Noncrystalline Solids and Glasses Using Solid State NMR," in *Progress in Nuclear Magnetic Resonance Spectroscopy*, Emsley JF, Feeney, J, Sutcliffe, ed., Vol. 24, Part 3, New York: Pergamon Press, 1992.

8. *Encyclopedia of Nuclear Magnetic Resonance*, D.M. Grant and R.K. Harris, eds., Vols. 1-8, New York: John Wiley & Sons, 1996.

9. Yesinowski JP, Sunberg RJ, Benedict JJ. pH control and rapid mixing in spinning NMR samples. J Magn Reson 1982;47:85-90

10. Yesinowski JP, Benedict JJ. ^{31}P NMR as a spectroscopic monitor of the spontaneous precipitation of calcium phosphates. Calcif Tissue Int 1983;35:284-286

11. Yesinowski JP. High-resolution NMR spectroscopy of solids and surface-adsorbed species in colloidal suspension: ^{31}P NMR spectra of hydroxyapatite and diphosphonates. J Amer Chem Soc 1981;103:6266-6267

12. Yesinowski JP, Wolfgang RA, Mobley MJ. "New NMR Methods for the Study of Hydroxyapatite Surfaces," in *Adsorption on and Surface Chemistry of Hydroxyapatite*, D.N. Misra, ed. New York: Plenum Publishing Corp., 1984.

13. Rothwell WP, Waugh JS, Yesinowski JP. High-resolution variable-temperature ^{31}P NMR of solid calcium phosphates. J Amer Chem Soc 1980;102:2637-2643

14. Aue WP, Roufosse AH, Glimcher MJ, Griffin RG. Solid-state phosphorus-31 nuclear magnetic resonance studies of synthetic solid phases of calcium phosphate: potential models of bone mineral. Biochem 1984;23:6110-6114

15. Note that Ref. 14 found that the csa was "considerably smaller" than this (no value given). However, the measured moments of the ^{31}P MAS-NMR spectra of hydroxyapatite at two different field strengths were significantly greater at the higher field, and the results tended to corroborate the values reported in Ref. 13 (J.P. Yesinowski, unpublished results).

16. Turner GL, Smith KE, Kirkpatrick RJ, Oldfield E. Structure and cation effects on phosphorus-31 NMR chemical shifts and chemical-shift anisotropies of orthophosphates. J Magn Reson 1986;70:408-415

17. Wu Y, Glimcher MJ, Rey C, Ackerman JL. A unique phosphate group in bone mineral not present in synthetic calcium phosphates: identification by phosphorus-31 solid state NMR spectroscopy. J Mol Biol 1994;244:423-435

18. Cho G, Yesinowski JP. Multiple-quantum NMR dynamics in the quasi-one-dimensional distribution of protons in hydroxyapatite. Chem Phys Lett 1993;205:1-5

19. Moran LB, Yesinowski JP. Chemical-shift-selective multiple-quantum NMR as a probe of the correlation length of chemical shielding tensors. Chem Phys Lett 1994;222:363-370

20. Cho G, Yesinowski JP. ^{1}H and ^{19}F Multiple-quantum NMR dynamics in quasi-one-dimensional spin clusters in apatites. J Phys Chem 1996;100:15716-15725

21. Pan Y. ^{31}P-^{19}F Rotational-echo, double-resonance nuclear magnetic resonance experiment on fluoridated hydroxyapatite. Sol State NMR 1995;5:263-268

22. Santos RA, Wind RA, Bronniman CE. ^{1}H CRAMPS and $1H$-^{31}P HetCor experiments on bone, bone mineral, and model calcium phosphate phases. J Magn Reson B 1994;105:183-187

23. Ramanathan C, Wu Y, Pfleiderer B, Lizak MJ, Garrido L, Ackerman JL. ADRF-CP surface-coil spectroscopy of synthetic calcium phosphates and bone mineral. J Magn Reson A 1996;121:127-138

24. Ackerman JL, Raleigh DP, Glimcher MJ. Phosphorus-31 magnetic resonance imaging of hydroxyapatite: a model for bone imaging. Magn Reson Medicine 1992; 25:1-11

25. Moore JR, Garrido L, Ackerman JL. Solid state ^{31}P magnetic resonance imaging of bone mineral. Magn Reson Med 1995;33:293-299

26. Yesinowski JP, Eckert H. Hydrogen Environments in Calcium Phosphates: ^{1}H MAS NMR at high spinning speeds. J Amer Chem Soc 1987;109:6274-6282

27. Belton PS, Harris RK, Wilkes PJ. Solid-state phosphorus-31 NMR studies of synthetic inorganic calcium phosphates. J Phys Chem Solids 1988;49:21-27

28. Yesinowski JP, unpublished results

29. Kormos DW, Waugh JS. Abstract factor analysis of solid-state nuclear magnetic resonance spectra. Anal Chem 1983;55:633-638

30. Rothwell WP. Investigations of structure and dynamics in solids by high resolution nuclear magnetic resonance. Ph.D. Thesis, Massachusetts Institute of Technology, 1980.

31. Miquel JL, Facchini L, Legrand AP, Marchandise X, Lecouffe P, Chanavaz M, Donazzan M, Rey C, Lemaitre J.Characterisation and conversion study into natural living bone of calcium phosphate bioceramics by solid state nmr spectroscopy. Clinical Materials 1990;5:115-125

32. Arends J, Christoffersen J, Christoffersen MR, Eckert H, Fowler BO, Heughebaert JC, Nancollas GH, Yesinowski JP, Zawacki SJ. A calcium hydroxyapatite precipitated from an aqueous solution. J Cryst Growth 1987;84:515-532.

33. Note that chemical shifts of ^{19}F are often reported using a variety of different compounds as references at "0 ppm", unlike the situation for ^{1}H and ^{31}P NMR. Conversions between different chemical shielding scales are readily made.

34. Yesinowski JP, Mobley MJ. ^{19}F MAS-NMR of fluoridated hydroxyapatite surfaces. J Amer Chem Soc 1983;105:6191-6193

35. Moran LB, Berkowitz JK, Yesinowski JP. ^{19}F and ^{31}P Magic-angle spinning nuclear magnetic resonance of antimony(iii)-doped fluorapatite phosphors: dopant sites and spin diffusion. Phys Rev B 1992;45:5347-5360

36. Cho G. NMR Investigations of chemical shielding, structure, and multiple-quantum dynamics of apatites. Ph.D. Thesis, Michigan State University, 1993.

37. Tropp J, Blumenthal NC, Waugh JS. Phosphorus NMR study of solid amorphous calcium phosphate. J Amer Chem Soc 1983;105:22-26

38. Roberts JE, Heughebaert M, Heughebaert JC, Bonar LC, Glimcher MJ, Griffin MJ. Solid state ^{31}NMR [sic] studies of the conversion of amorphous tricalcium phosphate to apatitic tricalcium phosphate. Calcif Tissue Int 1991;49:378-382

39. Kreinbrink AT, Sazavsky CD, Pyrz JW, Nelson DGA, Honkonen RS. Fast-magic-angle-spinning NMR of inorganic fluorides and fluoridated apatitic surfaces. J Magn Reson 1990;88:267-276

40. Smith KA, Burum DP. Application of fluorine-19 CRAMPS to the analysis of calcium fluoride/fluorapatite mixtures. J Magn Reson 1989;84:85-94

41. White DJ, Bowman WD, Faller RV, Mobley MJ, Wolfgang RA, Yesinowski JP. ^{19}F MAS-NMR and solution chemical characterization of the reactions of fluoride with hydroxyapatite and powdered enamel. Acta Odontol Scand 1988;46:375-389

42. Pan Y. "The Study of Fluoridated Hydroxyapatite by ^{31}P-^{19}F Rotational-Echo, Double-Resonance (REDOR) NMR," in *Solid State NMR Spectroscopy of Inorganic Materials*, J. Fitzgerald, ed., in ACS Symposium Series, New York:ACS Books, 1997, in press.

43. Iijima M, Nelson DGA, Pan Y, Kreinbrink AT, Adachi M, Goto T, Moriwaki Y. Fluoride analysis of apatite crystals with a central planar OCP inclusion: Concerning the role of F-ions on apatite/OCP/apatite structure formation. Calcif Tiss Int 1996;59:377-384

44. Ebifegha ME, Code RF, McNeill KG, Szyjkowski M. Nuclear-magnetic-resonance determinations of fluorine content and relaxation times in bone powder. Can J Phys 1986;64:282-288

45. Gelman N, Code RF. NMR Spin-echo study of ^{19}F environments in rat bone mineral. J Magn Reson 1992;96:290-301

46. Code RF, Armstrong RL, Hallsworth RS, Lemaire C, Cheng PT. Concentration. Phys Med Biol 1992;37:211-221

47. Hinedi ZR, Goldberg S, Chang AC, Yesinowski JP. A ^{31}P and ^{1}H MAS-NMR study of phosphate sorption onto calcium carbonate. J Colloid Interf Sci 1992;152:141-160

48. Fernandez VL, Reimer JA, Denn MM. Magnetic resonance studies of polypeptides adsorbed on silica and hydroxyapatite surfaces. J Amer Chem Soc 1992;114:9634-9642

49. Moran LB, Berkowitz JK, Yesinowski JP. A method for detection of spectral spin diffusion from minor peaks, and its application to ^{19}F MAS NMR of antimony(III)-doped fluorapatite. Solid State NMR 1992;1:307-311

50. Miquel JL, Facchini L, Legrand AP, Rey C, Lemaitre J. Solid state NMR to study calcium phosphate ceramics. Colloids and Surfaces 1990;45:427-433

51. Beshah K, Rey C, Glimcher MJ, Schimizu M, Griffin RG. Solid state carbon-13 and proton NMR studies of carbonate-containing calcium phosphates and enamel. J Sol State Chem 1990;84:71-81

52. Hinedi ZR, Chang AC, Yesinowski JP. Phosphorus-31 magic angle spinning nuclear magnetic resonance of wastewater sludges and sludge-amended soil. Soil Science Society of America Journal 1989;53:1053-1056

53. Nayeem A, Yesinowski JP. Calculation of magic-angle spinning nuclear magnetic resonance spectra of paramagnetic solids. J Chem Phys 1988;89:4600-4608

54. Herzfeld J, Roufosse A, Haberkorn RA, Griffin RG, Glimcher MJ. Magic angle sample spinning in inhomogeneously broadened biological systems. Phil Trans R Soc London B 1980;289:459-469

55. Roufosse AH, Aue WP, Roberts JE, Glimcher MJ, Griffin RG. Investigation of the mineral phases of bone by solid-state phosphorus-31 magic angle sample spinning nuclear magnetic resonance. Biochem 1984;23:6115-6120

56. Roberts JE, Bonar LC, Griffin RG, Glimcher MJ. Characterization of very young mineral phases of bone by solid state ^{31}phosphorus [sic] magic angle sample spinning nuclear magnetic resonance and x-ray diffraction. Calcif Tissue Int 1992;50:42-48

57. Rey C, Beshah K, Griffin R, Glimcher MJ. Structural studies of the mineral phase of calcifying cartilage. J Bone Miner Res 1991;6:515-525

58. Yesinowski JP, Evans JS. "High-Resolution ^{1}H MAS-NMR of Connective Tissue," in *Program and Abstracts, Third International Conference on the Chemistry and Biology of Mineralized Tissues, Chatham*, Massachusetts, October 1988.

59. Bonar LC, Shimizu M, Roberts JE, Griffin RG, Glimcher MJ. Structural and composition studies on the mineral of newly formed dental enamel: A chemical, x-ray diffraction, and ^{31}P and proton nuclear magnetic resonance study. J Bone Miner Res 1991;6:1167-1176

60. Funduk N, Kydon DW, Schreiner LJ, Peemoeller H, Miljkovi L, Pintar MM. Composition and relaxation of the proton magnetization of human enamel and its contribution to the tooth NMR image. Magn Reson Med 1984;1:66-75

61. Ackerman JL, Garrido L, Moore JR, Pfleiderer B, Wu Y. "Fluid and Solid State MRI of Biological and Nonbiological Ceramics," in *Magnetic Resonance Microscopy: Methods and Application in Materials Science, Agriculture and Biomedicine*, B. Blümich and W. Kuhn, New York: VCH, 1992.

7

INFLUENCE OF METAL IONS ON THE CRYSTAL GROWTH OF CALCIUM PHOSPHATES

Petros G. Koutsoukos

Institute of Chemical Engineering and High Temperature Chemical Processes
Department of Chemical Engineering, University of Patras, Patras, GREECE

INTRODUCTION

The development of living organisms and of mammals in particular, is closely related to the development of mineralization, i.e., to the mechanism of formation of hard tissue. The main inorganic chemical component of bone and teeth is calcium phosphate which is deposited through careful control by the development of a slight excess of the solubility product.[1] In the natural environment, the precipitation of calcium phosphates is part of the mechanism for the control of phosphate accumulation in lakes and estuaries in which there is a continuous influx of phosphates due to irrigation runoff.[2,3] Moreover, the removal of 85-90 % of the inorganic orthophosphates contained in wastewater may be achieved by precipitation of calcium phosphates by lime.[4] In addition, the formation of tenacious scale deposits of calcium phosphates has also been reported in boilers and industrial installations using recycled, chemically treated water.[5-7]

In all cases mentioned here, the formation of calcium phosphates takes place in supersaturated aqueous media. The first step is nucleation which proceeds up to the formation of a nucleus having the critical size, required for the further growth into crystallites of larger dimensions. Both nucleation and growth steps are very important in determining the nature and the morphology of the mineral phase forming. In the case of calcium phosphates a number of solids may form, depending on the solution supersaturation in the order of decreasing solubility: dicalcium phosphate dihydrate ($CaHPO_4{\cdot}H_2O$, DCPD), octacalcium phosphate ($Ca_8H_2(PO_4)_6{\cdot}5H_2O$, OCP) tricalcium phosphate ($Ca_3(PO_4)_2$, β-TCP) and hydroxyapatite ($Ca_{10}(PO_4)_6(OH)_2$, HAP).

From the thermodynamic point of view, the least soluble crystalline phase may be formed at the lowest degrees of supersaturation (with respect to the least soluble phase) while at higher degrees of supersaturation there is an increased possibility of the preferential formation of the most soluble phase through kinetic stabilization. This kinetic stabilization of unstable phases may be of great importance for the processes of formation of calcium phosphate minerals and may be affected by the interaction of the forming nuclei with foreign compounds, complexes or metal ions present in the supersaturated aqueous media. The phenomena of adsorption and lattice incorporation affect also considerably the preference of formation of one phase over another. As a rule, in biological mineralization, in the aquatic environment, in industrial use of water etc., the formation of calcium phosphates takes place in the presence of a number of different metal ions depending on the origin of the supersaturated fluids. Thus, for example, in the formation of biominerals, the metals present may be classified either as bulk metals including Na, K, Mg, and Ca or as trace metals in which the ions of Fe, Cu, Mn, Zn, Co, Cr, Sn, V, Ni and Sn are included.[8] In natural waters, on or under earth's surface the metals content depends on the composition of minerals in contact with water. Typical concentrations of metal ions in water and extracellular blood plasma are given in Table 1:

Table 1. Ion concentrations in seawater, groundwater and extracellular blood plasma.

Ion	Groundwater (M)	Seawater (mM)	Blood plasma (mM)
Na	0.42	470	138
Mg	0.38	50	1
Ca	0.60	10	3
K	-	10	4
Fe	2.1x10	10	1
Zn	-	10	2x10
Cu	-	10	1.5xx10
Co	-	10	2x10
Ni	-	10	-

The presence of metal ions in the supersaturated media especially of those with sizes similar to the size of calcium ions plays an important role not only from the equilibrium point of view but also from the kinetic, since ion substitution may well take place during the course of formation of the new crystals. In soils, the presence of trace metal ions is responsible for the formation of various phosphate minerals during the weathering process of primary phosphates, such as Plumbogummite ($PbAl_3(PO_4)_2(OH)_5H_2O$), Craudallite ($CaAl_3(PO_4)_2(OH)_5H_2O$), Gocyazite ($SrAl_3(PO_4)_2(OH)_5H_2O$), Gorceixite $BaAl_3(PO_4)_2(OH)_5H_2O$) and Lasungite ($SrFe_3(PO_4)_2(OH)_5H_2O$). Calcium may be found as a substitution in the above mineral families while OCP is also formed in several cases as a result of soil fertilization.

In this paper we discuss the aquatic chemistry of calcium phosphates in the presence of metal ions and its relationship with the solid phases formed, both from the equilibrium and from the kinetics point of view. The importance of the interactions of the metal ions present in the supersaturated medium with the nucleating phase will also be discussed.

SOLUTION EQUILIBRIA

The investigations on the precipitation of calcium phosphates, refer mostly to the formation of the thermodynamically most stable, HAP. The initial conditions however in most instances suggest that the aqueous phase is supersaturated with respect to all possible calcium phosphate phases. The solubility isotherms calculated over the pH range 4-10 are shown in Figure 1.

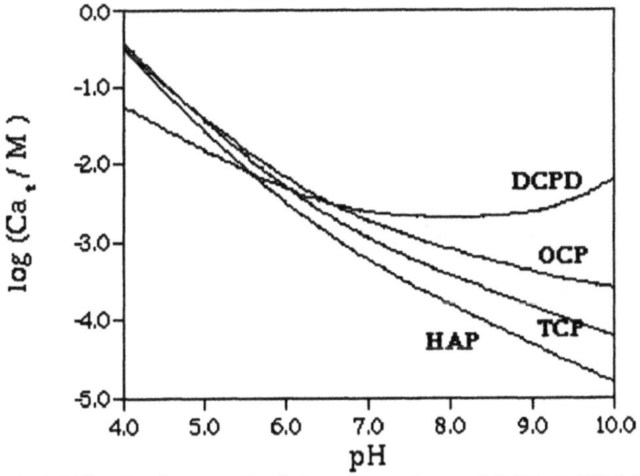

Figure 1. Solubility isotherms of calcium phosphates; 25 °C, 0.15M NaCl.

The thermodynamic solubility products of the calcium phosphate phases at 25 °C are:

for HAP[9]	$(Ca^{2+})^5(PO_4^{3-})^3(OH^-)$	$= 4.70\times10^{-59}$ (mol $l^{-1})^9$
for TCP [10]	$(Ca^{2+})^3(PO_4^{3-})^2$	$= 1.20\times10^{29}$ (mol $l^{-1})^5$
for OCP [11]	$(Ca^{2+})^4(PO_4^{3-})^3(H^+)$	$= 1.25\times10^{-47}$ (mol $l^{-1})^8$
for DCPD[12]	$(Ca^{2+})(HPO_4^{3-})$	$=1.26\times10^{-7}$ (mol $l^{-1})^2$
for OCPD[13]	$(Ca^{2+})(HPO_4^{3-})$	$= 2.49\times10^{-7}$ (mol $l^{-1})^2$

for metal phosphates in which Ca^{2+} ions have been replaced by other metals the following expressions may be written:

for $Mg_3(PO_4)_2 \cdot {}_8H_2O$[14]	$(Mg^{2+})^3(PO_4^{3-})^2$	$= 6.31\times10^{-26}$
for $MgHPO_4 \cdot {}_3H_2O$[14]	$(Mg^{2+})(HPO_4^{2-})$	$= 1.51\times10^{-6}$
for $SrHPO_4$[14]	$(Sr^{2+})(HPO_4^{2-})$	$= 1.20\times10^{-7}$
for $BaHPO_4$[14]	$(Ba^{2+})(HPO_4^{2-})$	$= 3.98\times10^{-8}$
for $LaPO_4$[14]	$(La^{3+})(PO_4^{3-})$	$= 3.71\times10^{-23}$ *
for $Cd_3(PO_4)_2$[14]	$(Cd^{2+})^3(PO_4^{3-})^2$	$= 5.50\times10^{-23}$ *
for $(UO_2)_3(PO_4)_2$[14]	$(UO_2^{2+})^3(PO_4^{3-})^2$	$= 2.00\times10^{-50}$ *
for UO_2HPO_4[14]	$(UO_2^{2+})(HPO_4^{2-})$	$= 6.76\times10^{-13}$ *
for $Fe_3(PO_4)_2 \cdot 8H_2O$[14]	$(Fe^{2+})^3(PO_4^{3-})^2$	$= 1.00\times10^{-36}$
for $(VO)_3(PO_4)_2$[14]	$(VO^{2+})^3(PO_4^{3-})^2$	$= 7.94\times10^{26}$
for Ag_3PO_4[14]	$(Ag^+)^3(PO_4^{3-})$	$= 2.82\times10^{-18}$
for $HgHPO_4$[14]	$(Hg^{2+})(HPO_4^{2-})$	$= 3.98\times10^{-13}$
for $Zn_3(PO_4)_2 \cdot 4H_2O$[14]	$(Zn^{2+})^3(PO_4^{3-})^2$	$= 5.01\times10^{-36}$
for $Pb_3(PO_4)_2$[14]	$(Pb^{2+})^3(PO_4^{3-})^2$	$= 2.95\times10^{-46}$
for $PbHPO_4$[14]	$(Pb^{2+})(HPO_4^{2-})$	$= 3.72\times10^{-12}$
for $GaPO_4$[14]	$(Ga^{3+})(PO_4^{3-})$	$= 1.00\times10^{-21}$ +
for $InPO_4$[14]	$(In^{3+})(PO_4^{3-})$	$= 2.34\times10^{-22}$ +

In the above expressions the values marked with * and + were determined for ionic strengths of 0.5 M and 1.0 M respectively, while parentheses denote activities of the enclosed ions. The computation of the ionic activities is necessary for the evaluation of the thermodynamic driving force for the formation of the corresponding solid phases, which is the change of chemical potential for going from the supersaturated solutions to equilibrium:

$$\Delta\mu = \mu_s - \mu_\infty \tag{1}$$

where μ_s, μ_∞ are the chemical potentials of the salt considered in the supersaturated solution and at equilibrium respectively. For a metal-phosphate salt of the type

$$M_m(PO_4)_n(HPO_4)_u$$

$$\mu_s = \mu_s^o + kT\ln(\alpha_{M_m(PO_4)_n(HPO_4)_u})_s \qquad (2)$$

$$\mu_\infty = \mu_\infty^o + kT\ln(\alpha_{M_m(PO_4)_n(HPO_4)_u})_\infty \qquad (3)$$

the mean ion activity concentrations are:

$$\alpha^{1/m+n+u}_{M_m(PO_4)_n(HPO_4)_u} = (M)^m(PO_4^{3-})^n(HPO_4^{2-})^u \qquad (4)$$

Equation (1) becomes therefore:

$$\Delta\mu = -\frac{kT}{m+n+u}\ln\frac{[(M)^m(PO_4^{3-})^n(HPO_4^{2-})^u]_s}{[(M)^m(PO_4^{3-})^n(HPO_4^{2-})^u]_\infty} \qquad (5)$$

The numerator and the denominator of the logarithmic term are the ionic product and the thermodynamic solubility product $K^o_{s,x}$ of the salt x respectively and $m+n+u=\nu$, the number of ions in the salt, hence:

$$\Delta\mu = -\frac{kT}{\nu}\ln\frac{IP}{K^o_{s,x}} \qquad (6)$$

and per mole,

$$\Delta G = -\frac{RT}{\nu}\ln\frac{IP}{K^o_{s,x}} = -\frac{RT}{\nu}\ln\Omega \qquad (7)$$

where $\Omega = IP/K^0_{s,x}$.

The computation of the ion activities may be done taking into account all calcium and other metals phosphate equilibria, together with the mass and charge balance equations which involve the analytical concentrations of the various species.[15] The relationship between the activity of a species i, γ_i, and the respective analytical concentration, c_i, is given by equation:

$$\alpha_i = c_i\gamma_i \qquad (8)$$

where γ_i is the activity coefficient of the species i. A number of expressions are available for the calculation for the activity coefficients of z-valent ions based on the extended form of the Debye-Hückel equations, such as that proposed by Davies:[16]

$$-\log\gamma_z = Az^2[I^{1/2}/(1+I^{1/2}) - 0.3I)] \qquad (9)$$

where A is a temperature dependent constant and I the ionic strength. The expression given in equation (9) is valid for ionic strength values up to 0.2 M for multiple charge ions. Other expressions and their validity at higher ionic strengths are discussed elsewhere.[17,18] The non-ideality of the electrolyte solutions is a problem which may introduce inconsistency in the calculations of the various equilibrium models depending on whether the ion association theory,[19] the specific-ion interaction theory[20,21] or the ion hydration theory[22] is used. For the computation of the complex ionic equilibria in which complex formation takes place between the various cations and the multivalent phosphate anions, a number of commercial software packages are available. The packages available are based either on the free energy minimization method or on the equilibrium constant approach.[23-26] In the former method the fundamental assumptions are that the free energy of the system is:

A. a function of the state of the system (state function),
B. a sum of the free energies of the constituents of the solution
C. at a minimum when the system is at equilibrium

This method is limited to rather simple systems where the free energies are available and reliable, while for more complex systems the equilibrium constant approach is preferable.[27] The equilibrium approach consists in the solution of a system of non-linear equations formed by the equations of the appropriate equilibria expressions involving a basis set of components from which the various derivative species are formed. The system of equations is solved by an iterative procedure with successive approximations for the ionic strength. The method of successive approximations is done either by substitution of the mass action expressions into the mass balance conditions or by Newton-Raphson methods which are used to find the solutions to a function which compares the difference between the total calculated component concentration and the total analytical component concentration.[28-30]

The Formation of Solid Phases in Aqueous Supersaturated Solutions

As may be seen from Figure 1, HAP is the thermodynamically most stable calcium phosphate phase, the formation of which is favored over a variety of equilibrium conditions. In the presence of metal ions the formation of HAP may yield isomorphous compounds the most important of which are summarized in Table 2.

Table 2. Isomorphous apatitic minerals formed in the presence of metal ions

Name	Formula	lattice parameters (pm) a	c
Barium HAP	$Ba_{10}(PO_4)_6(OH)_2$	1.019	0.77
Cadmium HAP	$Cd_{10}(PO_4)_6(OH)_2$	0.901	0.661
Lead HAP	$Pb_{10}(PO_4)_6(OH)_2$	0.990	0.729
Magnesium apatite	$Mg_{10}(PO_4)_6(OH)_2$	0.930	0.689
Strontium HAP	$Sr_{10}(PO_4)_6(OH)_2$	0.976	0.728

X-ray diffraction studies have shown that HAP crystallizes in the hexagonal Pb_{3m} space group with lattice constants a and c equal to 0.942 and 0.688 nm, respectively.[31-33] The arrangement of the unit cell lattice ions Ca^{2+}, PO_4^{3-}, and OH^- is shown in Figure 2. The projection of the same ions in the 001 plane is also shown in the same Figure.

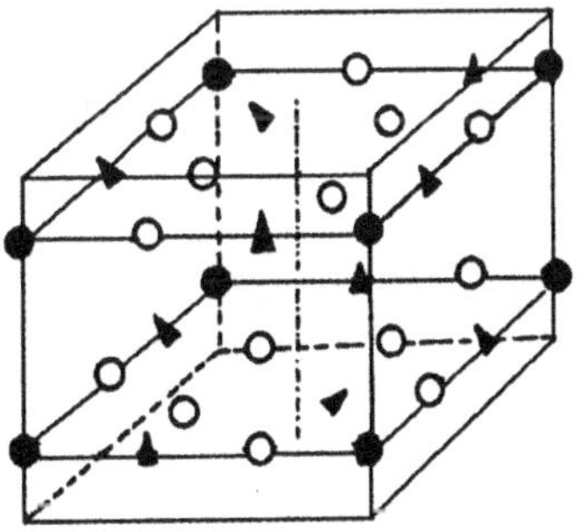

a

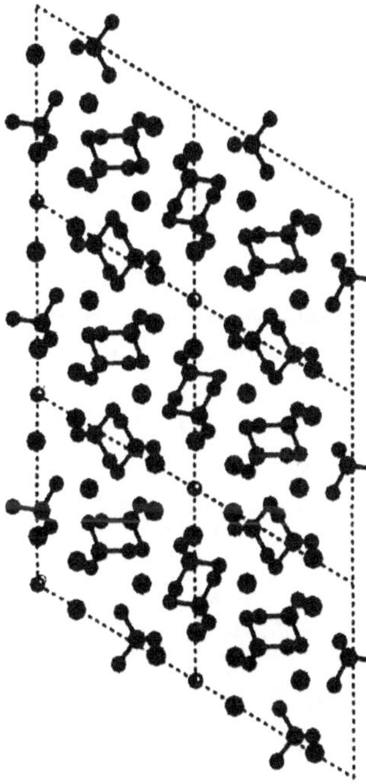

b

Figure 2. a. The unit cell of hydroxyapatite crystals: O Ca^{2+}; ▲ HPO_4^{3-}; ● OH^-
b. (001) plane. Calcium (single atoms) and phosphate tetrahedra are shown.

It is possible that HAP undergoes both iso and heteroionic substitutions. In the first type the substituting ion does not interrupt the crystal lattice while in the second, ions in the lattice are exchanged with similar ions present in solutions in contact with HAP.[34] Bivalent metal ions with ionic radii close to that of Ca^{2+} may replace at all proportions calcium in the apatite lattice.[35] The substitution with other ions of different charge than the corresponding to Ca^{2+} is also possible with a concomitant incorporation of additional ions for the attainment of charge balance.[36-38] Other metals such as Zn^{2+} are reported to substitute up to atom 25% in the HAP lattice when present in the supersaturated solution.[39] It should be noted that for bone HAP (non stoichiometric, carbonate containing HAP) studies with labeled ^{65}Zn have shown 100% substitution of Ca^{2+} by Zn^{2+} ions.[40,41] The effect of the presence of very low levels of Zn^{2+} in the supersaturated solutions has also been attributed to surface adsorption of Zn^{2+} onto the HAP surface.[42]

Among the most important heteroionic substitutions is the substitution of Ca^{2+} by Sr^{2+} ions in the apatite lattice which is known to interfere with the calcification mechanism.[43,44] Strontium is believed to substitute for Ca because of the similarity of its ionic radius (1.12 Å) with that of calcium (0.99).[45,46] A complete series of solid solutions of Sr-HAP have been prepared[47] and the unit cell lattice parameters were found to vary linearly with the Sr content in the HAP lattice.[48] These results have been confirmed by precipitation experiments[49] and by infra-red spectroscopic analyses.[50-52] Tracer experiments using, β-active ^{89}Sr suggested that strontium is first adsorbed onto the HAP surface, diffuses slowly into the crystal interior and finally it substitutes calcium through an HAP recrystallization procedure.[53] The extent of strontium incorporation into the HAP lattice was found to vary linearly with the Sr^{2+} content in the supersaturated solutions in which the solid was formed, while the similar variation in both a and c unit cell lattice parameters ($\Delta\alpha/\Delta c \approx 1$) shown in Figure 3 suggested that the Sr containing HAP at least for contents up to 10% retained the initial crystal morphology. This was confirmed by the morphological examination of the HAP crystals by scanning electron microscopy.[49]

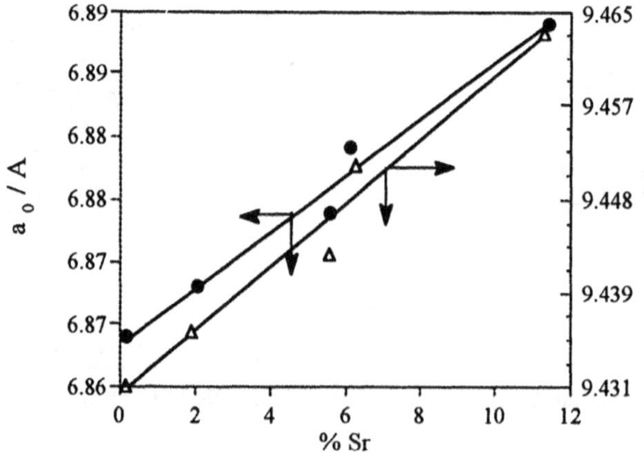

Figure 3. Variation of the a_0 and c_0 unit cell lattice parameters of hydroxyapatite grown in the presence of Sr^{2+} at constant supersaturation.[49]

The in vitro results of the dependence of the extent of Sr incorporation into the HAP lattice on the concentration of the corresponding ions in the fluid in contact with the HAP surface are in good agreement with the findings from experiments in rats which showed that Sr retention in rat femur was proportional to the diet calcium concentration levels.[54] Similar type of strontium incorporation has been reported for DCPD.[55] Toxic metals such as Cd and Pb may also substitute for calcium when present in fluids in contact with HAP through the formation of solid solutions.[56-58]

Of particular interest is the effect of the presence of magnesium in the fluids in contact with calcium phosphates because of the important biological implications of this metal ion. From the numerous reports concerning the effect of magnesium in the literature it seems that the presence of these ions is important in determining the nature of the solid phase forming in unstable supersaturated calcium phosphate solutions.[59-62] In vitro studies on the conversion of amorphous calcium phosphate into HAP have shown that Mg^{2+} ions play an important role in the hydrolysis of intermediate solid phases.[63,64] Unlike strontium, magnesium has not been reported to incorporate into the lattice of calcium phosphate phases although its presence is very important for the kinetics of crystallization of DCPD, OCP, and HAP.[55,66] The interaction of Mg^{2+} ions with the calcium phosphate crystalline solids may be described by adsorption. Since Mg^{2+} ions in solution are charged and the calcium phosphate/water interface is also charged with a charge depending on pH it is anticipated that adsorption of the metal ions on calcium phosphate is strongly influenced by electrostatic interactions. The investigation of adsorption of Mg^{2+} ions on HAP was done by measuring the adsorption isotherms. They showed relatively high affinity of Mg^{2+} ions for HAP surfaces at pH 8.0 where the HAP surface is predominantly negative. Electrophoretic techniques are very powerful tools for the characterization of the electrical double layer of the solids in which foreign metals have been incorporated in the bulk or on their surface, providing information about the magnitude of electrostatic interactions. The electrokinetic potential (ξ potential) which provides the information on the electrostatic interactions is calculated from the electrophoretic mobility, u, of the particles by applying the Helmholtz. Schmolukowski equation:

$$u = \frac{\varepsilon}{\eta}\zeta \qquad (10)$$

where $\varepsilon = \varepsilon_r^{es}\,\varepsilon_o$ (ε_r^{es}, ε_o the relative dielectric constants of the electrolyte solution and of vacuum respectively) and η the viscosity of the solution.[67-69] The validity of equation (10) when applied in the investigation of suspensions of inorganic particles rests on assumptions such as:

A. the particle surface is non conductive

B. the applied electric field strength and the field of the electrical double layer are additive

C. the electric charge is homogeneously distributed over the entire particle surface

D. the viscosity, dielectric constant and ion conductivities in the double layer, outside the shear plane are equal to their bulk values.

As very often the validity of the above assumptions for calcium phosphate particles in the presence of various, organic and inorganic additives is uncertain it is a common practice to present experimentally measured electrophoretic mobility values instead of zeta potentials.[70] In Figure 4 the electrophoretic mobility of stoichiometric HAP without and with the adsorption of 2.5 mmol Mg^{2+}/m^2 as a function of pH is presented. The electrophoretic mobility of HAP particles in the presence of adsorbed magnesium showed a shift of the isoelectric point (iep) of HAP to lower pH values, a fact which confirmed surface adsorption of the cations.[69]

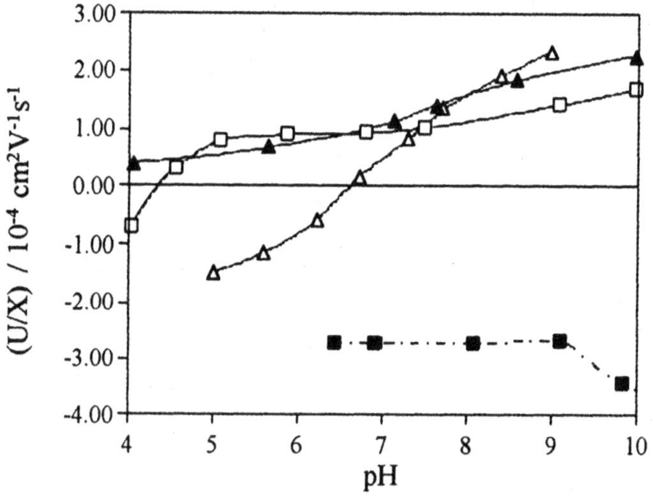

Figure 4. Electrophoretic mobility of HAP particles in the absence (Δ) and with adsorbed Mg^{2+}: (□)2.5 μmolm^{-2}; (▲)25 μmolm^{-2} and Cd^{2+} (■) ions.[55,113] 25 °C, 0.01M KNO_3

It should be noted that in the case of the presence of low concentrations of Cd^{2+} ions (1 mM) in equilibrium with stoichiometric HAP the adsorption of these ions caused a complete charge reversal as may also be seen in Figure 4.

Electrophoretic mobility measurements of HAP in the presence of 1 ppm Zn suggested the formation of a surface layer of zinc phosphate (hopeite) contrary to the magnesium ions which over a concentration range up to 1.5 mM at pH 7.0 showed a Langmuir type adsorption.[71] It should be noted that the concentration levels of Zn^{2+} and Mg^{2+} ions in equilibrium with HAP were sufficient for the formation of the corresponding mineral phases.

Monovalent metal ions can be incorporated into the lattice of calcium phosphates, in particular in the presence of carbonate. Thus sodium follows carbonate in the substitution into the HAP lattice. It has been reported that the incorporation of Na^+ ions into the calcium phosphate phases is accompanied by a retardation in the rates of formation of calcium phosphates.[72] The presence of other monovalent metal ions such as K^+, Cs^+ and NH_4^+ affected only the kinetics and in some cases they caused changes in morphology of the HAP forming.[73] Most pronounced was the effect of Li^+ on the formation of HAP. Small quantities of lithium were found to incorporate into the HAP lattice causing a reduction in the unit cell lattice parameters in accordance with the smaller size of this ion which most likely substitutes for calcium.[74] The effect of the rest of the monovalent ions, especially at relatively low concentrations may be attributed to adsorption on the calcium phosphate crystallites surface. Indirect proof of the adsorption was provided by measurements of the electrophoretic mobility in the presence of various 1:1 electrolytes. The shift of the iep. to more acidic pH values, shown in Figure 5, suggested adsorption of the cations. Potassium nitrate proved to be inert electrolyte as in this case the iep coincided with the point of zero charge (pzc) determined by potentiometric titrations.[55]

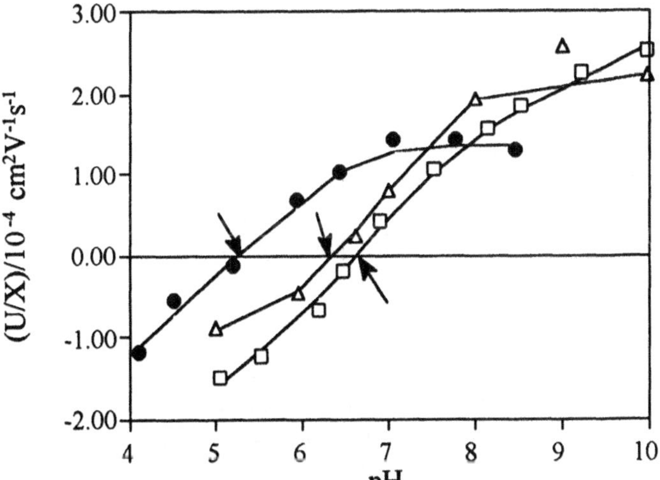

Figure 5. Electrophoretic mobility of HAP in the presence of various electrolytes: total concentration 0.01M; 37°C; (●) NaCl; (Δ)KCl; (□)KNO_3

It seems that metal ions with ionic radii differing considerably from the calcium radius tend to adsorb onto the calcium phosphate crystal surface. Besides the mono and divalent metal ions mentioned above, trivalent ions such as Ga^{3+} and Al^{3+} are reported to adsorb onto hydroxyapatite.[75,76] In several cases adsorption is treated in a simplistic manner by the adoption of the simple Langmuir model, which relates the adsorbed amount C_i of the adsorbate with the number of the active sites N on the adsorbent (calcium phosphate) with equations such as:

$$\frac{C_{eq}}{C_i} = \frac{1}{KN} + \frac{C_{eq}}{N} \tag{11}$$

where C_{eq} is the adsorbate equilibrium concentration and K is the affinity constant related with the specific rates of adsorption ad desorption.[77] Despite the fact that the adsorption data show a satisfactory fit according to this model it is doubtful whether the underlying fundamental assumptions are valid. The Langmuir model assumes monolayer coverage on energetically equivalent adsorption sites and lack of interactions between adsorbate molecules.[78] Despite the obvious deviations from these assumptions in the case of the water / calcium phosphate interface, the constants K and N computed from the fit of experimental adsorption data should be used with caution and for comparison reasons. Often, when the Langmuir adsorption isotherm cannot be used for fitting the experimental data the Freundlich isotherm may be used, which is successfully applied in dilute solutions:[79]

$$\Gamma = K' C_{eq}^{1/n} \tag{12}$$

In equation (12) Γ is the surface concentration of adsorbate (mol/m^2) and K' and n are empirical constants. This isotherm is more realistic for aqueous suspensions as it assumes energetically non-equivalent adsorption sites. The satisfactory fit of the adsorption data obtained from the corresponding experiments on an isotherm should not be taken as evidence of the actual mechanism of the adsorption process.[80] Metal adsorption processes should be examined after performing thorough calculations to ensure that the aqueous medium in equilibrium with the calcium phosphate solids is not supersaturated with respect to either metal phosphates or metal hydroxides. Thus experiments in which the influence of di-or higher valence metal ions on the mineralization of calcium phosphates in aqueous supersaturated solutions is examined, should be done at concentration levels precluding the formation of solid phases of the metal ions examined.[81]

Kinetic Considerations

A large number of kinetic studies of the crystal growth of calcium phosphates in supersaturated aqueous solutions both in the absence and in the presence of metal ions have been published.[23,82-84] The interaction of metal ions with the calcium phosphate crystals forming in supersaturated solutions may be considered as taking place during the crystal build-up by the growth units. The crystal growth process may be described by:[85]

A. surface diffusion models
B. volume diffusion models and / or
C. combined surface and volume diffusion models.

In these models the various steps involved are shown schematically in Figure 6.

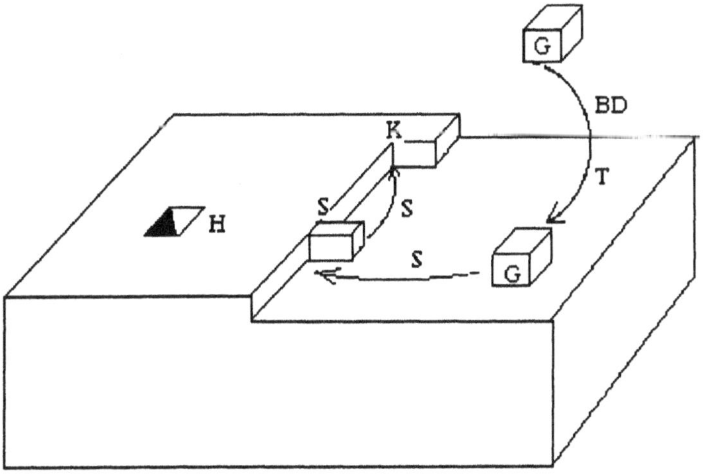

Figure 6. Surface of a growing crystal showing the possible locations for the attachment of the incoming growth units, and the steps involved D: Dislocation; S: Step; K: Kink; H: hole; T: terrace; G: growth unit; BD: Bulk diffusion; S: surface diffusion

The successive steps the growth from solution are:

A. Transport of the growth units to the growing crystal surface by bulk diffusion (path BD)
B. Adsorption of the growth units at the surface (site T)
C. Transport of the growth units to the crystal growth active sites (D, S, K, H)
D. integration of the growth units into the crystal steps

steps C. and D. may include additional steps such as transport through the hydrated crystal surface, dehydration of the growth units or add-ions (partial or total) and migration along the step

The rate of the crystal growth process is determined by the slowest of the above steps. In general, the crystal growth rates, expressed either as linear growth rates or as mass accumulation may be formulated according to equation (13):

$$R_G = K'_g \sigma^g \qquad (13)$$

In equation (13),[86] K'_g is the growth rate constant and g the overall "order" of the growth process. σ is the driving force for crystal growth given by:

$$\sigma = \Omega^{1/\nu} - 1 \qquad (14)$$

The rate constant K'_g is a function of the active growth sites which are considered to be evenly distributed over the entire surface area available for crystal growth:

$$K'_g = K_g S \qquad (15)$$

where K_g is the rate constant and S a function of the active sites for crystal growth. The exponent g is not related with the reaction order of the elementary reactions and lacks of any fundamental significance. It may be used however as indicative of the predominant mechanism i.e. of suggesting which of the steps (i) - (iv) is the rate determining step. If the growth process is controlled by bulk diffusion g = 1 and the rate expression may be written as:[87]

$$R_G = \frac{D}{\delta} A\sigma \tag{16}$$

where D is the diffusion coefficient of the solute, δ the thickness of the diffusion layer and A the crystal surface area. For surface diffusion controlled processes $g < 1$. In case g = 1 or 2 a spiral growth mechanism is anticipated while for $g > 2$ a polynuclear model fit of the kinetic data should be done.[88,89] It should be noted however that in most cases of crystal growth of sparingly soluble salts the rates measured were considerably lower in comparison with calculated rates assuming pure mass transport control, suggesting that growth rates are mostly controlled by surface processes (adsorption, diffusion and integration of the growth units in the crystal).[90] Experiments over a wide range of supersaturations and at various fluid dynamics conditions are needed in order to draw conclusions concerning the underlying mechanism. The measured rates as a function of the solution supersaturation should be fitted to equation (13) and conclusions concerning the mechanism may be drawn from the values of the "order" which can be calculated from the corresponding plots. It has been suggested that low "orders" may be obtained in the case of high surface concentrations of spirals.[91,92] The kinetics of the formation of calcium phosphates in aqueous supersaturated solutions both in the presence and in the absence of metal ions, is studied in experiments in which precipitation takes place either spontaneously or on well characterized calcium phosphate seed crystal substrates. According to the spontaneous precipitation approach the supersaturated solutions are prepared by mixing stock calcium and phosphate solutions in the presence and / or in the absence of inert electrolyte for ionic strength adjustment, the pH of the solution is adjusted as desired and the process is followed by different physicochemical methods e.g., optical density, pH conductance ionic activity measurements etc.[93-96] Considerable inconsistencies and disagreement between the results obtained by different workers who have published work using the spontaneous precipitation approach may be attributed to the fact that often precipitation takes place before it may be verified by the various sensors available and it is very sensitive to the details of the carrying out the experiments.[97] Spontaneous precipitation occurring at high supersaturations results in the formation of ACP which transforms to DCPD in the presence of Mg^{2+} ions.[96] Whether the formation of ACP or DCPD is favored, besides the presence of Mg^{2+} ions, depends also on the solution pH. Thus, at acidic pH the formation of brushite is favored while at pH > 6.5 ACP formed at 37 °C.[98] Alternatively, ACP formed by spontaneous precipitation may be transforming to HAP through the formation of OCP according to the Ostwald rule of stages.[95-100] The presence of Mg^{2+} however has been found to retard the formation of HAP in aqueous supersaturated solution.[101,102] Phase transformation, besides the presence of metal ions in the supersaturated solutions depends also on the solution supersaturation. Rapid desupersaturation may lead to rapid phase changes, which often cause inconsistencies in the experimental results of spontaneous precipitation experiments. The presence of heteronuclei, inherently present in the solutions used in most experiments enhances the problems of reproducibility not only within the same laboratory but also from one laboratory to another.

These problems have been overcome with the development of the seeded growth technique. The principle of this methodology is the employment of stable supersaturated solutions in which the precipitation process is initiated by the introduction of seed crystals of the mineral investigated. The choice of the supersaturation range may be done following the thorough examination of the stability of the supersaturated solutions employed so that spontaneous precipitation is precluded. In these experiments, the effect of parameters such as the presence of metal ions in the solution, temperature ionic strength and supersaturation may be examined quantitatively. The crystal growth process of the seed crystals introduced may be monitored by following pH changes or changes in total calcium, ΔCa_t, and total phosphate, ΔP_t.[103] In the presence of metal ions changes in total metal concentration, ΔM_t are also measured in order to determine whether the metal ion species in solution are incorporated into the lattice of the growing crystals, or if they are adsorbing onto the crystallites surface provided that the precipitation of any possible metal phosphate is precluded. The seeded growth technique provides the possibility of working practically at any desired supersaturation level. In the case of calcium phosphate system, in which the crystal growth process is accompanied by H^+ release in the solution, it is possible by using a glass electrode as a controlling sensor to keep the pH constant throughout the precipitation process by the addition of a standard base solution through the automatic control provided by an automatic titrator.[104] A limitation of the method applied to the calcium phosphate system is that at low supersaturations the pH, ΔCa_t and ΔP_t changes are very small and are complete over a short time. This fact introduces large errors in the calculation of the rates of crystal growth and in the identification of any precursor phases which may form. The characterization of the transient phases relies on the ratio $\Delta Ca_t/\Delta P_t$.[105] The development of the constant composition method[106,107] and its extension, dual constant composition[108] contributed in overcoming these problems. According to these two methodologies it is possible through the reaction probes (glass and calcium ion selective electrodes) in combination with the knowledge of the stoichiometry of the precipitating salt to maintain the activities of the lattice ions in the supersaturated solutions constant, during the crystal growth of the seed crystals. In the presence of metal ions it is possible to ensure the addition of sufficient quantities to compensate not only for dilution but also for the metal uptake by the calcium phosphates provided that it is known from separate experiments. Alternatively, if the interaction of the metal ion with calcium phosphate is not known, it may be deduced by the combination of the analysis for the metal present in the solution and by the amount of titrant solutions added in order to keep the concentration constant.[42,49,109] From the rates of titrants added for constant supersaturation maintenance both in the absence and in the presence of metal ions, it is possible to obtain accurate values for the rates of crystallization. By the adjustment of the concentration of the titrant solutions it is possible to measure accurately even very low rates of crystallization, unattainable by other methods. A typical example of titrants added in the absence and in the presence of Zn^{2+} ions is shown in Figure 7.

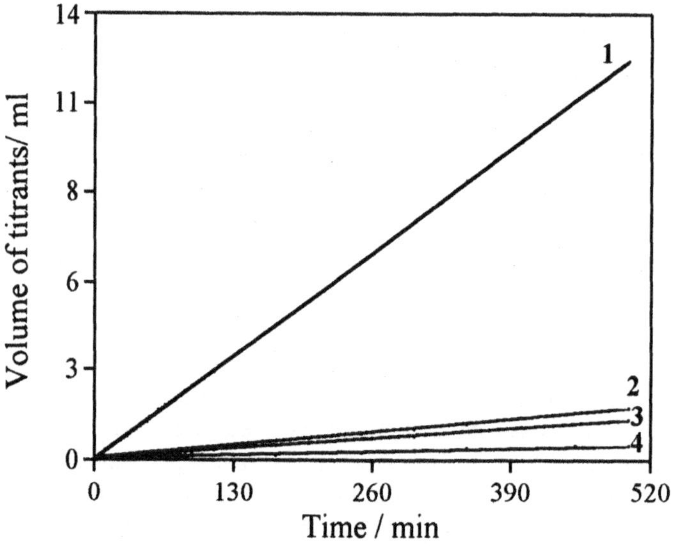

Figure 7. Crystal growth of HAP on HAP seed crystals in the absence and in the presence of Zn^{2+} ions at conditions of constant supersaturation: 1:Blank; 2: $1x10^{-8}$ M ; 3: $1x10^{-7}$ M; 4: $7x10^{-7}$ M; 37 °C, 0,1 M KNO_3 pH 7.40. Volume of titrants added as a function of time for the same amount of seed crystals.

Zn^{2+} ions inhibited the rate of crystallization of HAP.[42] A similar inhibition of calcium phosphate was found in the presence of Mg^{2+} ions. In this case, the order of inhibition of the crystal growth of the calcium phosphate phases was found:

HAP > OCP > DCPD

DCPD was not inhibited by the presence of the Mg^{2+} ions.[66] Both for Mg^{2+} and Zn^{2+} ions there has been no evidence for incorporation into the crystal lattice of the growing crystals. The kinetics analysis can be done therefore on the assumption that these metals adsorb on the calcium phosphates. This assumption has been confirmed and the corresponding adsorption isotherms or were determined. The rate of crystal growth in the presence of an adsorbate, R_a, which covers a fraction, θ, of the available surface may be written as a function of the corresponding rate, R_o, in the absence of any foreign dissolved species as[109,110]

$$R_a = R_0 - \theta R_0(1-\beta) \tag{17}$$

where β is a constant, which gives the limiting rate βR_0 in the presence of the adsorbate. The surface coverage θ is defined as the ratio:

$$\theta = \frac{\Gamma_m}{\Gamma_i} \tag{18}$$

where Γ_m is the maximum surface for a monolayer coverage by the adsorbate and Γ_i is the surface concentration of the adsorbate on the mineral surface. The adsorption is considered as a reversible process and the rates of adsorption, R_{ads}, and desorption, R_{des}, respectively are:

$$R_{ads} = k_{ads} C_a (\Gamma_m - \Gamma_i) \tag{19}$$

$$R_{des} = k_{des} \Gamma_i \tag{20}$$

where k_{ads}, k_{des} are the adsorption and desorption rate constants and C_a the adsorbate concentration in solution. At equilibrium:

$$R_{ads} = R_{des} \tag{21}$$

combination of equations (17)-(21) yields:

$$\frac{R_0}{R_0 - R_a} = \frac{1}{1-\beta} + \frac{1}{K(1-\beta)C_a} \tag{22}$$

where $K = k_{ads}/k_{des}$ is defined as the affinity constant and is used as a comparative measure for the relative influence of additives (both inorganic and organic) on the crystal growth of sparingly soluble salts. Plots of the rates measured in the absence and in the presence of adsorbates according to equation (22) are expected to be linear provided that the interaction between adsorbate - adsorbent may be described by a Langmuir isotherm. Equation similar to (22) may by obtained by assuming Freundlich type adsorption:[111]

$$\frac{R_0}{R_0 - R_i} = \frac{1}{K(1-\beta)C_a} \tag{23}$$

According to equation (22) plots of $\frac{R_0}{R_0 - R_i}$ as a function of $1/C_a$ are expected to be linear with an intercept of 1 (b = 0) corresponding to ideal Langmuir behavior, >1 implying that complete inhibition of the growth is not possible by the adsorbate or <1 for the cases in which complete inhibition is attained for adsorbate concentrations corresponding to less than monolayer coverage. A combination of the slope and the intercept obtained from the plots according to equation (22) yields the value of K for each adsorbate. Some values of the affinity constants for the influence of metal ions and of organic compounds with a capability of forming complexes with divalent metal ions, on the crystal growth of HAP are summarized in Table 3.

Table 3. Affinity constant values for the effect of additives on the crystal growth of HAP determined from the kinetics of crystal growth.

Additive	K	reference
Mg	2.5×10^4	73
Zn	1.6×10^6	71
Zn	3.0×10^7	42
Cd	1.4×10^7	113
Cd	1.7×10^7	112$^+$
Phytate	8.4×10^4	73
Ethane-l-Hydroxy -1,1-Diphosphonic Acid (EHDP)	20.8×10^5	73

$^+$ determined from adsorption isotherm

The values of K, although they lack thermodynamic significance they may be interpreted in a relative manner to conclude concerning the interaction of the adsorbate species with the mineral. It is interesting to note that Zn^{2+} and Cd^{2+} ions exhibit highest affinity. Since Cd^{2+} ions are known to enter the cadmium lattice it is possible that the high K values may signify a stronger interaction with the HAP surface (e.g., formation of a surface layer of cadmium substituted apatite).[43] The values for Mg^{2+} ions and the (organic) anions of phytate and EHDP are of similar order of magnitude. The organic anions showed higher affinity because of their capability to form polydentate anions. Mg^{2+} through adsorption is reported not to affect the formation of ACP while it retards the transformation of precursor into HAP by adsorption. The preferential adsorption at the faces affects the HAP crystallite morphology which become short and irregular.[73,114] The high value for the affinity constant of Zn^{2+} for HAP is of interest since there is evidence for surface adsorption of this ion[39,42] which may lead to reduction of the crystallinity of the HAP crystals.[115] It has been argued however that a continuous series of Zn-HAP solid solutions may be formed.[116-118] As a rule ions with a strong affinity for the solid surface act as crystal growth inhibitors.[119,120] It is believed that the interaction of the metal ions with the mineral surface of the various calcium phosphates is proportional to the stability of the corresponding surface complexes which may be calculated.[121] Moreover the knowledge of solution speciation is also very important for the analysis of the interaction of metal ions with the calcium phosphates. Thus, in the case of the inhibition of crystal growth of calcium phosphate dihydrate at pH 5-6 by Sn(II), the reduction in the rates measured was attributed in the presence of $Sn_3(OH)_4^{2+}$ ion pairs.[122] The free metal ion impurities or their ion-pairs formed in the aqueous supersaturated solutions affect crystal growth provided that they adsorb specifically onto the crystal / electrolyte interface. The specific adsorption is contrasted to simple electrostatic interaction and for ionic species is the result of the formation of surface complexes.[70,123] A most useful tool for the investigation of this type of interactions is the measurement of electrophoretic mobility of suspended particles in constant ionic strength inert electrolytes i.e., in electrolytes the ions of which do not exhibit specific adsorption on the solid investigated.

The presence of dissolved ionic impurities may also accelerate the formation of calcium phosphates in supersaturated solutions,[124] possibly through the initiation of the calcium phosphate precipitation by the formation of other impurity-phosphates, for which the solubility product is exceeded. Since in most cases of spontaneous precipitation, the process is initiated by the presence of heterogeneous impurity particles the driving force (change in Gibbs free energy) is lowered and nucleation of calcium phosphate and/or impurity phosphate may proceed simultaneously leading either to the formation of substituted calcium phosphates or to the coprecipitation of calcium and impurity phosphates.[125] In this case it is necessary to design experiments in which impurity phosphates or calcium phosphates cannot be formed in order to assess accurately the influence of the metal ions on the formation of calcium phosphates. It should be noted however at this point that the application of the constant composition or of the dual constant composition method is very helpful in separating the contributing effects of the various factors in the nucleation and crystal growth of calcium phosphates.

Finally, the presence of metal ions in the supersaturated calcium phosphate solutions may affect the aggregation of the nucleating crystallites or of the seed crystals introduced to initiate the crystal growth process. Irrespective of the nature of calcium phosphates formed in aqueous solutions the solids are ionic and they have a rather high surface charge which depends on the potential determining ions, which at first place are the ions constituting the lattice of the solids and indirectly the ions which through their equilibria in the aqueous phase may influence the concentration of the former type of ions (e.g., H^+ ions influence the relative concentration of $H_2PO_4^-$, HPO_4^{2-}, PO_4^{3-}).[70] Since the presence of charged metal ions in solutions may lead to the adsorption on the calcium phosphates surface it is anticipated that the surface potential will be influenced depending on the amount of impurity and on the extent of interaction with the solid. Reduction of the surface potential may lead to decrease of the repulsion between the calcium phosphate crystallites and thus aggregation may occur, which in most cases is irreversible and leads to the reduction of the total surface area available for crystallization. In such cases, in which agglomeration cannot be overlooked the problem should be treated by writing and solving the appropriate population balance equations.[126]

In conclusion, the role of metal ions present in the supersaturated solutions in which calcium phosphates are formed depend greatly on the interaction of the metal ion and its aquocomplexes with the solid phase forming, provided that the solutions are undersaturated with respect to the corresponding metal phosphates. Thus metal ions may through surface adsorption retard the crystal growth of certain phases or favor the formation of thermodynamically less stable phases inhibiting hydrolytic transformation to the more stable phases. In cases in which the ionic radii of the metal ions present in the solutions in which the solid calcium phosphate crystals are close to the calcium ion radius, incorporation may take place with a concomitant change of the unit cell lattice parameters of the growing crystals. Crystal growth experiments at constant supersaturation are excellent model experiments capable of distinguishing between the various mechanisms which may describe the interaction of foreign metal ions with calcium phosphates.

REFERENCES

1. da Silva JJRF, Williams RJP. *The Biological Chemistry of the Elements*, Clarendon Press-Oxford, Oxford 1994.

2. Phosphates in Detergents and the Eutrophication of America's Waters, In 23rd Report, Communications on Government Operations, US Printing Office, Washington DC.

3. Brown WE. Solubilities of phosphates and other sparingly soluble compounds, In *Environmental Phosphorus Handbook* Griffith EJ, Beeton A, Spencer JM, Mitchell DT (Eds.), Wiley and Sons, New York 1973.

4. Ferguson JF, Jenkins D, Eastman J. Calcium phosphate precipitation. In *44th Annual Conference of the Water Pollution Control Federation*, Ser. 5(2) Oct. 4 (1971).

5. Varsanik RG. The nature and control of calcium orthophosphate depositions in cooling water systems. Paper 142, Corrosion 75 NACE Toronto, Canada 1975.

6. *The NALCO Water Handbook*, Kemmer FN, Ed. Boiler water treatment 2nd Edition , MacGraw Hill, New York 1988.

7. BETZ Handbook of Industrial Water Conditioning, Boiler deposits: Occurence and control, 8th Edition, Betz lab Inc. Trevose, PA. 1982.

8. Hay RW. *Bioinorganic Chemistry*, Ellis Horwood Ltd, Chichester, 1987.

9. McDowell H, Gregory TM, Brown WE. Solubility of $Ca_5(PO_4)_3OH$ in the system $Ca(OH)_2$ -H_3PO_4-H_2O at 5, 15, 25 and 37 °C. J Res Nat Bur Stand 1977;81A:273-281

10. Gregory TM, Moreno EC, Patel JM, Brown WE, Solubility of $Ca_3(PO_4)_2$ in the system $Ca(OH)_2$-H_3PO4-H_2O at 5, 15, 25 and 37 °C. J Res Nat Bur Stand 1974;78A:667-674

11. Moreno EC, Brown WE, Osborn G. Stability of dicalcium phosphate dihydrate in aqueous solutions and solubility of octacalcium phosphate. Soil Sci 1960;24:99-102

12. McDowell H. Solubility of $CaHPO_4$ and ion pair formation Ph. D. Thesis, Howard University, Washington DC 1968.

13. Patel JM, Gregory TM, Brown WE. Solubility of $CaHPO_4.2H_2O$ in the quaternary system $Ca(OH)_2$-H_3PO_4-NaCl-H_2O at 25 °C. J Res Nat Bur Stand 1974;78A:675-681

14. Smith RM, Martell AE. *Critical Stability Constants v.4: Inorganic Complexes*, Plenum Press New York 1981.

15. Nancollas GH. *Interactions in Electrolyte Solutions*, Elsevier, Amsterdam, 1966.

16. Davies CW. *Ion Association*, Butterworth, London, 1962.

17. Lewis GN, Randall M. *Thermodynamics*, Revised by Pitzer JS and Brewer L, 2nd Ed. McGraw Hill, New York 1961.

18. Robinson RA, Stokes RH. *Electrolyte Solutions*, 2nd ed. Butterworth, London 1965.

19. Garrels RM, Tomson ME, A chemical model for seawater at 25 °C and 1 atm total pressure, Am J Science, 1962;260:57-66

20. Pitzer KS In *Thermodynamic modeling of geological materials: Mineral, fluids and melts.* Carmichael ISE, Engster HP eds. Reviews in Mineralogy, Miner Soc America, Washington DC, v. 17.

21. Wolery TJ, Jackson KJ. *Activity coefficients in aqueous salt solutions: hydration theory equations systems II*, Melchior DC, Basset RL Eds. ACS Washington DC 1990.

22. Ananthaswamy J, Atkinson G, Thermodynamics of concentrated electrolyte mixtures. S. A review of the thermodynamic properties of aqueous calcium chloride in temperature range 273.15-373.15, J Chem Eng Data, , 1985;30:120-128

23. Nancollas GH, Amjad Z, Koutsoukos P. Calcium phosphates-Speciation, solubility and kinetics considerations, In ACS Symposium Series No. 93, *Chemical Modeling in Aqueous Systems*, Jenne EA, Ed., ACS, Washington DC 1979.

24. Stumm W, Morgan JJ. *Aquatic Chemistry* 2nd Ed. J. Wiley, New York.

25. Morel FMM, *Principles of Aquatic Chemistry*, J. Wiley New York 1983.

26. Papelis C, Hayes KF, Leckie JO. HYDRAQL: A program for the computation of chemical equilibrium composition of aqueous batch systems including surface-complexation modeling of ion adsorption at the oxide/solution interface Teh. Rep. 306 Stanford Univ. Stanford CA 1988.

27. Nordstrom DK, Plummer LN, Wigley TML, Wolery TJ, Ball JW, Jenne EA, Basset RL, Crerar DA, Florence TM, Mattigod SV, McDuff RE, Morel F, Reddy MM, Sposito G, Thrailkill J. A comparison of computerized chemical models for equilibrium calculations in aqueous systems, In *Chemical Modeling in Aqueous Systems*, ACS Symposium Series No. 93 (E.A. Jenne Ed.), ACS Washington D.C. 1979.

28. Van Breeeman N. Calculation of ionic activities in natural waters Geochim. Cosmochim. Acta 1973;37:101-107

29. Mattigod SV, Sposito G. Chemical modeling of trace metal equilibria in contaminated soil solutions using the computer program GEOCHEM in Chemical Modelling in Aqueous Systems ACS Symposium Series No. 93 (EA Jenne ed.), ACS Washington D.C. 1979.

30. Pol T, Nancollas GH. EQUIL - A computational method for the calculation of solution equilibria, Anal Chem 1972;44:1940-1950

31. Eanes ED, Termine JD. Calcium in mineralized tissues, Met. Ions in *Biol. 6 Calcium in Biology*, TG Spiro, Ed., J. Wiley, 201-233 (1983).

32. Posner AS, Perloff A, Diorio AF. Refinement of hydroxyapatite structure, Acta Crystallogr 1958;11:308-309

33. Eanes ED, Posner AS. In *Biological Calcification - Cellular and Molecular Aspects*, H Schrarer, Ed. North Holland, Amsterdam 1970.

34. Narasaraju TSB, Phebe DE. Some Physico-chemical aspects of hydroxylapatite, J Mater Sci 1996;31:1-21

35. Klement R, Zureda F. Basic phosphates of bivalent metals. V. Phosphate and hydroxyapatite of cadmium, Z. Anorg Allgem Chem 1940;245:229-235

36. Termine JD, Lundy DR. Hydroxide and carbonate in rat bone mineral and its synthetic analogues, Calcif. Tissue Res 1973;13:73

37. Suzuki M. Physicochemical nature of hard tissue VIII. Determination of carbonate ions in the inorganic comporent of hard tissue by infrared reflectance spectroscopy, Hirosaki Iyaku 1973;24:450-456

38. Knappwost A. Fluoride-hydroxyl substitution in hydroxyapatite as an ion-exchange reaction and its use in the microanalysis of fluoride, Odontologisk Rev1957;8:30

39. Bigi A, Foresti E, Gandolfi M, Gazzano M, Roveri N. Inhibiting effect of zinc on hydroxylapatite crystallization, J Inorg Biochemistry 1995;58:49-58

40. Samachon J, Dennis H, Fowler R, Schmitz A. The reaction of ^{65}Zn with the surfaces of bone and bone mineral, Biochim Biophys Acta 1967;148:767-773

41. Samachon J, Schmitz A. Slow reactions of calcium and zinc ions with the surfaces of bone powder and inorganic bone, Biochim Biophys Acta 1969;192:231-237

42. Dalpi M, Karayanni E, Koutsoukos PG. Inhibition of hydroxyapatite formation in aqueous solutions by Zn and 1,2 Dihydroxy-1,2- bis (dihydroxyphosphonyl) ehtane, J Chem Soc Farad Trans 1993;89:965-969

43. Neuman WF, Bjornerstedt B, Mulryan BJ. Synthetic hydroxyapatite crystals II. Aging and strontium incorporation. Arch Biochem Biophys 1963;101:215-224

44. Bartley JC, Reber EF. Effect of roasting on radiostrontium in fresh ham, J Nutr 1961;75:21-26

45. Elliot JC. The problems of the composition and structure of the mineral components of the hard tissues. Clin Orthop 1973;93:313-345

46. Young RA. Some aspects of crystal structural modeling of biological apatites, Colloq Int 1973;CNRS No 230:21-44

47. Collin RL. Strontium-Calcium Hydroxyapatite Solid Solutions Precipitated from Basic Aqueous Solutions. J Amer Chem Soc 1960;82:5067-5069

48. Hayek E, Petter H. Hydrothermal synthesis of bivalent metal phosphate, Manatschh Chem 1960;91:356-360

49. Koutsoukos PG, Nancollas GH. Influence of strontium ion on the crystallization of hydroxyapatite from aqueous solutions. J Phys Chem 1981;85:2403-2408

50. Trombe JC, Montel G. Process for treating apatites applicable to other minerals,CR Acad Sci Ser C 1974;278:1227-1230

51. Fowler BO. Infrared studies of apatites. I. Vibrational assignements for calcium, strontium and barium hydroxylapatites utilizing isotopic substitution. Inorg Chem 1974;13:194-207

52. Urusov VS, Khudolozhkin VO. Energy analysis of cation ordering in the structure of apatite, Geokhimiya 1974:1509-1515

53. Ehrent A. Ph.D. Thesis, Tübigen University, Germany 1962.

54. Roushdy HM, Moloukhia MK, Abdel Fattah AT. Effect of dietary calcium level on the rate of deposition of radiostrontium in rat bones. Isot Rad Res 1981;13:19 -26

55. Zawacki SJ, Koutsoukos PG, Salimi MH, Nancollas GH. In *Geochemical Processes at mineral surfaces*, JA Davis, KF Hayes Eds. ACS Special Publication Series No. 323, ACS Washington DC 1976.

56. Duff EJ. Interactions of divalent cations with apatites Coll Int CNRS No 203, 1973:419-421

57. Müller M. The precipitation and the x-ray diffraction investigation of the mixed crystal system $Ca_{10}(PO_4)_6(OH)_2$-$Pb_{10}(PO_4)_6$-$(OH)_2$, Helv Chim Acta 1947;30:2069-2080

58. Narasaraju TSB, Singh RP, Rao VLN. New method of preparation of solid solutions of calcium and lead hydroxyapatites. J Inorg Nucl Chem 1972;34:2072-2074

59. Eanes ED, Posner AS. Alkaline earth intermediate phases in the basis solution preparation of phosphates. Calcif Tissue Res 1968;2:38-48

60. Boskey AL, Posner AS. Magnesium stabilization of amorphous calcium phosphate Mater Res Bull 1974;9:907-916

61. LeGeros RZ, Shirra WP, Miravite MA, LeGeros JP, Domingo GM, Quirolgico G. In *Physico-Chimie et cristallographie des apatites d' interet biologique.* Colloq Intern CNRS No 230 CNRS, Paris 1975.

62. Bachra BN, Fisher HRS. Effect of some inhibitors on the nucleation and crystal growth of apatite. Calcif Tissue Res 1969;3:348-357

63. Cuegan C. Contribution à l' étude cinetique de l' evolution de l' état amorphe a l' état apatitique des orthophosphate trimetalliques (Ca, Mg) Precipites, Thèse, Institut National Polytechnique de Toulouse, Toulouse, France, 1978

64. Montel G, Bonel G, Heughebaert JC, Trombe JC, Rey C. New Concepts in the composition, crystallization and growth of the mineral component of calcified tissues. J Crystal Growth 1981;53:74-99

65. Kibalczyc W, Christoffersen J, Christoffersen MR, Zielenkiewicz A, Zielenkiewicz W. The effect of magnesium ions on the precipitation of calcium phosphates. J Crystal Growth 1990;106:355-366

66. Salimi HM, Heughebaert JC, Nancollas GH. Crystsal growth of calcium phosphates in the presence of magnesium ions. Langmuir 1985;1:119-122

67. Blake RC, Shute EA, Howard T. Solubilization of minerals by bacteria: electrophoretic study of thiobacillus ferroxidans in the presence of iron, pyrite and sulphur. Appl Environ Microbiol 1994;60:3349-3357

68. Krekeler C, Ziehr H, Klein J. Influence of physicochemical bacterial surface properties on adsorption to inorganic porous supports. J Appl Microbiol Biotechnol 1991;35a:484-490

69. Van der Mel HC, Leonard AJ, Weerkamp AH, Rouxhet PG, Busscher HJ. Properties of oral streptococci relevant to adherence: zeta potential, surface free energy and elemental composition. Colloids & Surfaces 1988;32:297-305

70. Lyklema J. *Fundamentals of Interface and Colloid Science*, Vol. II, Academic Press London 1995.

71. Fuierer TA, LoRe M, Puckett SA, Nancollas GH. A mineralization adsorption and mobility study of hydroxyapapite surfaces in the presence of zinc and magnesium ions. Langmuir 1994;10:4721-4725

72. Neuman WF, Toribara TY, Mulryan BJ. Synthetic hydroxyapatite crystals. I. Sodium and potassium fixation. Arch Biochem Biophys 1962;98:384-390

73. Koutsoukos PG. The Precipitation of Hydroxyapatite in Aqueous Solutions at 37 °C, Ph. D. Thesis, SUNYAB, Buffalo NY 1980

74. Koutsoukos PG, Nancollas GH. The effect of lithium on the precipitation of hydroxyapatite from aqueous solutions. Colloids & Surfaces 1986;17:361-370

75. Donnely R, Boskey A. The effect of gallium on seeded HAP growth. Calcif. Tissue Int 1989;44:138-142

76. Christoffersen MR, Christoffersen J., The effect of aluminum on the rate of dissolution of calcium hydroxyapatite: A contribution to the understanding of aluminum-induced disease. Calcif Tissue Int 1985;37:673-676

77. Kresak M, Moreno EC, Zahradnik RT, Hay DI. Adsorption of aminoacids onto hydroxyapatite. J Colloid Interface Sci 1977;59:283-292

78. Adamson AW. *Physical Chemistry of Surfaces*, 5th edition, J. Wiley, New York 1990.

79. Xyla AC, Mikroyannidis J, Koutsoukos PG. The inhibition of calcium carbonate precipitation in aqueous media by organophosphorus compounds. J Colloid Interface Sci 1992;153:537-551

80. Sposito G. *The surface chemistry of soils*, Oxford University Press, N.Y. 1984.

81. Sposito G. Distinguishing adsorption from surface precipitation, In *Geochemical processes at mineral surfaces*, JA Davis, KF Hyaes Eds, ACS Symposium Series No 323, ACS Washington DC 1986,

82. Montel G, Bonel G, Heughebaert JC, Trombe JC, Rey C. New concepts in the composition, crystallization and growth of the mineral component of calcified tissues. J Crystal Growth 1981;53:74-99

83. Nancollas GH, Sawada K, Schuttringer E. Mineralization reaction involving calcium carbonates and phosphates, In *Biomineralization and Biological Metal Accumulation*, P Wesbroek, EW de Jong Eds. D. Dreidel Publi. Co. Dortrechf 1983.

84. Nancollas GH. In vivtro studies of calcium phosphate crystallization, In *Biomineralization*, S Mann, J Webb, RJP Williams. Eds., VCH Weinheim 1989.

85. Sangwal K. Kinetics and mechanism of crystal growth In *Elementary Crystal Growth*, K Sangwal Ed. Saan Publ. Lublin, 1994.

86. Garside J. in *Measurement of Crystal Growth Rates*. J Garside, A Mersmann, J Nyvlt Eds. Europena Federation of Chemical Engineering Working Party on Crystallization, Münich 1990.

87. Mullin JW. *Crystallization*, 3rd Ed. Butterworth-Heinemann, Oxford 1993.

88. Sõhnel O, Garside J. *Precipitation*, Butterworth Heinemann, Oxford, 1992.

89. Lundager Madsen HE. Calcium phosphate crystallization. III. Overall growth kinetics of tetracalcium monohydrogen phosphate. Acta Chem Scand A 1982;36:239-249

90. Nielsen AE, Christoffersen JC. The mechanism of crystal growth and dissolution. Life Sci. Res Rep 1982;23:37-77

91. Nielsen AE. Transport control in crystal growth from solution. Croat Chem Acta. 1987;60:531

92. O' Hara M, Reid RC. *Modeling Crystal Growth Rates from Solution*, Prentice Hall, New York 1973.

93. Walton AG, Bodin WG, Füredi-Milhofer H, Schwartz A., Nucleation of calcium phosphates from solution, Can. J. Chem 1967;45:2695-2701

94. Brecevic L, Füredi Milhofer H. Precipitation of calcium phosphate from electrolyte solutions. II. The formation and transformation of precipitates Calcif. Tissue Res 1972;10:82-90

95. Madsen HEL, Thorvardarson G. Precipitation of calcium phosphate from moderately acid solution. J Crystal Growth. 1984;66:369-376

96. Feenstra TP, DeBruyn PL. Formation of calcium phosphate in moderately supersaturated solutions. J Phys Chem 1979;83:475-479

97. van Kemenade MJJM, deBruyn PL. A kinetic study of the precipitation from supersaturated calcium phosphate solutions. J Coll Int Sci 1987;118:564-585

98. Boistelle R, Lopez Valero I. Nucleation and growth of calcium and magnesium phosphates, In *Geochemistry and Mineral Formation in the Earth Surface*, R Rodriguez-Clemente, Y Tardy Eds. CSIC-CNRS, Madrid 1987.

99. Meyer JL, Eanes ED. A thermodynamic analysis of the amorphous crystalline calcium phosphate transformation. Calcified Tissue Res 1978;25:5-68

100. Füredi-Milhofer H, Brecevic L, Purgaric B. Crystal growth and phase transformation in the precipitation of calcium phosphates, Faraday Discussions, Chem Soc 1976;61:184-193

101. Li P. In vitro and in vivo calcium phosphate induction on gel oxides, Ph.D. Thesis, RijksUniversiteit Leiden 1993.

102. Blumenthal NC. Mechanisms of inhibition of calcification. Clinical Orthopaedics and Related Research 1989;247:279-289

103. Moreno EC, Zahradnik RT, Glazman A, Hwu R. Precipitation of hydroxyapatite from dilute solutions upon seeding. Calcif Tissue Res 1977;24:47-

104. Nancollas GH, Mohan MS., The growth of hydroxyapatite crystals. Archs Oral Biol 1970;15:731-745

105. Nancollas GH. Phase transformation during precipitation of calcium salts, In *Biological Mineralization and Demineralization*. GH Nancollas Ed., Dahlem Konferenzen 1982, Berlin, Heidelberg New York, Springer-Verlag.

106. Tomson MB, Nancollas GH. Mineralization Kinetics: A constant composition approach. Science 1978;200:1059-1060

107. Koutsoukos PG, Amjad Z, Tomson MB, Nancollas GH. Crystallization of calcium phosphates. A constant composition study. J Am Chem Soc 1980;102:1553-1557

108. Zhang J, Ebrahimpour A, Nancollas GH. Dual constant composition studies of phase tranformation of dicalcium phosphate dihydrate to octacalcium phosphate. J Colloid Interface Sci 1992;152:132-140

109. Nancollas GH, Zawacki SJ. Inhibitors of crystallization and dissolution in *Industrial Crystallization 84*, SJ Jancic, EJ de Jong Eds. Elsevier, Amsterdam 1984.

110. Nancollas GH, Purdie N. The kinetics of crystal growth. Quart Rev Chem Soc 1964;18:1-20

111. Sabbides TG. Investigation of the calcium carbonate / water interface. Ph. D. Thesis, University of Patras, Patras, Greece 1994.

112. Christoffersen J, Christoffersen MR, Larsen R, Rostrup E, Tingsgaard P, Andersen O, Grandjean P. Interaction of cadmium ions with calcium hydroxylapatite crystals: A possible mechanism contributing to the pathogenesis of cadmium-induced bone diseases. Calcif Tissue Int 1988;42:331-339

113. Dalas E, Koutsoukos PG. Crytallization for hydroxyapatite from aqueous solutions in the presence of cadmium. J Chem Soc Faraday Trans I 1989;85:3159-3164

114. Abbona F, Baronnet A. An XRD and TEM study on the transformation of amorphous calcium phosphate in the presence of magnesium. J Crystal Growth 1996;165:98-105

115. Le Geros RZ, Taheri MM, Quirologico GB, Le Geros JP., Proceedings 2nd Int. Phosphorus, Boston, IMPHOS, Paris 1989.

116. Chiranjeevi Rao SV., A new method of preparation of solid solutions of calcium-zinc hydroxylapatites. J Indian Chem Soc 1976;LII:347-348

117. Pujari M, Patel PN. Strontium-copper-calcium hydroxylapatite solid solutions preparation, IR and lattice constant measurements. J Solid State Chem 1989;83:100-104

118. Panda A, Patel PN, Tripathy NK. Solid solutions of calcium-copper-zinc hydraxylapatiets: preparation and IR and lattice constant measurements. Ind J Chem A 1990;29A:70-72

119. Gilman JJ, Johnston WG, Sears GW. Dislocation etch pit formation in lithium fluoride. J Appl Phys 1958;29:747-755

120. Witkamp GJ. Crystallization of calcium sulfate and uptake of impurities. Ph. D. Thesis, T.V. Delf, The Netherlands, 1989.

121. Veintemillas-Verdaguer S. Chemical aspects of the effect of impurities in crystal growth. Prog Crystal Growth and Charact 1996;32:75-109

122. Nancollas GH, Wefel JS. The effect of stannous fluoride, sodium fluoride and stannous chloride on the crystallization of dicalcium phosphate dihydrate at constant pH. J Crystal Growth 1974;23:169-176

123. Lyklema J. Electrified interfaces in aqueous dispersion of solids. Pure and Appl Chem 1991;63:895-918

124. Melikhov IV, Lazic S, Vukovic Z. The effect of dissolved impurity on calcium phosphate nucleation in supersaturated medium. J Colloid Interface Sci 1989;127:317-326

125. Lazic S. pH effect on the coprecipitation of radioactive strontium with calcium phosphate. Radiochim Acta 1993;62:95-98

126. Berglund KA. Analysis and measurement of crystallization utilizing the population balance, In *Handbook of Industrial Crystallization.* AS Myerson, Ed., Butterworth-Heinemann, Boston 1993

8

STRUCTURE/PERFORMANCE RELATIONSHIPS OF PHOSPHOROUS AND CARBOXYL CONTAINING ADDITIVES AS CALCIUM PHOSPHATE CRYSTAL GROWTH INHIBITORS

John D. Sallis

Division of Biochemistry, University of Tasmania
Hobart, Tasmania, AUSTRALIA.

GENERAL COMMENTS

The identification of compounds suitable for controlling the development and growth of calcium phosphate crystals has been a research goal for many years. From a biological, medical and industrial standpoint, this area has attracted attention because of the important consequences that can develop if calcium phosphate crystallization proceeds in an uncontrolled manner. Common examples can be seen in industrial scaling and corrosion, dental calculus, calcified plaques of atherosclerosis, kidney calcinosis and to a lesser extent, an involvement in urinary tract stone diseases. Successful elimination of these problems has not yet been achieved but nevertheless, significant progress toward controlling events associated with calcium phosphate crystallization has been realized.

Numerous biological and synthetic molecules now have been identified which are capable of interfering with calcium phosphate development and growth as indicated in this report. In nature, various types of calcium phosphates exist ranging from a pure amorphous calcium phosphate state (ACP) to the hard needle-like crystalline form of hydroxyapatite (HAP, $Ca_{10}(PO_4)_6(OH)_2$) as found in bones and teeth.[1] Basic calcium phosphate is another form which includes a carbonate-substitute apatite. HAP can crystallize and grow in solutions where supersaturation is low or from the prior formation of kinetically favoured intermediates such as dicalcium phosphate dihydrate (DCPD, $CaHPO_4.2H_2O$), octacalcium phosphate (OCP, $Ca_4H(PO_4)_3.2.5H_2O$) and tricalcium phosphate (TCP, $Ca_3(PO_4)_2$) when the supersaturation is high. Opportunities are therefore presented at various stages to slow down or even eliminate the final development of HAP crystals and their subsequent growth through the presence of inhibitor compounds.

There are numerous and diverse attributes that determine a compound's potency as an inhibitor. Fundamentally, a compound's inhibitory activity to block the formation and crystallization of calcium phosphates relates to its degree of adsorption to the crystal surface, its conformation at that surface, the strength of its binding and the influence of possible sequestration of ion species. Anionic compounds, whether natural or synthetic are proven useful inhibitors of calcium phosphate crystallization because they conform or can be structured to conform to these basic characteristics.

In this review of structure/function relationships of inhibitor compounds, particular emphasis has been directed toward groups of compounds possessing only phosphate, only carboxylic acid moieties or, more importantly, when both moieties are present in the one compound. Structurally more complex inhibitors with additional functional groups (e.g., phosphoproteins) which are widely distributed in nature can also exert important and strong regulation over biological mineralization. They are noted briefly, but because of their multivaried roles and more complex modes of action, detailed discussion of this latter class of inhibitors has not been undertaken in the present review.

The widespread natural occurrence of calcium phosphates reflects their necessity to realize diverse biological roles. It is important then to recognize that not all compounds displaying potent inhibitory activity can be universally applied to correct situations of unwanted crystallization and/or unfavorable deposition. Comments have therefore been included where appropriate on any advantages or possible disadvantages in terms of environmental or metabolic tissue damage that might arise from an indiscriminant inhibitor application.

ASSESSMENT OF INHIBITORY ACTIVITY

No single technique has gained universal acceptance for standardizing assessment of a compound's inhibitory power because such compounds are expected to perform under many complex environmental conditions. The diversity of assessment techniques hinders any direct comparison of the activities of compounds, limiting delineation and a fuller understanding of mechanisms and restricting the fine tuning needed for the development of future ideal compounds. Fleisch[2] has commented on some of the assays commonly used and of these, perhaps the most favoured techniques are the following:

A. The pH-Stat assay[3] employing simple solution chemistry. Basically, from equimolar stock $CaCl_2$ and $NaHPO_4$ solutions, ACP can be instantly generated and then, over time, allowed to transform into the solid phase HAP form at a neutral pH. The time at which the sharp phase transition occurs is referred to as the "induction time" or "I_t". Measurement of this parameter can be accurately made by:

- determining calcium depletion from the medium
- laser beam nephelometry signalling changing solution turbidity or
- more commonly, by alkaline addition under the control of an automated pH-titrator to back titrate the acidity induced by expelled protons during the reaction. An inhibitor will lengthen the induction time.

B. The constant composition assay[4] is particularly valuable in following the kinetics of growth. Supersaturation is ensured at all times by continuous monitoring of ion species so that concentrations of precipitating ions can be maintained through regulated infusion at the same rate as they are being used. Spontaneous and seeded growth reactions can be more closely controlled in such a system.[5]

C. Measurement of zeta potentials and particle sizing[6,7] are also useful techniques to assist in evaluating the potential of a compound to inhibit nucleation or retard crystal growth. Sizing of crystals also provides an insight into aggregation and dissolution of crystals.

Through these types of *in vitro* experimental approaches, coupled with biological and *in situ* observations, many of the following inhibitor compounds have been identified.

NON-CARBOXYLATED PHOSPHATE COMPOUNDS

It is not surprising that phosphated compounds have enjoyed considerable appeal as inhibitory agents. A few representative compounds and their structures are shown in Figure 1.

Trisodium phosphate was trialed as a boiler antiscaling agent[8] as early as 1927 and this approach was extended in later years with the introduction of the more effective "glassy" phosphates i.e. polyphosphates.[9] Long-term use of polyphosphates however can lead to the greater risk of there reversion to an orthophosphate, thus inducing an increase in sludge as more calcium phosphate precipitates. In 1962, Fleisch and Bisaz[10] recognized that the simpler pyrophosphate (P-O-P) molecule also was an effective biological controller of HAP development and its action could be attributed to preventing the transformation of amorphous calcium phosphate into the crystalline HAP.[11] Being prone, *in vivo*, to rapid degradation by ubiquitous pyrophosphatases, however, can diminish its usefulness as an inhibitor but the development of pyrophosphate analogues in the form of phosphonates which are more biologically stable because of the P-C-P linkage, has continued to create interest. Both polyphosphates and polyphosphonates exert greater inhibitory activity than most phosphated compounds, binding tightly to HAP to inhibit crystal formation, aggregation and dissolution.[12,13] Aside from their useful application to prevent scaling, the bisphosphonate class of compounds has acquired importance for medical therapies in a number of bone disorders.[14,15] However, their use is not without problems because of the inherent non-biodegradable nature arising from the resilience of the core linkage of P-C-P.

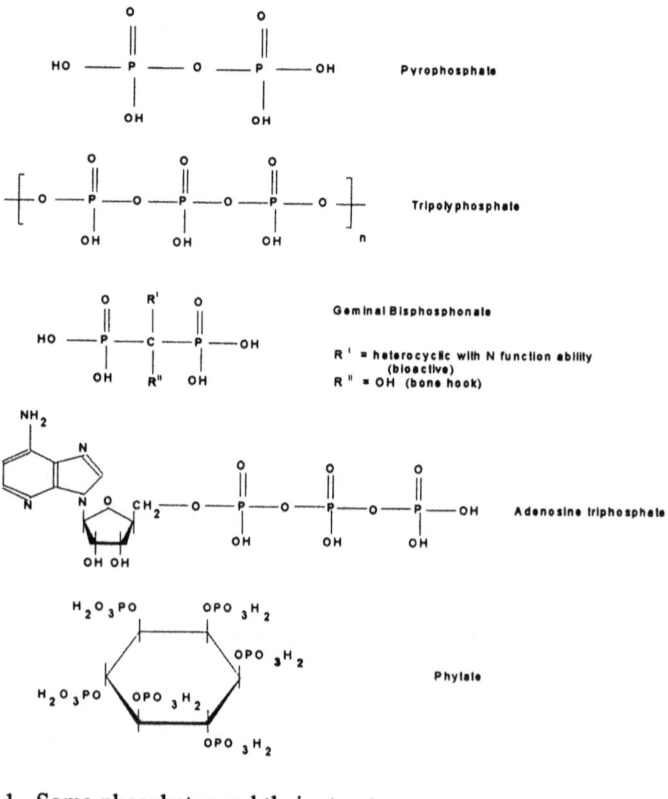

Figure 1. Some phosphates and their structures.

For a living environment where it may sometimes be necessary to ensure that the inhibitor compound will eventually be degraded, alternate classes of compounds which can be useful are the organophosphate esters (C-O-P linkage). Sugar and nucleotide phosphates are examples although generally it has been found that such compounds provide only weak inhibitory activity and because they are prone to metabolism, effective concentrations are harder to maintain.[16] Increasing the chain length of phosphate groups does improve inhibitory activity; polyphosphate > pyrophosphate; ATP (adenosine triphosphate) > ADP (adenosine diphosphate) and fructose-1,6-bisphosphate > fructose-6-phosphate. However, in comparing the former two series, it should be noted that the sterically bulky moiety of adenosine does exert considerable influence by way of lessening binding so that ATP << pyrophosphate or tripolyphosphate. The monophosphates and amino methylene trisphosphonic acid (AMP) are considered to have negligible inhibitory activity.

Using the constant composition technique, Amjad[17,18] also has compared the inhibitory influence of a range of polyphosphates which included hexametaphosphate, phytate (inositol hexametaphosphate) and pyrophosphate. For HAP, the inhibitory order was hexametaphosphate > phytate > pyrophosphate but for DCPD, phytate appeared superior to the other two compounds. The differing order of effectiveness is suggested to possibly reflect the differing rates of adsorption/desorption of the inhibitors. For linear chain type inhibitors, a plateau is reached beyond which there is no significant improvement in inhibition. This is not surprising with linear compounds as binding to active growth centres needs to be maintained so an irregular arrangement of active groups would be more likely to promote a closer packing of active groups. These considerations are, of course, but one aspect of the complex mechanisms that can operate.

In living organelles it is known that some compounds (e.g., nucleotides) will inhibit HAP embryo development but this can be reversed when the compound is removed by surface catalyzed hydrolysis. Further, the binding site available and the presence of metal ions will influence whether a mechanistic advantage has been gained by a chelation effect such as the removal of a calcium or iron species. An important characteristic which cannot be overlooked in any discussion of HAP inhibition is that it does present multiple binding site classes on its surface. Thus, the HAP crystal can provide access for many different types of compounds including both polyanions and polycations.[19] This means also, of course, that the number of sites offered to a compound can be quite variable as exemplified by the studies of Moreno et al.[20] and discussed later.

Based on the observed inhibitory activity of pyrophosphate, (1-hydroxyethylidene)-1,1-bisphosphonate (HEDP) and imidobisphosphate (PNP) Robertson and Fleisch[21] generalized that compounds possessing the basic structure H_2PO_4 - X - H_2PO_4 (where X = O, N or C) should be obvious choices as inhibitors and this has indeed proven to be true. In the development of overall potency however, further consideration then needs to be given to the influence that any group(s) other than phosphate present on the molecule might have on the compound's HAP binding.

For example, the hydroxyl group present in HEDP certainly augments the power of the compound. More recently, this feature has been exploited for other purposes. In the new generation bisphosphonates, the hydroxyl group associated with R II (see Figure 1) enables strong binding of HAP to bone while a heterocyclic with N-functionality provides control of the compounds bioactivity, in this case, its antiresorptive capacity.[22,23]

With new compounds continually being trialed, it is not surprising that some of the basic concepts on what might make a compound inhibitory, therefore, have undergone further refinement over time. Data related to organophosphate inhibition of HAP generated by Termine and Conn,[16] for example, highlighted that a minimum requirement of 2 phosphate esters does not necessarily imply a close grouping relationship (e.g., fructose-1,6-bisphosphate has inhibitory characteristics). Williams and Sallis,[24] in later studies with a range of phosphate and carboxyl compounds, suggested that a compound which possesses one phosphate together with a second acidic group (not necessarily a phosphate but perhaps a carboxyl) would serve as an even better inhibitor and such compounds are discussed in a later section.

Another important determinant of inhibitory power is, of course, the environmental pH under which the inhibitor is operating and so knowledge of this parameter can assist in predicting a compound's inhibitory usefulness over an extended pH range. The presence of dissolved ions near the crystal's surface also can't be overlooked. Pyrophosphate in its interaction with HAP or brushite adsorbs to the crystal surface and restricts growth despite an overall negativity in charge at neutral or alkaline pH. However, following chemisorption, a further interaction whereby the pyrophosphate can initiate a reaction with dissolved calcium will also influence inhibitory activity. Calcium pyrophosphate forms, precipitates on the surface which in turn leads to other interactions to regulate absorption and ultimately growth.[7]

The concentration of inhibitor in the immediate microenvironment is obviously an influencing factor but the crystal itself dictates its own response. Complete surface coverage with inhibitor is not necessarily a goal to achieve maximum effectiveness. If docking to growth sites can be controlled through the conformation of the inhibitor, then only a small fraction of the crystal surface needs inhibitor to interact, any excess of the substance being absorbed to non-growth sites.[19] In other situations, a complete blanketing of the crystal surface can be a requirement.[25]

NON-PHOSPHATED CARBOXYLATED COMPOUNDS

Carboxylated compounds through ionization of the carboxyl group(s) also can exert varying degrees of inhibition depending upon surface binding affinity and overall group charge. The structures of some carboxylic acids which have been investigated as possible inhibitors are shown in Figure 2.

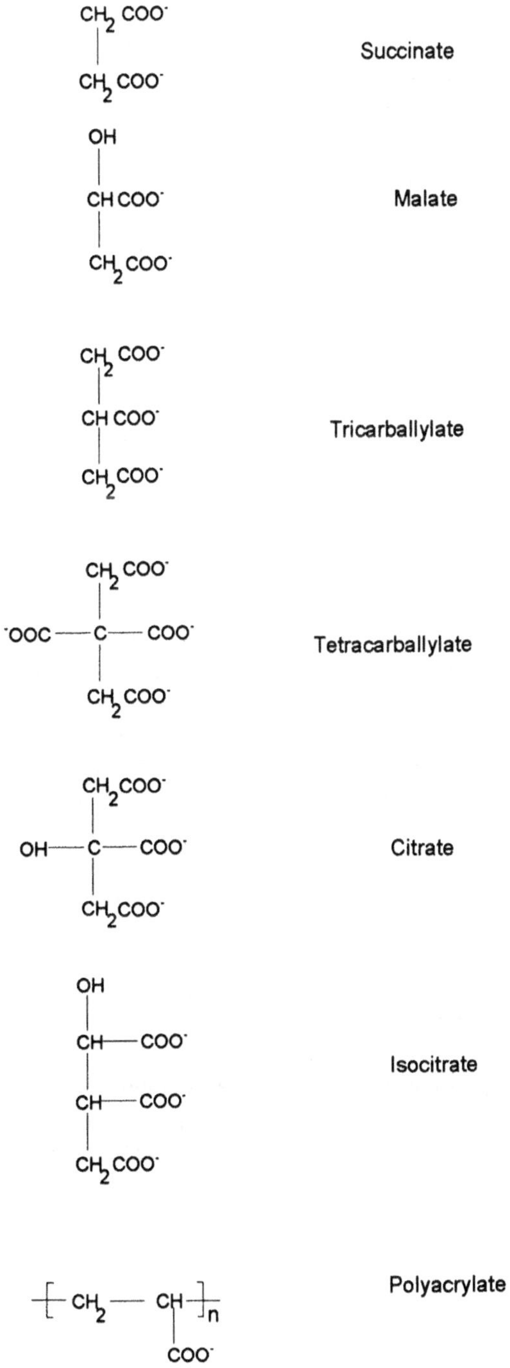

Figure 2. Structures of some representative carboxylates.

They include common di- and tri-carboxylates through to more extensively carboxylated compounds such as tetracarballylate (propane-1,2,2,3-tetracarboxylic acid), benzene hexacarboxylate, acrylic-maleate copolymers and polyacrylics. The latter named polycarboxylates are more effective inhibitors for DCPD and in preventing growth of hydroxyapatite than the simpler, lesser charged molecules.[26] Malate and succinate in fact show virtually no inhibition of HAP formation while citrate displays weak activity, no doubt a function of its calcium chelation ability. Aside from this aspect, these molecules also are essential metabolites for life cycles and so are rapidly metabolised.

In respect to citrate, it is interesting that tricarballylate and aconitate, dehydroxyl- and dehydro- derivatives of citrate respectively, are totally devoid of inhibitory activity which serves to emphasise the importance of the central hydroxy group in citrate.[27,28] Replacing the hydroxyl with a carboxyl to form the new compound, tetracarballylate offers superior inhibitory power. In studies with a range of α-hydroxyacids and dicarboxylates,[26] Amjad has reported that there is a definite correlation between their respective Calcium-carboxylate complex constants (K_{ML}) and their power as inhibitors of calcium phosphate dihydrate crystal growth.

In respect to longer chain macromolecules, inhibitor efficiency can be dependent upon the molecular weight of the compound. Hydroxyapatite crystal growth inhibition studies by Howie-Myers et al.[29] have demonstrated that low molecular weight substances, such as acrylate-maleate copolymers in the range 1000-2000 daltons, have a strong capacity to adsorb rapidly with their inhibitory power being greater and more efficient if a dissolved calcium salt is present. However, the higher molecular weight polyacrylates seem to perform better in the narrow range of 4000-6000 daltons when stronger binding occurs with calcium ion sequestration being an additional important aspect of their antiprecipitation action.

COMPOUNDS CONTAINING BOTH A PHOSPHATE AND CARBOXYLATE

A number of phosphorylated, carboxylated compounds have been investigated for their ability to inhibit calcium phosphate crystallization. Figure 3 shows the structures of some of the compounds which have been evaluated.

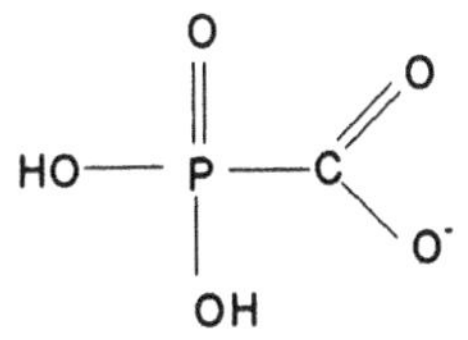

Phosphonoformate

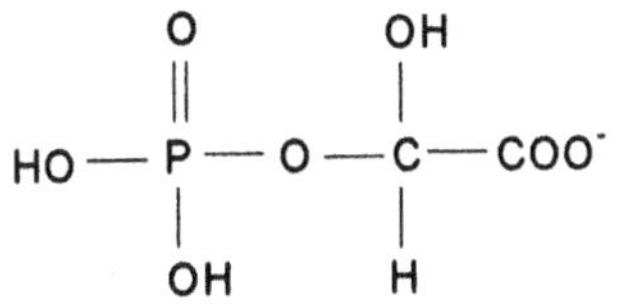

Hydroxy phosphonoacetate

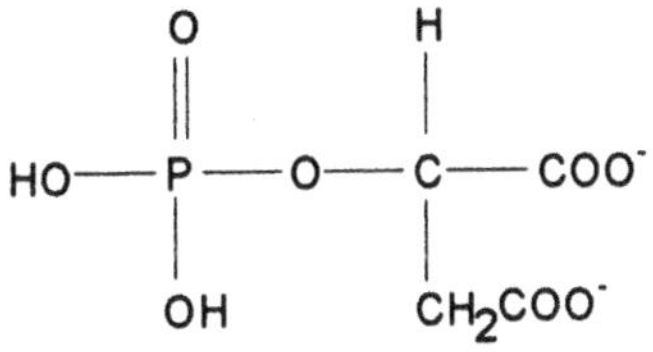

Phosphomalate

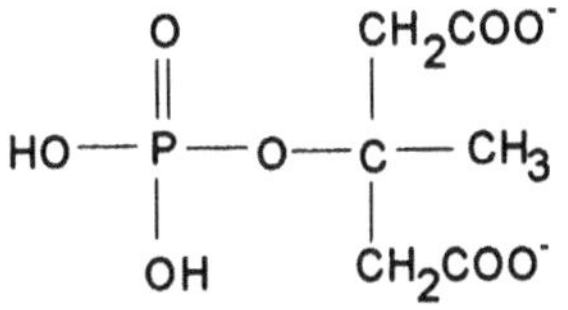

Phosphomethylglutarate

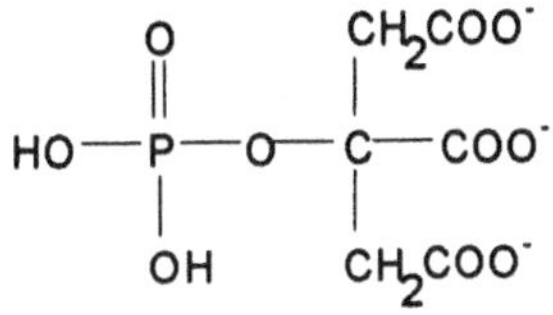

Phosphocitrate

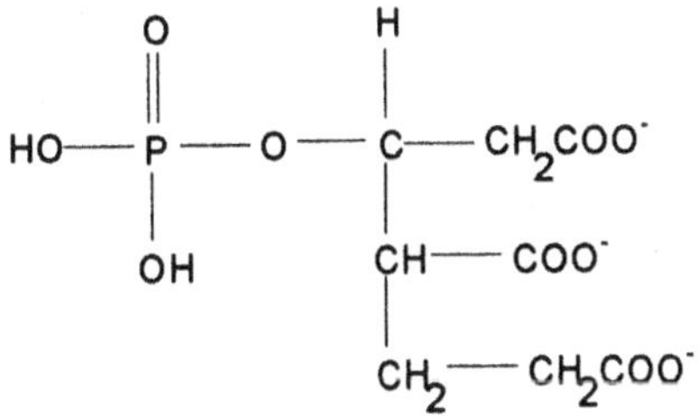

Phosphoisocitrate

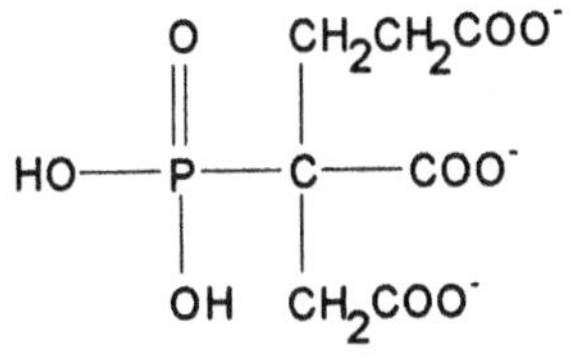

2-Phosphono butane 1,2,4, tricarboxylate

Figure 3. Phosphorylated carboxylic compounds.

These include compounds such as phosphonoformate, phosphonoacetate, hydroxyphosphonoacetate (HPA), phosphomalate, phosphomethylglutarate, phosphocitrate (PC) and 2-phosphonobutane-1,2,4-tricarboxylic acid (PTBC). It is well established that the presence of both anionic moieties in a compound's structure markedly increases inhibitory activity over that displayed by those compounds mentioned earlier as only having one of the moieties present. Studies by Williams and Sallis[24] using the pH-Stat technique compared, on a molar basis, the inhibitory activity of PC to that of citrate. The replacement of the central hydroxy group of citrate with a phosphate clearly had a profound effect in generating greater inhibitory activity, supporting the concept that a compound possessing a phosphate and additionally, either a carboxylic acid or phosphate moiety is favoured to exert strong inhibition. The actual ranking of the inhibitors of course is then dependent upon factors raised earlier and in the case of some of the Figure 3 compounds tested, the inhibitory activity grading is PC >> PIC > PM > PMG > PF. Additionally, Amjad[26] in constant composition studies with DCPD has reported that PBTC is slightly more active than HPA but comments further, that both are weak inhibitors.

MECHANISTIC INSIGHTS USING PHOSPHOCITRATE

As a representative of a low molecular weight phosphorylated carboxylic acid, more detailed mechanistic studies with PC have provided useful insights into the type of structure needed for inhibiting the formation and growth of calcium phosphates, keeping in mind the basic requirements outlined earlier i.e., a confirmation allowing surface interaction and a strong binding affinity with the HAP crystal. A comparison of its characteristics to that of either non-phosphated carboxylic compounds or non-carboxylated phosphate compounds reveals probable reasons for its more powerful action.

Figure 4 a, b, c, and d compare the pH dissociation curves for phosphate, pyrophosphate, citrate, and PC, respectively. At pH 7.4, citrate has three carboxyl groups fully ionized but replacement of the citrate hydroxyl group with a phosphate moiety now enables the new compound to contribute a greater negative charge density. For example, at pH 7.4, the second PO_4 hydroxyl group is contributing 50% of its ionization. Although of obvious importance to permit stronger crystal interaction, increased negativity alone is not solely responsible. A comparison of the pyrophosphate and PC dissociation curves shows similar characteristics, yet pyrophosphate is not nearly as inhibitory as PC. One possible explanation is that, at pH 7.4, a weaker surface fixing of the OH might be responsible for interaction with water on the outer surface of the crystal rather than internally.

Regardless of this possibility, the observation does highlight that PC's stereochemistry is critical to its action and additional studies with other propane carboxylate compounds supports this contention.[28] Repositioning of the phosphate moiety as in phosphoisocitrate leads to a lessening in inhibitory activity. Similarly, replacement of the β-COOH with a methyl group

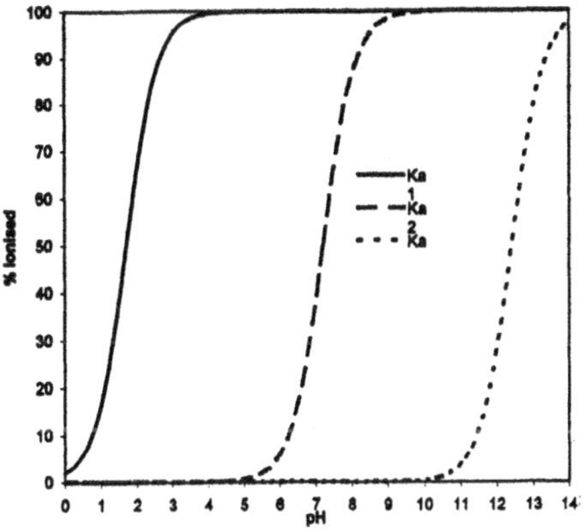

A o-phosphoric

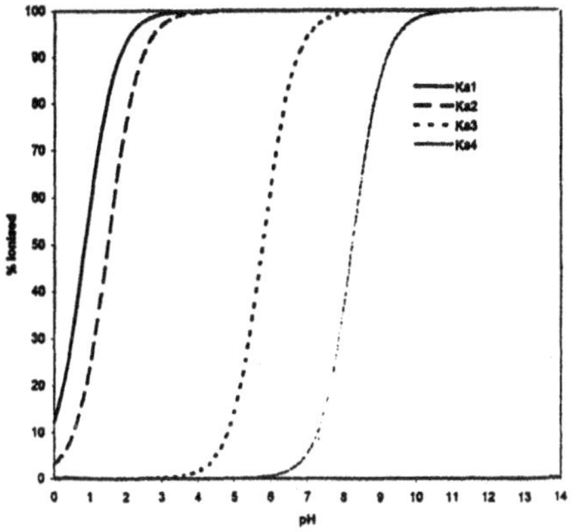

B pyrophosphoric

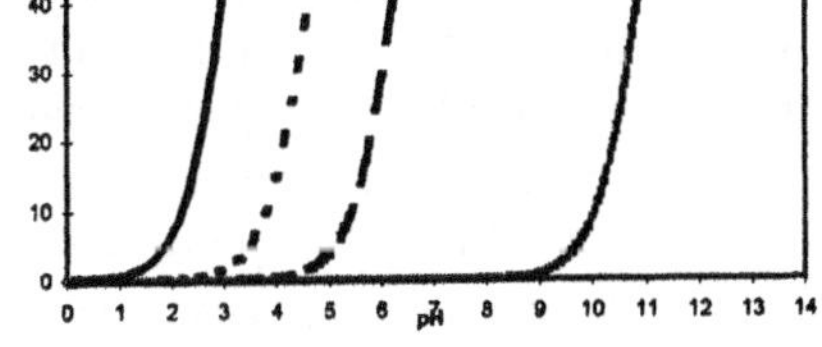

C citric

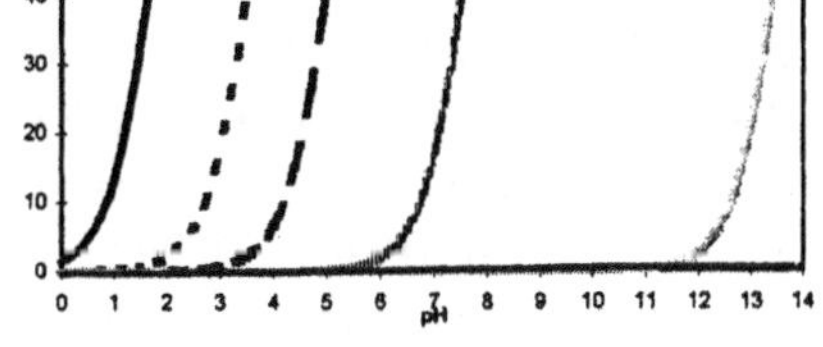

D phosphocitric

Figure 4. Comparison of dissociation curves for some selected inhibitors.[a]

(a) pKa values were obtained from the following source material:

Lide, DR. ed. (1992) Handbook of Chemistry and Physics. 8: 39-41 CRC Press, Boca Raton. FL. (graphs A - C).

Ward, LC; Shankar, R; Sallis, JD. Atherosclerosis 1987; 65: 117-124 (graph D).

has been observed to weaken activity as does the inclusion of additional side chaincarbons to position the α-COOH's further from the central carbon. Carboxylic acids with less than 3 carboxyl groups are known to be poor inhibitors when compared to the corresponding phosphorylated compounds on a molar basis. Tetracarballylate (TETCA, Figure 2) and longer chain carboxylates offer greater inhibitor power but compounds containing both a phosphate and carboxylic acid residue appear superior. A polymer of PC has yet to be synthesised so that its inhibitory power can be compared to that of a polyacrylate.

Of interest with PC, it seems that the stereochemical arrangement of chelating groups about the tetrahedral "sp^3" carbon is particularly suited to binding to active sites on the HAP molecule. Johnsson et al.[30] have demonstrated that PC has a much greater binding affinity than citrate. It would be interesting to substitute a phosphate for the presently attached β-carboxyl group to see whether greater inhibition occurs. If the anionic groups are bound to an "sp^2" hybridized carbon (i.e., double bonded carbon), the molecule no longer inhibits. For example, phosphonoacetate will inhibit but there is no inhibition with phosphoenol pyruvate again suggesting that the reacting groups are not binding sufficiently to the active sites.[24] Substitution of the original phosphate with a carboxyl group to produce TETCA (now one less ionizable group) also creates a less effective inhibitor. These studies highlight the necessity for a minimum number of anionic charged groups to be in the correct juxtaposition so that surface absorption occurs or even penetration into a crystal's lattice to bind and displace a cation such as calcium. In addition of course, an anionic group may occupy space normally filled with another anion group as growth is occurring. This has been predicted to occur in relation to PC's proposed action in preventing the development of another calcium crystallite, namely calcium oxalate monohydrate.[31] Clearly, the 3D spacing of the anionic groups is one key but also the ability of its anionic groups to ionize, (a function of pH) will dictate the controlling benefit that can be expected from the compound in a defined environment.

Not all data, however, support the concept that compounds with both carboxyl and phosphate moieties are superior as inhibitors to those with only one of the groupings present. In DCPD crystal inhibition studies, Amjad[26] has reported that hydroxy phosphono acetic acid (HPA) and PTBC are both weaker inhibitors than amino tris (methylene phosphonic acid) and hydroxyethylidine-1,1-bisphosphonic acid. Increasing numbers of phosphono groups and stronger calcium-phosphonate complexation are suggested as being responsible for DCPD inhibition. Perhaps these findings should not be considered surprising given the numerous inhibitory mechanisms which have been proposed by investigators. For phosphonates to inhibit strongly, surface absorption and coverage of growth sites is obviously important; for carboxylic acids, molecules that are highly charged are very effective but if maximum inhibition is to be achieved, the single most important parameter may well turn out to be the specific conformation of the molecule. It is interesting that on a molar basis, PC ranks higher as an inhibitor than HEDP.

Computer Modeling

A greater awareness of how a crystal surface is interacting with an inhibitor can now be gained through the use of computer modeling. Data generated from experimental techniques can be processed by modern software programmes to provide a visual image of probable interactions. One consequence of inhibitor binding is that a crystal's morphology (habit) can rapidly change as growth sites are shut down. The nature of the DCPD crystal, for example, is modified in the presence of citrate through preferential absorption to the (001) crystal plane. As a consequence, growth perpendicular to that plane is arrested.[32]

Combining molecular modeling techniques with information gained from scanning electron microscopy is a valuable aid to understanding an inhibitor's mechanism of action. Accurate measurements of the dimensions of each crystal face can be obtained from SEM images. Once data for the atomic lattice are derived and the crystal unit cell is computer generated, morphological images of the experimental crystals can be reproduced on the computer. To gain further information of the interactive nature of the inhibitor to the crystal surface, the geometry of the inhibitor needs first to be delineated before the data is computerized to provide an image. This can be accomplished either by *ab initio* calculations or obtaining the relevant data from x-ray analysis. Attempted positioning of the inhibitor to the crystal unit cell then can reveal valuable insights into energetic and spatial recognition, the type of charged groups involved, and the manner by which the experimental crystals have responded to the inhibitor. Through these approaches, it becomes possible to examine the positioning of functionally reactive groups and to predict the ideal structure to impose maximum inhibitory effect on the growth of a specific crystal. Striking evidence of an inhibitor's interaction with a crystal surface can sometimes also be acquired at the molecular level through application of atomic force microscopy.[33]

With respect to PC's action, two crystal types have recently been studied, namely the calcium oxalate monohydrate (COM) and struvite (magesium ammonium phosphate) crystallites. Using parameters of PC acquired from x-ray crystallography,[31] the interaction of PC to the surfaces of the COM crystal has been studied and related to morphological changes expressed by the crystals when grown in the presence of the inhibitor. The data indicated that PC not only bound better to the (-1 0 -1) face of this crystal type than other tested inhibitors but that displacement of the oxalate ion within the growing lattice was also possible. A similar study in progress with the struvite crystal indicates that PC probably replaces the phosphate ion in the lattice to cause growth inhibition.[34] To date, computational studies of PC and the HAP crystal have not been undertaken but in all likelihood, a similar mechanism of crystal surface-inhibitor interaction may emerge.

INHIBITORS WITH PHOSPHATE, CARBOXYLIC AND ADDITIONAL ELEMENTS

Phosphoproteins and particularly acidic polypeptides have also proven to be useful inhibitors of a wide variey of calcium salts including calcium phosphate. These compounds are rich in phosphate (as the amino acid residues, phosphoserine or phosphothreonine) together with carboxylic groups primarily contributed through aspartate and glutamate residues. Their distribution in nature is widespread, phosphoproteins being extracted from vertebrate tooth and bone tissue,[35] branchiopods,[36] and molluscan shells.[37] Osteopontin, a bone phosphoprotein, is rich in aspartate residues and has been described by Hunter et al.[38] as being much more powerful as an inhibitor of HAP formation than polyaspartic, polyglutamic or phosvitin. Their studies again highlight that both phosphate and carboxylic moieties promote the activity, as their individual removal by enzymic or chemical modification sharply decreased the inhibition obtained. Phosphate group influence in osteopontin appeared dominant, whereas in phosvitin there was less inhibition despite the presence of more phosphate groups. It is interesting that polyaspartate in its own right is more powerful an inhibitor than polyglutamate, reflecting perhaps that the carboxyl groups are closer together in space than in the polyglutamate molecule.[39]

As indicated earlier, not all compounds with both a phosphate and carboxyl group exert inhibition, particularly if additional functional groups are present. This is exemplified by the molecule phosphoserine where the presence of an α-amino group exerts influence on the adsorption of the compound to the crystal surface. Moreno et al.[20] have observed that serine, which has little inhibitory activity, does bind strongly to HAP but only at one specific surface site and that such sites are smaller in number than many of the other sites known to be present. With the amino group present, the α-carboxyl group does not seem to adsorb at all so despite replacing the hydroxyl with a phosphate (i.e., phosphoserine) which has the effect also of increasing hydrophilicity, no significant inhibitory power is gained when both anionic groups are present.

Statherin, another extensively investigated phosphoprotein, is found in human parotid and submandibular salivas. It has its own unique structure possessing a strong charge asymmetry and a high affinity for hydroxyapatite. This latter characteristic is associated with the N-terminus region which allows for strong electrostatic and hydrogen bonding interactions.[40]

It is worth commenting also that, besides proteinaceous compounds, natural carbohydrates such as those extracted from marine algae and or semisynthetic polysaccharides (polyanionic heparin analogues; pentosan polysulfate) also can offer potential to control mineralization. Although not phosphorylated, these compounds invariably have a sulphate moiety which like phosphate helps to provide useful inhibitory power through the dissociation of its available hydroxyl. The overall generation of negative charge in such compounds can be considerable, being related to the numbers of sulfate residues per monosaccharide unit. Studies of zeta potentials confirm the increase in negativity indicating that the compounds do offer favourable binding to HAP and DCPD surfaces through these exposed charged moieties.[41]

SOME ADVANTAGES AND DISDAVANTAGES OF AVAILABLE INHIBITORS

In general terms, inhibitors of biological mineralization are required to be non-toxic *in vivo* to living organisms or ecologically safe if in contact with water sytems or disposed of into the immediate environment. This means that any downstream products arising from their use must not create any unacceptable secondary responses. The imposition of tough new guidelines to ensure public health safety has inevitably led to reappraising the use of many previous efficient inhibitors. For example, the use of strongly acidic substances to assist metal corrosion protection and improve solubility of otherwise insoluble salts is not acceptable unless additional treatments are in place to neutralize the expelled waste waters. The additional costs imposed then have to be balanced against the economics for a continued use of the inhibitor. Bisphosphonates, one of the most effective inhibitors of scaling salt formation, are not without problems under normal industrial operating conditions as chlorination can lead to phosphate scale and also the phosphonates themselves may precipitate.[42] For *in vivo* medical treatments, adverse cellular responses can arise from their use dependant upon dosage and the type of compound employed.[15,43] The kidney is one organ affected and the strong affinity of these non-metabolizable compounds for bone means that in humans they can be present for a lifetime. The new generation of bisphosphonates, in which there has been improvement in dose sensitivity and reduction in toxicity, offers much broader scope for their future use. Similarly, whilst biopolyanionic compounds may offer treatment advantages to industry operating under alkaline conditions, some of the synthetic polymers if applied *in vivo* may not be so well tolerated (e.g., methyl acrylates are believed to be carcinogenic).

Thus, identified natural inhibitors (e.g., proteinaceous compounds with an abundance of phosphate and carboxyl groups) or synthetic compounds proven to be both non-toxic and effective inhibitors are keenly sought. Phosphocitrate (PC) as one such simple natural biological compound[44] offers promise. It not only strongly inhibits pure calcium phosphates but also, of course, a compound such as basic calcium phosphate (BCP), where calcite is impregnated into the molecule, the crystallization of pure calcite is known to be particularly sensitive to the presence of PC.[45] The fact that PC will biodegrade slowly over time in an aqueous environment to the safe by-products of phosphate and citrate is an additional advantage. The compound, however, has yet to be tested for its long-term stability under high temperature operating conditions such as might be encountered industrially. Whether the type and structural arrangement of the groups in the PC molecule is already ideal for providing maximal inhibitor effectiveness against calcium phosphates or whether additional modification would increase potency, also, still has to be researched. The importance of continued studies of both natural and synthetic substances in their structure-inhibitor function relationships cannot be underestimated. Maintaining control over calcium phosphate crystallization and growth is an important biological phenomenon. It can be anticipated then that considerable benefits would flow from a greater understanding of the performance characteristics of compounds containing both phosphate and carboxylic additives.

REFERENCES

1 Nancollas GH. The involvement of calcium phosphates in biological mineralization and demineralization processes. Pure & Appl Chem 1992;64:1673-1678.

2 Fleisch H. "Role of inhibitors and promoters of crystal nucleation, growth and aggregation in the formation of calcium stones." In: *Renal Tract Stone. Metabolic Basis and Clinical Practice,* JEA Wickam, AK Buck, eds, Chapt. 19, P295-306: Churchill Livingstone Publ., U.K., 1990.

3 Meyer JL, Eanes ED. A thermodynamic analysis of the amorphous to crystalline calcium phosphate transformation. Calcif Tissue Res 1978;25:59-68

4 Tomson MB, Nancollas GH. Mineralization kinetics: a constant composition approach. Science 1978;200:1059-1060

5 Amjad Z, Koutsoukos PG, Nancollas GH. The crystallization of hydroxyapatite and fluorapatite in the presence of magnesium ions. J Colloid Interface Sci 1984;101: 250-256

6 Sharma VK, Johnson M, Sallis JD, Nancollas GH. Influence of citrate and phosphocitrate on the crystallization of octacalcium phosphate. Langmuir 1992;8:676-679

7 Vasudevan TV, Somasundaran P, howie-Myers CL, Elliot DL, Ananthapadmanabhan KP. Interaction of pyrophosphate with calcium phosphates. Langmuir 1994;10:320-325

8 Hall R-E. Trisodium phosphate in boiler water to reduce scale. US 1, 613, 656, 1927.

9 Harris JC. In: *Phosphorus and its Compounds,* JR Van Wazer, ed., Vol. II, pp.1775-1822: Intersci. Pub. Inc., N.Y., 1961.

10 Fleisch H, Bisaz S. Isolation from urine of pyrophosphate, a calcification inhibitor. Am J Physiol 1962;203:671-675

11 Fleisch H, Russell RGG, Bisaz S, Termine JD, Posner AS. Influence of pyrophosphate on the transformation of amorphous to crystalline calcium phosphate. Calcif Tissue Int 1968;2:49-59

12 Francis MD. The inhibition of calcium hydroxyapatite crystal growth by polyphosphonates and polyphosphates. Calcif Tissue Res 1969;3:151-162

13 Jung A, Bisaz S, Fleisch H. The binding of pyrophosphate and two diphosphonates by hydroxyapatite crystals. Calcif Tissue Res 1969;11:269-280

14 Francis MD, Martodam RR. "Chemical, biochemical and medicinal properties of diphosphonates". In: *The Role of Phosphonates in Living Systems,* RL Hilderbrand, ed., pp.55-96: CRC Press Inc. Boca Raton, Fl, 1983.

15 Fleisch H. Bisphosphonates in bone disease. 1-144: Stampfli and Co. Ltd. Publ., Berne, 1993.

16 Termine JD, Conn KM. Inhibition of apatite formation by phosphorylated metabolites and macromolecules. Calcif Tissue Res 1976;22:149-157

17 Amjad Z. The influence of polyphosphates, phosphonates and poly (carboxylic acids) on the crystal growth of hydroxyapatite. Langmuir 1987;3:1063-1069

18 Amjad Z. Constant composition study of crystal growth of dicalcium/phosphate dihydrate. The influence of polyphosphates, phosphonates and phytate. Can J Chem. 1988;66:2281-2287

19 Mueller E, Sikes CS. Adsorption and modification of calcium salt crystal growth by anionic peptides and spermine. Calcif Tissue Int 1993;52:34-41

20 Moreno EC, Kresak M, Hay DI. Adsorption of molecules of biological interest onto hydroxyapatite. Calcif Tissue Int 1984;36:48-59

21 Robertson WG, Fleisch H. The effect of imidodiphosphate (P-N-P) on the precipitation and dissolution of calcium phosphate *in vitro*. Biochem Biophys Acta 1970;222:677-680

22 Ebetino FH, Kass SM, Crawford RJ. Bisphosphonates: Molecular modelling, structure-activity relationships and the rational design of new analogs. Phosphorus, Sulfur and Silicon 1993;76:151-154

23 Rogers MJ, Xiong XJ, Brown RJ, Watts DJ, Russell RGG, Bayless AV, Ebetino FH. Structure-activity relationships of new heterocycle-containing bisphosphonates as inhibitors of bone resorption and as inhibitors of growth of dictyostelium Dicoideun amoebae. Molec Pharmacol 1995;47:398-402

24 Williams G, Sallis JD. Structural factors influencing the ability of compounds to inhibit hydroxyapatite formation. Calcif Tissue Int 1982;34:169-177

25 Boskey AL, Dick, BL. The effect of phosphatidylserine on *in vitro* hydroxyapatite growth and proliferation. Calcif Tissue Int 1991; 49:193-196

26 Amjad Z. "Evaluation of phosphono-hydroxyphosphono-, α-hydroxycarboxylic and polycarboxylic acids as calcium phosphate dihydrate crystal growth inhibitors." In: *Mineral Scale Formation and Inhibition*, Z Amjad, ed. Chapt. 17, pp.207-217: Plenum Press, N.Y., 1995

27 Nancollas GH, Tomson MB. Precipitation of calcium phosphates in the presence of polycarboxylic acids. Faraday Discuss Chem Soc 1976;61:2976

28 Sallis JD, Brown MR, Parker NM. "Phosphorylated and nonphosphorylated carboxylic acids: Influence of group substitutions and comparison of compounds to phosphocitrate in respect to inhibition of calcium salt crystallization." In: *ACS Symposium Series 444 - "Surface Reactive Peptides and Polymers*, CS Sikes, AP Wheeler, eds, pp.149-160: Am. Chem. Soc. Pub. Washington D.C.,1991.

29 Howie-Meyers CL, Yu K, Elliott D, Vasudevan T, Aronson MP, Ananthapadmanabhan KP, Somasundaran P. "Crystal growth inhibition of hydroxyapatite by polycarboxylates". In: *Mineral Scale Formation and Inhibition*, Z Amjad, ed., Chapt. 15, pp.169-182: Plenum Press, N.Y., 1995.

30 Johnsson M, Richardson CF, Sallis JD, Nancollas GH. Adsorption and mineralization effects of citrate and phosphocitrate on hydroxyapatite. Calcif Tissue Int 1991;49:134-137

31 Wierzbicki A, Sikes CS, Sallis JD, Madura JD, Stevens ED, Martin KL. Scanning electron microscopy and molecular modeling of calcium oxalate monohydrate crystal growth inhibition. Calcif Tissue Int 1995;56:297-304

32 Brecevic LJ, Füredi-Milhofer H. Precipitation of calcium phosphates from electrolyte solutions V. The influence of citrate ions. Calcif Tissue Int 1979;28:131-136

33 Sikes CS, Mueller EM, Madura JD, Drake B, Little BJ. Polyamino acids as antiscalants, corrosion inhibitors, and dispersants: atomic force microscopy and mechanisms of action. Corrosion 1993;465:1-21

34 Sallis JD, Wierzbicki A, Smith ME. "*In vitro* and modeling studies of struvite-inhibitor interactions." In: *Urolithiasis 96*, CYC Pak, MI Resnick, GM Preminger, eds, pp.232-233. Millet the Printer, Inc., Dallas, 1996.

35 Butler WT. Mineralized tissues: an overview. Methods Enzymol 1987;145:255-261

36 Tuross N, Fisher LW. Characterization of the proteins in an invertebrate hydroxyapatite shell. Connect Tissue Res 1989;22:318-

37 Marsh M-E, Sass RL. Distribution and characterization of mineral binding phosphoprotein particles in Bivalvia. J Exp Zool 1983;234:237-242

38 Hunter GK, Kyle CL, Goldberg HA. Modulation of crystal formation by bone phosphoproteins: structural specificity of the osteoprotein-mediated inhibition of hydroxyapatite formation. Biochem J 1994;300:723-728

39 Hay DI, Moreno EC, Schlesinger DH. Phosphoprotein-inhibitors of calcium phosphate precipitation from salivary secretions. Inorg Perspect Biol-Med 1979;2:271-285

40 Raj PA, Johnsson M, Levine MJ, Nancollas GH. Dependence on sequence, charge, hydrogen binding potency and helical conformation for adsorption to hydroxyapatite and inhibition of mineralization. J Biol Chem 1992;267:5968-5976

41 Boeve ER, Cao LC, Deng G, de Bruijn WC, Schroder FH. Effect of two new polysaccharides on growth agglomeration and zeta potential of calcium phosphate crystals. J Urology 1996;155:368-373

42 Sikes CS, Wheeler AP. Regulators of biomineralization. Chem Tech 1988;620-626

43 Adami S, Zamberlan N. Adverse effects of bisphosphonates - A comparative review. Drug Safety 1996; 14:158-170

44 Williams G, Sallis JD. "The sources of phosphocitrate and its influential role in inhibiting calcium phosphate and calcium oxalate crystallization." In: *Urolithiasis, Clinical and Basic Research*, LH Smith, WG Robertson, B Finlayson, eds, pp.569-577: Plenum Press, N.Y., 1981.

45 Sallis JD, Juckes W, Anderson, ME. "Phosphocitrate: potential to influence deposition of scaling salts and corrosion." In: Mineral Scale Formation and Inhibition, Z. Amjad, ed, Chapt. 8, pp87-98: Plenum Press. NY. 1995

9

CRYSTAL GROWTH OF CALCIUM PHOSPHATES IN THE PRESENCE OF POLYMERIC INHIBITORS

Saburo Shimabayashi and Tadayuki Uno

The University of Tokushima, Faculty of Pharmaceutical Sciences, Tokushima, Tokushima 770, JAPAN

INTRODUCTION

Biological hydroxyapatite (HAP, $Ca_{10}(PO_4)_6(OH)_2$) is the main inorganic component of mammalian hard tissues such as bones and teeth[1-5] and synthetic HAP is being developed as a biomedical material for artificial bones and teeth. HAP is also used in chromotography for separation and purification of biopolymers from their mixture or raw extract by elution with a phosphate buffer. Recently, HAP has been used as an immunoadsorbent.[4] Other calcium phosphates are also important in agriculture as fertilizers, in wastewater treatment processes, in chemical industry, and in other various fields.

HAP is directly crystallized and grows in its mother solution at low supersaturation. Conversely, it is generally agreed that the precipitating phase which initially appears from a highly supersaturated solution at a neutral pH is an amorphous calcium phosphate (ACP). The precipitate often contains both octacalcium phosphate (OCP) and dicalcium phosphate dihydrate (DCPD). These calcium phosphates are recrystallized to HAP over time in the aqueous phase through the mechanism of so-called Ostwald's ripening. The mechanism of the precipitate formation of ACP and that of recrystallization to HAP is, however, so complicated that they have not been fully elucidated.

Crystallization of HAP in biological and industrial systems is affected by many kinds of inhibitors. There are several possible mechanisms by which an additive can inhibit the reaction of crystal growth. A. An additive may form a stable complex with one of the precipitating ions, calcium or phosphate ion. B. The additive may increase the ionic strength of the supersaturated solution. Both these factors can result in decrease in the degree of supersaturation of the solution. C. The additive may be adsorbed on the crystal surface and/or crystal nuclei (in the case of homogeneous nucleation), on the seed surface in the case of heterogeneous nucleation either generally or at active growth sites. The inhibitory effect by polymers is mainly due to the adsorption and subsequent blocking of the growth sites. That is, the influence of polymers may be explained in terms of its adsorption either on all crystal faces or on selective faces resulting in reduction of the crystallization rate and/or in changes of crystal habit and morphology of the developing crystals. In this chapter, the effect of various polymers (i.e., biological and synthetic, and/or nonionic and ionic) on crystal growth of HAP will be discussed, taking the effects of the low molecular weight model compounds into consideration.

METHODS

The calcium ion activity and pH (i.e., hydrogen ion activity) after mixing calcium ion (e.g., from $CaCl_2$) with phosphate ion (e.g., from K_2HPO_4) in an aqueous phase as a function of time can give information on the transformation of ACP to HAP in the presence or absence of the inhibitor. Calcium ion activity steeply decreases twice as the precipitate forms and ripens. The first decrease in calcium ion concentration is observed immediately after the mixing due to the formation of ACP, and the second is observed after the induction time as the ACP transforms to HAP. The pH also decreases as ACP transforms to HAP because OH^- is consumed from the solution to form HAP. The induction time (T_{trans}) can be determined from the intersection of the tangents drawn to the time course of pH and calcium ion activity just before and after the second steep decrease. The induction period obtained from these observations of pH and Ca^{2+} activity are genarally in fair agreement.

The constant composition crystallization technique[5] is often applied to the study of the formation and transformation to ACP to HAP. Stable supersaturated solutions with respect to HAP with a given molar ratio of total calcium ion to total phosphate ion (= 1.67) are prepared by adjusting the pH of a premixed subsaturated solution of $CaCl_2$ with KH_2PO_4 to a value of 7.40 by slow addition of a dilute solution of KOH. Following the addition of HAP seed to the above solution, the crystal growth rate can be monitored by the addition of titrant solutions from mechanically coupled automatic burets mounted on the modified pH-stat, controlled by H^+ or Ca^{2+} sensitive electrodes immersed in the sample solution. The titrant solutions in the burets consist of $CaCl_2$, KH_2PO_4, KOH, and the inhibitor. The ratio of molar concentration of the titrant corresponds to the stoichiometry of the HAP phase. KCl is added to the calcium phosphate supersaturated solution in order to maintain its ionic strength constant.

X-ray powder diffraction spectrometry (Cu α+ radiation) is often used to examine the precipitates (the angle $2\delta = 31.8$ degrees is specific for HAP). Samples for the diffraction measurements are obtained by filtration of the precipitate from the solution. Using x-ray diffractometry, the degree of crystallinity of the precipitated ACP can be determined. By using a scanning electron microscope (SEM) and/or transmission electron microscope (TEM), the shape and crystal habit of HAP can be determined.

Polymers inhibit the crystal growth of HAP due to adsorption. Determination of the amount of adsorption, therefore, at various conditions of temperature, pH, and species and ionic strength, is important to understand the inhibiting mechanism as well as the affinity of the inhibitor for the surface of HAP.

Mean diameter of secondary particles of HAP (i.e., aggregate of primary particles) can be determined by means of a Coulter counter. The diameter depends on the degree of supersaturation in a mother solution and on species and adsorption of the polymeric inhibitor. Sometimes the adsorbed polymer forms an intermolecular complex with calcium on the surface of HAP (i.e., surface complex formation), which affects the mean diameter of secondary particles and the crystallinity of primary particles of HAP.

Standard crystalline HAP can be synthesized by both wet and dry methods. Typical preparation procedures are summarized in Table 1.[6-9] HAP crystals thus obtained are used as the seed in the study of HAP formation with heterogeneous nucleation. The seed competitively adsorbs the polymeric inhibitors concomitantly with Ca^{2+} and phosphate ions from the supersaturated solution.

Table 1. Preparation and Formation of HAP by Wet and Dry Methods[6-9]

Methods	Reagents and Conditions
precipitation[6]	$Ca(NO_3)_2$+$(NH_4)_2HPO_4$ and H_3PO_4+$Ca(OH)_2$
reflux[6]	hydrolysis of $CaHPO_4 \cdot 2H_2O$ for 1 month
hydrothermal[6]	CaO+$Ca_2P_2O_7$ at 1000°C
high-temperature solid-state reaction[6]	$CaCO_3$+$Ca_2P_2O_7$ with N_2 at 1000°C for 24 hr
conversion of chlorapatite[6]	chlorapatite+steam at 1000°C for 180-450 hr

precipitation in silica gel[7]	double diffusion method in an aqueous phase
precipitation in gelatin gel[8]	one-dimensional double diffusion method; ACP/DCPD→ OCP→ HAP
coating on titanium plate[9]	thermal dissiciation of Ca-EDTA complex in phosphate solution under hydrothermal condition

ACP: amorphous calcium phosphate
DCPD: dicalcium phosphate dihydrate
EDTA: ethylenediaminetetraacetate
HAP: hydroxyapatite
OCP: octacalcium phosphate

POLYMERS CONTAINING PHOSPHATE/PHOSPHONATE GROUPS

It is known that phosphoproteins[3] (salivary prorin-rich phosphoprotein, for example) regulate biological mineralization through the adsorption on an HAP crystal seed and, therefore, through blocking the active growth sites for HAP, resulting in a retardation of the crystal growth of HAP. Table 2[3,10-21] shows phosphoproteins which affect crystallization of HAP and/or precipitate formation of calcium phosphates.

Table 2. Proteins Affecting Crystallization and Precipitation of Calcium Phosphates (except for collagen and gelatin)[3,4,10-21]

Polymer	Studies of
albumin, bovin serum[10]	effect on HAP formation through ACP
albumin, human serum[11]	dispersing effect for HAP particles at equilibrium and during crystal growth
phosphoproteins[3,4]	inhibition of apatite formation from physiological fluids by proteins, such as phosvitin, etc.
acidic proline-rich proteins (PRP), human salivary[12,13]	inhibition of apatite crystal growth and dissolution
proline-rich phosphoproteins (PRP 1, 2, 3, and 4)[14,15]	adsorption on HAP from binary mixtures of adsorbates; effect of dephosphorylation (phosphoserine vs. serine). see also ref. 25
proline rich phosphoprotein[16]	effect of fluoride on HAP formation; acceleration of formation of fluoridated HAP by 1 ppm F
amelogenin, porcine[18]	adsorption and inhibitory activity on the HAP growth; effect of partial enzymatic degradation
salivary protein, human[19,20]	inhibitor of spontaneous precipitation of CaPi
statherin, human salivary[21]	inhibition of CaPi precipitation in mouth and salivary glands

To elucidate the role of phosphate groups and the effects of condensed phosphates (i.e., hexametaphosphate, triphosphate, and diphosphate) on the formation of HAP various studies were made.[22] The sample precipitate was obtained by mixing 9.85 mmol/ℓ K_2HPO_4 and 4.93 mmol/ℓ $CaCl_2$ in an aqueous solution of 154 mmol/ℓ NaCl containing a given amount of the additive. Periodically, aliquots of the test solution were withdrawn, filtered and the residue analyzed by x-ray powder diffraction spectroscopy. The spectral band specific for HAP did not register for the length of the induction time, after which, its intensity increased with time. The diffraction intensity decreased with an increased additive concentration. Hexametaphosphate was found to be the most effective inhibitor while diphosphate the least.

The induction time, T_{trans}, was determined by using pH- and Ca^{2+} ion sensitive electrodes in tests where the precipitate was formed by mixing 1.25 mmol/ℓ $CaCl_2$ with 2.50 mmol/ℓ K_2HPO_4 in the presence of 154 mmol/ℓ NaCl and a given amount of the additive. The value of T_{trans} increased with a concentration of the additive. The order of effectiveness was reported to be:[22]

hexametaphosphate > triphosphate > diphosphate

The dispersant effect of condensed phosphate on the precipitated calcium phosphate phase was studied by means of the Coulter counter method. It was observed that the mean diameter of the particles decreased while particle-number concentration (in the unit of ℓ^{-1}) increased with additive concentration. The order of effectiveness was, again:[22]

hexametaphosphate > triphosphate > diphosphate

Dispersing effect of condensed phosphate on ripened particles of synthesized HAP was also studied by means of a peptization - redispersion and an optical density methods (λ = 550 nm) as well as the Coulter counter method. The effective concentration of the condensed phosphate to retard the transformation of ACP to HAP was found to be between 10^{-4} and 10^{-3} mmol/ℓ; that to sufficiently disperse the precipitate particles was between 10^{-3} and 10^{-2} mmol/ℓ; and that to disaggregate ripened HAP particles was between 3 and 10^2 mmol/ℓ. The order of effectiveness of the condensed phosphate was common regardless of the measuring method.

These results suggest that the condensed phosphate ion maybe tightly adsorbed on the surface of precipitate and nuclei of HAP by virture of the terminal phosphate group, which is similar in structure to the orthophosphate ion, i.e., the lattice ion of HAP. The condensed phosphate ion competes for the adsorption site with orthophosphate ion and interrupts the crystal growth of HAP by replacing the orthophosphate ion on the nucleus surface (isomorphous substitution) by blocking the active site of the crystal growth. As a result, inhibition of the crystal growth requires only a small amount of adsorption to occupy the active growth sites. On the other hand, significant adsorption is needed to completely disperse the matured/immatured HAP particles because the dispersion occurs by virtue of electrostatic repulsion between condensed phosphate ions adsorbed on the particles. Needless to say, electric charges of the ion also contribute to the retardation.

Considering the phosphoproteins, it can be concluded that they are adsorbed on the surface of HAP nucleus mainly by isomorphous substitution between phosphate group of the protein and inorganic orthophosphate ion on the HAP surface. Electrostatic attractive force between phosphate group of the protein and positively charged sites on the surface, such as surface Ca^{2+} and/or sites defective in phosphate ion also play a role. Negative charges of the adsorbed protein are repulsive to phosphate ions approaching the HAP nucleus from the bulk solution. Phosphoproteins are, therefore, strong mineralization inhibitors.

Because the size of phosphonate group is quite similar to that of phosphate group, phosphonates also inhibit the HAP formation. Table 3 shows the effect of phosphates[15,22-26] together with that of phosphonates.[27,28] These phosphonates affect crystallization and precipitation of the calcium phosphates. Peroxydiphosphate also inhibits the crystal growth of HAP after the endothermic adsorption concomitant with its dehydration.[29]

Table 3. Phosphates and Phosphonates Affecting Crystallization and Precipitation of Calcium Phosphates[13,19-25]

Polymer	Studies of
Phosphates[22-26]	
condensed phosphates[22] -hexametaphosphate -triphosphate -diphosphate	dispersion/ crystallinity of CaPi
phosphorylated polyvinylalcohol[23,24]	aggregation and crystallinity; ACP/HAP formation
phosphoserine[15,23-25]	effect of adsorption on crystallization of CaPi
phosphorylated cellulose[24,26]	precipitate formation and transformation from ACP to HAP, biological matrix model
Phosphonates[27,28]	
phophonates[27] -aminotri(methylenephosphonate) -hydroxyethane 1,1-diphosphonate	precipitation of inhibitor of ACP and HAP
N,N,N',N' ethylenediaminetetra (methylene phosphonic acid)	inhibition of HAP formation

CaPi: amorphous and crystalline calcium phosphates

MODEL COMPOUNDS FOR PHOSPHOPROTEINS (PHOSPHOSERINE, PHOSPHORYLATED POLYVINYLALCOHOL, AND SULFATED POLYVINYLALCOHOL)

Phosphorylated polyvinylalcohol (Phos. PVA) was prepared as a polymeric model compound for phosphoproteins. Phosphoserine (Pser) was assumed to be a monomeric model compound because phosphoproteins usually contain PSer residues.[23,24,30,31] The effect of these model compounds on the formation of HAP and on the transformation of ACP to HAP has been reported.[27,28] It is possible to examine the differences in efficacy between the phosphorylated and unphosphorylated compounds of a given molecular size (i.e., PSer vs. Ser, and Phos. PVA vs. PVA), and between the low and high molecular weight compounds at a given ester phosphate concentration (i.e., PSer vs. Phos. PVA).

The precipitate for an x-ray powder diffractometry was prepared by mixing 10 mmol/ℓ K_2HPO_4 with 5 mmol/ℓ $CaCl_2$ in the presence of 154 mmol/ℓ NaCl and a known concentration of additive at 25 °C. No diffraction peaks were detected from a precipitate immediately after precipitation. However, specific diffraction peak developed ($2\delta = 31.8$ °), and increased with time leveling off after approximately 20 minutes following the initiation of the precipitate formation. This result indicates that initial precipitate is ACP which subsequently crystallizes to HAP. The peak height decreased with a concentration of PSer or Phos. PVA, while Ser and PVA showed no effect on the peak height up to 1.2 mmol/ℓ and 40 mg/ℓ, respectively. To compare the effect of Phos. PVA with that of PSer on a common scale, the relationship between [Pi] and relative peak height, h/h_o, is shown in Figure 1, where [Pi] is a concentration of the added ester phosphorus in mole/ℓ, and h and h_o are the peak height in the presence and absence of phosphorylated compound at 30 min after the precipitate formation.

According to Figure 1, the concentration of PSer should be about 20 times higher than that of Phos. PVA to achieve the same effect on h/h_o.

Figure 2 shows the relationship between the induction time, T_{trans}, and the phosphorus concentration of the ester phosphate, [Pi]. The precipitate formation for this measurement was done by mixing 2.5 mmol/ℓ K_2HPO_4 with 1.25 mmol/ℓ $CaCl_2$ in the presence of 154 mmol/ℓ NaCl and a given amount of the additive at 25°C. As shown in Figure 2, the concentration of PSer should be ca. 20 times higher than that of Phos. PVA to observe the same induction time. It is interesting that the induction time obtained in these experiments is the same as that obtained from the x-ray powder diffractometry, shown in Figure 1.

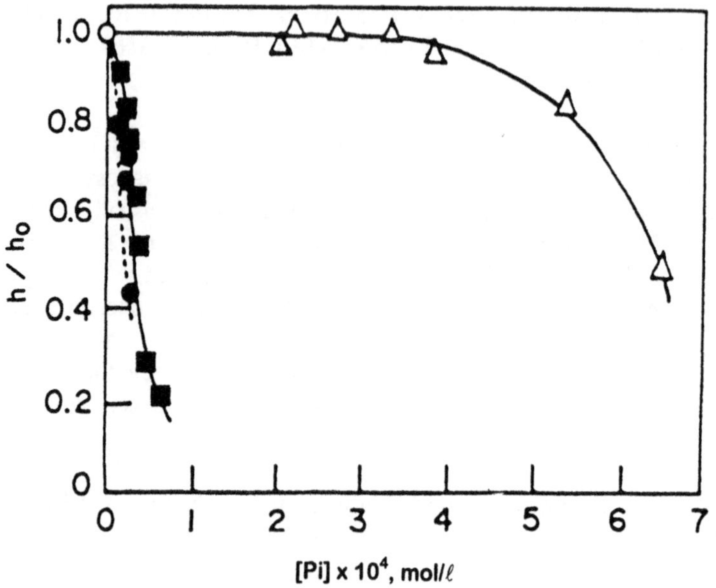

Figure 1. Relative diffraction strength, h/h_0, as a function of the phosphorus concentration, [Pi], of the added ester phophate. Degree of polymerization of the original PVA before preparation to Phos. PVA is 450. The degree of phosphorylation of Phos. PVA is 8.17 %. Additives are none (○), Pser (Δ), H_2Phos. PVA (□), and Na_2Phos. PVA(●).

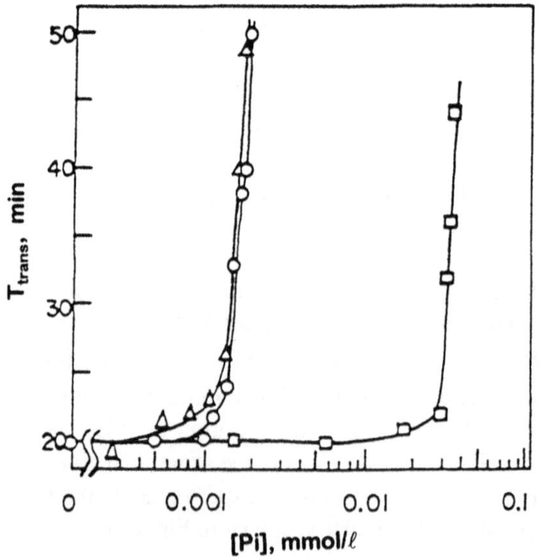

Figure 2. Relationship between the induction time, T_{trans}, and phosphorus concentration, [Pi], of the added ester phosphate Degrees of polymerization and phosphorylation are the same as those for Figure 1. Additives are Pser (□), Na_2Phos. PVA (○), and H_2Phos. PVA (Δ).

Although both phosphorylated (PSer and Phos. PVA) and unphosphorylated compounds (Ser and PVA) are assumed to be adsorbed by HAP, the former inhibits crystallization but the latter does not. This is due to the affinity of the ester phosphate group for active sites of HAP crystal being higher than that of the hydroxyl group of Ser and PVA. Thus, the ester phosphate groups of these compounds as well as condensed phosphates and phosphoproteins exhibit a significant role in the regulation of the crystal growth of HAP. These organic compounds also inhibit the adsorption of anionic proteins due mainly to the high affinity for the HAP surface.

Some of the phosphate groups along the polymer chain of the adsorbed Phos. PVA participate in the adsorption in contact with HAP as train segments, while others remain on the polymer loops or tails protruding from the HAP surface. Negative charges of the ester phosphate groups in the thick adsorption layer effectively repel the inorganic phosphate ions approaching toward the growth site of HAP, resulting in effective inhibition of the crystal growth. The effect of Phos. PVA is, therefore, more pronounced than that of PSer which is adsorbed separately on the HAP surface. These facts suggest that polymeric phosphorylated compounds are more important than monomeric and/or low-molecular phosphorylated compounds in the regulation of the crystal growth of HAP.

Chemical structure and size of sulfate ion (SO_4^{2-}) are quite similar to those of phosphate ion (PO_4^{3-}). That is, these are tetrahedral oxoacid ions and the interatomic distance between S and O (= 0.149 nm) is almost the same as that between P and O (= 0.155 nm). Therefore, phosphate ion on the surface of HAP is easily exchanged with inorganic sulfate ion (SO_4^{2-}) or sulfate group of organic compounds ($R\text{-}SO_4^-$) through isomorphous substitution, although their valencies are different. The effect of sulfated compounds[24,30-36] on the formation of calcium phosphates are shown in Table 4, where the effect of polysaccharides[24,33,37] are also included.

Studies with Na_2SO_4,[30,38,39] sodium dodecylsulfate (SDS),[30,39] and sodium chondroitin-6-sulfate (Na_2Chs)[40] show that they are easily adsorbed on HAP, and phosphate ion was concomitantly released from HAP, as would be expected. However, Na_2SO_4 and Na_2Chs did not show any effect on the induction time and the degree of crystallization regardless of concentration, while mean diameter of secondary particles of the precipitate decreased with increased Na_2Chs.[32] On the other hand, in the presence of high concentration of SDS, the crystallization was retarded.[30] Thus, the effects are specific, depending on the chemical species and/or their affinity for HAP seed surface.

Table 4. Polymers Affecting Crystallization and Precipitation of Calcium Phosphates.

Polymer	Studies of
Sulfates[24,30-36]	
polyvinylsulfate[24,30,31]	formation of crystallinity of HAP; isomorphous substitution with Pi; effect of degree of sulfation
chondroitin-6-sulfate[32]	precipitation, crystallinity; ACP and HAP formation
chondroitin-4-sulfate[33]	CaPi formation in artificial matrices
polyvinylpyrrolidone-dodecylsulfate complex[34-36]	formation of surface complex; aggregation of HAP particles
Polysaccharides[24,33,37]	
agar gel[33]	CaPi formation; ACP,HAP
hyaluronidate[33]	CaPi formation; ACP,HAP
cotton fiber[24]	CaPi formation; ACP transformation to HAP
chitin[37]	orientation of HAP crystals in Lingula unguis (inarticulate braciopod) shell

Pi: inorganic phosphate ion

The effects of Na_2SO_4, Na_2Chs, Phos. PVA, and sulfated polyvinylalcohol (Sulf. PVA) on the retardation of crystallization from ACP to HAP were studied at various concentrations of the additives. Figure 3 shows the relationship between relative induction time, $T_{trans}/T_{trans,0}$, and concentration of sulfate or phosphate group of the additive, where T_{trans} and $T_{trans,0}$ are the induction time in the presence and absence of the additive. The data for Phos. PVA and the experimental conditions for Figure 3 are the same as that for Figure 2. The effect of an additive on $T_{trans}/T_{trans,0}$ was in the order of Phos.PVA, Sulf. PVA, Na_2CHs = Na_2SO_4, where the retardation effect of Sulf. PVA with high degree of sulfation was stronger than that with low degree of sulfation.

Mechanism for the retardation and inhibition by Sulf. PVA is similar to that by Phos. PVA after adsorption by virtue of sulfate group through the isomorphous substitution with the surface phosphate ion on the embryos/nuclei of HAP. However, the affinity of the sulfate groups for the sites of phosphate ion on the nucleus is weaker than that of phosphate groups of Phos. PVA, because the size and valence of sulfate group are smaller than those of the phosphate group. Therefore, the effect of Sulf. PVA is weaker than that of Phos. PVA, even though the degree of esterification, charge density along the polymer chain, and probably the thicknessof the adsorption layer are larger in Sulf. PVA than Phos. PVA. The difference in effcts between high and low degree of sulfation of Sulf. PVA is explained in terms of efficiency of rejection against inorganic phosphate ion in the adsoption layer. That is, the higher the degree of sulfation, the higher the electric charge density and the repelling efficiency. Polyvinylalcohol (PVA) has no charged groups along the polymer chain and the amount of adsorption by HAP is low. Therefore, the retardation effect by PVA was minimal.

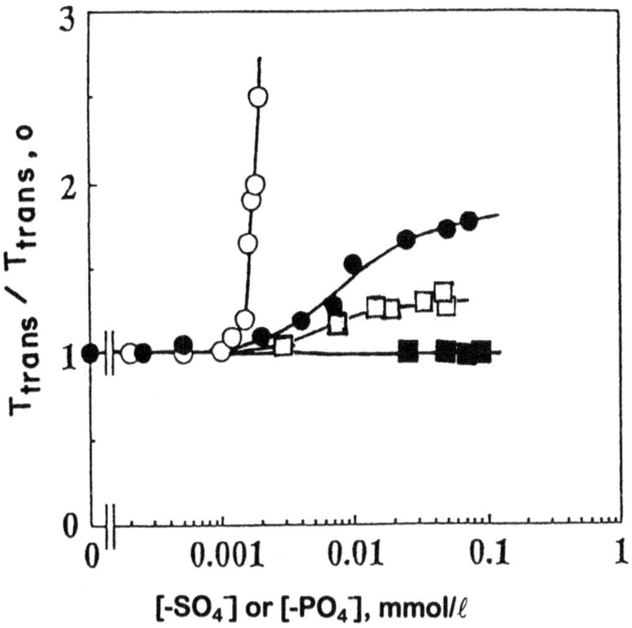

Figure 3. Effect of concentration of sulfate or phosphate group on the relative induction time, $T_{trans}/T_{trans,0}$The data for Phos. PVA was quoted from Figure 1. Potassium salt of Sulf. PVA was used, where the degree of polymerization of PVA as a starting material was about 1500. The additives are 8.17 % phosphorylated Phos. PVA (○), 92.2 % sulfated Sulf.PVA (●), 32.1 % sulfated Sulf. PVA (□), and Na_2SO_4 and Na_2Chs (■).

Na_2Chs is an acidic mucopolysaccharide which has sulfate, carboxylate, and hydroxyl groups. Charge density along the polymer chain of Na_2Chs is lower than that of Sulf. PVA, resulting in the fact that phosphate ion could more easily penetrate into the adsorbed layer of Na_2Chs than that of Sulf. PVA. In addition, the size of carboxylate group is smaller than that of sulfate group. This means the affinity of carboxylate group for HAP is lower than that of sulfate group. That is, it is more easily desorbed from the seed surface when phosphate ion as one of the lattice ion approaches and contacts with the surface to compete for the growth/adsorption sites with the carboxylate/sulfate groups of Na_2Chs. The desorption of inorganic sulfate ions from the site and selective adsorption of inorganic phosphate ions to the sites on the nuclei of HAP easily occur after the competition between these ions due to the lack of polymer in the adsorbed layer. This fact results in accelerated growth of the HAP crystal.

OTHER POLYELECTROLYTES

Polyelectrolytes affecting the crystallization and precipitation are listed in Table 5.[5,11,41-49] In this section, the effects of calcium ion and molecular weight of the polymer will be discussed briefly.

Table 5. Synthetic Polymers Affecting Crystallization and Precipitation of Calcium Phosphates[5,11,41-49]

Polymer	Studies of
poly(acrylic acid)[41]	anomalous adsorption behavior as compared with its salt
poly(acrylate)[43-47]	inhibition of crystal growth of DCPD/HAP and ACP precipitation
poly(acrylate)[5]	effect of M.W. on inhibition of crystal growth/formation
poly(acrylate)[11,48,49]	particle size distribution of HAP crystallite
poly(acrylamide)[5,44,45]	inhibiting effect on ACP precipitation and HAP crystal growth
poly(methacrylate)[44,45]	inhibition of the precipitate formation of ACP
poly(2-acrylamido-2-methyl propane sulfonate)[5]	HAP crystal growth inhibitor
poly(diallydimethyl ammonium chloride)[5,44,45]	negligible inhibiting effect on ACP precipitation and HAP crystal growth
acrylate-maleate copolymer[47]	crystal growth inhibition of HAP, and effect of M.W. and Ca^{2+} on it

M.W.: molecular weight of the polymer

It was found[47] that low molecular weight acrylate/maleate cotelomers are effective inhibitors against dental calculus. In these systems, addition of calcium ion in the form of dissolved calcium salt was necessary to activate the polymers as crystal growth inhibitors (Figure 4). At a given calcium level, dependence on the molecular weight was observed for these inhibitors against the crystal growth. The optimum activity was found between 1000 and 2000 in molecular weight.

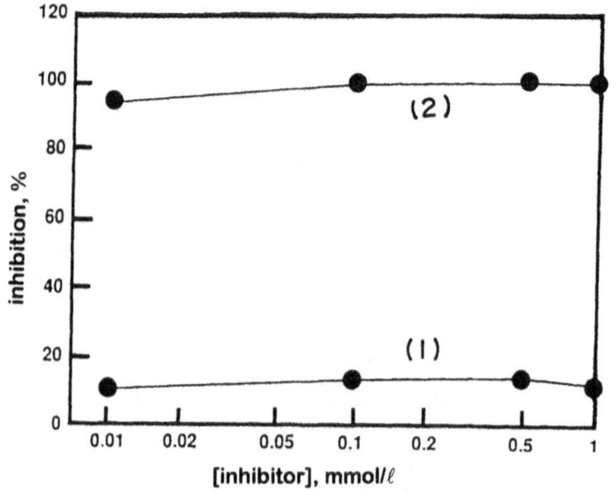

Figure 4. Seeded crystal growth inhibition of HAP by acrylate/ maleate cotelomer in the presence and absence of added calcium as a function of the reagent concentration (1) in the absence of Ca^{2+}, (2) in the presence of 1.5 mmol/ℓ $CaCl_2$.

Addition of calcium ion can make the surface more positive. Therefore, the adsorption of the polyanion should increase probably by the bridging mechanism, i.e., the polymer chain-Ca^{2+}-HAP. This fact results in high affinity of the polymer for HAP and high amount of its adsorption. The reason for the existence of an optimum molecular weight range is attributed to the maximum adsorption amount of polymer on HAP in that molecular weight range. At the low molecular weight, the adsorption amount is low because of the limited number of functional carboxylate groups. At very high molecular weight, the polymer tend to form complexes in solution, which lead to the precipitation rather than the adsorption at the HAP surface. Non-flat conformation, that is, adsorption with protruding loops and tails, of high molecular weight polymer could lead to incomplete surface coverage. Thus, the delicate balance between adsorption amount, conformational effect, and complex formation with Ca^{2+} results in an optimum molecular weight range for the maximum inhibiting activity against crystal growth of HAP.

In the case of polyacrylate,[5,44] it was found that the inhibiting effect increases with its concentration, while molecular weight of 2000, in the molecular weight range of 800 to 10000, is at the optimum to inhibit the crystal growth of seeded HAP at a concentration of 0.36 ppm at pH 7.40. The assumption that the inhibitory effect of the additive is due to both adsorption on and subsequent blocking of the growth sites was tested using Langmuir adsorption model. According to this model, the decrease in crystallization rate might be related to the crystal surface area covered by the adsorbed additive molecules.

Given that J and J_1 are the rate of crystallization in the absence and presence of the additive, respectively, the Langmuir adsorption isotherm (Equation (1)) requires a linear relationship between the reciprocal of the relative reduction in rate, $(J - J_1)/J$, and the reciprocal of the additive concentration, [A] (see Equation (2)). Equation (2) can be obtained directly from Equation (1).

$$\theta = k[A]/(1+k[A]) \qquad (1)$$

where θ is fraction of surface occupied by the adsorbent and k is an affinity constant, $J/(J - J_1)$ is equal to $1/\theta$ according to the assumption. Therefore, following equation (2) is obtained,

$$J/(J - J_1) = 1 + K/[A] \qquad (2)$$

where $K = k^{-1}$. The relationship between $J/(J - J_1)$ and $[A]^{-1}$ for polyacrylate[5] is shown in Figure 5. The linear relationship suggest that the mechanism of inhibition is almost the same as that proposed for the Langmuir adsorption isotherm. The crystal growth is completely inhibited ($\theta = 1$ and/or $J_1 = 0$) at 0.5 ppm polymer. This result suggests that the active growth sites are completely covered with the inhibitor of 0.5 ppm. However, strictly speaking, Equation (1) requires that [A] should be a free concentration but not an additive one and that [A] should become infinity or $[A]^{-1} = 0$ to attain $\theta = 1$. A specific correction factor should be added to Equation (2) to compensate for the bulky and thick adsorption layer of the charged polymer. In the studies of the influence of low-molecular-weight compounds such as glycolic acid, hydroxyphosphonoacetic acid (HPA), and 2-phosphonobutane-1,2,4-tricarboxylic acid (PBTC) on the crystal growth of DCPD,[46] and of magnesium ion on the crystal growth of HAP and fluorapatite (FAP)[17], the good linear relationships have been observed and the value of $J/(J - J_1)$ becomes unity when $[A]^{-1}$ is zero or [A] is infinity.

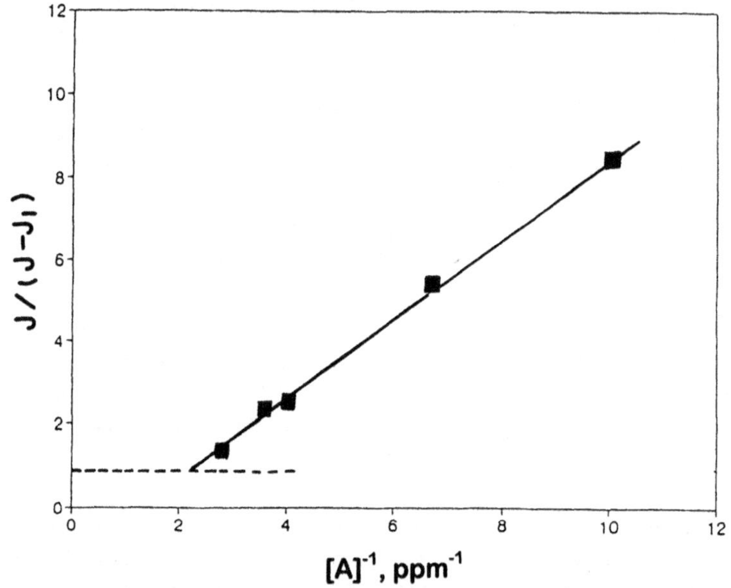

Figure 5. Testing the Langmuir adsorption isotherm (Equation (2)) Molecular weight of polyacrylate is 2100. The solution pH was adjusted at 7.40 by KOH. The constant composition technique was applied.

ACCELERATION OF HAP CRYSTAL GROWTH

The effects of collagen on the formation of crystalline calcium phosphate (CaPi) are quite complicated, depending on their origin, type, pretreatment, solubility in water, etc., as shown in Table 6.[8,50-56]

Table 6. Collagens Affecting the Crystallization and Precipitation

Polymer	Studies of
collagen, bovine incisor[50]	remineralization of the incisor lesion; CaPi precipitates on residual crystals but not on the organic matrices
collagen, rat tail tendon[8]	precipitation of CaPi in gel; collagen itself does not promote the calcification
collagen, type I[51]	inhibitory effect on HAP proliferation by steric blockage

collegen, bones of mice and rats[52]	orientation in the hole and gap zone; needle like crystals aggregate to plate-like crystals (SAXS).
collagen, turkey leg tendon[53]	HAP crystal growth/orientation; visualized by selected-area dark field electron microscope
collagen, turkey leg tendon[54]	comparing CaPi formed inside the gap region with that formed in the space between collagen fibrils; Mg effect; xray diffraction
collagen, pepsinated soluble[55]	acceleration of the transformation from TCP to HAP; incorporation of collagen into the precipitate (HAP-collagen and OCP-collagen complex formation)
collagen, type II (bovine)[56]	binding of inorganic phosphate ion in excess amount of monolayer coverage (pH 4-8). Phosphate ion is adsorbed/ absorbed either on the surface or inside the helical structure
gelatin gel[8]	precipitation formation of CaPi in the gel

SAXS: small angle x-ray scattering
TCP: tricalcium phosphate

Some studies report that collagen inhibits or does not participate in the formation/crystallization of HAP.[8,50,51] On the other hand, it was observed[48] that water-insoluble collagen adsorbs as much as the phosphate ion from an aqueous phase at pH 4 -8, 25 - 50 °C. This fact suggests that CaPi are easily formed in the neighborhood of collagen fibers if both phosphate and calcium ions are present in the system. Following the formation of ACP or amorphous TCP, the transformation to HAP is accelerated in the presence of soluble collagen, which is incorporated into the precipitate to form HAP-collagen complex.[55]

According to the small angle x-ray scattering (SAXS),[52] needle-like crystals, which are oriented along the long axis of collagen fibrils, aggregate with each other to grow to plate-like crystals (3 nm thick x ~ 30 nm long) with time. The thickness of 3nm corresponds to maximum space available in the hole of collagen fibrils. The crystal formed inside these zones is small in size but high in Mg^{2+} content while that formed in the space between collagen fibrils is long in size but low in Mg^{2+} content.[54] This corresponds to the fact that Mg^{2+} inhibits the crystal growth of HAP.[17]

Arsenault[53] studied in detail the distribution and orientation of biological HAP crystals in calcified turkey leg tendon. At the early stage of HAP formation, rod-shaped HAP crystals were localized and dispersed within gap zones bordering both C- and N-terminal regions of collagen molecules. At later stage of the deposition, the crystals were more extensive, occupying greater areas of the gap zones; in addition, HAP crystals were found to occur in the overlapping zones of collagen fibrils. The orientation of HAP crystals was gradually and continuously rotated along the axial period of collagen fibrils.

The effect of partially phosphorylated cellulose (Phos. Cell, PO_4^{3-} content = 0.34 mmol/g) on the formation of HAP was studied in order to compare the role of phosphate group on a water-soluble polymer (Phos. PVA) with that of a water-insoluble polymer (Phos. Cell)[24,26] taking into consideration the effect of phosphoproteins which regulate and/or participate in biological mineralization.

Phos. Cell captured Ca^{2+} as its counter ion in the supersaturated solution and showed the duplicity on the induction time, T_{trans}, depending on its concentration (Figure 6). At the low concentrations, the induction time decreased with the concentration, that is, it accelerated the transformation to HAP by virtue of the thick and dense clusters of calcium phosphates formed around the Phos. Cell fibers (Figure 7-(1)). These clusters are sites for heterogeneous nucleation of HAP.

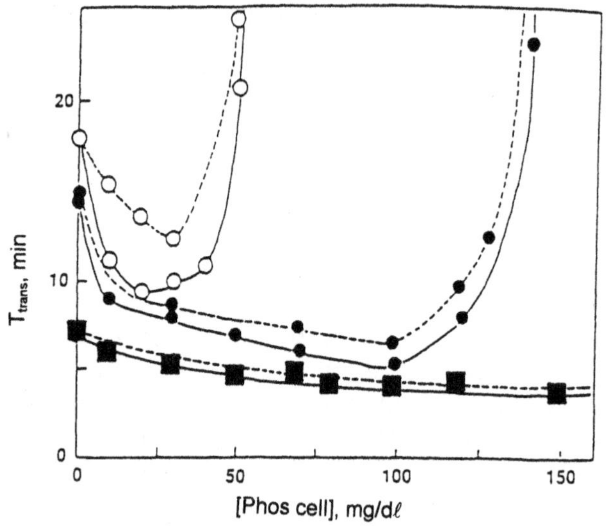

Figure 6. Relationship between the induction time and concentration of phosphorylated cellulose in the presence and absence of bovine serum albumin, BSA

[BSA]/(mg/100 mℓ) = 0 (dottes line) and 10 (solid line), Initial concentrations of Ca^{2+} and orthophosphate ion/(mmol/ℓ) = 1.25 and 2.50 (○),2.50 and 5.00 (●), and 5.00 and 10 (■).

On the other hand, at the high concentrations, the induction time increased with the concentration of Phos. Cell, that is, it suppressed the transformation of ACP to HAP (Figure 6). This effect is explained mainly in terms of the consumption of Ca^{2+} and decrease in its thermodynamic activity by the adsorption/counter-ion binding to Phos. Cell. Ca^{2+} ion was dispersed/separately bound to the phosphate group along the fiber, therefore, the cluster was small in size and less available for the HAP crystal seed (Figure 7-(2)). Although bovine serum albumin (BSA) is not directly adsorbed by Phos. Cell,[26] the transformation to HAP was retarded in the presence of BSA, as shown by dotted lines in Figure 6. This effect was explained in terms of the adsorption of BSA onto the HAP nuclei formed on the surface of Phos. Cell fibers. In fact, BSA is one of the inhibitors in an aqueous phase after adsorption on the nucleus surface by repelling the phosphate ion approaching to the growth site, as shown in Table 2.

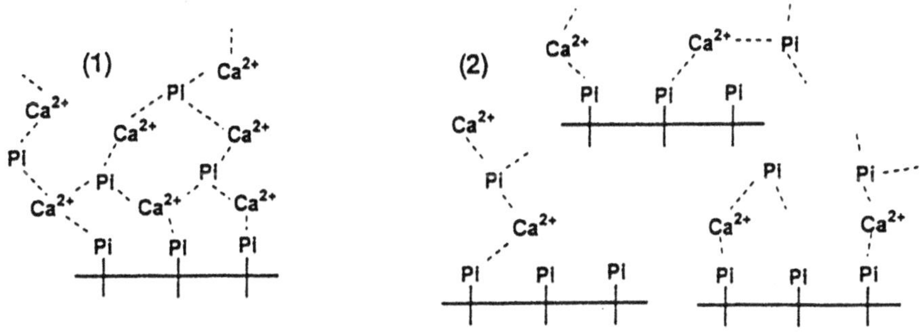

Figure 7. Schematic illustration for the formation of ACP/HAP on the surface of phosphorylated cellulose, Phos. Cell.

(1) at low concentration of Phos. Cell, clusters large in size are formed, where the clusters behave as embryos/nuclei for CaPi/HAP formation.

(2) at high concentration of Phos. Cell, clusters small in size are formed much in number along the fibers, where Ca^{2+} of thesupersaturated solution is consumed due to the cluster formation/ counter-ion binding, resulting in decrease in the thermodynamic activity of Ca^{2+} and, therefore, in the degree of supersaturtion.

Thus, Phos. Cell showed the duplicity in retardation/acceleration of the formation of HAP. On the other hand, water-soluble Phos. PVA as well as phosphoproteins was a crystal inhibitor for HAP through the adsorption/isomorphous substitution on the HAP nucleus. Collagen also showed the duplicity. It might be interesting to assume Phos. Cell as a model matrices for biological mineralization/calcification. It seems surface of the water-insoluble polymer is the place for crystallization, while the water-soluble polymer itself behaves as the inhibitor.

AGGREGATION/DISPERSION OF HAP PARTICLES

The inhibiting effect against the crystal growth is initiated by adsorption of the inhibitor on the nucleus. The adsorption also affects the particle size distribution of the products. Sodium chondroitin-6-sulfate[32] did not retard the ACP transformation but decreased the particle size. The effects of polyacrylate, human serum albumin, and human saliva were comprehensively investigated from the viewpoints of adsorption, aggregation/dispersion, and retardation of the transformation (see Tables 2 and 5).[11]

Although polyvinylpyrrolidone (PVP) is slightly adsorbed by HAP in a single dose, it is adsorbed in the presence of sodium dodecylsulfate, SDS.[34-36] This is explained in terms of hydrophobic interaction as follows: SDS is adsorbed on the surface of HAP by isomorphous substitution of sulfate group, as mentioned before, and hydrophobic groups of the adsorbed SDS are protruding from the surface. They capture PVP present in an aqueous phase by virtue of hydrophobic interaction, resulting in formation of a surface complex between SDS and PVP on HAP. The complex bridges between the HAP particles, following formation of the aggregate and secondary particles of HAP. Its mean diameter, determined by a Coulter counter, is shown as a function of concentrations of SDS and PVP in Figure 8. The mean diameter at the optimum condition is about 3 times larger than that in the absence of the additives. This fact suggests that even non-adsorbable polymers could flocculate the HAP particles in the presence of the second additive. Similar effects of the surface complex on the formation of secondary particles were also observed in the system of HAP-SDS-hydroxypropylcellulose (HPC) in an aqueous phase.[34-36]

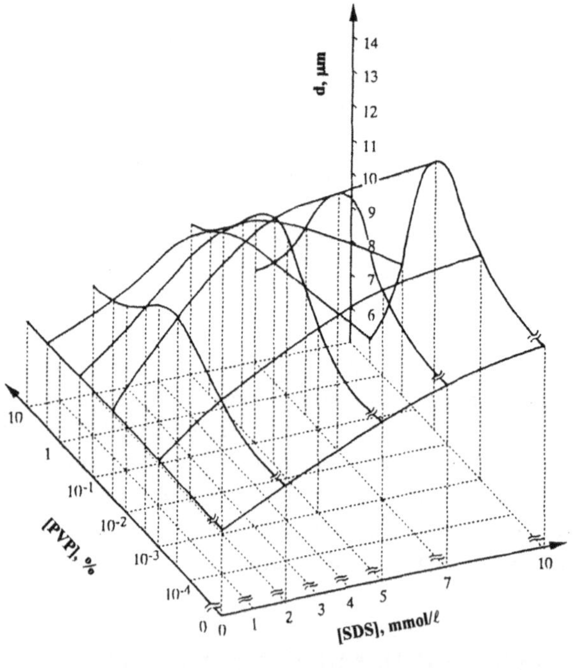

Figure 8. Mean diameter of secondary particles of matured HAP in the presence of polyvinylpyrrolidone (PVP) and sodium dodecylsulfate (SDS)
HAP concentration = 50 g/ℓ in 0.9 % NaCl. Mean diameter was measured after dilution with an aqueous solution of 0.9 % NaCl, but the data shown in the Figure were those extrapolated to zero time elapsed after the dilution. The concentrations of PVP and SDS are those before dilution.

Non-adsorbable polymer PVP retarded the crystallization with the aid of SDS, forming the SDS-PVP complex, because the complex behaves like a sulfated polymer[30] (see Table 4). On the other hand, the fact that the complex accelerates the aggregation of HAP particles suggests that nonionic hydrophobic compounds that hardly adsorb on HAP can still contribute to the formation of biological hard tissues after hydrophobic interaction and/or complex formation with ionic amphiphilic compounds which are easily adsorbable to HAP. In fact, biological hard tissues are assumed as composite material of biological HAP and organic compounds. In this regard, systems of PVP-SDS-HAP and HPC-SDS-HAP are quite interesting as models for studying composite inhibitors.

SUMMARY

Effects of polymeric inhibitors on formation of ACP and subsequent transformation to HAP were discussed, taking into account various experimental techniques and polymeric materials in an aqueous phase.

A. The retarding effect of a polymer was stronger than that of a corresponding monomer by virtue of the polymer size and charged adsorption layer, which are steric and electrostatic barriers against ions approaching to the active growth sites.

B. Phosphoproteins and phosphorylated polymers exhibited strong retardation effect. Sulfated polymers also showed the retardation effect. These effects were explained in terms of isomorphous substitution of the phosphate or sulfate groups with orthophosphate ion at the active growth sites on the surface of the nuclei/embryos.

C. Water-soluble Phos. PVA and Sulf. PVA retarded the transformation, while water-insoluble Phos. Cell showed the duplicity of acceleration/retardation, depending on its concentration.

D. Collagen exhibited mixed effects, depending on its origin, type, pretreatment, and solubility in water. Matured HAP crystals were observed around the water-insoluble collagen fibrils.

REFERENCES

1. Shimabayashi S, Uno T. "Hydroxyapatite-biopolymer interactions." In Polymeric Materials Encyclopedia vol 5, Joseph C. Salamone, ed. Boca Raton, Florida: CRC Press, 1996.

2. Shimabayashi S. Physicochemical studies on the interactions of hydroxyapatite with biologically significant ions. Yakugaku Zasshi 1987;107:323-337

3. Termine JD, Bone and tooth mineralization: matrix effect and crystal developement. Prog Crystal Growth Charact 1980;3:65-75

4. Paul W, Sharma CP. Modified hydroxyapatite microspheres as immunoadsorbent for plasma perfusion. J Colloid Interface Sci 1995;174:224-229

5. Amjad Z. Performance of polymeric additives as hydroxyapatite crystal growth inhibitors. Phosphorus Res Bull 1995;5:1-12

6. Young RA, Holcomb DW. Variability of hydroxyapatite precipitations. Calcif Tissue Int 1982;34:S17- S32

7. Kibalczyc W, Sokolowski T, Wiktorowska B. Calcium hydroxyapatite crystals grown in silica gel. Crystal Res Tech 1982;11: K105-K106

8. Pokric B, Pucar Z. Precipitation of calcium phosphates under conditions of double diffusion in collagen and gels of gelatin and agar. Calcif Tissue Int 1979;27:171-176

9. Fujishiro Y, Fujimoto A, Sato T, Okuwaki A. Coating of hydroxyapatite on titanium plate using thermal dissociation of calcium- EDTA chelate complex in phosphate solution under hydrothermal conditions. J Colloid Interface Sci 1995;173:119-127

10. Shimabayashi S, Hashimoto N, Uno T. Formation of hydroxyapatite in the presence of bovine serum albumin in an aqueous phase. Phosphorus Res Bull 1995;5:19-24

11. Nancollas GH, Budz JA. Analysis of particle size distribution of hydroxyapatite crystallites in the presence of synthetic and natural polymers. J Dent Res 1990;69:1678-1685

12. Aoba T, Moreno EC, Hay DI. Inhibition of apatite crystal growth by the amino-terminal segment of human salivary acidic proline-rich proteins. Calcif Tissue Int 1984;36:651-658

13. Hay DI, Carlson ER, Schluckebier, Moreno EC, Schlesinger DH. Inhibition of calcium phosphate precipitation by human salivary acidic proline-rich proteins: structure-activity relationship. Calcif Tissue Int 1987;40:126-132

14. Moreno EC, Kresak K, Kane JJ, Hay DI. Adsoption of proteins, peptides and organic acids from binary mixtures onto hydroxyapatite. Langmuir 1987;3:511-519

15. Moreno EC, Kresak M, Hay DI. Adsorption of molecules of biological interest onto hydroxyapatite. Calcif Tissue Int 1984; 36:48-59

16. Margolis HC, Varughese K, Moreno EC. Effect of fluoride on crystal growth of calcium apatites in the presence of a salivary inhibitor. Calcif Tissue Int 1982;34:S33-S40

17. Amjad Z, Koutsoukos PG, Nancollas GH. The crystallization of hydroxyapatite and fluorapatite in the presence of magnesium ion. J Colloid Interface Sci 1984;101:250-256

18. Aoba T, Fukae M, Tanabe T, Shimizu M, Moreno EC. Selective adsorption of porcine amelogenins onto hydroxyapatite and their in- hibitory activity on hydroxyapatite growth in supersaturated solutions. Calcif Tissue Int 1987;41:281-289

19. Hay DI, Schluckebier SK, Moreno EC. Equilibrium dialysis and ultrafiltration studies of calcium and phosphate binding by human salivary proteins(Implications for salivary supersaturation with respect to calcium phosphate salts). Calcif Tissue Int 1982;34: 531-538

20. Moreno EC, Varughese K, Hay DI. Effect of human salivary proteins on the precipitation kinetics of calcium phosphate. Calcif Tissue Int 1979;28:7-16

21. Hay DI, Smith DJ, Schluckebier SK, Moreno EC. Relationship between concentration of human salivary statherin and inhibition of calcium phosphate precipitation in stimulated human parotid saliva. J Dent Res 1984;63:857-863

22. Shimabayashi S, Moriwaki A, Nakagaki M, Effect of condensed phosphate on the precipitate formation and dispersion of calcium phosphate in water. Chem Pharm Bull 1985;33:4641-4648

23. Shimabayashi S, Tanizawa Y. Formation of hydroxyapatite in the presence of phosphorylated polyvinylalcohol as a simplified model compound for mineralization regulator phosphoproteins. Chem Pharm Bull 1990;38:1810-1814

24. Shimabayashi S, Hashimoto N, Kawamura H, Uno T. "Formation of hydroxyapatite in the presence of phosphorylated and sulfated polymers in an aqueous phase." In *Mineral Scale Formation and Inhibition*, Z. Amjad, ed. New York: Plenum Press, 1995.

25. Aoba T, Moreno EC. Adsorption of phosphosrine onto hydroxyapatite and its inhibitory activity on crystal growth. J Colloid Interface Sci 1985;106:110-121

26. Shimabayashi S, Hashimoto N, Uno T. Effect of phosphorylated cellulose and bovine serum albumin on crystallization of hydroxyapatite. Phosphorus Res Bull 1993;3:7-12

27. Amjad Z. Effect of precipitation inhibitors on calcium phosphate scale formation. Can J Chem 1989;67:850-856

28. Gaffar A, Moreno EC, Varughese K, Kresak M. In vitro evaluation of editempa on hydroxyapatite formation and its effacts on dental enamel. Calcif Tissue Int 1982;34:S8-S16

29. Moeno EC, Kresak M, Gaffar A, Adsorption of peroxydiphosphate onto hydroxyapatite: kinetics, thermodynamics, and crystal growth inhibition. J Colloid Interface Sci 1994;168:173-182

30. Shimabayashi S, Kawamura H, Uno T. Effect of sulfate group on the formation of hydroxyapatite in an aquous phase. Phosphorus Res Bull 1993;3:1-6

31. Shimabayashi S, Koh N, Uno T. Effect of potassium polyvinylsulfate of various degrees of sulfation on the formation of hydroxyapatite in an aqueous phase. Phosphorus Res Bull 1995;5: 13-18

32. Shimabayashi S, Nakagaki M. Effect of chondroitin sulfate on the precipitate formation of calcium phosphate in water. Chem Pharm Bull 1985;33:3589-3594

33. Udich HJ, Hoft H, Bornig H. Studies of calcium phosphate formation in artificial matrices. Biomed Biochim Acta 1985;44: 547 -559

34. Shimabayashi S, Yoshida Y, Arima K, Uno T. Complex formation through hydrophobic interaction between polyvinylpyrrolidone and sodium dedecylsulfate on the surface of hydroxyapatite(Modification to hydrophobic surface by the adsorption of sodium dodecylsulfate). Phosphorus Res Bull 1994;4:89-94

35. Shimabayashi S, Nishine S, Uno T. Formation of intermolec ular complex between polyvinylpyrrolidone and sodium dodecylsulfate through hydrophobic interaction on the surface of hydroxyapatite in an aqueous phase (Flocculation of hydroxyapatite particles by the surface complex). Phosphorus Res Bull 1996;6: accepted/in press

36. Shimabayashi S, Uno T, Nakagaki M. Review: Formation of surface complex between polymer and surfactant and its effect on the dispersion of solid particles. Colloids Surfaces 1997; accepted/ in press

37. Iijima M, Moriwaki Y. Orientation of apatite and organic matrix in Lingula unguis shell. Calcif Tissue Int 1990;40:237-242

38. Shimabayashi S, Matsumoto M. Non-stoichiometric dissolution of hydroxyapatite in the presence of simple salt. Nippon Kagaku Kaishi 1993;1118-1122

39. Shimabayashi S, Matsumoto M. Effect of sulfate ion and dodecyl sulfate ion on non-stoichiometric dissolution of hydroxyapatite. Nippon Kagaku Kaishi 1994;26-30

40. Shimabayashi S, Itoi K. Interaction of hydroxyapatite with sodium chondroitin sulfate and calcium chondroitin sulfate in an aqueous phase. Chem. Pharm. Bull 1989;37:1437-1440

41. Misra DN. Adsorption of polyacrylic acids and their sadium salts on hydroxyapatite: effect of relative molar mass. J Colloid Interface Sci 1996;181:289-296

42. Misra DN. Adsorption of low-molecular-weight sodium polyacrylate on hydroxyapatite. J Dent Res 1993;71:1418-1422

43. Amjad Z. Inhibition of hydroxyapatite crystal growth by phosphonated hydroxycarboxylic acids, nonphosphonated hydroxycarboxylic acids, and poly(acrylic acid). Phosphorus Res Bull 1996;6: in press

44. Amjad Z. Influence of polyelectrolytes on the precipitation of amorphous calcium phosphate. Colloids Surfaces 1990;48:95-106

45. Amjad Z, Hooley J. Influence of polyelectrolytes on the crystal growth of calcium sulfate dihydrate. J Colloid Interface Sci 1986;111:496-503

46. Amjad Z. "Evaluation of phosphono-, hydroxyphosphono-, â+- hydroxycarboxylic and polycarboxylic acids as calcium phosphate dihydrate crystal growth inhibitors." In *Mineral Scale Formation and Inhibition*, Z. Amjad, ed. New York: Plenum Press, 1995.

47. Howie-Meyers CL, Yu K, Elliott D, Vasudevan T, Aronson MP, Ananthapadmanabhan KP, Somasundaran P. "Crystal growth Inhibition of hydroxyapatite by polycarboxylates(Role of calcium and polymer molecular weight)." In Mineral Scale Formation and Inhibition, Z. Amjad, ed. New York: Plenum Press, 1995.

48. Amjad Z. The inhibition of dicalcium phosphate dihydrate crys- tal growth by polycarboxylic acids. J Colloid Interface Sci 1987; 117:98-103

49. Amjad Z. Calcium sulfate dihydrate scale formation on heat exchanger surface: the influence of scale inhibitors. J Colloid Interface Sci 1988;123:523-536

50. Klont B, ten Cate JM. Remineralization of bovine incisor root lesions in vitro: the role of the collagenous matrix. Caries Res 1991; 25:39-45

51. Blumenthal NC, Cosma V, Gomes E. Regulation of hydroxyapatite formation by galatin and type I collagen gels. Calcif Tissue Int 1991;48;440-442

52. Fratzl P, Fratzl-Zelman N, Klaushofer K, Vogl G. Koller K. Nucleation and growth of mineral crystals in bone studied by small-angle X-ray scattering. Calcif Tissue Int 1991;48:407-413

53. Arsenault AL. Crystal-collagen relationships in calcifies turkey leg tendons visualized by selected-area dark field electron microscopy. Calcif Tissue Int 1988;43:202-212

54. Bigfi A, Foresti E, Grerorini R, Ripamonti A, Roveri N, Shah J. The role of magnesium on the structure of biological apatites. Cacif Tissue Int 1992;50:439-444

55. Matsushita S, Motooka I, Kanaji Y, Studies on hydrolysis of â+-tricalcium bis(phosphate) in solubilized collagen solutions. Nippon Kagaku Kaishi 1994;409-415, 1995;285-290, 1995;885-890.

56. Koutsoukos PG, Nancollas GH. The adsorption of inorganic phos- phate by collagen. Colloids Surfaces 1986;17:81-90

10

CALCIUM PHOSPHATES FOR MEDICAL APPLICATIONS

Christian Rey

Laboratoire des Matériaux - Physico-chimie des Phosphates, INPT, ENSCT, UPRESA CNRS 507131400 Toulouse, FRANCE

INTRODUCTION

In the last decade, the biomedical applications of calcium phosphate have increased significantly.[1-5] This success is mainly due to the biological properties of these materials and the total absence of toxicity of their constituents. Much progress has been made in recent years to facilitate their use. Several different types of Ca-P materials including resorbable and non-resorbable ceramics, prosthetic coatings, cements, injectable cements, composite materials, and drug carriers are proposed that are suited to a wide range of applications: bone defect filling, bone reconstruction (especially in maxilo-facial applications), bone replacement (especially small bones and middle ear bones), drug carrier (antibiotics, anticancerous drugs, growth factors), coatings of metal prostheses (hip and knee joints), and even nerve guide.[6] Considering the large amounts of natural Ca-P materials of human or animal origin that are available, why is there a need for synthetic calcium phosphates? In addition to cultural and religious objections, there are also well founded medical reasons.

Since the development of graft surgery, the substitution and replacement of hard tissues has always been a concern. The use of bone tissues of the patient (autograft) is inconvenient due to the limited amount of material that can be taken from iliac crest, fibulae or other locations. Thus, bone tissues from different origins are often used. These grafts are usually taken from other persons (allografts) or from animals (xenografts).[7] The replacement of a bone or of part of must be distinguished from organ grafts (*e.g.*, kidney, heart). In organ grafts, the cells are generally kept alive and the reestablishment of blood supply restores the functions of the whole organ. In bone tissues, however, the blood supply cannot be efficiently restored and the tissue is condemned to die, be reabsorbed, and replaced. In most cases the replacement of bone fragments by foreign bone tissue leads to poor junction, rehabitation, and remodeling problems. Moreover, the use of foreign human or animal tissues can be detrimental, due to possible contamination by viruses, bacteria, etc., along with possible antigenic problems.[8] In addition, bone composition and properties can vary considerably within a population and are difficult to control. Thus remodeling of the implant can be very different from one person to another and from one implant to another. Synthetic products offer controlled composition and properties that can be modified and optimized for bone replacement. They can be easily stored, shaped, and sterilized, and do not need the heavy organization of bone organ banks.

Logically the first modern materials used for bone substitution attempted to imitate bone constituents, especially bone mineral which represents about 70% of the mass of dry bone tissue. This chapter will present the bone model and describe different types of Ca-P materials, their preparation methods, properties and biological activity.

THE BONE MODEL

Bone mineral can be described as a heterogeneous compound in constant evolution[9] between the moment of its formation and the time it is remodeled or turned-over. Bone mineral composition varies depending on species, strain, type of bone, diet, etc. However, its chemical composition can be represented approximately by a single chemical formula:[10]

$$Ca_{8.3}\square_{.7}\ (PO_4)_{4.3}(HPO_4\text{ and }CO_3)_{1.7}\ \ (2OH\text{ and }CO_3)_{0.15}\square_{1.7}$$

In which □ represents a vacancy. The amount of HPO_4^{2-} and CO_3^{2-} may vary considerably, however the sum of divalent ions is quite constant in most bone tissues. The amount of vacancies in cationic sites and in monovalent anionic sites is always very high and close to the maximum attainable for an apatitic structure. Although bone mineral is often described as a "hydroxyapatite", it can be seen from the above formula that it contains very few OH^- ions that have never been detected by any technique.

The initial stage of formation of bone mineral crystals is extremely difficult to study, due to the tiny size of the crystals and their very high reactivity. Several hypotheses involving the existence of mineral precursors such has amorphous calcium phosphate (ACP), dicalcium phosphate dihydrate or brushite (DCPD) and octacalcium phosphate (OCP) have been proposed, based on the study of calcium phosphate formation, in vitro. However, despite isolated studies claiming the formation of one or the other of these phases, none of them has been reproducibly detected.[11] The local arrangements of atoms analyzed by spectroscopic techniques reveal, however, specific environments different from those existing in pure well crystallized apatites. These environments named "non-apatitic" are specific to bone mineral crystals and fresly precipiated apatites in physiological conditions.[12]

In early stages of formation, bone mineral crystals exhibit very specific features: a very high amount of non-apatitic environments, a low carbonate and a high HPO_4^{2-} content.[10] During aging however the amount of non-apatitic environments and that of HPO_4^{2-} ions decrease, whereas the proportion of carbonate increases, and bone mineral evolves towards a regular carbonated apatite with a better crystallinity (maturation). The ability of a poorly crystalline apatite to mature is related to the proportion of non-apatitic environments.[13] The non-apatitic environments can be easily and reversibly exchanged or removed from bone mineral without alteration of the stable carbonate-apatite part. This reactivity suggests that non-apatitic environments are in easily accessible domains, possibly corresponding to a surface location on the crystals. It has also been shown that the alteration of non-apatitic environments is related to massive desorption of non-collagenic protein.[14]

Apatites are not however the only Ca-P compounds involved in biological systems. Amorphous Ca-P has been shown to be more abundant in certain invertebrates. Other potent precursors of apatites like DCPD and OCP may also present an interest for biomedical applications. However, Ca-P compounds not related at all with living systems also figure among biomaterials. a- or b-Tricalcium phosphate (a-TCP or b-TCP) and especially tetracalcium phosphate (TTCP) are used often. These salts can be well accepted by living organisms and have found specific applications (Table 1).

PREPARATION OF Ca-P BIOMATERIALS

Referring to any Ca-P compound vaguely related to apatite by its composition or structure as "hydroxyapatite" causes confusion among biomaterials users. Stoichiometric hydroxyapatite is represented by the chemical formula:

$Ca_{10}(PO_4)_6(OH)_2$

and characterized by an apatitic structure. Bone, for example, is often referred to as hydroxyapatite although, as indicated in the previous section, OH^- ions cannot be detected. Similarly, plasma sprayed coatings of hydroxyapatite generally contain large amounts of amorphous compounds and oxyapatites, but are still named "hydroxyapatite coatings". For clarity sake, pure stoichiometric hydroxyapatite will be symbolized by S-HA, and non-stoichiometric hydroxyapatite by NS-HA.

Apatites may form over a wide range of conditions. Even apatites quite analogous to bone mineral composition, structure and crystal size can easily be obtained. However, these compounds cannot be formed into usable objects by the standard techniques used for ceramics because of their thermal instability. Also, like many poorly crystalline materials, they are often difficult to characterize with precision. Therefore, most Ca-P compounds used as biomaterials are well crystallized, stable calcium phosphates which are quite different from the bone mineral model. Recently however, more and more materials synthesized to mimic bone mineral have been studied, especially among Ca-P cements and coatings of prosthetic devices.

The most important industrial Ca-P material remains hydroxypatite, which is widely used around the world and is the most produced and the least expensive industrial Ca-P biomaterial. HA might not be the most suitable for some applications, though. A few trials have also been made with derived apatites, especially carbonated and fluoridated compounds. One of the oldest materials, also available on an industrial scale, is tricalcium phosphate, which is probably the most used bioresorbable Ca-P biomaterial.

Table 1. Ca-P compounds used as biomaterials.

Name and Formula	Abbreviation	Type of compound	Occurrence
Stoichiometric Hydroxyapatite $Ca_{10}(PO_4)_6\ (OH)_2$	S-HA	Ceramics PSC[a] Composites DC[b]	close to enamel tissue
Non-stoichiometric apatites (d)	NS-HA	LTC[c] Composites DC	bone tissue ectopic calcification end term of Ca-P cements
α- or β-Tricalcium Phosphate $Ca_3(PO_4)_2$	α- or β-TCP	Ceramics Composites Cements PSC DC	high temperature phase impurities in PSC
Dicalcium phosphate dihydrate $CaHPO_4,\ 2H_2O$	DCPD	Cements	forms also during Ca-P cement setting
Anhydrous Dicalcium	ADCP	Cements	

Phosphate $CaHPO_4$			
Octacalcium phosphate $Ca_8(PO_4)_4(HPO_4)_2 \bullet 5H_2O$	OCP	Cements	forms also during Ca-P cement setting
Tetracalcium phosphate $Ca_4(PO_4)_2O$	TTCP	Cements	impurity in PSC
Amorphous calcium phosphate (e)	ACP	Cements DC LTC	biomineralizations PSC

(a): Plasma sprayed coatings
(b): Drug carrier
(c): Low temperature coating
(d)(e) several chemical formulae have been proposed.[1]

The emergence of low-temperature forming processes, especially mineral cements and low temperature coatings, has resulted in many other reactive calcium salts being introduced (Table 1). The structure and physical-chemical characteristics of most of these compounds are described in other chapters of this book, as well as in other books.[1] Processes used for the formation of calcium phosphates, their main uses, and biological characteristics will be discussed in this chapter.

Hydroxyapatites

Many patents and publications describe the formation of hydroxyapatite. Several mechanisms of formation have been explored including precipitation, conversion of other calcium salts, solid-state reactions, and sol-gel preparation. Nevertheless, some variability may appear depending on the synthesis method.[15] Precipitation methods are the most widely used since they can produce large amount of S-HA with good reproducibility.

Precipitation of HA

Three problems limit the purity of HA obtained by aqueous precipitation: A. the ability of the lattice to accept substituents and vacancies, B. the apatite crystal formation, and C. the surface properties.

All ionic sites of the apatitic structure accept substituents. It is generally the size of the ionic substituent that determines its incorporation in a given lattice. For Ca-P apatites, the monovalent anionic site substitution of OH^- by F^-, Cl^-, Br^- are known but I^- is not. These sites may also be occupied by carbonate ions, as in enamel and bone. Apatite without any anions in these sites have never been described. The trivalent anionic phosphate sites may be occupied by many other trivalent ions such as vanadate or arseniate, and also by bivalent or tetravalent ions like carbonate, HPO_4^{2-}, or silicate. These sites cannot accept vacancies, probably because the trivalent anions are quite large and vacancies would destabilize the lattice.

Cationic sites can also be occupied by other divalent cations like Pb^{2+}, Mg^{2+}, Zn^{2+}, Cu^{2+}, Mn^{2+}, and by monovalent ions such as Na^{+} and K^{+} or trivalent ions like the lanthanium ion. The cationic sites can also accept vacancies: up to a maximum of 2 sites out of the 10 existing in stoichiometric apatites. This property of the apatitic structure allows the adaptation of its physical and chemical properties to many different organisms, from bacteria to vertebrates, and many different calcified tissues, such as enamel, bone, dentine, and fish scales. However, in the preparation of pure S-HA, this property is a drawback because many ions from the solution may be incorporated, including toxic ions, or mineral ions with a positive biological activity. Thus, the preparation of pure apatite has to take these properties into account.

Preparations made from sodium or potassium phosphate may lead to the incorporation of these cations and result in impure solids exhibiting some thermal instability. The best precipitation methods use ammonium phosphate and calcium nitrate salts. Although the counter ion, ammonium, has been reported to be incorporated in small amounts in the precipitate,[16] it can be easily eliminated by heating above 500 °C. The main substituents preventing the formation of pure apatites are then carbonate and hydrogen phosphate ions, which are, unfortunately, always present at one stage or the other of the preparation. These ions are largely responsible for the non-stoichiometry of precipitated apatites.

The second problem limiting the purity of the samples is the mode of formation of apatites. Stoichiometric hydroxyapatites do not form spontaneously in solution. Their formation in highly supersaturated solutions is preceded by the precipitation of an amorphous phase possessing a Ca/P ratio much lower than that of apatite.[17] This first precipitate then transforms into an apatitic phase whose Ca/P ratio tends progressively towards that of S-HA (1.67). This evolution of the first precipitate (maturation) depends on the temperature and pH. As the temperature increases, the stability of the first amorphous precipitate decreases and stoichiometry can be reached rapidly. This phase cannot, however, be eliminated. The pH is an important factor in the preparation of apatites as it determines the HPO_4^{2-} content of the final product. Stoichiometric apatites are generally obtained at alkaline pH and in boiling solutions.

Surface properties also play a role in the preparation of apatites. Generally, precipitated apatites have a very high specific surface area (between 50 and 90 m^2/g) that may adsorb many ions or constituents. In fact, adsorbed ions may eventually be incorporated into the lattice during growth or maturation of the crystals. Surface constituents generally slow down the maturation process and the evolution towards stoichiometry. The surface CO_3^{2-} and HPO_4^{2-} ions play a special role as they may alter the global stoichiometry of the solid.

Precipitation has been performed in various conditions.[1,15,18] The preparation of S-HA seems to be better achieved in Ca-rich solutions than in phosphate-rich solutions. The use of ammonium phosphate and calcium nitrate avoid incorporation of foreign ions. Alkaline pH and high temperature facilitate the evolution towards stoichiometry.

Of course, as in any other chemical preparation, contamination of apatites may also come from the vessels, containers, and instruments used. Contamination by trace elements in stainless steel reactors may occur or from debris due to abrasion by apatite particulates. Concerning samples for biological applications, these potential contaminants must be considered, as even trace elements may have a biological activity. In general, only toxic trace elements are subject to regulations and are determined (As, Pb, Cd, Hg), however other potentially active biological elements should also be known.

Conversion of calcium compounds

The second method for the preparation of apatite biomaterials is the conversion of a calcium compound in phosphate solution.[1,19-21] Calcium hydroxide and calcium carbonate are most often used. These heterogeneous reactions are generally more difficult to control than the simple precipitation, because they involve a dissolution/reprecipitation phenomenon. The characteristics of the starting material (specific surface area, particle size, porosity) have to be carefully verified to obtain reproducible results. These preparations do however offer some advantages, such as the possibility of transforming porous calcium carbonate coral skeletons into porous Ca-P ceramics.[52] In addition, the reactants are abundant and inexpensive. The reaction may, in some conditions, be performed at low temperature and do not produce any by-products that have to be disposed of.

The hydrolysis of several Ca-P salts (DCPD, DCPA, OCP, TTCP, TCP, ACP) may also give apatites.[1,22] These methods, also difficult to control, lead essentially to NS-HA, though formation of S-HA might be achieved in alkaline solutions.

Solid State Reactions

Several high temperature solid-solid reactions may lead to hydroxyapatite. These reactions are not generally used to produce large amounts of apatites,[23] but they are involved in the direct production of ceramic pieces (reactive sintering). Albeit several other processes have been proposed to synthesize hydroxyapatite,[24-25] they seem too expensive to produce such a basic compound. Their only advantage would be to lead directly to HA materials or coatings, or to give more sophisticated Ca-P compounds, possibly usable in cements.

Non-stoichiometric apatites

Non-stoichiometric apatites are quite easy to synthesize. In fact they can be obtained by any precipitation method over a wide range of pH, temperature and concentration.[1] Their composition and crystal characteristics are however more difficult to control. The most interesting of these apatites are HPO_4^{2-}- and CO_3^{2-}-containing apatites which are close to bone mineral. Although bone mineral has a variable composition, diverse apatites mimicking the evolution of bone mineral in young and old animals can be prepared.[26] These apatites also show similar crystal shape and dimensions and their composition can be represented by a chemical formula analogous to that of bone mineral.

Other types of hydroxylated apatites can be prepared[27-28] which offer higher crystallinity or a special morphology. The control of the crystal size and shape is a major concern for some applications. Generally apatite crystals exhibit a needle shape. Very elongated crystals can be obtained either by hydrolysis of other calcium salts or by hydrothermal synthesis. Preparation leading to plate-like crystals has also been reported in non-aqueous media.[28]

Tricalcium phosphate

Tricalcium phosphate exists under two crystallographic forms: β-tricalcium phosphate and α-tricalcium phosphate. The α-form is unstable at low temperature and is obtained by quenching the β-form heated above 1125 °C (transition temperature). The β-form may however be stabilized by several ionic impurities, such as Mg^{2+} ions, frequently associated with Ca salts.[29] α-tricalcium phosphate is more soluble and more reactive than β-TCP, and it can be rapidly hydrolyzed into apatite in aqueous media.[30] Both varieties can be used in Ca-P cement preparations.

The β-form makes an excellent resorbable biomaterial. However, it cannot be obtained by precipitation. This phase should not be confused with whitlockite, a natural iso-structural compound that contains Mg^{2+} ions substituted for Ca^{2+}, and HPO_4^{2-} ions substituted for PO_4^{3-}. The solubility of these two compounds are very different: β-tricalcium phosphate is more soluble than apatite and can be hydrolyzed into apatitic phases; whitlockite on the contrary seems less soluble and has never been shown to convert into apatite. The presence of Mg in β-tricalcium phosphate is thus an important parameter affecting its biological properties, especially its ability to be resorbed.

β-tricalcium phosphate can be obtained by thermal decomposition of a non-stoichiometric apatite with a Ca/P ratio of 1.5[31] or by solid-state reaction.[32] The industrial product may sometimes be blended precipitated powders with a correct overall composition but grain heterogeneity. Prolonged heating at 1000 °C is generally sufficient to insure homogenization of the chemical composition.

In order to control the resorbability of Ca-P ceramics, mixtures of β-TCP and S-HA (biphasic calcium phosphate or BCP) have been proposed.[33] It has been shown that the ratio of β-TCP/S-HA in the final product is related to the resorption rate, *in vivo*. Such mixtures can be obtained either by blending pure powders, or more simply by heating a non-stoichiometric apatite with an adequate Ca/P ratio (see section on thermal stability of apatites).

Other Ca-P compounds

Other calcium phosphate compounds are used in biomaterial preparations, especially as constituents of Ca-P cements (Table 1). DCPD forms very easily, at room temperature, in any Ca-P-containing solution at acidic pH (4-5).[34] While OCP is more difficult to obtain, the preparation method given by LeGeros[35] seems to be efficient and reproducible. TTCP is a high-temperature phase that can be obtained by solid-state reaction at high temperature.[36]

Amorphous Ca-P may play an important role in the future because of their high reactivity. Several ways of formation exist: precipitation in highly supersaturated solutions, from alkaline to neutral pH,[37] quenching of fused Ca-P compounds,[38] and non-aqueous precipitations.[39] They may exhibit very different Ca/P ratios, from 1 to 2 or more. All are easily hydrolyzed into apatitic compounds. Amorphous Ca-P begins, generally, to crystallize around 600 °C. Despite different composition, they all appear to be structurally related, independent of the preparation method.[40]

Several other Ca-P compounds can have biological applications. Glasses and crystalline metaphosphates have been suggested, but they have not yet reached important developments despite interesting biological results.[41]

FORMING OF Ca-P MATERIALS

Ca-P ceramics were first formed using classical methods involving high temperature treatments. These techniques are still widely used today. These methods, however, do not allow the forming of reactive and potentially interesting biomaterials. Thus, specific low temperature methods have been developed permitting compounds more similar to biological models to be obtained.

High temperature forming

The two most important techniques in the forming of Ca-P materials are natural sintering and plasma spraying. The forming of Ca-P at high temperature is, however, essentially related to their thermal stability. Two ranges of decomposition of apatites and Ca-P compounds can be distinguished: a low-temperature irreversible decomposition (150-1000 °C) and a high-temperature reversible decomposition (above 1000 °C).

Reactions Occurring at moderate temperatures

The first reaction occurring in Ca-P materials is the condensation of hydrogenophosphate ion HPO_4^{2-}. This reaction begins at temperatures as low as 150-300 °C in NS-HA and other HPO_4-containing compounds:

$$2HPO_4^{2-} \rightarrow P_2O_7^{4-} + H_2O \qquad (1)$$

The P_2O_7 ions are believed to remain in the apatite lattice, in a very distorted form.

Carbonate ions also begin to decompose at low temperature (100-300 $^{\circ}C$) especially in HPO_4-containing apatites, probably because of proton transfer and bicarbonate ion decomposition:

$$2HPO_4^{2-} + 2CO_3^{2-} \rightarrow 2PO_4^{3-} + H_2O + 2CO_2 \qquad (2)$$

At higher temperatures (500-1000 °C) carbonate ions begin to decompose by themselves :

$$CO_3^{2-} \rightarrow CO_2 + O^{2-} \qquad (3)$$

The O^{2-} ion, formed in the apatite structure, is generally rapidly hydrolyzed into OH^-. However, when apatites are rich in carbonate ions calcium oxide also forms as a separate phase.

At about 700 °C pyrophosphate ions react with OH^- ions to restore PO_4^{3-} ions :

$$P_2O_7^{4-} + 2\,OH^- \rightarrow 2PO_4^{3-} + H_2O \qquad (4)$$

All these reactions are irreversible. At around 800 °C massive restructuration of the non-stoichiometric apatite generally occurs depending essentially on the Ca/P ratio of the initial product. When the atomic Ca/P ratio is between 1.5 and 1.667 β-tricalcium phosphate and stoichiometric hydroxyapatite are formed. When the atomic Ca/P ratio is higher than 1.667 then CaO and stoichiometric hydroxyapatite are detected. It is only for a ratio equal to 1.667 that S-HA is observed as a single phase. This decomposition reaction is a very sensitive way to determine the stoichiometry of apatites and it has been recommended by most standards in Europe and in the US. However, the role of impurities in this decomposition is not clear, especially that of Mg^{2+}, which is the major impurity in Ca-P compounds.

A final reaction may occur at about 850 °C, the decomposition of OH^-:

$$2\,OH^- \rightarrow H_2O + O^{2-} \qquad (5)$$

This reaction depends, of course, on the water partial pressure and is completely reversible. The oxyapatite eventually formed:

$$Ca_{10}(PO_4)_6\,O \qquad (6)$$

has been proposed as a potential biomaterial. This compound, very unstable, is however not known in a pure form and only exists as an oxyhydroxyapatite:[42]

$$Ca_{10}(PO_4)_6\,OH_x\,O_{1-x/2} \qquad (7)$$

High-Temperature Decomposition

At temperatures higher than 1200 °C apatites may began to decompose in a reversible manner into tricalcium phosphate and tetracalcium phosphate. This is attributed to the instability of the oxyapatite formed:

$$Ca_{10}(PO_4)_6O \rightarrow 2Ca_3(PO_4)_2 + Ca_4(PO_4)_2O \quad (8)$$

At higher temperatures (above 1600 °C) a liquid phase forms. Then CaO is the only solid phase to subsist. Slow cooling allows these high-temperature phases to react again and restore S-HA.

Sintering of Ca-P compounds

Due to these decomposition reactions, apatites are generally sintered in air at 1100-1300 °C.[43-44] Sintering is frequently associated with growth of crystals and a change of their shape. Tricalcium phosphate can be sintered in a similar range of temperatures,[45-46] as can most Ca-P phases stable at high temperatures. Additives are often used to stabilize the desired phases, thus sintering of β-TCP may involve Mg^{2+} or SO_4^{2-} ions. Mixtures of phases can also be sintered, although frequently segregation is observed. Thus, in S-HA - β-TCP mixtures β-TCP is often found at the surface of the material and may considerably change its surface properties.[47]

The production of porous ceramics able to be rehabited by living tissues has been extensively explored. Many porogenic agents have been tried. The best agents shall give connected pores of controlled sizes and most makers are now using calibrated foams[48] impregnated with HA suspension, dried, and sintered.

Calcinated cancellous bone of animal origin can also be considered as a sintered material. The removal of organic matter and the decomposition of biological apatite can be performed while the bone architecture is preserved.[49] Due to bone mineral variability, however, a careful selection has to be made to guarantee a product with a constant quality.

Plasma-sprayed coatings

Most orthopedic applications need good mechanical performance that cannot be obtained from Ca-P materials alone. Metals are still, and will probably long remain, the main constituents of load bearing prosthetic devices. Efforts have been made to improve the biological activity of metallic surfaces by coating with Ca-P compounds.

The coating of metallic prosthesis by plasma spraying is one of the major applications of Ca-P biomaterials.[50-51] Plasma spraying has been developed essentially for industrial coating of metallic pieces working at high temperatures. This technique is based on the use of plasma flames reaching very high temperatures and high velocities that transport the hydroxyapatites particles and project them onto the surface to be coated. The individual particles are brought to a very high temperature during the short transit time in the flame and are quenched on the metal surface where they adhere.

The application of this technique to hydroxyapatite has raised several difficulties due to the surface decomposition of HA particles at high temperatures and quenching that prevents the reverse formation of HA. Plasma sprayed HA coatings generally contain many phases[52] (Table 2) heterogeneously distributed in the coating (Figure 1). These phases may affect its chemical, mechanical and biological properties. Plasma spraying may also alter the Ca/P ratio by loss of phosphorous in the flame. The presence of CaO creates a special problem, as it can be rehydrated as $Ca(OH)_2$ with a considerable increase in molar volume associated with the formation of cracks and possibly the detachment of the coating from the metal surface. Foreign phases more soluble than HA can not only facilitate the degradation and resorption of the coating, but also increase its biological activity, as will be discussed later.

Table 2. Phases that can be formed in plasma spray processing of hydroxyapatite.

Phase	Formula	Chemical Origin
Most soluble phases:		
Calcium oxide	CaO	Decomposition of HA
Calcium carbonate	$CaCO_3$	Reaction of CaO with CO_2 of air during storage
Calcium hydroxide	$Ca(OH)_2$	Reaction of CaO with water during storage
Amorphous calcium phosphate	?	Decomposition of HA
Relatively soluble phases:		
α-tricalcium phosphate	$Ca_3(PO_4)_2$	Decomposition of HA, metastable phase
β–tricalcium phosphate	$Ca_3(PO_4)_2$	Decomposition of HA, stable
Tetracalcium phosphate	$Ca_4(PO_4)_2O$	Decomposition of HA
Insoluble phases:		
Hydroxyapatite	$Ca_{10}(PO_4)_6(OH)_2$	
Oxyapatite	$Ca_{10}(PO_4)_6O$	Loss H_2O from HA

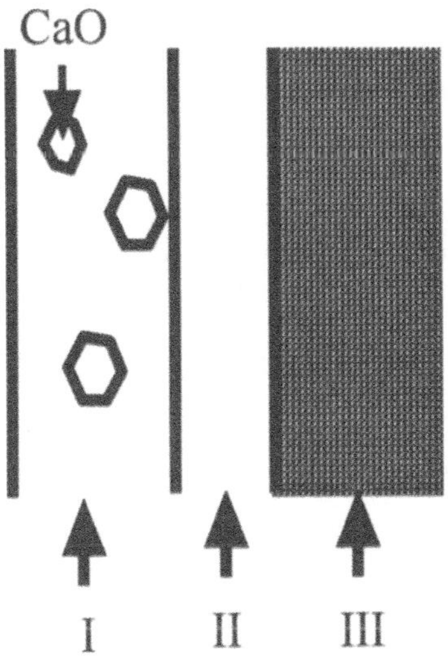

Figure 1. Schema of different decomposition domains of plasma-sprayed HA.

I : fused layer CaO islands in a liquid having a low Ca/P ratio
II : intermediate layer, high temperature phases TCP, TTCP, CaO
III : apatite layer, partly decomposed into oxyapatite

Several other apatites may be a better choice for plasma spray coating. Fluorapatites[53] and Fluorhydroxyapatites[54] have been found to be more stable and lead to fewer decomposition products. Curiously, chlorapatite or fluorchlorapatites, which can melt without decomposition, have not yet been tested.

Low temperature forming

Several low-temperature processes for forming of Ca-P biomaterials have been proposed in the recent years to overcome the limitations and the inadaptation of high-temperature processes. An already large and still developing domain is bone cements based on Ca-P materials.

Cements

Several formulae of self-setting cements have been studied (Table 3). The setting of a cement is obtained through the crystallization of a mineral phase. Two types of Ca-P cements may be distinguished depending on the chemical reactions involved: the crystallization reaction may come from two reactants or simply from one. Most cements are based on the formation, in an aqueous environment, of a crystalline Ca-P phase by reaction of at least two constituents, generally a calcium-rich phase (or alkaline phase) and a phosphate rich phase (or acid phase). The first Ca-P cements to be proposed involved the reaction of DCPD and tetracalcium phosphate to give hydroxyapatite: [55]

$$2\,Ca\,(HPO_4)\bullet 2H_2O + 2\,Ca_4\,(PO_4)_2\,O \rightarrow Ca_{10}\,(PO_4)_6\,(OH)_2 + 4H_2O \qquad (10)$$

Since then several other formulae have been proposed (Table 3) leading to different end-products.[56-60] The end-product of most cements is apatite, however the reactions are often quite complex and transitory phases like DCPD or OCP are generally observed. The setting time may vary from a few minutes to over an hour depending on the characteristics of the powders and the use of additives, which can modify either their dissolution rate or the formation rate of transitory phases or apatites.

Table 3. Ca-P cements.

Main reacting compounds		Main final constituent	reference
DCPD	TTCP	HA	55
DCPD	α-TCP	OCP	60
MCPM[a]	β-TCP	DCPD	57
MCPM	$Ca(OH)_2$	OCP/HA	56
MCPM	α-TCP/$CaCO_3$	NS-HA	58
MCPM	α-TCP/TTCP	NS-HA	59

[a]MCPM : monocalcium phosphate monohydrate

Ca-P cements may also involve organic phases. Generally however these cements have to be very carefully controlled. Variations in the particle size, hydration degree, mixing conditions, or temperature, may considerably alter their setting conditions and performance.

The second type of Ca-P cement behaves more like a plaster,[30,61] and contains a setting agent. Interesting results have been obtained with amorphous Ca-P compounds associated with a nucleating agent. The setting time can also be controlled by the amount of nucleating agent added or the presence of crystal growth inhibitors.[61]

Both categories may be used as bone substitute and injectable material to minimize surgical trauma. These cements however still lack mechanical strength, and cannot be recommended for load bearing applications. All cements are resorbable materials and are replaced by bone after variable periods of time, depending on their final composition and textural characteristics.

Low temperature coatings

As was previously reported high-temperature plasma spraying has several drawbacks. It is difficult to cover homogeneously porous surfaces or devices with a complex geometry and also difficult to control the decomposition of the C-P phase. In addition, the technique cannot be used for unstable Ca-P compounds more similar to bone mineral and physiologic calcium phosphates.

Most coatings made at low temperatures use supersaturated solutions to bring about apatite precipitation.[62-64] The aim is to promote precipitation onto the surface of the metallic prosthesis. Several techniques have been explored. The simplest way is to control the supersaturation of the solution so that it is adapted to the nucleation properties of the metallic substrate. Then, after a variable induction time, the metal is covered by a layer of calcium phosphate that can be both chemically and physically very similar to bone mineral crystals. This technique is very slow though, and does not seem suitable for mass production. Another variation is to grow, at high rates, a precursor of the apatite phase and then to hydrolyze it. Thus dicalcium phosphate dihydrate, one of the fastest growing Ca-P phases, has been proposed as a precursor for an apatite coating. Precipitation can also be encouraged at the metal surface by locally changing the supersaturation. Several techniques have been proposed. Abe et al.[62] set up a process using Ca-rich bioglasses that maintain a high supersaturation ratio in the vicinity of the material to coat. Electrolysis of metastable supersaturated solutions produces the same effect when the metal to coat is at the cathode.[63-64] The OH^- generated by electrolysis increases supersaturation and the calcium phosphate precipitate is laid down on the metal. The main drawback of these techniques is the slow rate and the poor adhesion of the coating to the metal surface. They do offer however, the ability to control the composition, structure and crystal morphology of the mineral phase. Thus, non-stoichiometric apatites containing carbonate and/or HPO_4^{2-} and bone-like mineral crystals can be formed. These layers with a very high reactivity show good biological properties and are generally bioresorbable. Thus, they may play only a temporary role during the healing process.

Other coating processes of Ca-P compounds have been investigated including ion sputtering[65] and pulsed laser deposition.[66] Unfortunately they produce a decomposition of the Ca-P compound and are unable to produce biological apatites. Pulverization of solutions on a heated substrate[67] and electrophoretic deposition[68] have also been proposed, but none of these techniques has yet received commercial development.

Associations With Organic Matrixes

Ca-P compounds can confer to composite materials their bioactive properties. In addition, composite materials can show a mechanical behavior closer from bone mineral tissue[69] than Ca-P ceramics. Several systems, both non-bioresorbable and bioresorbable, have been studied.

Non-bioresorbable composites

These materials result from the association of a non-bioresorbable polymer with a non-bioresorbable Ca-P compound, generally S-HA. Polymethyl methacrylate (PMMA) is widely used as bone cement for the fixation of metal prosthesis. PMMA, or its derivatives, have been mixed with various Ca-P compounds in order to make bioactive polymeric cements.[70-71] These attempts have failed however for two main reasons: the amount of Ca-P particles is too low to maintain the cement fluidity and the surface of Ca-P particles is covered by polymer and cannot induce bioactivity by contact. Some formulae containing highly soluble Ca-P salts may release calcium and phosphate ions. These compounds were suggested to be bioactive.[72]

The association of polymers with apatites may accomplish the additional goal of matching of bone mechanical properties. In polyethylene composites, the mineral/organic ratio can be very large. The surface of HA particles is exposed and promotes a bioactive behavior.[73] These materials have found different applications in ear surgery and reconstructive surgery.

Biodegradable composites
Bioresorbable polymers such as polylactic, polyglycolic, polybutyric acids have been used either associated with S-HA granules or with bioresorbable Ca-P compounds like PTC.[74-75] Although these composite materials may have interesting mechanical properties, their biological activity is less evident. Their resorption can lead to a massive release of particles and produce inflammation. The primary challenge in the association of Ca-P compounds and bioresorbable polymers is to adapt resorption of both constituents to avoid inflammation due to particles release. Associations between chitosan,[76] cellulosic compounds,[77] collagen,[78] and Ca-P materials have been studied. Some may give injectable materials. Most successful associations contained sintered particulates large enough to avoid inflammatory reactions loosely bound by macromolecules. Such compounds maintained the ceramic constituent in bone defects and could easily be handled by surgeons. They cannot, however, truely be considered composite materials.

Bone itself constitutes a composite material. Used as a graft, several drawbacks have been mentioned, such as infection and antigenic risks. The tissue can however be treated by supercritical carbon dioxide to remove cells, fat, and antigenic tissues.[79] The material obtained, after purification, contains essentially collagen and bone mineral and retains its mechanical properties. The field of organic-inorganic composite materials does not currently represent a major utilization of Ca-P compounds. However, these materials should develop in the future, due to wide possibilities of associations and to their potential ability to combine a biological activity and mechanical properties close to those of bone.

CHARACTERIZATION OF Ca-P BIOMATERIALS

The characterization of Ca-P biomaterials covers several aspects. In addition to the traditional physical-chemical and mechanical characterization, implantable materials have to prove their innocuousness and, if possible, their bioactivity. As in many multidisciplinary domains, chemical and physical properties relevant for the use of the biomaterial must also to be identified.

Chemical and Physical Characterization

Three levels of characterization may be distinguished: A. chemical composition, B. structural identification and C. texture of the material. Beyond these, surface and bulk characterization have to be considered.

Chemical composition

The chemical composition of biomaterials can be determined by classical chemical analysis. Several methods for calcium and phosphate determination have been proposed. More accurate techniques have been developed especially for S-HA, based on their stability at high temperatures as mentioned earlier. Less than 1% β-TCP can be detected by x-ray diffraction. The detection of CaO can even be more sensitive, using pH measurements of suspensions obtained just after calcination. Thus the purity of S-HA can be assessed with a good precision. In the case of bone mineral and non-stoichiometric apatites, the composition can be more complex, as carbonate and HPO_4^{2-} ions can be present as well as other substituents. Although the carbonate content can be chemically analyzed, an accurate determination of the amount of HPO_4^{2-} ions in the presence of carbonate is not possible.[1] Moreover, several locations of carbonate and HPO_4^{2-} ions exist that can only be distinguished by IR or Raman spectroscopy, but they cannot yet be determined quantitatively with a satisfactory accuracy.

Minor elements are frequently associated with apatites due to their incorporation as Ca^{2+}, OH^-, or PO_4^{3-} substitutes, or their attachment to the crystal surface. These elements may play an important role in the properties of apatites, and their determination is often forgotten except for toxic elements, which most standards recommend to determine. Mg, Sr, and F, for example may, at low concentration, have an action on cell activity and they should be analyzed systematically, as well as other essential elements such as Mn, Zn, and Cu. Several accurate methods can be used to find the trace element content. But, in addition, their location on the surface or in the interior of the materials and crystals should be determined, as the activity of these elements also depends on their availability or release.

Structural identification

Crystalline compounds can be conveniently identified and quantified by x-ray diffraction.[52] Therefore, most standards have selected this technique. Several dilemmas in standard identification remain including the evaluation of amorphous phases and the determination of the nature of apatitic phases.

Amorphous Ca-P materials are frequently observed as the first mineral phase in the precipitation from highly supersaturated solutions in neutral or alkaline solutions or in plasma sprayed coatings. Although several determinations of the amorphous content of biomaterials or bone mineral have been proposed none is very accurate. In fact, 5 to 10 % of the amorphous phase in a poorly crystalline sample, can remain undetected.

X-ray diffraction allows a good identification of well crystallized phases and in some cases may distinguish between different apatitic phases, however the technique is not sensitive enough to identify solid solutions like oxy-hydroxyapatites in plasma sprayed coatings. In the case of poorly crystalline compounds peak broadening prevents also any precise identification of apatite phases. In addition, it is often difficult to discriminate between closely related structures like OCP and apatite. Thus, x-ray diffraction information has to be supplemented with other techniques better suited to poorly crystalline samples such as IR and Raman spectroscopy and solid state NMR.

Unlike diffraction methods, spectroscopic techniques are sensitive to the short range order of solids. It has been shown that crystalline compounds with the same long-range order might have different short range orders depending on the manner in which they have been obtained. Thus, spectroscopic techniques reveal finer structural details than diffraction techniques and are a valuable complement.

Textural characterization

Composition and structure determine the behavior of Ca-P biomaterials, however, the morphology, the size of the crystals, their arrangement in the materials, and the porosity are also important parameters. These factors must be investigated in order to complete the bulk physical-chemical information on the material.

Microscopy (SEM, TEM, optical microscopy, atomic force microscopy) and derived techniques such as image analysis, have provided most of the results in this field. However, some data may also be obtained by other techniques.

Porosity determines whether a material will be rehabited by a tissue or not. It has been proven in several studies that a pore size between 300 and 600 mm allows cellular invasion and tissue reconstruction. In addition to these macropores, micropores may also play an important role as they may be involved in protein adhesion. Mechanical properties are also determined by porosity. One of the main challenges for biomaterials manufacturers is to prepare porous Ca-P materials with high mechanical strength.

The size and shape of the crystals appears as an equally important factor for mechanical properties, dissolution, and biodegradation. Most Ca-P crystalline materials are anisotropic and do not present the same properties depending on the crystallographic planes exposed. These effects are not always known for all calcium phosphates, but its seems quite probable that bioactivity depends on this factor. Diffraction methods may here complement information given by microscopic techniques.

In heterogeneous compounds containing several phases, such as plasma-sprayed deposits, multiphasic ceramized samples and cements, the distribution of phases determines the properties of the materials. In plasma-sprayed deposits, amorphous vitreous phases are thought to occur on the surface and at the interface with the metal. The thickness of this amorphous layer is difficult to determine. In sintered

HA-TCP mixtures, the TCP has been found to gather on the surface of the ceramics, modifying the surface properties. In most cases, however, precise placement of the phases is not possible. Similarly when a Ca-P cement forms in vivo, the setting reactions are different at the surface of the material than in the bulk. This leads to different compositions and, probably, to different properties as resorption begins. Such determinations of phase distribution within a material are among the most difficult analyses and are seldom performed.

This problem of phase distribution and heterogeneity between the surface of a material and the bulk introduces another type of physical chemical characterization that should be performed, surface characterization. This point is especially important for non-bioresorbable materials, as the surface will establish a permanent contact with living tissues.

Surface characterization

Most standards do not recommend surface characterization, but it is essential for understanding failures or malfunctions of biomaterials. The bioactivity of different biomaterials has been shown to be directly linked to surface reactions and modifications in the body.

Several physical chemical techniques are available to determine the surface composition of biomaterials. The chemical composition can be determined by XPS or ESCA. These techniques are sensitive to the first atomic layers on a surface of a material. They may warn of any surface location of toxic ions or indicate the presence of beneficial ions that may appear only as trace elements in the bulk chemical composition. Physical techniques such as surface tension and electrokinetic potential measurements may also complement our knowledge of the surface properties. They may also be related to biological properties.[80]

Another important aspect is the alteration of surface characteristics after successive remodeling at the bone-biomaterial interface. This difficult and crucial subject has never been addressed but might well be a determining factor in the longevity of permanent implants.

Mechanical characterization

The mechanical behavior of an implant determines its biological use. Several mechanical characterizations can be made depending on the nature of the implant, before implantation and after implantation. For permanent implants, fatigue tests should also be performed.

Homogeneous implants can be tested by standardardized methods. The Ca-P ceramics are characterized, like all ceramic compounds, by their brittleness. Several attempts have been made to improve bending strength and fracture toughness by controlling the sintering process or the incorporation of additives.[81] These modified Ca-P ceramics could find some application in dentistry and for small bones replacement or in rachis operation. Organic-mineral composites can match more closely their mechanical behavior to that of compact bone,[69] but they have not yet been used in major load bearing applications. The mechanical properties of porous ceramics are directly related to their porosity.[22] After implantation, as the amount of pore diminishes due to bone ingrowth their ultimate compressive strength increase. They are used essentially for bone regeneration in medium size defects although wider uses have been proposed. Ca-P cements show also, generally poor mechanical properties that depend on the preparation conditions,[82] but seem also to improve with the implantation time. They have the same applications as porous implants. For important bone substitution, however there is no other solution than metals. Despite their development, plasma-sprayed Ca-P coated prosthesis generally show a poor metal-mineral adhesion.[83] This is probably related to the decomposition of S-HA at high temperature and the formation of CaO nodules at the metal-coating interface.

Several other characterizations of biomaterials have been proposed, for example dissolution properties. These data are however difficult to use in predicting the biomaterial behavior in its biological environment. Besides the results depend strongly on experimental parameters that are not always well controlled (specific surface area, texture of the material, confusion between degradation and dissolution). These experiments may, however, be useful to compare biomaterial of the same class (coatings, bulk ceramics).

BIOLOGICAL BEHAVIOR

Several levels of biological behavior can be studied. Cell culture systems have been adopted by most standards for specific accurate determinations of cell toxicity or bone cell behavior on the material surface. Additionally, animal trials remain a necessity for testing biomaterials, especially bonding with bone tissues.

Cell culture

Three main types of cells have been recognized in bone tissue: A. osteoblast, which build an organic matrix and mineralize it; B. osteoclasts, which dissolve and degrade bone tissue; C. osteocytes, which are considered to be resting osteoblasts embedded in the tissue. Only the first two types can be cultured.

Osteoblast cell culture

The culture of osteoblastic cells has now been mastered by many teams[80,85] and gives valuable information on their adhesion, multiplication, and expression on the Ca-P material surface. This system is not suitable for very reactive compounds that can alter the composition of the cell culture media. Thus DCPD, CaO or TTCP-containing Ca-P materials cannot, generally, be tested, due to the change of pH and Ca and P concentration they produce. Less reactive samples can be tested. Generally S-HA materials are found to be less effective than the treated polyethylene which is traditionally used as cell culture support. The response improves however with time, and osteoblastic cells are able to build a bone matrix with all constituents. The reasons for the poor adaptation of osteoblastic cells to a S-HA surface can be linked to the complex events following contact of the mineral surface with the media (see the following section). All Ca-P surfaces seem to exhibit similar behavior, although differences may be seen depending on the substrate and its equilibration reactions with the media. Unlike treated polyethylene, Ca-P materials interact strongly with the cell media, especially modifying their mineral ion content.

Osteoclast cell culture

The culture of osteoclastic cell appears to be technically more difficult.[86] These cells do not replicate and can be kept alive only for a very few days. Their attachment to the Ca-P surface and their activity can however be evaluated during this short period. These giant multinucleated cells adhere strongly to bone and create resorption pits under the cell in which bone tissue is destroyed due to the release of acid and enzymes. S-HA supports have been found to be only weakly altered by osteoclast cells, probably because of the poor adhesion of the cells. However, NS-HA supports are well accepted and have even been patented. The activity of osteoblasts and osteoclasts are strongly correlated in living tissues. Because they constitute only part of the body response, animal implantations are necessity.

Animal implantation

The development of human skeleton is quite unique. No other mammal shows such a slow growth rate. In addition, the distributions of loads, due to a standing position, is also rarely found in other animals. Moreover, it should be noticed that bone composition and turn over depend on the species. Thus, the choice of an animal model is a crucial step. Though, one must recognize that none will behave like the human system. Some standards recommend rabbit for implant materials. Regarding prosthetic devices, however, tests are being made on bigger mammals: generally, sheep, goats, or dogs. The type of bone in which the implant is incorporated is also variable, depending on the proposed use of the biomaterial or device.

Histological and mechanical studies on implanted materials have proven the osteoconductivity of Ca-P compounds, i.e., their ability to enhance bone growth when they are in contact with living bone tissue.[87] The contact between Ca-P compounds and bone tissue is a direct apposition and does not involve any fibrous tissue embedding the implant, as happens with other non-bioactive materials. This situation defines bioactive behavior, and leads to the improvement of the mechanical strength of the bone-biomaterial interface.

Implantations can also be made in non-osseous sites, especially to test the ability of a material to induce bone mineral formation. Ca-P materials are generally not considered as osteoinductive materials (i.e., materials able to induce bone formation in a non-osseous tissue), however in association with active compounds they could be.

Bioactivity

Bioactivity is a common feature of all Ca-P compounds. Schematically, bioactivity may be described as a four-step process: A. modification of the material surface due to chemical interactions and cell activity, B. nucleation of carbonate-apatite crystals from body fluids, C. protein adsorption, and D. osteoblast adhesion and expression.

Most synthetic Ca-P materials are not in equilibrium with biological fluids. Thus, several reactions occur on the mineral surface. For apatites, very first reaction is the hydrolysis of the surface PO_4^{3-} groups into HPO_4^{2-} and the uptake of carbonate and sodium ions. Generally the hydrolysis of surface phosphate groups is associated with a decrease of the calcium concentration at the surface.[88] These reactions correspond simply to exchanges between the solution and the first atomic layers of the apatite surface. The extent of these reactions may vary according to the composition of the mineral and how it was produced. Inflammatory cells in close proximity to the material, may also create an acidic environment. Poorly crystalline apatites with many defects are particularly sensitive to these changes.

As biological fluids are generally supersaturated with respect to apatite, the Ca-P materials also serve as a template for the formation of carbonate-apatite crystals. This event has been shown to be the major factor behind the osteoconductivity of all biomaterials and has been evidenced in different classes of bioactive orthopedic implants (polymers and bioactive glasses). The crystals formed are very much like bone mineral crystals. They are carbonated apatites and exhibit non-apatitic environments.[89]

The precipitation of the neo-formed crystals can also be boosted by release of calcium, phosphate ions, or rise in the pH that may locally increase the supersaturation ratio in the vicinity of the implant surface. This phenomenon can be favored by hydrolysis reactions of the material constituents. The relatively soluble impurities that are present in plasma sprayed coatings, for example, may act this way and facilitate the formation of the carbonate-apatite layer. The presence of an amorphous phase in plasma spray deposits, has been shown to improve their biological activity.[90] These impurities then exhibit ambivalent behavior: they are mainly responsible for the degradation of the coating, but in return they enhance its biological activity. Enlarging the surface area is another way to increase the amount of reactive, neo-formed crystals. In this respect, low-temperature deposition offers extended capabilities.

The ability of a biomaterial to generate this layer has been considered as a measure of its biological activity, although such a relationship has not yet been established on a quantitative basis. Despite the analogy of the crystals formed with bone mineral crystals, the processes of formation and organization appear to be rather different. Bone mineral crystals are always formed in an organic collagen matrix with which they are closely associated. Although the layer formed on bioactive biomaterials frequently incorporates bone proteins, there is no organization like that of bone. It appears more like an uncontrolled mineralization.

The functions of the neo-formed layer have not yet been clearly established. This layer creates a very reactive surface with a high specific surface area, which is able to attach bone specific proteins, especially osteopontin and osteocalcin.[91] The neo-formed layer appears, as a support, and the adsorption of proteins to it is necessary to accommodate osteoblast cell attachment and favor their multiplication. Although such proteins might bind directly to the HA surface, the adsorption affinity could be quite different depending on the apatitic substrate. Thus, synthetic apatites rich in labile non-apatitic environments exhibit a high surface reactivity and have a higher affinity for some proteins than stoichiometric HA. Depending on the apatitic substrate and the rate of formation of the layer, the presence of labile non-apatitic environments and the reactivity of the neoformed layer may vary.

The formation of carbonate-apatite crystals has also been observed on non-Ca-P materials and this *in vivo* coating process has been recognized as determining bioactivity. These very particular modes of action have been studied and developed to give rise to a new class of biomaterials: Ca-P inducing materials.

Ca-P inducing materials

The formation of a carbonate apatite layer has been shown to be the characteristic of all bioactive biomaterials for bone substitution. It occurs on several biomaterial surfaces such as collagen, Polyactiv®, Bioglass®, apatites, titania gels, corals, either *in vivo* or *in vitro*, in cell culture media.[89] On other substances, such as titanium, for example, this layer may also be formed by simple heterogeneous nucleation in supersaturated fluids.

Table 4. Classification of biomaterials owing to their ability to generate a carbonate apatite neoformed layer.

Class of biomaterials	Nature of biomaterials	Active ions	Induction process
Active biomaterials	Bioglasses	Ca^{2+} Ca^{2+}, CO_3^{2-}	Diffusion out of the Glass
	Corals (aragonite)	Ca^{2+}	Dissolution
	Plaster (gypsum)		Dissolution
	Ca-P cements[a]	Ca^{2+}, PO_4^{3-}, OH^-	Dissolution and hydrolysis
	Alkaline hydrogels	OH^-	Hydrolysis
		Nucleating ability	
Passive biomaterials	Sintered apatites	Excellent	Epitaxy
	Titanium	Average	Phosphate binding
	Neutral hydrogels	Good	Phosphate binding
	Collagen	Good	Phosphate binding
	Polyactiv	Good	Calcium binding

[a]Many different formulae of cements exist that may release ions at different rates and proportions, other ions may also be released (F^-, CO_3^{2-}, for example)

Two ways of forming this layer can be recognized depending on the interaction of the biomaterial with the body fluids. The biomaterial may release (by dissolution or degradation) active ions such as OH^-, F^-, Ca^{2+}, and phosphate, capable of locally inducing supersaturation of the body fluids and the precipitation of apatite. Such biomaterials may be qualified by "active" with regard to their ability for apatite layer production (Table 4). Depending on the ions released, several types of activity can be identified. So, one material may in fact act in several different ways. A second group of biomaterials is formed by "passive" materials that only release minor amount of ions and do not interfere with the process of formation of the apatitic layer but act as heterogeneous nucleators of apatite. Polyactiv® seems to belong to the group of passive materials, despite its name, as it acts as a promoter of precipitation, but does not itself bring any mineral ions.

Bone Bonding

The quality of bone-biomaterial bonding results of two processes: a mechanical effect, the interlocking of bone tissue with the biomaterial, and chemical effect, the interaction of bone and biomaterials components.

Interlocking is facilitated by the exitence of surface irregularities on the material. These are invaded by bone and allow a stable anchoring to the tissue. This step is crucial as it insures the stability of the implant and its bone integration. The existence of a porous surface and/or a partly biodegradable material, always irregularly resorbed, add more irregularities and should be considered as positive. The osteoconductivity of Ca-P biomaterials facilitate the rehabitation of the implant surface, however, bone development is also linked to bone physiology. The importance of micromovements of the implant has been stressed by different authors. Movements of low amplitude (less than 50 micrometer) do not affect osseointegration. Movements between 50 and 100 micrometer lead to partial bone ingrowth only, whereas micromovements of the implant superior to 200 micrometer were characterized by the formation of fibrous encapsulation tissue.[92-93] The integration to living bone of an implant material is also related to the mechanical stimulation of bone tissue. This is determined by biomechanical considerations developed in other reports and is based on the simple observation that non-stimulated bone tissue tends to lose its volume and strength. The matching of the Young modulus of bone with that of the implant is considered as an essential quality to insure a mechanical continuity. Pure Ca-P materials by themselves cannot fulfill this objective. Only organo-mineral composites show a close mechanical analogy with bone tissues.

The second effect that determine the integration of the material to bone tissue is a surface interaction. The neoformed layer certainly assures a continuity between bioactive biomaterial and bone mineral, preventing the interposition of a fibrous encapsulating tissue. The adhesion of this layer to the biomaterial is thus an important parameter in the stability of the implant-bone interface. It has been shown that a continuity exists between the apatite crystals of a biomaterial and the crystals of the neoformed layer. This particular relationship does not exist with other materials and appear as a very important factor justifying the use of apatitic biomaterials.[94] Other Ca-P substrates may give weaker interactions, as there will always exist a mismatch between the apatitic lattices and the substrate lattice.

One last problem of importance is the behavior of the neo-formed layer during bone turn-over in the vicinity of the implant. This phenomenon is difficult to study and it has never been addressed. The permanent nature of the implant bone interface essentially relies on the ability of the implant material to regenerate the neo-formed layer again and again on successive turn-overs or remodeling. Implants composed of "passive" bioactive biomaterials will not have their potential for generating the neo-formed layer altered. Biomaterials belonging to the "active" group on the contrary are consumed or modified during the process of the carbonate apatite layer formation. Thus, they become less and less effective after each turn-over. This is not a real drawback for the bioresorbable implants that constitute a large part of the active group. It might however be a problem for bioglasses and some implants of the hydrogel series.

Ca-P Alteration and Disaggregation

The degradation of Ca-P materials may involve biological and physical-chemical factors. Biological degradation of Ca-P materials[95] is linked to the production of acids by several cells such as osteoclasts or inflammatory cells. This production is dependent on the intensity of the inflammatory response following implantation.

In addition, chemical degradation may occur, depending on the material make up. Some physical parameters may also determine the degradation rate, such as porosity, the heterogeneity of phase distributions, and the size of the Ca-P crystals. Biodegradation has been found to be roughly related to the solubility of Ca-P compounds[96]. Thus relatively soluble Ca-P materials such as tricalcium phosphate (α or β) and poorly crystalline apatites are resorbable, whereas stoechiometric apatites are considered non-bioresorbable.

The most common reaction that can occur in all Ca-P materials is dissolution at grain boundaries and release of the constituent particulates into the body. Depending on the nature, size and the amount of particulate released, the phenomenon may lead to an inflammatory response, adding cell degradation to chemical degradation. Dissolution of sintered HA at grain boundaries depends on its purity and on the sintering temperature: pure HA sintered at high temperatures shows very little degradation and is generally considered as non-bioresorbable although some surface alterations have been reported. In mixtures of HA and other constituents like in BCP (Biphasic Calcium Phosphates made of S-HA and β-TCP) or plasma sprayed coatings, the release of HA particles is essentially related to the degradation of the more soluble constituents. The slow release of particles could play a role in the degradation by promoting a mild inflammatory response. Also, although ceramic S-HA is a non-bioresorbable material, isolated S-HA crystals can be dissolved due to cell action (osteoclasts or inflammatory cells)

The second obvious phenomenon is the complete dissolution of phases that cannot reach their solubility equilibrium in physiological media. All the most soluble phases including DCPD, TTCP or CaO which appear as impurities in plasma spray coatings or in cements are involved in this process. However, stoichiometric pure apatite (hydroxyapatite, oxyapatites, fluorapatite) with very low solubility products cannot dissolve in body fluids without specific cell activity.

Dissolution may also be followed by reprecipitation if the local supersaturation is high enough. When relatively soluble phases, such as tricalcium phosphate, tetracalcium phosphate, calcium oxide, hydroxide or carbonate, or an amorphous phase are present, an apatite precipitate may be locally induced. This phenomenon might be particularly rapid for compounds or impurities inducing a local alkaline pH, such as calcium hydroxide or tetracalcium phosphate, and depends on the surface exposed. This reaction may involve ions from the body fluids or ions from the soluble material itself. In this case, it can be identified as a hydrolysis reaction leading to a more insoluble and stable Ca-P.

Hydrolysis may be particularly fast for OCP, DCPD and α-TCP. Although the hydrolysis rate might be very low at physiological temperatures, for less soluble Ca-P materials like β-TCP and for samples with low specific surface areas, this reaction may in the long term, induce disaggregation of the material.

Another important alteration process is related to the presence of phases with different solubilities. Transfer of matter may occur, in this case, between the more soluble phase and the less soluble one, leading to disaggregation. This phenomenon would be particularly efficient at the boundary between the two phases where strong ionic gradient shall be expected.

The release of debris, the nucleation of free crystals in body fluids and the generation of alkaline pHs, all of which may induce strong adverse responses of the organism, can be related to the existence of multi-phasic Ca-P compounds. It appears then particularly necessary to characterize the phases present in the Ca-P materials as precisely as possible. In the case of plasma sprayed coatings especially, the presence of a hydrolyzable, relatively soluble amorphous phase, at the interface between a plasma sprayed coating and the metal prosthesis may be one of the causes of the weakening of the coating-metal interface after long implantation periods.

The coatings obtained at low temperature are generally quite homogeneous, but they are deficient apatites, hence more soluble than stoichiometric HA. They cannot constitute a permanent coating, although several ways of decreasing their resorbability are under study.

Inflammatory behavior of Ca-P biomaterials

Natural Ca-P compounds are known to induce inflammatory reactions in synovial fluids, calcified tendon, and more general ectopic calcifications. Ca-P biomaterials may have a similar behavior. Studies agree to relate the intensity of the inflammatory response to the size of the crystals, or more specifically, to the surface area of the Ca-P compound. The smallest crystals induce the strongest response[97]. Other factors are, however, certainly involved, such as the chemical composition and the surface characteristics. It seems that the inflammatory response is stronger with more insoluble materials. Particles with sharp edges have also been demonstrated to promote stronger inflammation. These experiments were however made in soft tissues in animal models. Generally, mild inflammatory responses are observed in bone tissues.

Associations with active molecules

In order to enhance or orient biological activity, Ca-P materials have been associated with active molecules including antibiotic, anti-osteoporotic, antimitotic agents, and growth factors. Ca-P compounds have also been suggested as potential drugs carrier due to their ability to progressively release active compounds as they are dissolved and degraded. Several types of association can be distinguished from impregnation of porous materials to mixing or adsorption.

The simplest way to associate an active molecules with Ca-P materials is to mix the dry powders and to compress them. This technique seems however inefficient for Ca-P compounds as the compression of a apatite powder does not give a cohesive pellet. The results is a disaggregation, a massive release of mineral particulates and an inflammatory response. The possibility of segregation of the active molecules as well as their irregular release are potential drawbacks of this technique. Only elaborated techniques such as dynamic high speed compression (compression by explosion) could possibly be effective, but studies have not yet been completed. Besides this technique is not practical for industry.[98] The addition of a binding agent (generally, a bioresorbable polymer) may, however, allow the control of the release rate of the active compound. The risk to induce an important inflammatory response when degradation is not controlled should be kept in mind though.

When biomaterials have been constituted by a high temperature forming technique (sintering, plasma spray), association can be made by adsorption. However, due to the very low specific surface area of the materials, the amount of active molecules attached to the biomaterial is low. The surfaces are immediately exposed to body fluids and the release of the substance cannot really be controlled. This technique may however be interesting for low temperature coatings made of poorly crystalline apatites with very high surface specific area. Then, the progressive development of the layer may allow an attachment of the active molecules within the thickness of the layer and the release would be determined by its dissolution or degradation rate.

In the case of porous materials the active molecules can be simply impregnated into the implant. The release is then related to diffusion of the active molecules out of the pores. It seems, however, difficult to control and decreases rapidly with the implantation time. Active molecules can also been incorporated within cements obtained at low temperature. Though they must be stable in the acidic or alkaline pH which is usually generated during the setting.

Associations with living cells

The activation of bone growth could be promoted by the attachment and expression of osteoblastic cells on Ca-P materials. The extraction of osteoblast cells is, however, difficult. Most studies in this field have been made with progenitor cells from bone marrow.[100-101] These cells have the properties to differentiate into osteobastic cells and express their characteristics. Associations with different types of materials have been made including sintered S-HA particles and porous ceramics. The feasibility of such association has been demonstrated both *in vitro* cell culture and *in vivo.* A development of these techniques could lead to the reconstruction of bone tissues *in vitro*, in a similar way as skin tissue can be rebuilt from a few cells from a patient.

CONCLUSION

Although significant progress has been made in the last decade, Ca-P compounds cannot yet be considered as ideal biomaterials. They do, however, offer strong advantages on other types of materials for biodegradable implants. Various Ca-P compounds are available under different forms applicable to surgical procedures, and adaptable to many operative situations. The main progress to come is in the area of mechanical properties. The improvement may involve organo-mineral composites. For non-biodegradabe implants, the use of Ca-P compounds offer minor advantages and they will probably be replaced by more efficient fillers in composites for load bearing applications. However, for bone-bonding, Ca-P compounds cannot be replaced. This is beacuse as it has now been established that bioactivity of an orthopedic biomaterial is always related to the formation of carbonate-apatite. Most efforts in the future will probably be to better understand the formation conditions and the properties of this layer in order to control, and possibly orient, its biological activity.

REFERENCES

1. Elliott JC. *Structure and Chemistry of the Apatites and Other calcium Orthophosphates*, Elsevier, Amsterdam, 1994.

2. de Groot K. Clinical applications of calcium phosphate biomaterials: a review. Ceram Int 1993; 19:363-366

3. Yamamuro T, Hench L, Wilson J.*CRC Handbook of Bioactive Ceramics Vol II Calcium Phosphate and Hydroxylapatite Ceramics*. CRC Press, Boca Raton, 1990.

4. Legeros RZ, *Calcium Phosphates in Oral Biology and Medicine Monographs in Oral Science Vol 15.* H Myers ed. Karger, Basel, 1991.

5. Rey C, Freche M, Heughebaert M, Heughebaert JC, Lacout JL, Lebugle A, Szilagyi J, Vignoles M. "Apatite Chemistry in Biomaterial Preparation, Shaping and Biological Behaviour". In: *Bioceramics 4.* W. Bonfield, G.W. Hastings, K.E. Tanner eds. Butterworth-Heineman, London, 1991.

6. Barthelemy R, Montoya R, Lebugle A, Guilhem A. Nerve regeneration through synthetic biodegradable nerve guide. Rev Europ Tech Biomed 1989; 11:347-350

7. Glimcher MJ, Kato F, Ninomaya S, Hirotani H, Suzuki H, Mijayi N, Eguro H, Landis W, Shapiro F. "The Biology of Bone Healing and the Repair of Autograft, Allograft and Xenograft Metatarseal-phalangeal Joint Transplants in Rabbits". In: *Osteochondral allografts.* GE Friedlaender, HJ Mankin, Sell KW, eds. Little, Brown and Co., Boston, 1983.

8. Friedlaender GE, Strong DM, Sell KW. Studies on the antigenicity of bone I- Freeze dried and deep-frozen allografts in rabbits. J Bone Joint Surg 1976; 58: 854-858

9. Rey C, Renugopalakrishnan V, Collins B, Glimcher MJ. Fourier transform infrared spectroscopy study of the carbonate ions in bone mineral during aging. Calcif. Tissue Int 1991; 49:251-258

10. Legros R, Balmain N, Bonel G. Age related changes in mineral of rat and bovine cortical bone. Calcif Tissue Int 1987; 41:137-144

11. Glimcher MJ. "The Nature of the Mineral Component of Bone and the Mechanism of Calcification". In: *Disorders of Bone and Mineral Metabolism.* FL Coe, MJ Favus, eds. Raven Press, NewYork, 1992.

12. Rey C, Lian L, Grynpas M, Shapiro F, Zylberberg L, Glimcher MJ. Non-apatitic environments in bone mineral : FT-IR detection, biological properties and changes in several disease states. Connective Tissue Res 1989; 21:267-273

13. Rey C, Kim HM, Glimcher MJ. "Maturation of Poorly Crystalline Synthetic and Biological Apatite". In *Hydroxyapatite and Related Materials.* Brown PW and Constantz B, eds. CRC Press, Boca Raton, 1994.

14. Rey C, Strawich E, Glimcher MJ. Non-apatitic environments in Ca-P biominerals; implications in reactivity of the mineral phase and its interactions with the organic matrix constituents. Bull Inst Océanographique de Monaco N spécial, 1994; 14:55-64

15. Young RA, Holcomb DW. Variability of hydroxyapatite preparations. Calcif Tissue Inter 1982; 34:517-532

16. Doi Y, Moriwaki Y, Aoba T, Okazaki M, Takahashi J, Joshin K. Carbonate apatites from aqueous and non-aqueous media studied by ESR, IR and X-ray diffraction: effect of NH_4^+ions on crystallographic parameters. J Dent Res 1982; 61:429-434

17. Eanes ED, Gillessen IH, Posner AS. Intermediate states in the precipitation of hydroxyapatite. Nature 1965; 208:365-367

18. Bonel G, Heughebaert JC, Heughebaert M, Lacout JL, Lebugle A. "Apatitic Calcium Phosphates and Related Compounds for Biomaterials Preparations". In: *Bioceramics: Material characteristics versus in vivo behavior.* P Ducheyne and J. Lemons, eds. Ann New York Acad Sci, New York, 1988.

19. Mayer I, Wahnon S, Cohen S, Preparation of hydroxyapatites via the MSO_4 suphates (M=Ca,Sr,Pb and Eu). Mater Res Bull 1979; 14:1479-1483

20. Clark JS, Turner RC. Reactions between solid calcium carbonate and orthophosphate solutions. Can J Chem 1955; 33:665-671

21. Roy DM, Linnehan SA. Hydroxyapatite formed from coral skeleton carbonate by hydrothermic exchange. Nature 1974; 247:220-227

22. Kanazawa T, *Inorganic Phosphate Materials, Materials Science Monograph 52.* Elsevier, New York, 1989.

23. Fowler BO. Infrared studies of apatites II. Preparation of normal and isotopically substituted calcium, strontium, and barium hydroxyapatite and spectra-structure-composition correlations. Inorg Chem 1974; 13:207-204

24. Takahashi H, Yashima M, Kakihana M, Yoshimura M. Synthesis of stoichiometric hydroxyapatite by a "gel" route from the aqueous solution of citric and phosphonoacetic acids. Eur J Solid State Inorg Chem 1995; 32:829-835

25. Brendel T, Engel A, Ruessel C. Hydroxyapatite coatings by a polymeric route. J. Mater Sci: Mater in Med 1992; 3:175-178

26. Rey C, Hina A, Tofighi A, Glimcher MJ. Maturation of poorly crystalline apatites: chemical and structural aspects in vivo and in vitro. Cells and Materials 1995; 5:345-356

27. Mortier A, Lemaitre J, Rodrique L, Rouxhet PG. Synthesis and thermal behavior of well-crystallized calcium-deficient phosphate apatite. J Solid State Chem 1989; 78:215-219

28. Nagata F, Yokogawa Y, Toriyama M, Kawamoto Y, Suzuki T, Nishizawa K. Hydrothermal synthesis of hydroxyapatite crystals in the presence of methanol. J Ceram Soc Japan 1995; 103:69-72

29. Schroeder LW, Dickens B, Brown WE. Crystallographic studies of the role of Mg as a stabilizing impurity in b-$Ca_3(PO_4)_2$: II Refinement of Mg-containing b-$Ca_3(PO_4)_2$. J Sol State Chem 1977; 22:253-262

30. Monma H, Ueno S, Kanazawa T. Properties of hydroxyapatites prepared by the hydrolysis of tricalcium phosphate. J Chem Tech Biotechnol 1981; 31:15-24

31. Heughebaert JC, Montel G. Préparation de l'orthophosphate tricalcique pur. Bull Soc Chim Fr 1970; 8:1923-1924

32. Gregory TM, Moreno EC, Patel JM, Brown WE. Solubility of b-$Ca_3(PO_4)_2$ in the system $Ca(OH)_2$-H_3PO_4-H_2O at 5, 15, 25 and 37 °C. J Res Natl Bur Stand 1974; 78:667-674

33. Daculsi G, LeGeros RZ, Nery E, Lynch K, Kerebel B. Transformation of biphasic calcium phosphate ceramics in vivo: ultrastructural and physico-chemical characterization. J Biomed Mater Res 1988; 23: 883-894

34. Aia MA, Goldsmith RL, Mooney RW. Precipitating stoichiometric $CaHPO_4$, $2H_2O$. Ind Eng Chem 1961; 53:55-57

35. Legeros RZ. Preparation of octacalcium phosphate (OCP): a direct fast method. Calcif Tissue Int 1985; 37:194-197

36. Ciesla K, Rudnicki R. Synthesis and transformation of tetracalcium phosphate in solid state. Part I. Synthesis of roentgenographically pure tetracalcium phosphate from calcium dibasicphosphate and calcite. Pol J Chem 1987; 61:719-727

37. Heughebaert JC, Montel G. Conversion of amorphous tricalcium phosphate into apatitic calcium phosphate. Calcif Tissue Int 1982; 34: 103-108

38. Gross KA, Berndt CC. "The Amorphous Phase in Plasma Sprayed Hydroxyapatite Coatings". In: *Bioceramics 8*. J Wilson, L Hench, D Greenspan, eds. 1995.

39. Layrolle P, Lebugle A. Characterization and reactivity of nano sized calcium phosphates prepared in ahydrous ethanol. Chem Mater 1994; 6:1996-2004

40. Nelson LS, Holt C, Harries JE, Hukins DWL. Amorphous calcium phosphates of different composition give very similar EXAFS spectra. Physica B 1989; 158:105-106

41. Kasuga T, Ichino A, Abe Y. Preparation of calcium phosphate fibers for applications to biomedical field. J Ceram Soc Japan 1992; 100:1088-1089

42. Trombe JC, Montel G, Some feature of the incorporation of oxygen in different oxidation state in the apatite lattice 1. On the existence of Calcium and strontium oxyapatites, J Inorg Nucl Chem 1978; 40:15-21

43. Puajindanetr S, Best S, Bonfield W. Preparation, charaterization and sintering of hydroxyapatite. British Ceram Trans 1993; 3:96-99

44. Ababou A, Bernache-Assolant D. "Sintering of Hydroxyapatites with Different Additives". In *Fourth Euro-Ceramics Volume 8 Bioceramics.* A Ravaglioli, ed. Gruppo Editoriale Faenza Editrice Spa, Faenza, 1995.

45. Jarcho M, Salsbury RL, Thomas MB, Doremus RH, Synthesis and fabrication of β-tricalcium phosphate ceramics for potential prosthetic applications, J Mater Sci 1979, 14:142-150

46. Famery R, Richard N, Boch P. Preparation of a-and b-tricalcium phosphate ceramics with and without Mg addition. Ceram Int 1994; 20:327-336

47. Binder-Royer A, Etude de l'influence de la composition de l'hydroxyapatite frittée sur ses propriétés mécaniques. Thèse, Institut National Polytechnique de Grenoble, 1990.

48. Sherman AJ, Tuffias RH, Kaplan RB, Refractory ceramic foams: a novel high temperature stucture, Ceram Bull 1991,70:6

49. Mittelmeier H, Mittelmeier W, Processus for the preparation of bone substitute material based on natural bone. CA patent, CA 1,242,653

50. Khor KA, Cheang P, Wang Y. Thermal spray processing of hydroxyapatite powders and coatings. J of Materials 1997, February; 51-57

51. Wolke JGC, de Blieck-hogervorst JMA, Dhert WJA, Klein CPAT, de Groot K. Studies on the thermal spraying of apatites bioceramics. J Thermal Spray Technol 1992; 1:75-82

52. LeGeros JP, LeGeros RZ, Burgess A, Edwards B, Zitelli J, "X-ray Diffraction Method for the Quantitative Characterization of Calcium Phosphate Coatings". In: *Characterization and Performance of Calcium Phosphate Coating for Implants, ASTM STP 1196.* E. Horowitz, JE Parr, eds, ASTM, 1994.

53. Dhert WJA, Thomsen P, Klein CPAT, de Groot K, Rozing PM, Ericson LE. Fluorapatite coated implants in experimental arthritis: the response of rabbit trabecular bone. J Mater Sci: Mater Med 1994; 5:59-66

54. Ranz X, Rey C, Antolotti N, Harmand MF, Moroni A, Orienti L, Viola G, Bertini S, Scrivani A. "Properties of Plasma sprayed bioactive fluorhydroxyapatite coatings". In: *Bioceramics 10,* L Sedel C Rey, eds. Elsevier, Oxford, (in press).

55. Brown WE, Chow LC. "A New Calcium Phosphate Water-setting Cement". In: *Cement Research Progress.* P Brown ed. Amer Ceram Soc, Westfield, 1986.

56. Driessens FCM, Botlong MG, Bermudez O, Planell JA, Ginebra MP, Fernandez E. Effective formulations for the preparation of calcium phosphate bone cements. J Mater Res: Mater Med 1994; 5:164-170

57. Mirtchi AA, Lemaitre J, Terao N. Calcium phosphate cements: study of the b-tricalcium phosphate - monocalcium phosphate system. Biomaterials 1989; 10:475-480

58. Constantz BR, Ison IC, Fulmer MT, Poser RD, Smith ST, VanWagoner M, Ross J, Goldstein SA, Jupiter JB, Rosenthal DI. Skeletal repair by in situ formation of the mineral phase of bone. Science 1995; 267:1796-1798

59. Lacout JL, Mejdoubi E, Hamad M. Crystallization mechanism of calcium phosphate cement for biological uses. J Mater Sci: Mater Med 1996; 7:371-374

60. Kurashina K, Hurita H, Hirano M, Blieck de JMA, Klein CPAT, Groot de K. Calcium phosphate cement: in vitro and in vivo studies of the a-Tricalcium phosphate-dicalcium phosphate dibasic-tetracalcium phosphate monoxide system. J Mater Sci: Mater Med 1995; 6:340-347

61. Lee D, Rey C, Aiolova M. Orthopaedic and dental ceramic implants. US Patent 1996, 08/729,343

62. Abe Y, KokuboT, Yamamuro T. Apatite coating on ceramic metals and polymers utilizing a biological process. Materials in Medicine 1990; 1:233-238

63. Royer P, Rey C. Calcium Phosphate coatings for orthopaedic prosthesis. Surf Coat Tech 1991; 45: 171-177

64. Shirkhanzadeh M. Calcium phosphate coatings prepared by electrocrystallization from aqueous electrolytes. J Mater Sci: Mater Med 1995, 6:90-93

65. Lacefield WR, Rigney ED, Lucas L, Ong J, Gautenberg JB. Sputter deposiiton of calcium-phosphate coatings onto metallic implants. Mat Sci Monog 1961; 69:167-76

66. Cotell CM. Pulsed laser deposition and processing of biocompatible hydroxylapatite thin films. Appl Surf Sci 1993; 69:140-148

67. Freche M. Contribution à l'étude des phosphates de calcium, croissance sur le phosphate dicalcique anhydre, croissance de dépôts élaborés par pulvérisation chimique. Thèse d'Etat, INPT, Toulouse

68. Ducheyne P, Radin S, Heughebaert JC, Heughebaert M. Calcium phosphate ceramics coatings on porous titanium: effect of structure and composition on electrophoretic deposition, vacuum sintering and in vitro dissolution. Biomaterials 1990; 11:244-254

69. Bonfield W. Composites for bone replacement. J Biomed Eng 1988; 10:522-526

70. Delpech V, Lebugle A. Calcium phosphate and polymer interfaces in orthopaedic cement. Clin Mater 1990; 5:209-216

71. Castaldini A, Cavallini A. Setting properties of bone cement with added synthetic hydroxyapatite. Biomaterials 1985; 6:50-60

72. Antonucci JM, Eanes ED, Skrtic D. Bioactive polymeric dental materials based on amorphous calcium phosphate. Polym Prep 1994; 35:460-461

73. Wang M, Porter D, Bonfield W, Processing characterization and evaluation of hydroxyapatite reinforced polyethylene composites, Britsh Ceram trans 1993, 3:91-95

74. Verheyen CCPM. Resorbable materials with bone-bonding ability. PhD-Thesis, University of Leiden, 1993.

75. Doyle C, Tanner ET, Bonfield W. In vitro and in vivo evaluation of poluhydroxybutyrate and of poluhydroxybutyrate reinforced with hydroxyapatite. Biomaterials 1991; 12:841-847

76. Mattioli-Belmonte M, Biagini G, Muzzarelli RAA, Castaldini C, Gandolfi MG, Krajevski A, Ravaglioli A, Fini M, Giardino R. Osteoinduction in the presence of chitosan-coated porous hydroxyapatite. J Bioact Compat Polym 1995; 10:249-257

77. Grimaldi G, Weiss P, Millot F, Passuti N, Daculsi G. In vitro evaluation of a new injectable calcium phosphate material. J Biomed Mater Res (in press)

78. Serre CM, Papillard M, Chavassiuex P, Boivin G. In vitro induction of calcifying matrix by biomaterials constituted of collagen and/or hydroxyapatite: an ultrastructural comparison of three types of biomaterials. Biomaterials 1993; 14:97-106

79. Frayssinet P, Asimus E, Autefage A, Fages J, Histological evaluation of xenogenic bone treated by supercritical CO_2 implanted into sheep. J Mater Sci: Mater in Medicine 1995; 6: 473-478

80. Moller K, Meyer U, Szulczewski DH, Heide H, Priessnitz B, Jones DB. The influence of zeta potential and interfacial tension on osteoblast-like cells. Cells and Materials 1994; 4:263-274

81. Dimitrova-Lukacs M, Lukacs P, Sajo I. "Composite Bioceramics in the System Hydroxyapatite (Fluorapatite)-Yttria Stabilized Zirconia". In: *Materials in Clinical Applications*. P. Vincenzini ed. Techna, Faenza, 1995.

82. Chow LC, Markovic M, Takagi S, Cherng M. Effect of cement liquid on the physical properties of the cement. Innov Techn Biol Med 1997; 18: 11-14

83. Bermudez O, Boltong MG, Driessens FCM, Planell JA. Compressive strength and diametral tensile strength of some calcium orthophosphate cements: a pilot study. J Matyer Sci: Mater Med 1993;4:389-393

84. Pillar RM, Filiaggi MJ. "Mechanical Characterization of Plasma-sprayed Hydroxyapatites-titanium Alloy Interfaces". In: *Bioceramics 4*. W. Bonfield, G.W. Hastings, K.E. Tanner, eds. Butterworth-Heinemann, London, 1991.

85. Davies JE. "The Use of Cell and Tissue Culture to Investigate Bone Cells Reactions to Bioactive Materials". In: *CRC Handbook of Bioactive Ceramics*. T Yamamuro, LL Hench, J Wilson, eds. CRC Press, Boca Raton, 1990.

86. Gomi K, Lowenberg B, Shapiro G, Davies JE. Resoption of sintered synthetic hydroxyapatite by osteoclasts in vitro. Biomaterials 1993; 14:91-96

87. Spivak, J.L. Ricci, N.C. Blumenthal, H.Alexander. A new canine model to evaluate the biological response of intramedullary bone to implant materials and surfaces. J Biomed Mater Res 1990; 24:1121-1149

88. Rey C, Hina A, Amrah-Bouali S, Ranz X. "Surface Reactions of Calcium-phosphate Bioceramics, Comparison with Bone Mineral Surface Chemistry". In: *Fourth Euro-Ceramics . Symposium Bioceramics*. A. Ravaglioli, ed. Gruppo Editoriale Faenza Editrice Spa, Faenza, 1995.

89. Rey C, Hina A. Neoformed apatite layers on bioactive orthopedic biomaterials. mode of formation and and analogies with bone mineral. Innov Techn Biol Med 1997; 18:1-10

90. van Blitterswijk CA, Bowell YP, Flach JS, Leenders H, van den Brink I, de Bruijn J. "Variation in Hydroxyapatite Crystallinity : Effects on Interface Reactions". In: *Hydroxyapatite Coatings in Orthopaedic Surgery*. RGT Geesink and MT Manley, eds. Raven Press, New York, 1993.

91. Kawagushi H, McKee MD, Okamoto H, Nanci A, Tissue response to hydroxyapatite implanted in an alveolar bone defect: Immunocytochemical and lecitin-gold characterization of the hydroxyapatite-bone interface. Cells and Materials 1993; 3:337-350

92. Hollis JM, Hofmann OE, Stewart CL Flahiff CM, Nelson C. Effect of micromotion on ingrowth into porous coated implants using a transcortical model. Fourth World Biomaterials Congress, Berlin 1992.

93. Pillar R, Deporter D, Watson P. "Tissue-implant Interface: Micromovements Effects". In: *Advances in Science and Technology, Vol 12, Materials in Clinical Applications.* P. Vincenzini ed. Techna, Faenza, 1995.

94. Bonfield W, Lublinska ZB. "High Resolution Microscopy of a Bone Implant Surface". In: *The Bone-Biomaterial Interface.* JE Davies, ed. Toronto University Press, Toronto, 1991.

95. Frayssinet P, Rouquet N, Tourenne F, Fages J, Hardy D, Bonel G. Cell-degradation of calcium phosphate ceramics. Cells and Materials 1993; 3:383-394

96. LeGeros RZ, LeGeros JP, Daculsi G, Kijkowska R. "Calcium Phosphate Biomaterials: Preparation, Properties and Biodegradation". In *Encyclopedic Handbook of Biomaterails and Bioengineering.* DB Wise et al. eds. Marcel Dekker, New York, 1995.

97. Prudhommeaux F, Schiltz C, Lioté A, Hina A, Champy R, Bucki B, Ortiz-Bravo E, Meunier A, Rey C, Bardin T. Variation in the inflammatory properties of basic calcium phosphate crystals according to crystal type. Arth and Rheum 1996; 39:1319-1326

98. Trecant M, Daculsi G, Leroy M, Dynamic compaction of calcium phosphate biomaterial, J Mater Sci: Mater in Med 1995, 6: 545-551

99. Martinetti R, Denissen H, Roncari E, Pinasco P, Mangano C. "Development of Hydroxyapatite Implants as Release Systems for Drugs". In: *Fourth Euro-Cermics, Volume 8 Bioceramics.* A Ravaglioli ed. Gruppo Editoriale Faenza Editrice Spa, Faenza, 1995.

100. Frayssinet P, Autefage A. Hybrid materials for use as bone substitutes. Rev Rhum (english edition) 1993; 60:302-310

101. Goshima J, Goldberg VM, Caplan AI. The osteogenic potential of of culture expanded rat marrow mesenchymal cells assayed in vivo in calcium phosphate ceramic blocks. Clin Orthop Rel Res 1991; 262:298-311

11

IMPORTANCE OF CALCIUM PHOSPHATES IN THE DEVELOPMENT OF CALCIUM UROLITHIASIS

Saeed R. Khan

Department of Pathology, College of Medicine
University of Florida, Gainesville, Florida 32610 USA

INTRODUCTION

Calcium oxalate (CaOx) and calcium phosphate (CaP) are the two most common crystalline constituents of urinary stones and CaOx is their major component.[1-9] Appropriate urinary supersaturation is necessary for crystallization. But mammalian urine is normally insufficiently supersaturated for homogeneous nucleation of CaOx. As a result, suggestions have been made that some urinary elements such as macromolecules or crystals act as the substrate for heterogeneous nucleation of CaOx. CaP is often considered a promoter of CaOx crystallization in the urine because, A. it is common in both urine and urinary stones, B. it frequently occurs together with calcium oxalate (CaOx), C. it is occasionally present at the nidus of CaOx-CaP stones, and D. it can support nucleation of CaOx *in vitro*. The possibility of CaP nucleating CaOx crystals and promoting the formation of CaOx stones will be explored in this chapter.

CRYSTALLINE COMPOSITION OF HUMAN URINARY STONES

Results of various studies of mineral composition of urinary stones from different countries and regions indicate a presence of more than 30 types of crystalline components.[2-9] Pure stones consisting of only one type of crystals are rare. In the most recent study in which a total of 10,617 stones were analyzed by optical microscopy and infrared spectroscopy,[9] only 6.7% were pure stones; 22.6% of stones contained two components, 52.9% three components, 13.7% four components while 3.9% contained more than four components. Some reports have suggested a much higher frequency of pure, single component stones. The discrepancy may be a result of, A. application of analytical techniques such as x-ray diffraction that have poor sensitivity to detect minor amounts of CaP and, B. inadequate sampling of various sections of a stone.

Based on the data in text references 2-9, Table 1 lists the calcific components of human urinary stones. CaOx is the most common crystalline element in the stones while CaP is second in frequency of occurrence. CaOx stones are more common in men while CaP stones are typical in women. Otnes et al. analyzed 245 stones from Norwegian patients[2] and found pure CaOx stones were more common in men (40%) than in women (7%), while pure CaP stones were more frequent in women (11%) than men (2%). CaOx stones were also more common in upper urinary tract while CaP stones frequented the bladder.

Table 1. Frequency of occurrence of various crystals in mixed or pure stones[2-9]

Crystalline component	Mixed stones	Pure stones
Calcium Oxalate Monohydrate (COM)	43-83.4%	3.58-24.9%
Calcium Oxalate Dihydrate (COD)	42.1-59%	0.23-4.7%
Calcium Oxalate Trihydrate	0.06%	
Apatite (HAP)	23.1-79.39%	0.3-6.6%
Brushite	0.5-2.3%	0.02-1.2%
Whitlokite	0.3-5.1%	
Octacalcium Phosphate	0.2-1.3%	
COM/COD	21.12-33.2%	
COM/COD/HAP	5.2-40.5	

Calcium Oxalate exists in three different forms; calcium oxalate monohydrate (COM), calcium oxalate dihydrate (COD) and calcium oxalate trihydrate (COT). All three forms have been detected in human urinary stones.[5] But COM is the most common, being present in as many as 83% of all stones analyzed. COD has been found in as many as 59% of the stones while COT in only 0.06% of the stones. COM is also the most common constituent of pure stones, present in 52% of them. CaP is found as apatite, brushite (calcium hydrogen phosphate dihydrate), whitlokite (β-tricalcium phosphate), octacalcium phosphate or amorphous calcium phosphate. As a constituent of stones, apatite is ubiquitously being present in approximately 80% of the stones making it the second most common crystalline component of the urinary stones. It may actually be present in all the urinary stones and not be recognized because of its small size and amount. Apatite occurs in two forms, hydroxyapatite (HAP) or carbonate apatite. Other types of calcium phosphate are very rare in stones.

CRYSTAL MORPHOLOGY

Scanning electron microscopic examination of urinary stones and crystalluria particles has provided extensive information about the structure of crystals and their relationship with each other inside a stone.[10] COM crystals have a monoclinic, platelike habit (Figure 1). They are radially organized inside the stones (Figure 2) and seen as dumbbell-shaped aggregates in urine. COD has a distinctive bipyramidal tetragonal form (Figure 3). Apatites occur as small 2-6 μm across spherulitic aggregates (Figure 4) of needle-shaped or platelike crystals. Occasionally spherulites coalesce (Figure 5) to make larger units which form thin but solid laminations (Figure 2) between layers of COM. Such apatite layers appear harder and less brittle.

Smaller, fine-grained powdery crystals of apatite (Figure 4) fill the intracrystalline spaces between larger crystals of COM and COD. Since stones with COD crystals have numerous interstices their CaP contents are generally higher than that of COM stones. Similar crystals of apatite have also been seen occupying the nidus of mixed CaOx/CaP stones. Occasionally, CaP crystals in the amorphous form or with spherulitic habit are detected at the core of radially organized CaOx crystals (Figures 6,7) or coating their surfaces (Figure 5,8).

Brushite crystals have monoclinic habit (Figure 9) and exist in radial arrangements[10] a morphology reminiscent of radially organized COM crystals.

Figure 1. Platelike crystals of COM inside a human urinary stone.

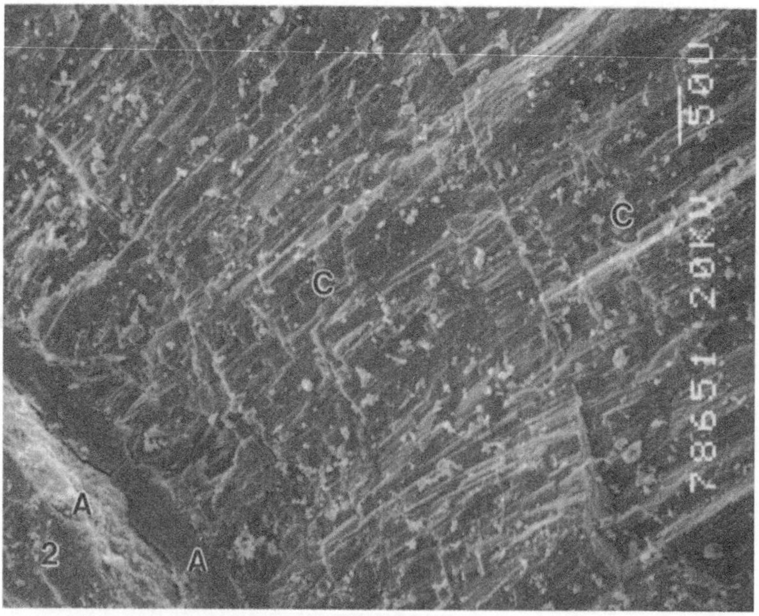

Figure 2. Fractured surface of a CaOx stone showing radially organized COM crystals (C). A thin layer of apatite (A) is also present.

Figure 3. Pyramidal COD crystals in a human CaOx stone.

Figure 4. Spherulitic crystals of apatite (A) inside a human CaOx stone are seen surrounding platelike crystals (C) of COM.

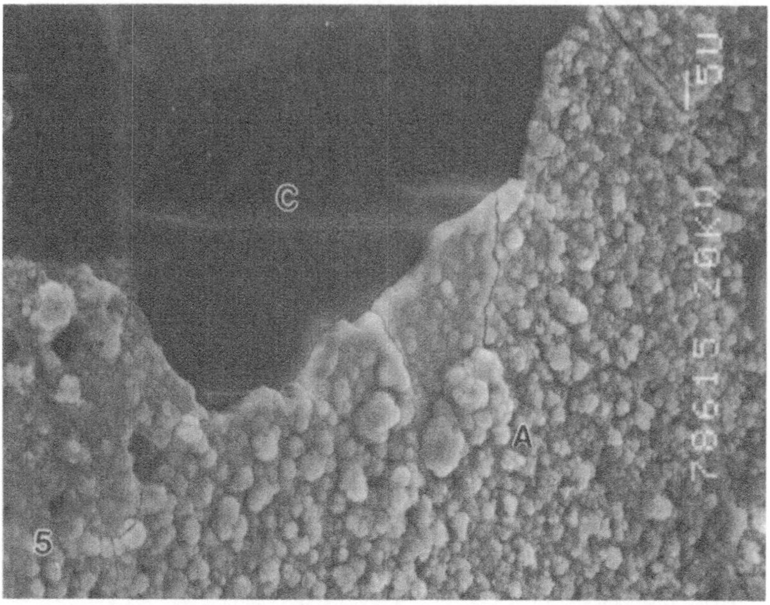

Figure 5. Apatite crystals (A) coalesce to form a thin coating over a CaOx crystal (C).

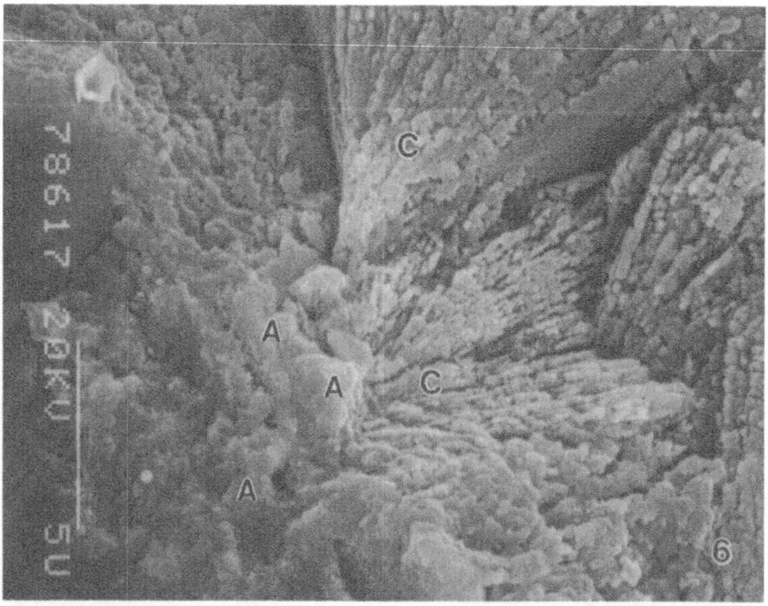

Figure 6. Surface of a fractured stone showing apatite (A) crystals at the core of radially organized COM (C) crystals.

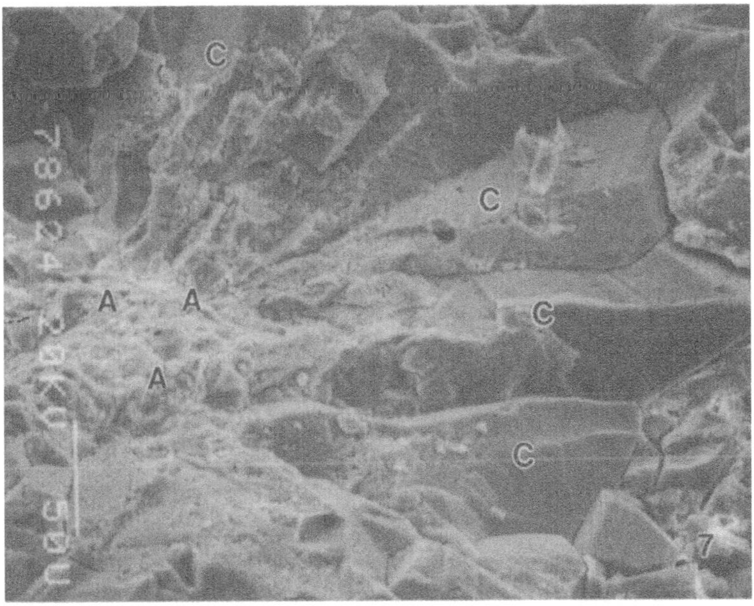

Figure 7. Surface of a fractured stone showing apatite (A) crystals at the core of COD (C) crystal aggregate.

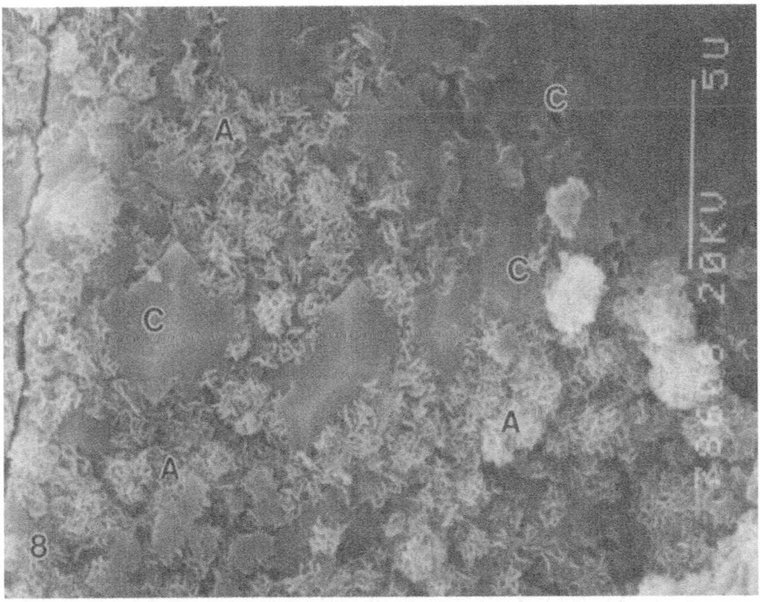

Figure 8. Thin platelike crystals of calcium phosphate (A) coating the surface of a COD crystal (C).

Figure 9. Outer surface of a calcific stone showing brushite crystals.

CRYSTALS IN HUMAN URINE

Human urine is often supersaturated beyond the metastable limit with respect to a variety of stone forming salts including calcium phosphate and calcium oxalate.[1,11] As a result both calcium phosphate and calcium oxalate are common in urine from both normal and stone forming people.[12-14] CaOx is generally encountered in urine with a pH between 5.0 and 6.2 while CaP dominates crystalluria at a pH above 6.2.[12] Urinary crystals of CaOx are either COM or COD with their distinctive morphologies as discussed above. CaP is generally present in the form of poorly crystalline hydroxylapatite (HAP)[14] with spherulitic aggregates of needle-shaped or platelike crystals. Frequently, urinary CaP appears amorphous. Brushite is also found in the urine but under specific circumstances.

Precipitation of various forms of CaP is critically dependent upon pH of the milieu.[15] Forms such as brushite and octacalcium phosphates are preferentially formed in more acidic circumstances while HAP is more likely to develop under basic conditions. Brushite is rare as crystalluria particle but has been seen in the urine from stone patients with primary hyperparathyroidism or idiopathic calcium urolithiasis. It is also interesting to note that only the urine from calcific stone formers is supersaturated for brushite.[16,17] Brushite was the first precipitate to appear at a pH below 6.9, when urinary supersaturation was experimentally increased by addition of calcium chloride.[18] Since brushite is rarely found in urinary stones it was proposed that brushite nucleates CaOx and then transforms into the more stable HAP. Octacalcium phosphate, an extremely rare constituent of urinary stones, has similarly been proposed to nucleate CaOx and then transform into apatite.

Urine from both normal healthy subjects and stone formers contain CaOx and CaP crystals but stone formers produce larger crystals and crystal aggregates.[12] In addition studies of Werness et al.[13,17] have shown that stone formers as a group void more crystals in their urine and more often than normal controls and that HAP is the most common crystal in urine from both normal controls and stone formers. In one study urinary crystals were examined in 4,835 total voiding obtained from 16 normal controls, 45 patients with primary hyperparathyroidism, 12 with primary hyperoxaluria and 89 with idiopathic calcium urolithiasis.[13] Seventy four percent of voiding from normal controls, 21% from stone patients with primary hyperparathyroidism, 8% from primary hyperoxaluria patients and 34% from patients with idiopathic urolithiasis were negative for crystals. Fifty percent urinary voiding with crystalluria of normal controls, 70% of stone patients with primary hyperparathyroidism and 30% of patients with idiopathic calcium urolithiasis contained HAP as the only crystal. All patients with hyperoxaluria had COM as the major crystal usually mixed with small amount of COD crystals. Patients with primary hyperoxaluria produced only CaOx crystals. In another study Werness et al. compared crystalluria between 96 idiopathic stone formers and 20 normal controls.[17] Sixty three percent of the normal voiding had no crystals as compared to 37% of voiding from stone patients. HAP was present in urine from 80% of normal controls and 96% of stone patients. It was a major crystal in urine from 75% of normal controls and 83% of stone patients and was the only crystal in urine from 60% of normal controls and 32% of stone patients. 45% of the stone patients had both HAP and CaOx together in at least one voiding. Only 3% of the normal controls showed such a mixed crystalluria.

EPITAXY BETWEEN URINARY CRYSTALS

Since oriented growth of one crystalline phase over another appears common in stones of mixed composition, it has been suggested that epitaxy may be a major mechanism involved in development and growth of such urinary stones.[19] It is difficult to study actual faces in contact with one another to determine epitaxy. Most of the research has been limited to calculations of lattice misfits for the common stone components. A lattice misfit of no greater than 10-20% can support epitaxial growth.[20] But a misfit of less than 5% is critical for epitaxy in urinary stones.[21] At 15% misfit, a large number of matches are possible. The number of possible matches is dramatically reduced when allowing 10% mismatch and is nearly zero for most common stone components at 5% mismatch.[21] Based on these data, epitaxy could not be responsible for the apatite/COM/COD stone, the most frequent combination, because of the total absence of matches between COM and apatite or apatite and COD at or below 5%. However, when atomic interactions between crystals were examined using molecular overlays, the (100) face of COM and (-110) face of apatite totally supported each other indicating the possibility of an epitaxial overgrowth of one over the other.[22]

IN VITRO STUDIES

Calcium Phosphate As Promotor Of Calcium Oxalate Crystallization

To understand the relationship between crystallization of CaOx and CaP, a number of experimental *in vitro* studies have been carried out. Using solution depletion seeded crystallization techniques, Meyer et al. showed that addition of HAP seeds to a metastable solution of CaOx induced crystallization of COM on the seed.[23] Brushite was also shown to induce nucleation of COM but induction time was longer than with HAP.[24] On the other hand COM was able to induce crystallization of brushite but not of HAP from a metastable solution of CaP in conditions suitable for their respective nucleation. Recent studies using constant composition crystallization system, confirmed the ability of HAP to induce nucleation of COM and inability of COM to induce nucleation of HAP.[25] Calcium oxalate trihydrate could however nucleate HAP.

A very interesting study was performed by Ebrahimpour et al. who pretreated the HAP seeds with human serum albumin and then incubated them in a metastable CaOx solution at a constant supersaturation.[26] Pretreatment of seeds with albumin resulted in reduction in crystallization induction time and increase in rate of subsequent COM crystallization indicating that HAP surfaces covered with albumin are more effective in inducing formation of COM.

Achilles et al. recently evaluated the ability of CaP crystals to induce nucleation of CaOx crystals utilizing a dynamic flow model of crystallization in gels.[27] First a metastable solution of calcium phosphate was coursed over the surface of 1 wt % agar gel at 37 °C for 4 to 8 hours. This resulted in the formation of about 200 μm diameter hemispherulitic particles of CaP in the gel matrix. These particles contained poorly crystalline carbonate apatite in the center and well crystallized radially arranged platelike crystals of octacalcium phosphate on the periphery. CaOx crystals were then generated by flowing a metastable solution of calcium oxalate over the gel surface containing the CaP crystals. Scanning electron microscopic examination showed them to be platelike crystals of COM which appeared to grow in continuation with the platelike crystals of octacalcium phosphate. This study is one of the few to actually demonstrate oriented overgrowth of COM on CaP crystals.

Our own studies have shown that HAP is capable of inducing CaOx crystallization in a buffer as well as artificial urine of a constant but low CaOx relative supersaturation (Figure 10). Addition of HAP seeds to metastable CaOx in buffer or artificial urine[28] at relative supersaturation of 6 or 10 resulted in the formation of typical COM crystals. Crystal induction took longer in artificial urine than in the buffer.

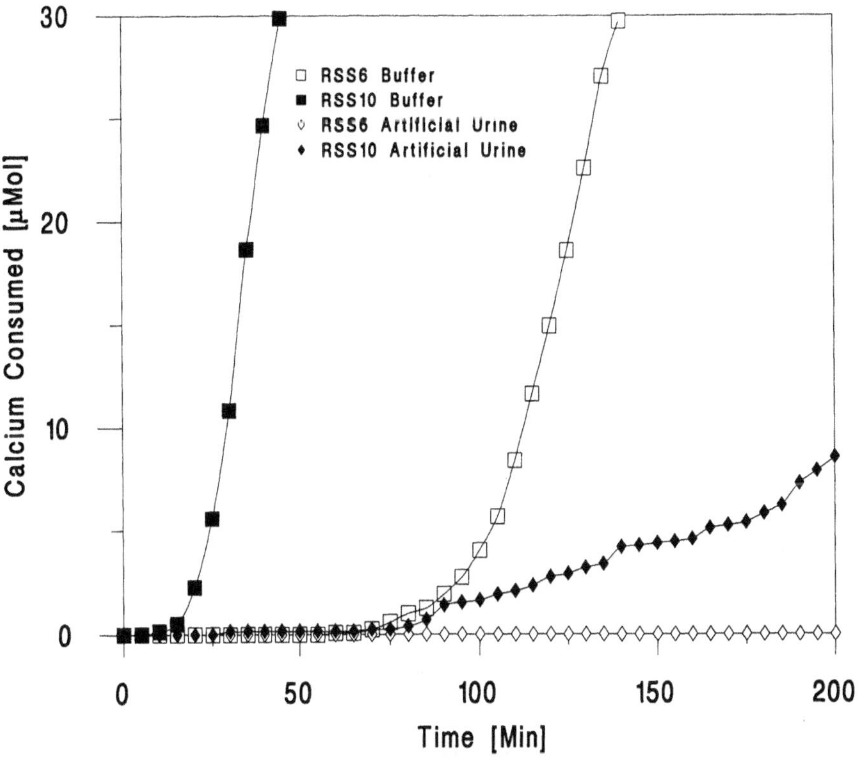

Figure 10. Results of a crystallization experiment in which hydroxyapatite crystals were incubated in a buffer or artificial urine at a constant but low CaOx supersaturation.

Cellular Membranes And Their Lipids As Nucleator Of Calcium Oxalate

Cellular degradation products are common in human urine[29] and small amounts of lipids appear normally therein. Moreover, there is a significant increase in the amount of lipids in the urine of stone formers particularly the phospholipids[30] which have a high affinity for calcific crystals. It is generally agreed that cellular membranes and their lipids promote nucleation of CaP.[31] To understand whether cellular membranes can also promote nucleation of CaOx, brush border membrane vesicles isolated from the renal cortex of male rats were incubated in a metastable solution of CaOx.[32] In a constant composition system crystallization started within 8 minutes of incubation in a metastable solution of CaOx relative supersaturation of 12. Retentate obtained after filtration of the solution was examined by scanning and transmission electron microscopy. The investigation displayed COM crystals forming in close association with the membranous vesicles.

Achilles et al. utilized their dynamic flow model of crystallization in gels[27] to investigate the ability of phospholipids to induce nucleation of CaOx crystals.[33] Their results showed that d-isomer of phosphatidylcholine can induce nucleation of COM.

CRYSTAL DEPOSITION IN THE URINARY SYSTEM

Urinary stones can be found anywhere in the urinary system including kidneys, ureters and bladder. They may form at one location and relocate to another. For example, stones forming in the kidneys can lodge in the ureter. Nephrolithiasis or kidney stone formation refers to deposition of crystals in the kidneys and starts in lumina of renal tubules where it is often called intranephronic calculosis. Crystal deposition outside the tubular lumina is called nephrocalcinosis. Nephrocalcinosis is common in humans and can be found in almost all the kidneys of older humans.[34] Nephrolithiasis on the other hand is relatively uncommon. Intranephronic calculi have been suggested to precede the development of kidney stones.[35,36] Malek and Boyce examined fifty two kidneys from CaOx stone patients and found intranephronic calculi in every one of them.[36] Randall found calcification plaques in renal papillae of 19.6% of 1154 pairs of kidneys he examined at autopsy, and suggested that such plaques could be precursors of kidney stones.[37] The plaques were of two types. Type 1 appeared to include subepithelial deposits while type 2 were intraluminal. Cifuentes-Delatte demonstrated that 28.5% of 500 spontaneously passed stones contained nidi consisting of Randall's plaques.[38] Plaques themselves consisted of calcified tubules in which CaP was easily identifiable.

Experimental Nephrolithiasis

Studies of CaOx nephrolithiasis in rats[39] have demonstrated that hyperoxaluria alone can induce CaOx crystalluria and nephrolithiasis without the assistance of any CaP phase. Acute hyperoxaluria promotes a diffuse deposition of CaOx crystals throughout the kidney while a sustained low level chronic hyperoxaluria causes preferential deposition of CaOx crystals in the renal papilla.[40] Cellular degradation products in the form of membranous vesicles are consistently seen in association with crystal deposits. A low grade hyperoxaluria, which by itself can not generate CaOx crystallization, promotes CaOx crystalluria[41] and accelerates the onset of CaOx nephrolithiasis in male rats, when induced in association with membranuria brought about by the administration of gentamicin sulphate, an aminoglycoside antibiotic.

CaP nephrolithiasis can be experimentally induced in rats by various treatments which cause increased urinary excretion of calcium[42] or phosphate[43] or decreased excretion of magnesium.[44] CaP preferentially deposits in the renal tubules at the cortico-medullary junction and crystals are closely associated with cellular degradation products.[45]

To understand the relationship between CaP and CaOx crystal deposition in the mammalian kidneys we induced experimental CaP and CaOx nephrolithiasis in male and female Sprague-Dawley rats.[46] Renal deposition of CaP was induced by nutritional modifications giving rats a semi-purified diet, AIN-76A. This diet meets nutritional requirements but results in CaP deposition in kidneys of female rats. CaOx deposition was produced by administration of ethylene glycol, a hyperoxaluria-inducing lithogen, and ammonium chloride, a urine acidifying agent. It was hypothesized that induced CaP crystallization followed by hyperoxaluria would result in CaOx deposition on preexisting crystal deposits of CaP. One group of male and female rats was put on CaOx inducing regime and another on CaP inducing regime. Another group received CaP inducing diet for first 2 weeks followed by CaP inducing diet in association with hyperoxaluria inducing regime of ethylene glycol and ammonium chloride. Last group received CaP inducing diet followed by hyperoxaluria inducing protocol. Urine was analyzed for crystalluria and CaOx and CaP relative supersaturation. Kidneys were examined for crystal deposition.

No significant differences were noticed in CaOx relative supersaturation between male and female rats. CaP relative supersaturation was generally lower in males than females. Both male and female rats receiving CaOx inducing treatments produced CaOx crystals in their urine. Only female rats on CaP inducing protocol developed renal deposits of CaP and only male rats on CaOx inducing treatment developed renal deposits of CaOx. CaP deposits were located at the cortico-medullary junction of the kidneys while CaOx deposits were located in collecting ducts of the renal medulla.

In a separate experiment a group of female rats received AIN diet and regular water for 4 weeks and then switched to regular diet and hyperoxaluria inducing treatment of 0.75% ethylene glycol and 2% ammonium chloride in drinking water for next 18 days. Both CaP and CaOx deposited in rat kidneys. Crystals of CaP were located at the cortico-medullary junction and were more abundant than CaOx crystals. The later were randomly dispersed as small deposits in the kidneys. But no direct association was evident between CaOx and CaP crystals. Thus even when CaP crystals were present in the kidneys to act as substrate for nucleation of CaOx and conditions were suitable for CaOx precipitation from a urine with higher urinary oxalate and obviously metastable for CaOx, CaOx crystals did not form in association with CaP.

Results indicate gender-based differences in the crystalline composition of stones. Male rats were prone to form CaOx, while female rats were susceptible to produce CaP. Moreover CaP and CaOx did not deposit in the same location within the kidneys. It is obvious that CaP is not necessary for renal deposition of CaOx and even preexisting CaP deposits do not provoke an overgrowth of CaOx. Gender dependent crystallization of CaOx or CaP may be the reason why more men develop CaOx stones while women produce CaP stones.[2-9]

Foreign Body Stones

Stones comprising CaOx and CaP were produced in male rats by implanting a plastic foreign body in their bladder and inducing hyperoxaluria. Foreign body implantation resulted in the development of stones on the foreign body as well as the formation of small daughter stones.[47] Stones contained CaOx, CaP and magnesium ammonium phosphate (MAP), COM being the most common crystal.

Rats produce MAP crystals in the absence of infection. Examination of fractured stone surfaces showed concentric layers of radially arranged COM crystals which at places alternated with randomly distributed MAP and CaP crystals. The innermost layer of COM crystals appeared to be deposited directly on the foreign body or on a layer of CaP crystals which in turn was laid down on a thin layer of amorphous biological material. The CaP layer present on the foreign body consisted of thin platelike crystals. The layer of platelike CaP crystals also coated the surface of other stone crystals such as MaP and CaOx. Many of the CaP plates were radially organized into spherical units which frequently filled intracrystalline spaces. X-ray microanalysis of CaP crystals showed calcium, phosphorus and magnesium in varying proportions. Our other studies of foreign body stone formation have demonstrated that encrustation begins with laying down of organic material on the foreign body surface. TEM showed the organic material to contain amorphous, fibrillar, membranous and cellular substances and that crystals deposited on this layer. Deposition of CaP was not found essential for laying down CaOx. Our current SEM study also demonstrated a thin amorphous layer on the foreign body surface.

ORGANIC MATRIX OF URINARY STONES

Microscopic Analysis

All urinary crystals and stones contain a ubiquitous organic matrix which represents approximately 2-5% of the dry weight of stones composed primarily of calcium. Demineralization by treatment with 0.25M ethylenediamine tetra-acetic acid (EDTA) at pH 8 and 4 °C in the presence of aldehyde fixatives causes dissolution of crystals and removal of the mineral resulting in the appearance of crystal ghosts.[48-51] Despite the total loss of their mineral contents, stones (Figure 11) and crystals (Figure 12) for the most part maintain their shapes because matrix material is pervasive throughout the crystals and stones and is distributed as a network providing architectural integrity to the calculi. Demineralization of COM stones shows the matrix to be present as a succession of concentric layers like growth rings of a tree (Figure 11). Examination of the matrix at higher magnification also reveals radial striations in the individual layers. The striations are formed by individual COM ghosts that have narrow rectangular boxlike shape (Figure 12). COD crystal ghosts of maintain pyramidal shape and apatite crystal ghosts keep their spherulitic morphology.

Transmission electron microscopic examination provides more detailed morphological information and demonstrates that crystal ghosts are delineated by an organic outer coat (Figures 13, 14). COM ghosts have a rectangular outline made of organic material (Figure 14) while ghosts of spherulitic CaP show organic material radially organized around a central space (Figures 13,14). There appears to be a merging or fusion of matrix material at the junction between the ghosts of CaP and CaOx crystals (Figure 14). Thus CaOx/CaP crystal interface is actually occupied by an organic matrix associated with the two types of crystals.

Transmission electron microscopy also reveals that organic matrix includes fibrillar and membranous substances. Cellular degradation products including membranous vesicles can often be clearly identified.[49] Light microscopic cytochemical analysis using Sudan black B and electron microscopic cytochemical analysis using malachite green displayed presence of lipids in the matrices of calcium oxalate stones.[52,53]

Figure 11. Fractured surface of a demineralized COM stone showing concentrically laminated ubiquitous organic matrix. Despite a total loss of the mineral, stone maintains the shape (Compare with Figure 2).

Figure 12. Higher magnification of Figure 11 showing fractured ghosts of rectangular COM crystals. Arrows point to one such ghost.

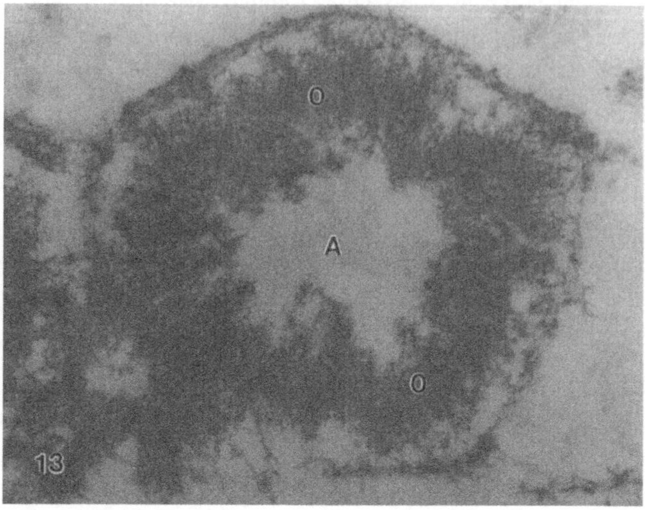

Figure 13. Section through a demineralized spherical CaP crystal showing radial arrangement of organic material (O) around a central space (A).

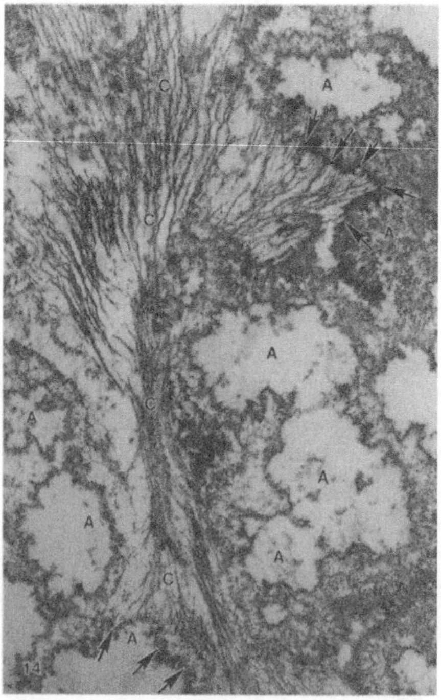

Figure 14. Transmission electron micrograph of demineralized COM crystals (C) surrounded by spherulitic calcium phosphate crystals (A). Organic matrix of the COM crystals is coalescing (arrows) with the organic matrix of the CaP crystals.

Biochemical Analysis

The main biochemical components of stone matrix are glycosaminoglycans, lipids and proteins.[51] Some of the components are preferentially incorporated through specific interactions with the calcific crystals while other materials become a part of the matrix just because stones form in urine in presence of such substances.[49] Calcium oxalate stone matrix contains 3.8$\pm$0.5 mg total lipid and 0.42$\pm$0.12 mg phospholipid/g stone.[52] Matrix of calcium oxalate crystals which are induced in human urine by addition of sodium oxalate also contains lipids and phospholipids[53] in addition to many proteins and glycosaminoglycans. More urinary phospholipids are incorporated into crystals than the neutral lipids or glycolipids. Matrix of CaOx crystals induced in urine from stone formers contain more phospholipids and total lipids than the matrix of crystals induced in urine from normal controls.

CONCLUSION

Based on the many observations, that CaP is the most common crystal in human urine, is ubiquitous in human urinary stones, is seen at the center of mixed CaOx/CaP urinary stones and can promote CaOx precipitation *in vitro*, it has been proposed that in the absence of appropriate CaOx supersaturation, CaP is the heterogenous nucleator of CaOx and nidus for renal stone formation. A cursory look at the data presented above tend to support the proposition. But a critical examination of the data provides another interpretation.

Experimental studies have demonstrated that CaP and CaOx crystals deposit at different locations in the kidneys, CaP at the cortico-medullary junction and CaOx in the renal papilla. Perhaps urinary environment is more suited for CaP precipitation in earlier segments of the renal tubules and for CaOx crystallization in the later segments. Evidence has been provided that urine in thin descending limb of the loop of Henle is metastable for CaP and can support crystallization of HAP.[54,55] Recent studies of Luptak et al. have also indicated that urinary conditions in the earlier segments of the nephron may be more suitable for precipitation of CaP while those in collecting ducts more amenable for CaOx crystallization.[56] It has been proposed that HAP formed in earlier segments of the renal tubules migrates to the collecting ducts of the renal papilla where urine is more concentrated and conditions more suitable for heterogeneous nucleation of CaOx. Results of *in vivo* experiments however indicate that crystals formed in the proximal segments do not necessarily move down the nephron. Moreover, crystals formed in the urine and seen either as crystalluria particles or as constituents of urinary stones are always associated with organic material which appears to coat the crystal surfaces. Thus, only the organic matrix material covering the crystals may be accessible for heterogeneous nucleation. Therefore, even though a crystal may be able to act as substrate for nucleation as shown by *in vitro* studies, it may not be available to do so *in vivo*. Studies of Campbell et al.[57] and Ebrahimpour et.al.[26] demonstrate that CaP crystals coated with certain macromolecules are better substrate for CaOx nucleation than the native crystals themselves, clearly indicating the importance of the organic material associated with the crystals in heterogenous nucleation of CaOx.

Theoretical model of epitaxy has provided evidence that apatite and COM are totally supportive of each other at the atomic level. *In vitro* studies have however, shown HAP promoting COM crystallization but the reverse does not happen. Moreover theoretical models are based on the atomic structure of inorganic crystals. Urinary crystals are not pure. They contain an abundance of organic material. How would the lattice of a poorly crystalline HAP match with protein associated CaOx crystals? Even though epitaxy between well formed inorganic crystals is theoretically possible it has been rarely seen in stones and pronounced tendency of stone components to grow on unrelated substrates[1,47] suggest that epitaxy is not necessary for the formation of mixed stones.

Crystalluria is an indication of supersaturation and not necessarily of nephrolithiasis. It is common in both stone formers as well as normal people. Although all patients with hyperparathyroidism demonstrate hydroxyapatite crystalluria, only some produce kidney stones. Presence of CaP and CaOx crystals in the urine and urinary stones does not necessarily mean nucleation of one by the other, only that urine can be appropriately supersaturated with respect to both types of crystals. CaP crystals present in the CaOx stones may simply represent crystalluria particles trapped between the larger CaOx crystals. The situation is similar to that existing in struvite stones where magnesium ammonium phosphate hexahydrate crystals are seen in association with carbonate apatite not because one nucleated the other but because urinary conditions were suited for crystallization of both during stone formation.

What about the presence of CaP in the stone nidus? In the stones with CaP as nidus, stone center may be formed by aggregation and retention of CaP crystals which are then enclosed by a mantle of organic matrix material. The retained mass of CaP crystals coated with organic material provide platform for further crystal deposition similar to foreign body encrustation.[47] Randall's plaques which are calcifications at the papillary tips[37,38] and frequently represent calcified renal tubules[38] may similarly act as encrustation platforms and become nidi for formation of renal stones. Organic matrix of urinary stones contain membranous as well as fibrillar and amorphous substances. Organic matrix,[58] its lipids as well as cellular membrane vesicles obtained from the renal proximal tubular epithelial cells,[32,59] are all capable of inducing CaOx crystals from metastable solution of CaOx.

Even though CaP is a good nucleator of CaOx *in vitro* from a macromolecule free metastable solution, it is the biological material coating the crystal surfaces which is most likely involved in nucleation of CaOx *in vivo*. We propose that calcium phosphate crystals formed in the loop of Henle become associated with urinary proteins and membranous cellular degradation products which in turn act as nucleator of CaOx when the crystals reach collecting duct system where urine is metastably supersaturated with respect to CaOx. Many of the urinary proteins including albumin and osteopontin[60-62] and membrane lipids[31,63] have high affinity for calcific crystals and cellular membranes and their lipids have been shown to promote nucleation of both the CaP and CaOx crystals. Thus, it is the biological material associated with the CaP crystals which appears to be involved in heterogeneous nucleation of CaOx.[32,59] Development of multicrystalline kidney stones does not appear to be a result of heterogeneous nucleation of one crystal by the other but because conditions are suitable for precipitation of various constituent crystals. Urinary environment can support crystallization of both CaP and CaOx as is indicated by mixed CaOx-CaP crystalluria seen in idiopathic CaOx stone formers.[13-17]

ACKNOWLEDGMENTS

This work was supported in part by National Institutes of Health grants #PO1-DK20586 and RO1-DK41434.

REFERENCES

1. Finlayson B. Renal lithiasis in review. Urol Clin North Am 1974; 1:181-212

2. Otnes B. Urinary stone analysis, methods, materials and value, Scand J Urol Nephrol 1983; Supplement 71: 1-109

3. Leusmann DB, Blaschke R, Schmandt W. Results of 5035 stone analyses: a contribution to epidemiology of urinary stone disease, Scand J Urol Nephrol 1990; 24:205-210

4. Murphy BT, Pyrah LN. The composition, structure and mechanism of the formation of urinary calculi, Br J Urol 1962; 34:129-159

5. Heijnen W, Jellinghaus W, Klee WE. Calcium oxalate trihydrate in urinary calculi, Urol Res 1985; 13:281-283

6. Herring LC, Observations on the analysis of ten thousand urinary calculi. J Urol 1962; 88: 545-562

7. Mandel NS, Mandel GS. Urinary tract stone disease in the United States veteran population. II. Geographic analysis of variations in composition. J Urol 1989; 142:1516-1521

8. Brien G, Schubert G, Bick C, 10,000 analyses of urinary calculi using x-ray diffraction and polarizing microscopy. Eur Urol 1982; 8: 251-256

9. Daudon M, Donsimoni R, Hennequin S, Fellahi S, Le Moel G, Paris M, Troupel S, Lacour B, Sex- and age-related composition of 10, 617 calculi analyzed by infra red spectroscopy. Urol Res 1995; 5: 319-326

10. Khan SR, Hackett RL, Identification of urinary stone and sediment crystals by scanning electron microscopy and x-ray microanalysis, J Urol 1986; 135:818-825

11. Robertson WG, Peacock M, Nordin BEC, Activity products in stone-forming and non-stone-forming urine, Cln Sci 1968; 34:579-594

12. Robertson WG, Peacock M, Nordin BEC, Calcium crystalluria in recurrent renal-stone formers, Lancet 1969; 2:21-24

13. Werness PG, Bergert JH, Smith LH, Crystalluria. J Crystal Growth 1981; 53:166-181

14. Tozuka K, Yonese Y, Konjika T, Sudo T, Study of calcium phosphate crystalluria. J Urol 1987; 138:326-329

15. Nancollas GH, Lore M, Perez L, Richardson C, Zawacki SJ, Mineral phases of calcium phosphate. Anat Rec 1989; 224:234-241

16. Berland Y, Boistelle R, Olmer M, Urinary supersaturation with respect to brushite in patients suffering calcium oxalate lithiasis. Nephrol Dial Transplant 1990; 5: 179-184

17. Werness PG, Wilson JWL, Smith LH, Hydroxyapatite and its role in calcium urolithiasis, in *Urinary stone*, Eds., Ryall, R., Brockis, J.G., Marshall, V., and Finlayson, B., Churchill Livingstone, London, New York, 273-277, 1984.

18. Pak CYC, Eanes ED, Ruskin B, Spontaneous precipitation of brushite in urine: evidence that brushite is the nidus of renal stones originating as calcium phosphate. Proc Nat Acad Sci 1971; 68:1456-1460

19. Lonsdale K, Epitaxy as a growth factor in urinary calculi and gallstones. Nature 1968; 217:56-58.

20. Frank FC, Van der Merwe JH, One dimensional dislocations (I) static theory, (II) misfitting monolayers and oriented overgrowth, Proc R Soc Lond [Biol.] 1949; 198A:205-211

21. Mandel N, Mandel G. "Epitaxis in renal stones" In *Renal Tract Stone, Metabolic Basis and Clinical Practice*, Eds., Wickham, JEA, Buck AC, Churchill Livingstone, New York, 87-101, 1990.

22. Mandel G, Mandel N. "Crystal-crystal interactions" In *Kidney stones: Medical and surgical management*, Eds. FL Coe, MJ Favus, CYC Pak, GL Preminger. Raven press, Boston, 115-127, 1996.

23. Meyer JL, Bergert JH, Smith LH, Epitaxial relationships in urolithiasis: the calcium oxalate monohydrate - hydroxyapatite system. Clin Sci Mol Med 1975; 49: 369-374

24. Meyer JL, Bergert JH, Smith LH, Epitaxial relationships in urolithiasis: the brushite - whewellite system. Clin Sci Mol Med 1977; 52:143-148

25. Koutsoukos PG, Sheehan ME, Nancollas GH, Epitaxial considerations in urinary stone formation. Invest Urol 1981; 18:358-363

26. Ebrahimpour A, Perez L, Nancollas GH, Induced crystal growth of calcium oxalate monohydrate at hydroxyapatite surfaces. The influence of human serum albumin, citrate, and magnesium. Langmuir 1991; 7:577-583

27. Achilles W, Jockel U, Schaper A, Burk M, Riedmiller H, In vitro formation of urinary stones: generation of spherulites of calcium phosphate in gel and overgrowth with calcium oxalate using a new flow model of crystallization. Scann Microsc 1995; 9:577-586

28. Burns JR, Finlayson B, A proposal for a standard reference artificial urine in *in vitro* urolithiasis experiments. Invest Urol 1980; 18: 167-169

29. Prescott LF, The normal urinary excretion rates of renal tubular epithelial cell, leucocytes, and red blood cells. Clin Chem 1966; 31:425-429

30. Khan SR, Glenton PA, Increased urinary excretion of lipids by patients with kidney stones. Br J Urol 1996; 77:506-511

31. Boskey AL, "Phospholipids and calcification," In *Calcified Tissues*. Edited by, DWL Hukins. Boca Raton, CRC Press, p 215, 1989.

32. Khan SR, Whalen PO, Glenton PA, Heterogeneous nucleation of calcium oxalate crystals in the presence of membrane vesicles. J Crystal Growth 1993; 134:211-218

33. Achilles W, Kroning M, Schaper A, Khan SR, Riedmiller H, "Specific nucleation by phospholipids of spherulitic calcium phosphate and calcium oxalate crystals in a gel matrix," In *Urolithiasis, Consensus and Controversies*. Edited by PN Rao, JP Kavanagh, H.-J. Tiselius, University Hospital, Manchester, p 275, 1994.

34. Burry AF, Axelson RA, Trolove P, Saal JR, Calcification in the renal medulla, a classification based on a prospective study of 2,261 necropsies. Human Pathol 1971; 7:435-449

35. Drach GW, Boyce WH, Nephrocalculosis as a source for renal stone nuclei. Observations on humans and squirrel monkeys and on hyperparathyroidism in the squirrel monkey. J Urol 1972; 107: 897-904

36. Malek RS, Boyce WH, Intranephronic calculosis: its significance and relationship to matrix in nephrolithiasis. J Urol 1973; 109: 551-555

37. Randall A, The etiology of primary renal calculus. Intl Abst Surg 1940; 71: 209-240

38. Cifuentes-Delatte L, Minon-Cifuentes JLR, Medina JA, Papillary stones, calcified renal tubules in Randall's plaques, J Urol 1985; 133: 490-494

39. Khan SR, Hackett RL, Calcium oxalate urolithiasis in the rat: is it a model for human stone disease? A review of recent literature. Scann Elect Microsc 1985; 2:759-774

40. Khan SR, Pathogenesis of oxalate urolithiasis: lessons from experimental studies with rats. Am J Kid Dis 1991; 17:398-401

41. Hackett RL, Shevock PN, Khan SR, Cell injury associated calcium oxalate crystalluria. J Urol 1990; 144:1535-1538

42. Duffy JL, Suzuki Y, Churg J, Acute calcium nephropathy, early proximal tubular changes in the rat kidney. Arch Path 1971; 91:340-351

43. Ritskes-Hoittinga J, Lemmens AG, Danse LHJC, Beynen AC, Phosphorus-induced nephrocalcinosis and kidney function in female rats. J Nutr 1989; 119:1423-1431

44. Oliver MD, MacDowell M, Whang R, Welt LG, The renal lesions of electrolyte imbalance, IV. The intranephronic calculosis of experimental magnesium depletion. J Exp Med 1966; 124: 263-277

45. Nguyen HT, Woodard JC, Intranephronic calculosis in rats. Am J Pathol 1980; 100: 39-56

46. Khan SR, Glenton PA, Deposition of calcium phosphate and calcium oxalate crystals in the kidneys. J Urol 1995; 153:811-817

47. Khan SR, Hackett RL, Developmental morphology of calcium oxalate foreign body stones in rats. Calcif Tissue Intl 1985; 37:165-173, 1985

48. Khan SR, Hackett RL, Microstructure of decalcified human calcium oxalate urinary stones, Scann Elect Microsc 1984; 2:935-941

49. Khan SR, Hackett RL, Role of organic matrix in urinary stone formation: an uktrastructural study of crystal matrix interface of calcium oxalate monohydrate stones. J Urol 1993; 150:239-245

50. Khan SR, Hackett RL, Backscattered electron imaging of urinary stone matrix, Scanning 1993; 15, Suuplement III:84-85

51. Ryall RL, "Urinary macromolecules in calcium oxalate stone and crystal matrix: good, bad, or indifferent?" In *Calcium Oxalate in Biological Systems*, Edited by SR Khan, CRC Press, Boca Raton, p265, 1995.

52. Khan SR,"Lipid matrix of urinary calcium oxalate crystals and stones." In *Calcium Oxalate in Biological Systems*, Edited by SR Khan, CRC Press, Boca Raton, p 291, 1995.

53. Khan SR, Atmani F, Glenton P, Hou Zc, Talham DR, Khurshid M. Lipids and membranes in the organic matrix of urinary calcific crystals and stones. Calcified Tissue Intl 1996, In press

54. DeGanello S, Asplin J, Coe FL, Evidence that tubule fluid in the thin segment of the loop of Henle normally is supersaturated and forms a poorly crystallized hydroxyapatite that can initiate renal stones. Kidney Intl. 1990; 37: 472

55. Kok D, "Intratubular crystallization events", In *Renal Stones- aspects on their formation, removal, and prevention*, Proc. VIth European Symspoium on Urolithiasis, Ed., Tiselius, H.-G, Akademitryck AB, Edsbruk, 26-29,1996.

56. Luptak I, Bek-Jensen H, Fornander AM, Hojgaard I, Nilsson MA, Tiselius HG, Crystallization of calcium oxalate and calcium phosphate at supersaturation levels corresponding to those in different parts of the nephron. Scann Microsc 1994; 8: 47-51, 1994

57. Campbell AA, Ebrahimpour A, Perez L, Smesko SA, Nancollas GH, The dual role of polyelectrolytes and proteins as mineralization promoters and inhibitors of calcium oxalate monohydrate. Calcif Tissue Itl 1989; 45:122-128

58. Khan SR, Shevock PN, Hackett RL, In vitro precipitation of calcium oxalate in the presence of whole matrix or lipid components of the urinary stones. J Urol 1988; 139:418-422

59. Khan SR, Heterogeneous nucleation of calcium oxalate crystals in mammalian urine. Scann Microsc 1995; 9: 597-616

60. Atmani F, Opalko FJ, Khan SR, Association of urinary macromolecules with calcium oxalate crystals induced *in vitro* in normal human and rat urine. Urol Res 1996; 24:45-49

61. McKee MD, Nanci A, Khan SR, Ultrastructural immunodetection of osteopontin and osteocalcin as major matrix components of renal calculi. J Bone Min Res, 1995; 10:1913-1922

62. Boskey AL, Maresca M, Ullrich W, Doty SB, Butler WT, Prince CW, Osteopontin-hydroxyapatite interactions *in vitro*: inhibition of hydroxyapatite formation and growth in a gelatin gel. Bone Miner, 1993; 22: 147-152

63. Mandel N, Crystal-membrane interaction in kidney stone disease. J Am Soc Neprol, 1994; 5:S37-S42

12

CALCIUM PYROPHOSPHATE CRYSTAL FORMATION AND DISSOLUTION

Kenneth P.H. Pritzker

Pathology and Laboratory Medicine, Mount Sinai Hospital
University of Toronto, Toronto, Ontario, CANADA

INTRODUCTION

Pyrophosphates have their historical mode of formation embedded in their name. These condensed phosphates are commonly synthesized by the application of heat to orthophosphates.[1,2] Amongst condensed phosphates, pyrophosphate (P_2O_7), can be considered to be the smallest structure and shortest linear polyphosphate.

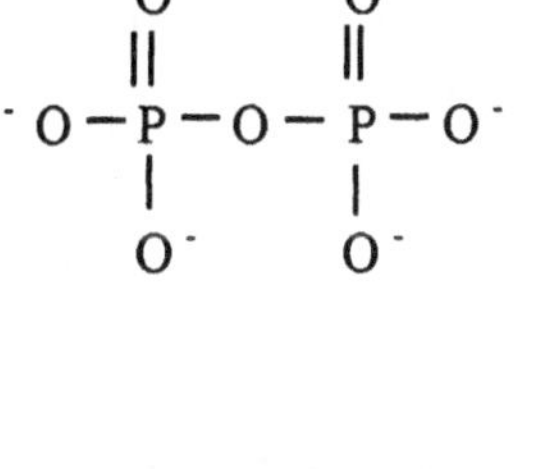

The scientific study of pyrophosphate salts and in particular calcium pyrophosphates can be divided into three periods. The first period, starting in the 1840's, sought to understand the fundamental chemistry of ion coordination in solutions, and multiple phases of hydrated single and double salts.[3] The second period which flourished in the 1950's and 1960's was directed towards identifying the chemical properties of condensed phosphates which could have superior characteristics to orthophosphate in industrial products such as fertilizer.[4-7] The third period, characterized by the study of pyrophosphates in biology, began with the identification by Fleisch and Neuman that pyrophosphate, a substance found in urine inhibited mineralization.[8-10] Subsequently, McCarty and colleagues discovered calcium pyrophosphate dihydrate (CPPD) crystals as pathologic deposits in joints.[11] CPPD crystals released from cartilage into the synovial fluid, imitated features of acute gouty arthritis. Ever since, calcium pyrophosphate crystal studies have been driven by the need to understand mechanisms of CPPD crystal formation and dissolution under biologic conditions.

Although this chapter is concerned primarily with the chemistry of formation and dissolution of the biologic calcium pyrophosphate crystals, it is worthwhile to point out relevant insights in pyrophosphate crystal chemistry developed by early investigators.

Bassett et al.[3] reviewed the chemical literature from 1847 and performed extensive studies on hydrated salts of pyrophosphate and cites the following observation:

"If **equivalent** amounts of MCl_2 in $Na_4P_2O_7$ solutions are mixed, the amorphous precipitate persists indefinitely without becoming crystalline."

"We *(Bassett et al.)* find that at ordinary temperature, mixtures will keep for months or even years without any detectable orthophosphate formation in the absence of acid. This is largely true also of acid mixtures of room temperature but not for temperatures above 50°C."

Bassett et al.[3] further observed that in solutions with excess pyrophosphate, some amorphous pyrophosphates could transform to crystals. In solutions with excess metal, crystals formed from amorphous precipitates only from those metal compounds such as magnesium pyrophosphate in which the precipitate was relatively soluble in water. In Bassett's studies of many types of hydrated metal pyrophosphate salts, most compounds formed were poorly crystallized. Therefore, under their experimental conditions, these compounds could be characterized by polarized light microscopy only with difficulty.

Bassett et al. noted[3] that "sodium pyrophosphate crystallized from solutions at ordinary temperatures as a decahydrate in well formed clear crystals". Substitution of Na^+ by Mg^{2+} resulted in salts that were only one third as soluble as the sodium salts. Although previous workers studied many metal pyrophosphate salts, apparently only very limited studies of calcium pyrophosphates were undertaken.

Key insights from these early studies include:

A. The common mode of metal pyrophosphate preparation involved equivalent (equimolar) P_2O_7.
B. Metal pyrophosphate salts forming from solution tend to be amorphous or poorly crystallized.
C. Metal pyrophosphate salts are sparingly soluble compared to orthophosphates.
D. At physiologic temperatures, pyrophosphate precipitates appear to be stable in aqueous solutions for long periods (days, weeks).
E. Excess pyrophosphate added to solution mixtures permits transformation of amorphous precipitates to hydrated crystals. Under some conditions, the crystals transform with time indicating metastable crystal forms.
F. For magnesium pyrophosphates, excess magnesium added to solution mixtures permits solubilization of amorphous precipitates and subsequent formation of hydrated magnesium pyrophosphate crystals.
G. Under specific conditions, in the presence of two of more metals, double salts of pyrophosphate can form.
H. The specific metal pyrophosphate product formed varies with the primary and secondary ions present.

Calcium pyrophosphate crystal chemistry was stimulated by the search for condensed calcium phosphates which were present or could be synthesized as fertilizers. More than 25 calcium pyrophosphate crystal compounds were characterized by Brown, Lehr and colleagues.[5-7] Condensation of phosphates and pyrophosphates had the potential economic advantage of transport of phosphates at higher concentration with greater stability and less weight. Theoretically, these advantages were offset by the decreased solubility of some calcium pyrophosphate compounds. In practice, soil added to calcium pyrophosphate crystals in aqueous solution facilitated the dissolution of the compounds.[12] This effect was attributed to pyrophosphatase enzymes elaborated by soil bacteria. Paradoxically, this knowledge resulted in scientific apprehension (later discovered to be unfounded) about bacterial contamination affecting experimental calcium pyrophosphate crystal formation and dissolution in solutions. As a consequence, experiments were performed in which CPPD crystals were sterilized at temperatures > 100°C prior to dissolution studies.[13] Unfortunately, the annealing of CPPD crystals at these temperatures does alter the crystal structure and crystal surface properties.[14]

More than 30 known calcium pyrophosphate crystal phases have been recognized (Table 1). However, only CPPD triclinic (CPPD(T)) and CPPD monoclinic (CPPD(M)) crystals deposit in joints under pathologic conditions.[15,16] This observation stimulated research on the basis for the specificity of calcium pyrophosphate crystal phases. These studies centred on CPPD crystal formation, crystal dissolution and their biologic properties are now the principal themes of calcium pyrophosphate crystal research.

Table 1. Calcium Pyrophosphate Crystal Phases

Calcium Pyrophosphate	Abbreviated Name	Structure Identified	Reference
Amorphous $Ca_2P_2O_7$	CaPPi		30,62
α $Ca_2P_2O_7$ (M)		+	35
β $Ca_2P_2O_7$ (tetragonal)		+	36
γ $Ca_2P_2O_7$ (?)			7
$Ca_2P_2O_7.2H_2O$ (M)	CPPD(M)		4,5
$Ca_2P_2O_7.2H_2O$ (T)	CPPD(T)	+	4
$Ca_2P_2O_7.2H_2O$ (H)	CPPD(H)		37
$Ca_2P_2O_7.4H_2O$ (O)	CPPT(O)		5
$Ca_2P_2O_7.4H_2O$ (M)	CPPT(M)	+	38
$CaH_2P_2O_7$ (O)	CDPP		5,37
$Ca_3H_2(P_2O_7)_2.4H_2O$ (T)			5
$Ca_3(H_2P_2O_7)_3.4H_2O$ (O)			
α $CaNa_2P_2O_7.4H_2O$ (M)			28,52
β $CaNa_2P_2O_7.4H_2O$ (M)			28,52,106
γ $CaNa_2P_2O_7.?H_2O$			5
$CaNaNH_4.P_2O_7.3H_2O$			
α $CaMgP_2O_7$. ? H_2O (?)	$CMPP_1$		28,33
β $CaMgP_2O_7$. ? H_2O (?)	$CMPP_2$		33
$CaK_2P_2O_7$ (M)			5
$CaK_2P_2O_7.4H_2O$ (H)			5
$Ca_3K_2(P_2O_7)_2.2H_2O$ (?)			5
$Ca_3K_2(P_2O_7)_3.6H_2O$ (H)			5
$CaK_4H_2(P_2O_7)_2$ (M)			5
$CaK_2H_4(P_2O_7)_2$ (M)			5
$CaKHP_2O_7$ (M)			5
$CaKHP_2O_7.2H_2O$ (O)			5
$Ca_2KH_3(P_2O_7)_2.3H_2O$ (M)			5
$Ca_3KH(P_2O_7)_2.4H_2O$ (M)			5
$CaNH_4HP_2O_7$ (M)		+	5,107
$Ca(NH_4)_2P_2O_7.H_2O$ (M)			5
$Ca_3(NH_4)_2(P_2O_7)_2.6H_2O$ (M)		+	5,108
$Ca_5(NH_4)_2(P_2O_7)_3.6H_2O$ (H)			5
$Ca(NH_4)_2H_4(P_2O_7)_2$ (M)			5
$Ca_2NH_4H_3(P_2O_7)_2.3H_2O$ (M)			5
$Ca_2NH_4H_3(P_2O_7)_2.H_2O$ (T)			5
$Ca_3(NH_4)_4H_6(P_2O_7)_4.3H_2O$ (O)			5

Table 1. Calcium Pyrophosphate Crystal Phases (continued)

Other Selected Metal Pyrophosphates	Abbreviated Name	Structure Identified	Reference
α $Mg_2P_2O_7$		+	109
β $Mg_2P_2O_7$		+	24
$(Na_4)_4P_2O_7.10H_2O$ (M)		+	110
$Mn_2P_2O_7.2H_2O$ (M)		+	111
$K_4P_2O_7.3H_2O$		+	112
$NH_4P_2O_7$ (M)		+	113
$Cd_2P_2O_7$ (T)		+	114
β $2Zn_2P_2O_7$ (M)		+	115
α SrP_2O_7		+	35
$KAlP_2O_7$		+	116
$VOH_3P_2O_7$		+	117

To facilitate an understanding of solubility and thermodynamic properties of calcium pyrophosphate crystals, Brown and Gregory reviewed the literature on calcium pyrophosphate crystal chemistry.[17] Using known equilibrium constants, they described calcium pyrophosphate ionic properties in aqueous solutions and predicted the solubility products for calcium pyrophosphate dihydrate crystals. They pointed out the complexity of pyrophosphoric acid dissociation system and the numerous ionic complexes that can form in the presence of calcium ions. From consideration of known pyrophosphate equilibrium data and by fitting the empirical CPPD crystal dissolution data observed by McCarty et al.,[18] they predicted the CPPD crystal solubility product, $K_{CPPD} = 3 \times 10^{-18}$. Many assumptions were made in this prediction including the reliability of published equilibrium data, the possibility of calcium pyrophosphate complexes other than those specified in their predictions and the experimental limits of measurement of CPPD crystal dissolution. Further, they recognized that "some disquieting evidence based on its solubility in plasma and synovial fluid suggests that K_{CPPD} may be orders of magnitude larger than 10^{-18}". Today, we can suggest that the "disquieting evidence" is related to the pyrophosphatase activity of alkaline phosphatase present in these fluids.

Calcium Pyrophosphate Crystals: Standard Preparations of Crystal Structures

Anhydrous calcium pyrophosphate crystals ($Ca_2P_2O_7$) can be prepared by heating dicalcium phosphate $CaHPO_4$ crystals.[7] $Ca_2P_2O_7$ forms in the region of 240 to 500 °C; $Ca_2P_2O_7$ forms > 750 °C and $Ca_2P_2O_7$ forms in the temperature range of 1140 to 1350 °C. $Ca_2P_2O_7$ can be formed from heating calcium hydroxyapatite dicalcium phosphate mixtures > 750 °C [2,19] if the calcium:phosphorous ratio is maintained at < 1.5. Similarly, $Ca_2P_2O_7$ can be seen as intermediate products of thermal condensation in the $Ca(H_2P0_4)_2.H_20$-$CaHPO_4.2H_2O$ crystal system.[20] $Ca_2P_2O_7$ can be formed by heating $Ca(H_2PO_4)_2.H_2O$ > 400 °C in the presence of SiO_2.[21]

$Ca_2P_2O_7$, sintered with the dopant $Na_4P_2O_7.10H_2O$ to reduce the sintering temperature and improve mechanical properties has been developed as a biomaterial substitute for bone grafts.[22] As well, calcium pyrophosphate has been incorporated into a triphasic ceramic (calcium apatite, calcium pyrophosphate, tricalcium phosphate) for potential biologic applications.[23]

Anhydrous magnesium pyrophosphate crystals ($Mg_2P_2O_7$) can be formed at much lower temperatures and undergo phase transitions at 68 °C.[24]

Two types of experimental protocols have been used to synthesize pure calcium pyrophosphate dihydrate crystals. The first protocol designed to maximize crystal yield, involves preparation of a precursor crystal followed by solubilization in aqueous solution and recrystallization. For hydrated calcium pyrophosphate, the precursor crystal compound commonly used is calcium dihydrogen pyrophosphate (calcium acid pyrophosphate (CDPP, $CaH_2P_2O_7$).[4,13,25-27] The second method designed to understand the relative role of initial ion concentrations involves preparation of solution mixtures containing primary and secondary ions.[28-30] Crystallization is achieved either directly or via initial amorphous solid intermediates which transform to crystals upon standing in solution mixtures for up to 21 days.

Mandel et al. outlined preparation of hydrated calcium pyrophosphate crystals following the first method.[27] CPPD(T) was formed by dissolving 1.1 gm of CDPP in 100 mℓ distilled water, followed by incubation of the solution without stirring for 7 days. The first crystal phase precipitating is metastable calcium pyrophosphate tetrahydrate, orthorhombic, (CPPT(O)) which transforms to CPPD(T) within five days. CPPD(M) was formed by adding CDPP crystals to $MgCl_2$ heated to 80 °C with a subsequent addition of calcium acetate. Of note, the presence of a secondary cation, magnesium, was required for CPPD(M) formation. Stable CPPT(O) crystals were synthesized by addition of CDPP to 0.05 M NH_4OH solution (pH = 9.7).[27]

Under vacuum conditions, the CPPT(O) crystals underwent solid state transformation to CPPD hexagonal (CPPD(H)) crystals.[27] Burt and colleagues synthesized CPPD(T) and CPPD(M) using modified Mandel methods[31,32] and characterized the relative solubilities of the two crystals. CPPD(M) is more soluble than CPPD(T) at all pH's and temperatures measured.

As described below, Pritzker, Cheng and colleagues[28,30,33] synthesized CPPD(T) and CPPD(M) or CPPD(T)/CPPD(M) mixtures as well as other calcium pyrophosphates in aqueous solutions at room temperature (20 °C) or 37 °C. The solution mixtures contained various concentrations and concentration ratios of primary and secondary ions. These studies demonstrated the importance of kinetics (ion concentrations) in the determination of the crystal phase formed. On a biologic time scale and under quasibiologic conditions, these studies demonstrated that metastable CPPD crystals such as CPPD(M) could be stable. For example, no dissolution or transformation could be observed from CPPD(M) present in aqueous solutions containing Mg^{2+} 1 mM at 150 days.[34]

Table 2. P-O-P bond Angle in Selected Pyrophosphate Crystals

Calcium Pyrophosphate	P-O-P Bond Angle (°)	Other Metal Pyrophosphate
$Ca_2P_2O_7.2H_2O$ (T)	123.1	
	123.2	$KALP_2O_7$
	123.6	$VOH_3P_2O_7$
$Ca(NH_4)_2P_2O_7$	125.2-138.7	
$Ca_2P_2O_7.4H_2O$ (O)	125.9	
	127.5	$Mn_2P_2O_7.2H_2O$
α $CaNa_2P_2O_7.2H_2O$	128.7	
$Ca_3(NH_4)(P_2O_7)_2.6H_2O$	128.9	
β $CaNa_2P_2O_7.2H_2O$	129.6	
α $Ca_2P_2O_7$	130.0	
	130.2	$Na_4P_2O_7.10H_2O$
	130.3	$K_4P_2O_7.3H_2O$
β $Ca_2P_2O_7$	131.0-138.0	
	131.0-138.0	β $Sr_2P_2O_7$
	132.0	$Cd_2P_2O_7$
	132.9	$(NH_4)_4P_2O_7$
	133.5	α $Sr_2P_2O_7$
	144	α MgP_2O_7
	180	β MgP_2O_7
	180	ZnP_2O_7

Calcium Pyrophosphate Crystals: Structural Studies

Despite the difficulties in synthesizing calcium pyrophosphate crystals, sufficiently large crystals have been formed to determine the crystal structures of several calcium pyrophosphates including anhydrous $Ca_2P_2O_7$[35] and $Ca_2P_2O_7$,[36] CPPD(T),[37] CPPT(M),[38] calcium disodium pyrophosphate tetrahydrate,[39] and calcium disodium tetrahydrate.[40] The structure for the biologic crystal phase CPPD(M) has not yet been determined. Mandel and Mandel summarized the properties of pyrophosphate bonds in the known calcium pyrophosphate crystal structures.[41] A key characteristic is the flexibility of P-O-P bond in crystals which can vary from 123-180° depending on the coordinating cation. A comparison of P-O-P bonds in calcium pyrophosphate and other selected metal pyrophosphate crystals is presented in Table 2. The P-O-P bond flexibility facilitates interaction with cations having different ionic radius and hence different bonding requirements. In CPPD(T) the P-O-P angle is narrow (123°) and the calcium ions are 7 coordinate implying a thermodynamically stable complex.[37,41] Hydrated calcium pyrophosphate crystals, $Ca_2P_2O_7.H_2O$, have narrower P-O-P bond angles then anhydrous phases. Computer modeling of the P-O-P bond interaction with manganese, zinc and calcium has confirmed the experimental observations related to P-O-P bond angles thereby providing a tool to study more complex interactions of pyrophosphate in biomineralization.[42] It is of considerable interest that P-O-P bond in $Mg_2P_2O_7$ is more open than $Ca_2P_2O_7$. As well, the O atom within the P-O-P bond of metal pyrophosphate structures such as β-$Zn_2P_2O_7$ which has linear pyrophosphate group has considerable thermal motion perpendicular to the plane of the P-O-P vector. If this observation can be extrapolated to soluble $CaMgP_2O_7$ complexes in aqueous solution, this suggests that Mg^{2+} facilitates widening of the P-O-P bond angle. A further extrapolation suggests that narrow P-O-P bond angle inhibits and an open P-O-P bond angle facilitates enzymatic pyrophosphatase hydrolysis (Mg^{2+} is a cofactor for inorganic pyrophosphatase[43]).

Calcium Pyrophosphate Crystal Formation in Gels

Calcium pyrophosphate crystals were first grown in gel systems in an attempt to form CPPD(M) crystals sufficiently large to characterize atomic and molecular structure by single crystal x-ray diffractometry.[44] An extremely simple test tube system in which $CaCl_2$ solution was layered over a silica or gelatin gels doped with sodium pyrophosphate was used. Multiple layers of white precipitates, each layer containing crystals of a different phase, formed within the gels. For reasons subsequently understood (ion concentrations, absence of secondary ion such as magnesium), this system failed to generate CPPD(M) crystals. Nonetheless, this data indicated that kinetic conditions were critically important to the calcium pyrophosphate crystal phase formed and suggested that the crystal phase may be influenced by the gel matrix.

Calcium pyrophosphate crystals have considerable biologic specificity.[15] In tissues, the crystal phases found are restricted to CPPD(T), CPPD(M) or CPPD(T)/CPPD(M) mixtures.[45,46] Calcium pyrophosphate crystals form only in articular tissues, usually hyaline (cartilage) fibrocartilage, or soft tissues which have undergone cartilaginous metaplasia. There is even some species restriction. Calcium pyrophosphate crystals deposit in the cartilage of human and nonhuman primates whereas under similar conditions in rabbit cartilage, calcium apatite crystals form.[47]

As animal models for CPPD crystal formation are not available, in vitro gel systems were developed to model CPPD crystal formation in hyaline articular cartilage matrix, a natural biologic gel[14,41,48-56] (Table 3).

In early experiments using articular cartilage from human tibial plateau, wells were excavated in articular cartilage using a corkborer. The wells were filled with solid calcium chloride and sodium pyrophosphate respectively, a preparation analogous to an immunodiffusion gel plate system.[48] Indeed, a linear crystal precipitate formed within the cartilage midway between the two wells. These crystals were identified as $CaNa_2P_2O_7.2H_2O$. The crystal phase did not change despite degradation of the cartilage by various physical and chemical modalities. This study indicated that ion concentrations were more important than the matrix composition in determining the calcium pyrophosphate crystal phase formed. Mandel classified the experimental calcium pyrophosphate gel systems into 3 categories.[41] The Type A system is a system in which calcium chloride solution is layered on top of a gel doped with pyrophosphate. The Type B system is a system in which the gel contains both calcium and pyrophosphate. The ion concentrations and gradients can be controlled by infusion of calcium and pyrophosphate from separate reservoirs. Type C is a modification of the Type B system in which calcium pyrophosphate crystals are seeded in the gel.

Table 3. CPPD Crystal Formation in Gels

System	Matrix	Ions + -	Temp	Crystal Product	Ref.
A	Silica	Ca^{2+} Cl^- Na^+ $P_2O_7^{-4}$	20°C	CPPD(T) CPPD(O) $CaNa_2P_2O_7.4H_2O$	44
	Gelatin Bloom #275	Ca^{2+} Cl^- Na^+ $P_2O_7^{-4}$	20°C	CPPD(T) CPPD(O) $CaNa_2P_2O_7.H_2O$ $CaNa_2P_2O_7.H_2O$ $CaNa_2P_2O_7.H_2O$	44
Articular Cartilage	Articular Cartilage (Human)	Ca^{2+} Cl^- Na^+ $P_2O_7^{-4}$	10°C 37°C	$CaNa_2P_2O_7.4H_2O$	48
A	Gelatin	Ca^{2+} Cl^- Na^+ $P_2O_7^{-4}$	20°C	CPPD(T) CPPT(O)	49
A	Polyacrylamide	Ca^{2+} Cl^- Na^+ $P_2O_7^{-4}$	37°C	CPPD(T) CPPD(O)	
A	Type I Collagen	Ca^{2+} Cl^- Mg^{2+} Na^+ $P_2O_7^{-4}$	37°C	CPPD(T) CPPT(O)	50
B	Gelatin Bloom # 275	Ca^{2+} Cl^- Na^+ $P_2O_7^{-4}$	5°C 23°C	Amorphous PPi CPPD(T) CPPD(M) CPPT(O) $CaNa_2P_2O_7$ $CaNa_2P_2O_7$ $CaNa_2P_2O_7$	51,5 2
B	Gelatin	Ca^{2+} Cl^- Na^+ $P_2O_7^{-4}$ Fe^{3+} Fe^{2+}	23°C	CPPD(T) CPPD(M) CPPD(O)	51
C	Type I Collagen	Mg^{2+} Cl^- Ca^{2+} SO_4^{-2} Na^+ $P_2O_7^{-4}$ K^+	37°C	Amorphous CaPPi CPPD(T) CPPD(M)	55
C	Gelatin MSU Crystals Ca Apatite Crystals	Ca^{2+} Cl^- Na^+ $P_2O_7^{-4}$	23°C	CPPD(T) CPPD(M) CPPT(O)	53

Harries et al. investigated a Type A system using both gelatin and polyacrylamide gels. The polyacrylamide gels were used because in contrast to the gelatin, these gels were stable at 37 °C. At relatively low calcium concentrations $[Ca^{2+}]$ = 0.2 M, CPPD(T) crystals formed under mildly alkaline conditions.[49]

Hunter et al. used the Type A system with a collagen gel to determine the effects of glycosaminoglycans on CPPD crystal formation.[50] Chondroitin sulphate which has a greater affinity for calcium ions than keratan sulphate was also a more potent inhibitor of CPPD(T) crystal formation. Using a Type B system in which the calcium pyrophosphate concentrations in solution above the gel can be kept constant, Mandel and colleagues demonstrated that different crystal phases formed in different gel domains with different ion concentrations.[48,51,52,54-56] Further, the patterns of CPPD crystal formation differed markedly when Type I collagen was substituted for gelatin in the gels.

In a Type C system in which the matrix is seeded by monosodium urate or calcium apatite crystals, CPPD crystals demonstrated epitaxial nucleation and growth of on monosodium urate.[53] By trapping pyrophosphate ions, calcium apatite crystals facilitated CPPD crystal formation in adjacent matrix domains. The complexity of the crystal phases formed by gels and the variation of crystal phase with the type of gel matrix illustrates that diffusion in extracellular matrices is far from the simple process predicted by classical diffusion theory.[57,58]

CPPD(M) can form in lipid rich conditions. CPPD(M) crystals have been identified in processed cheese in which the melting salts contain pyrophosphate and the preparation condition was water content 56%, fat:dry weight ratio 1:2, pH 5.9 and storage temperature 70 °C.[59]

Calcium Pyrophosphate Crystal Formation in Aqueous Solutions

The observations of CPPD crystal deposition in human arthritic tissues led to the recognition that whatever other conditions were favourable, the crystal phases CPPD(T) and CPPD(M) must be capable of formation under biologic ionic conditions: $[Na^{+}]$ = 140 mM, $[Ca^{2+}]$ = 1 mM, $[Mg^{2+}]$ = 0.5 mM and inorganic phosphate, [Pi] = 1 mM. Of these ions, Ca^{2+} is known to be maintained relatively constant throughout the extracellular space in a very narrow concentration range about $[Ca^{2+}]$ = 1 mM. Inorganic pyrophosphate, [PPi], is present in plasma in a concentration range of [PPi] = 1-5 μM, a concentration below the minimum concentration (20-70 μM) required for CPPD crystal formation (Table 4). However, in contrast to $[Ca^{2+}]$, PPi is produced by chondrocytes (and other cells) and diffuses outward to the circulation.[60] Although not directly observed because of technical difficulties, it is presumed that PPi diffuses through cartilage matrix, becomes bound and increases in concentration.[15,45] Both PPi and pyrophosphatase are elaborated by chondrocytes on the outer cell membrane. Further, specific to cartilage, the extracellular matrix composition inhibits diffusion of large molecules such as enzymes with pyrophosphatase activity much more than small anions such as PPi.[61] In effect, by differential diffusion from the cell, PPi escapes from pyrophosphatase. Subsequently, PPi becomes bound to the outer territorial and to the extraterritorial extracellular matrix of cartilage.[15] When the [PPi] is sufficiently elevated, nucleation with [Ca^{2+}] occurs resulting in CPPD crystal formation and aggregation in the cartilage matrix.[16]

Extensive aqueous solution mixture experiments have been performed designed to understand the initial ionic conditions required for biologic calcium pyrophosphate crystal formation.

Hearn and Russell prepared calcium pyrophosphate dihydrate crystals in aqueous solution mixtures at 37 °C after mutual preparation of CPPT(O) and amorphous CaPPi.[62] Their studies demonstrated:

A. [PPi] 40 μM is required for calcium pyrophosphate crystal formation.
B. At pH 7.2 7.4, CPPD crystal formation increases rapidly.
C. In the range [Mg^{2+}] = 0.5 mM to 150 mM, increasing [Mg^{2+}] increased the [PPi] required for calcium pyrophosphate crystal formation.

Cheng, Pritzker and colleagues[28-30,63,64] in an extensive series of experiments explored systematically the conditions favoring CPPD crystal formation in aqueous solution mixtures simulating the range of biologic ionic conditions (Table 4). These studies demonstrate conclusively that CPPD(T), CPPD(M), and CPPD(T)/CPPD(M) mixture could form at 37 °C under ionic conditions present in the extracellular matrix.[30,33] The crystal phase formed was exquisitely dependent on the concentration of the ions present. Biologically relevant observations from these studies included:

A. [Ca^{2+}] and [Mg^{2+}] in the normal biologic ranges favors CPPD(T) and CPPD(M) mixtures.
B. Decreased [Mg^{2+}] favors CPPD(T) crystal formation.
C. Elevated [Mg^{2+}] or [Ca^{2+}] favors CPPD(M) crystal formation.
D. When [Mg^{2+}] > [Ca^{2+}], the [PPi] required for crystal formation is increased.
E. When [Mg^{2+}] > 1.0 mM, calcium magnesium pyrophosphate crystals form. These calcium magnesium crystal phases have not been found in vivo.

These observations provide evidence that CPPD crystals form only in the extracellular matrix. Isolated case reports of intracellular CPPD crystal formation can be reconciled with the crystal chemistry by postulating a defect in pyrophosphate catabolism in specific intracellular compartments such as mitochondria. Together with a Ca^{2+} leak into the intracellular compartments, it may be feasible under these exceptional pathologic circumstances to have CPPD crystal formation.[65] Reports of amorphous MgPPi or CaMgPPi granules within intracellular vesicles[66-71] are compatible with solid pyrophosphate precipitation in the presence of elevated [Mg^{2+}]. At [Ca^{2+}] = 1 mM and [Mg^{2+}] > 1 mM, there is a region where crystal formation occurs only when [PPi] exceeds 0.25 mM. These [PPi] concentrations are substantially above that predicted for intracellular reactions. This accounts for the absence of intracellular CPPD crystal formation in normal metabolism and even during most cell injury where rapid influx of calcium can be observed.

Table 4. Minimum [PPi] required for CPPD crystal formation in aqueous solution mixtures 37°C, $[Na^{+}]$ 140 mM.

Minimum [PPi] mM	$[Ca^{2+}]$ mM	[Ca]:[PPi]	$[Mg^{2+}]$ mM	[Pi] mM	Crystal Phase	Ref
0.500	0.5	1	-	-	CPPD(O) +CPPD(T)	28
0.150	0.5	3.3	0.3	-	CPPD(T)	33
0.050	1.0	20	0.3	-	CPPD(T)	33
0.150	1.0	6.6	0.7	-	CPPD(T) trace+ CPPD(M)	33
0.250	1.0	4	0.7	-	CPPD(M)	33
0.100	1.0	10	0.5	0.3	CPPD(M)	30
*0.005	1.0	200	0.5	1.0	----------	*
0.100	1.5	15	0.5	0.3	CPPD(M)	30
0.225	1.5	6.6	1.0	-	CPPD(O)	62
0.175	1.5	8.6	0.5	-	CPPD(O)	62
0.150	1.5	10	1.1	-	CPPD(T)/ CPPD(M)	33
0.050	1.5	30	0.3	-	CPPD(T)	33
0.040	1.5	37.5	-	-	CPPD(O)	62

* Ambient extracellular biologic ionic conditions

In further studies, the effects of inorganic phosphates [PPi] on CPPD crystal formation were examined.[30] The Pi:PPi ratio was found to be a critical determinant to the crystal product formed. At [Pi]:[PPi] < 3, CPPD crystals precipitated. With increasing [Pi]:[PPi] ratios between 3 and 100, amorphous calcium pyrophosphate, amorphous calcium pyrophosphate/amorphous calcium phosphate mixtures and amorphous calcium phosphate progressively formed. When [Pi]:[PPi] > 100, calcium apatite crystals precipitated. These studies demonstrate that [Pi] inhibits CPPD crystal formation and also showed the concentration range in which [PPi] inhibits calcium apatite crystal formation.

The biologic significance of these observations includes:

A. CPPD and calcium apatite form under mutually exclusive conditions. Therefore, tissues where both crystal phases are found in proximity, the crystals formed at different times under differing conditions.
B. Amorphous calcium pyrophosphates do form under pathologic condition in vessel walls and other heterotopic calcifications. Under these circumstances [PPi] as well as [Mg^{2+}] act as inhibitors of calcium apatite crystal formation. Hence, precipitation of an amorphous phase is favoured.

The extracellular matrices in which the CPPD crystals deposit contain glycosaminoglycans in which carboxylate and sulphate anionic groups are prominent. Evidence from crystal formation in gels suggested that glycosaminoglycans with high carboxylate concentration were inhibitory to CPPD crystal formation.[50] The effects of carboxylates were modeled in solution by comparing the inhibitory effects of mono, di and tricarboxylates on CPPD crystal formation.[64] Monocarboxylates (acetate, glycuronate) failed to inhibit CPPD crystal formation. Progressively greater inhibition was observed with dicarboxylate (malate) and tricarboxylate (citrate). Biologically, the effects of carboxylate groups may be blocked by binding of these groups to other ions, thereby decreasing inhibition and facilitating CPPD crystal formation. CPPD crystal formation has been associated with metabolic diseases such as haemochromatosis in which increased iron is deposited in tissues.[63] To model these conditions, the effects of [Fe^{2+}] and [Fe^{3+}] were studied in CPPD crystal formation in solution. Contrary to inhibition, [Fe^{2+}] > 1 μM and [Fe^{3+}] > 10 μM inhibited CPPD crystal formation. These observations indicated that CPPD crystal formation in haemochromatosis occurs by mechanisms other than the direct effects of the ions on crystal nucleation inhibition. Secondary hyperparathyroidism or [Fe^{2+}] depletion therapy, which removes inhibitors of crystal formation, may facilitate CPPD crystal deposition in the disease.

The studies of CPPD crystal formation and solution have provided insight into the control of crystal size and habit (aspect ratios). In contrast to calcium apatite in which initial crystal size is 10-40 nm, calcium pyrophosphate crystals form in a minimum size range of 100-500 nm in length. The difference in minimum crystal size or "ripening" is determined by diffusion through the initial solution as well as the relative kinetics of crystal growth and dissolution. Compared to calcium apatite, the larger size of calcium pyrophosphate crystals indicate that once nucleated, ion kinetics strongly favour crystal formation over dissolution.[72] At very high equivalent concentrations, large crystals are formed (Table 5). At lower absolute concentrations, increasing [PPi^{2+}]:[Ca^{2+}] ratios favors formation of larger crystals (Figures 1 and 2). Excess pyrophosphate facilitates complex ion formation and dissolution.[73] This effect, by preventing formation of intermediate amorphous calcium pyrophosphate precipitates, facilitates the availability of the primary ions for crystal growth. Conversely, increasing [Ca^{2+}]:[PPi] ratio tends to limit the growth of calcium pyrophosphate crystals.

Table 5. Calcium pyrophosphate crystals. Control of crystal size.

Crystal Phase	Crystal Length	$[Ca^{2+}]$ mM	[PPi] mM	[PPi]/ $[Ca^{2+}]$	$[Mg^{2+}]$ mM	$[Na^{+}]$ mM	pH	Ref
CPPD(T)	5-10 m							(33)
	50-100 m	1.0	.15	-	1.0	-	7.4	(96)
	> 200 m	1.0	30	30	-	120	5.5	(41)
							7.0(?)	
		50	50	1.0	-	-		
CPPD(M)	3-10 m							(34)
	40-60 m	1.5	0.5	0.3	0.75	140	7.4	
		1.5	3.0	2.0	0.75	140	7.4	
	> 200 m	1				40-		(28)
$CaNa_2P_2O_7$.			10-30	10-30	10-30	120	8.5	
$4H_2O$								
	> 200 m	1	30-			120-		(28)
$CaNa_2P_2O_7$.			100	30-	30-100	400	7.0(?)	
$4H_2O$				100				
CPPT(O)	> 200 m	160	170	1.06				(41)
					-	-	9.7	

Calcium Pyrophosphate Crystal Morphology

Historically, calcium pyrophosphate crystal phases are detected by their habit in light microscopy and identified by their characteristic birefringence under compensated polarized light or by powder x-ray diffraction pattern.[74-76] Indeed, the observation of blue positively birefringent rhomboid crystals 1-20 μm in length by polarized light microscopy with compensation constitutes the principal clinical method for identifying calcium pyrophosphate dihydrate crystals in synovial fluids.[46,75,77] While other methods including atomic force microscopy and spectroscopy have been explored,[78] definitive identification of calcium pyrophosphate crystals is dependent on powder diffraction analysis using either x-rays or an electron microscope source.[74,76] To observe calcium pyrophosphate crystal growth and dissolution, scanning electron microscopy, a technique which can visualize each crystal face is extremely useful.[76,79] With field emission scanning microscopy, sufficient resolution can be obtained to observe defects in the crystal surface, twinning, epitaxy as well as etch pits characteristic of crystal dissolution[80] (Figures 1, 2, 3 and 4).

Figure 1.Calcium pyrophosphate dihydrate monoclinic crystals. Synthesized from aqueous solution mixtures. Field emission scanning electron microscopy. Magnification bar = 2 μ.

Figure 2.Calcium pyrophosphate dihydrate triclinic crystals. Synthesized from aqueous solution mixtures. Field emission scanning electron microscopy. Magnification bar = 50 μ.

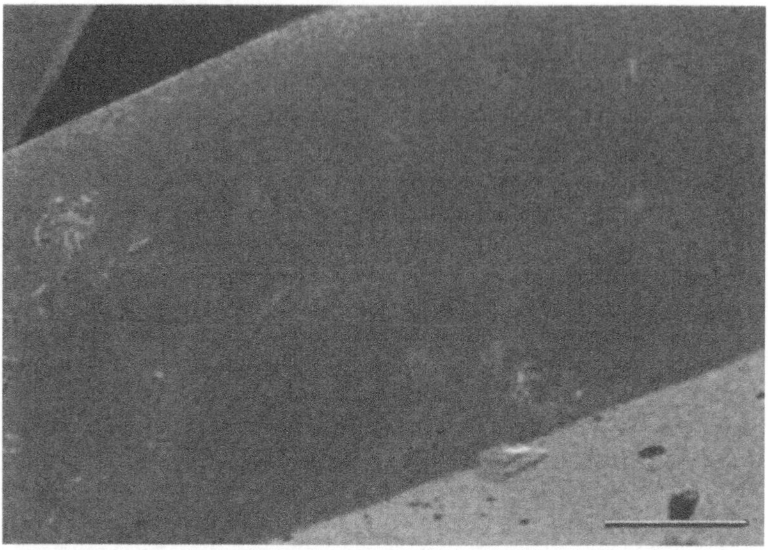

Figure 3. Crystal 010 face of calcium pyrophosphate dihydrate triclinic crystals. Field emission electron microscopy. Magnification bar = 2 μ.

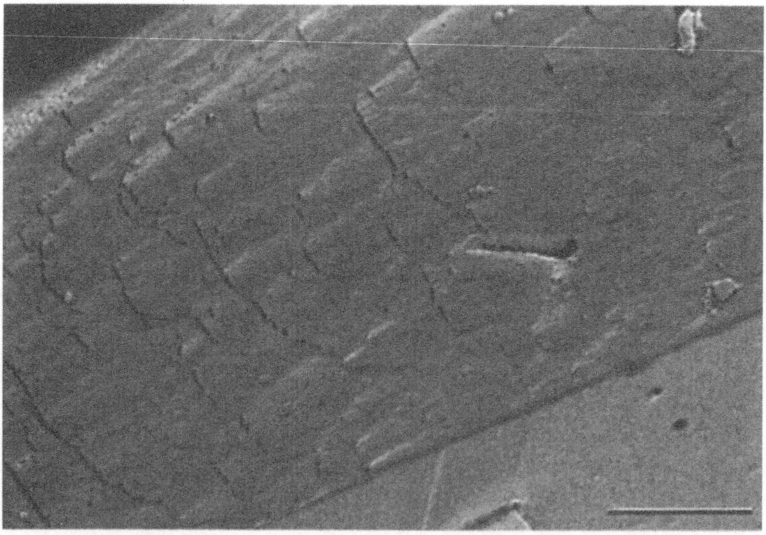

Figure 4. Optical 010 face of calcium pyrophosphate dihydrate crystal after incubation in presence of 10 units alkaline phosphatase for 72 hours. Note selective etch pit formation on the 010 face.Field emission electron microscopy. Magnification bar = 2 μ.

Calcium Pyrophosphate Crystal Dissolution

Conceptually, calcium pyrophosphate crystal dissolution could occur either by dissociation of Ca^{2+} from PPi or by hydrolysis of PPi to orthophosphate. In the absence of promoters, both reactions are unfavorable thermodynamically.[17,81-87] The hydrolysis of P-O-P in a 1% solution at 37 °C, pH = 7 has been estimated to require 10-100 years.[83]

With calcium pyrophosphate dihydrate crystals, the P-O-P bond is extremely stable. An experimental attempt to disrupt the P-O-P bond in calcium pyrophosphate dihydrate crystals by exposure of CPPD crystals to massive fluxes of free radicals in a nuclear reactor was unsuccessful.[88]

In contrast, under biologic conditions, calcium pyrophosphate crystals can be cleared from the joint space by sequential phagocytosis and dissolution.[89-92] Calcium pyrophosphate crystal solubility can be increased by decreasing pH and/or by increasing [Mg^{2+}], citrate or albumin concentrations.[13] However, the major biologic catalyst for calcium pyrophosphate crystal dissolution are enzymes with pyrophosphatase activity including alkaline phosphatase.[79,93-97]

Extensive study of CPPD crystal dissolution by alkaline phosphatase enzyme activity at pH = 7 has demonstrated that for dissolution to occur, the enzyme must be proximate or bound to the crystal surface.[93,96] Further both enzyme binding and crystal dissolution are stereoselective with the most energetic short end faces (optical 010 faces) favored (Figures 3 and 4). This stereoselectivity can be modulated by adjusting the ambient ion concentration.[96] Stereoselectivity of crystal faces for pyrophosphatase enzyme binding and crystal dissolution may have many important biologic implications such as the differential effects in inflammatory activity between CPPD(T) and CPPD(M) crystal phases.[31,32]

Enzymatic calcium pyrophosphate crystal dissolution is facilitated by Mg^{2+} and polyamines[79,93,94] and inhibited by orthovanadate > cadmium > calcium ions.[97] Ascorbate facilitates CPPD crystal dissolution only at high concentrations, [ascorbate] > 1 mM. Bisphosphonates at high concentrations inhibit CPPD crystal dissolution but not pyrophosphatase activity indicating that these agents act by binding directly to the CPPD crystal surface.[97]

Calcium Pyrophosphate - Calcium Phosphate Crystal Interactions

Calcium pyrophosphate - calcium phosphate crystal interaction studies have been stimulated by two diverse lines of scientific inquiry, biologic evolution (origins of life) and pyrophosphate inhibition of calcium phosphate crystal formation. Selected adsorption of biopolymers on bioactive inorganic substances such as silicon dioxide, clays or calcium apatite have engendered considerable study as primitive bioactive processes which facilitate formation of biologic compounds.[98] A central problem in these investigations has been the formation of high energy phosphate compounds. In the presence of polyphosphate synthesized by abiogenic processes and metal ions, pyrophosphate can form.[99] Further, pyrophosphate can form on calcium apatite surfaces in the presence of thioesters such as N,S-diacetylcysteamine as condensing agents.[100] The adsorption of pyrophosphate to calcium apatite is also considered to be the principal mechanism by which pyrophosphate inhibits calcium apatite crystal growth.[8,9,101-104]

X-ray photo electron spectroscopy (ESCA) studies of pyrophosphate, calcium apatite interaction have provided evidence for a solid calcium pyrophosphate phase on a surface of apatite crystals.[105] Given the stereoselective adsorption properties of calcium pyrophosphate,[96] this suggests that a calcium pyrophosphate layer on calcium apatite may modify the crystal interactions with organic molecules.

Epilogue

The study of calcium pyrophosphate solid phases has provided important concepts in several fields of both basic and applied science. This includes physical chemistry (ion coordination, thermodynamics of hydrolysis and crystal dissolution), biologic evolution, agriculture (fertilizers) biomaterials, and human pathology (calcium pyrophosphate crystal arthropathy, dental caries, osteoporosis). Only recently, the specific reactivity of calcium pyrophosphate crystal surfaces has been recognized. Future investigations on the regulation of calcium pyrophosphate crystal formation, crystal dissolution, and crystal surface interaction with inorganic and organic substances will continue to be important to understanding diverse chemical and biologic processes.

Acknowledgement

We thank our colleagues, Dr. Stanley Nyburg, Dr. Pei-Tak Cheng, Dr. Marc Grynpas for helpful discussions, the Canadian Arthritis Society and the Medical Research Council of Canada for grant support and Ms. Anna Vaccaro for preparation of the manuscript.

REFERENCES

1. Bassett H. The phosphates of calcium. Part V. Revision of the earlier space diagram. Journal of Chemical Society 1958;601:2949-2961

2. Mortier A, Lemaire J, Rouxhet P. Formation of β-tricalcium di(orthophosphate) from mixtures of hydroxyapatite and calcium hydrogen orthophosphate. Thermochimica Acta 1987;113:133-139

3. Bassett H, Bedwell WL, Hutchinson JB. Studies of phosphates. Part IV. Pyrophosphates of some bivalent metals and their double salts, and solid solutions with sodium pyrophosphate. J.Amer.Chem.Soc. 1936;1412-1429

4. Dubost M-P. Etude des phosphates de calcium. Bulletin de la societe chimique de la France 1959;5:810-815

5. Brown EH, Lehr JR, Smith JP, Frazier AW. Preparation and characterization of some calcium pyrophosphates. Journal of Agricultural and Food Chemistry 1963;11:214-222

6. Brown EH, Lehr JR, Frazier AW, Smith JP. Calcium ammonium and calcium potassium pyrophosphate systems. Journal of Agricultural and Food Chemistry 1964;12:70-73

7. Lehr JR, Brown EH, Frazier AW, Smith JP, Thrasher RD. Crystallographic properties of fertilizer compounds. Chemical Engineering Bulletin 1967;6:1-166

8. Fleisch H, Neuman WF. Mechanisms of calcification: role of collagen, polyphosphates, and phosphate. American Journal of Physiology 1961;200:1296-1300

9. Fleisch H, Russell RGG, Straumann F. Effect of pyrophosphate on hydroxyapatite and its implications in calcium homeostasis. Nature 1966;212:901-903

10. Fleisch H. Inhibitors of calcium phosphate precipitation and their role in biological mineralization. Journal of Crystal Growth 1981;53:120-134

11. McCarty DJ, Jr., Kohn NN, Faires JS. The significance of calcium phosphate crystals in the synovial fluid of arthritic patients: The "pseudogout syndrome". I. Clinical aspects. Annals of Internal Medicine 1962;56:711-737

12. McCarty DJ. Dissolution of calcium pyrophosphate by soil. 1997. Personal Communication

13. Bennett RM, Lehr JR, McCarty DJ. Factors affecting the solubility of calcium pyrophosphate dihydrate crystals. Journal of Clinical Investigation 1975;56:1571-1579

14. Cheng P-T, Pritzker KPH. Effect of heat on crystals. Arthritis and Rheumatism 1981;24:637-638

15. Pritzker KPH. Calcium pyrophosphate crystal arthropathy. A biomineralization disorder. Human Pathology 1986;17:543-545

16. Pritzker KPH, Cheng P-T, Renlund RC. Calcium pyrophosphate crystal deposition in hyaline cartilage: Ultrastructural analysis and implications for pathogenesis. Journal of Rheumatology 1988;15:828-835

17. Brown WE, Gregory TM. Calcium pyrophosphate crystal chemistry. Arthritis and Rheumatism 1976;19:446-462

18. McCarty DJ, Solomon SD, Warnock ML, Paloyan E. Inorganic pyrophosphate concentrations in the synovial fluid of arthritic patients. J Lab Clin Med 1971;78:216-229

19. Mortier A, Lemaitre J, Rouxhet PG. Temperature-programmed characterization of synthetic calcium-deficient phosphate apatites. Thermochimica Acta 1989;143:265-282

20. Vaimakis TC, Pomonis PJ, Sdoukos AT. A detailed study of the condensation of the $Ca(H_2PO_4)_2 \cdot H_2O$-$CaHPO_4 \cdot 2H_20$ system under thermal treatment. Thermochimica Acta 1990;168:103-113

21. Tynsuaadu K. Influence of silicic acid and glauconite on thermal dehydration of $Ca(H_2PO_4)_2.H_2O$. Journal of Thermal Analysis 1990;36:1785-1793

22. Lin F-H, Lin C-C, Lu C-M, Liu H-C, Sun J-S, Wang C-Y. Mechanical properties and histological evaluation of sintered β-$Ca_2P_2O_7$ with $Na_4P_2O_7.10H_2O$ addition. Biomaterials 1995;16:793-802

23. Nordstrom EG. Reaction of bone to HA, carbonate-HA, hydroxyapatite + calcium orthophosphate and to hydroxyapatite + calcium ortho- and pyrophosphate. Bio-Medical Materials and Engineering 1992;2:115-121

24. Calvo C. Refinement of the crystal structure of β-$Mg_2P_2O_7$. Canadian Journal of Chemistry 1965;43:1139-1146

25. Bennett RM, Lehr JR, McCarty DJ. Crystal shedding and acute pseudogout. Arthritis and Rheumatism 1976;19:93-97

26. Andrews R, Phelps P. Release of lysosomal enzymes from polymorphonuclear leukocytes (PMN) after phagocytosis of monosodium urate (MSU) and calcium pyrophosphate dihydrate (CPPD) crystals: effect of colchicine and indomethacin. Arthritis and Rheumatism 1971;14:368(Abstract)

27. Mandel GS, Renne KM, Kolbach AM, Kaplan WD, Miller JD, Mandel NS. Calcium pyrophosphate crystal deposition disease: preparation and characaterization of crystals. Journal of Crystal Growth 1988;87:453-462

28. Cheng P-T, Pritzker KPH, Adams ME, Nyburg SC, Omar SA. Calcium pyrophosphate crystal formation in aqueous solutions. Journal of Rheumatology 1980;7:609-616

29. McCarty DJ. Diagnostic mimicry in arthritis - Patterns of joint involvement associated with calcium pyrophosphate dihydrate crystal deposits. Bulletin on the Rheumatic Diseases 1974;25:804-809

30. Cheng P-T, Pritzker KPH. Pyrophosphate, phosphate ion interaction: Effects on calcium pyrophosphate and calcium hydroxyapatite crystal formation in aqueous solutions. Journal of Rheumatology 1983;10:769-777

31. Burt HM, Jackson JK. Characterization and membranolytic effects of triclinic calcium pyrophosphate dihydrate crystals. Journal of Rheumatology 1987;14:968-973

32. Winternitz CI, Jackson JK, Burt HM. The interaction of monoclinic calcium pyrophosphate dihydrate crystals with neutrophils. Rheumatology International 1996;16:101-107

33. Cheng P-T, Pritzker KPH. The effect of calcium and magnesium ions on calcium pyrophosphate crystal formation in aqueous solutions. Journal of Rheumatology 1981;8:772-782

34. Pritzker KPH, Mendes M, Mandel E. Calcium pyrophosphate monoclinic crystal size. 1997. Unpublished data

35. Calvo C. The crystal structure of α-$Ca_2P_2O_7$. Inorganic Chemistry 1968;7:1345-1351

36. Webb NC. The crystal structure of β-$Ca_2P_2O_7$. Acta Crystallographica 1966;21:942-948

37. Mandel NS. The crystal structure of calcium pyrophosphate dihydrate. Acta Crystallographica 1975;B31:1730-1734

38. Davis NL, Mandel GS, Mandel NS, Dickerson RE. Structure of monoclinic dicalcium pyrophoshate tetrahydrate. Journal of Crystallographic and Spectroscopic Research 1985;15:513-521

39. Cheng P-T, Pritzker KPH. α-calcium disodium pyrophosphate tetrahydrate. Acta Crystallographica 1980;B36:921-924

40. Cheng P-T, Nyburg SC. α-calcium disodium pyrophosphate tetrahydrate, $CaNa_2P_2O_7•4H_20$. Crystal Structure Communications 1979;8:313-317

41. Mandel N, Mandel G. Calcium pyrophosphate crystal deposition in model systems. Rheumatic Diseases Clinics of North America 1988;14:321-340

42. Taylor MG, Simkiss K, Leslie M. Computer modelling of phosphate biominerals. Transfer of parameters for interatomic potentials for different polymorphs of divalent metal diphosphates. J.Chem Soc.Faraday Trans. 1994;90:641-647

43. Schwarm H, Vigenschow H, Knobloch K. Kinetic Characterization and Partial Purification of the Membrane-Bound Inorganic Pyrophosphatase from Rhodopseudomonas palustris. Biol Chem Hoppe-Seyler 1986;367:127-133

44. Pritzker KPH, Cheng P-T, Adams ME, Nyburg SC. Calcium pyrophoshate dihydrate crystal formation in model hydrogels. Journal of Rheumatology 1978;5:469-473

45. Pritzker KPH, Cheng P-T, Hunter GK, Grynpas MD, Kessler MJ, Renlund RC. In vitro and in vivo models of calcium pyrophosphate crystal formation. In: Butler WT, ed. The chemistry and biology of mineralized tissues. Birmingham: Ebsco Media, 1985:381-384

46. Pritzker KPH. Calcium pyrophosphate dihydrate crystal deposition and other crystal deposition diseases. Curr Op Rheum 1994;6:442-447

47. Yosipovitch ZH, Glimcher MJ. Articular chondrocalcinosis, hydroxyapatite deposition disease, in adult mature rabbits. Journal of Bone and Joint Surgery 1972;54-A:841-853

48. Pritzker KPH, Cheng P-T, Omar SA, Nyburg SC. Calcium pyrophosphate crystal formation in model hydrogels. II. Hyaline articular cartilage as a gel. Journal of Rheumatology 1981;8:451-455

49. Harries JE, Dieppe PA, Heap P, Gilgead J, Mather M, Shah JS. In vitro growth of calcium pyrophosphate crystals in polyacrylamide gels. Annals of the Rheumatic Diseases 1983;42:100-101

50. Hunter GK, Grynpas MD, Cheng P-T, Pritzker KPH. Effect of glycosaminoglycans on calcium pyrophosphate crystal formation in collagen gels. Calcified Tissue International 1987;41:164-170

51. Mandel NS, Mandel GS. A model for human calcium pyrophosphate crystal deposition disease: Crystallization kinetics in a gelatin matrix. Scanning Electron Microscopy 1984;IV:1779-1792

52. Mandel NS, Mandel GS, Carroll DJ, Halverson PB. Calcium pyrophosphate crystal deposition. An in vitro study using a gelatin matrix model. Arthritis and Rheumatism 1984;27:789-796

53. Mandel GS, Halverson PB, Mandel NS. Calcium pyrophosphate crystal deposition: the effect of monosodium urate and apatite crystals in a kinetic study using a gelatin matrix model. Scanning Microscopy 1988;2:1189-1198

54. Mandel GS, Halverson PB, Mandel NS. Calcium pyrophosphate crystal deposition: The effect of soluble iron in a kinetic study using a gelatin matrix model. Scanning Microscopy 1988;2:1177-1188

55. Mandel GS, Halverson PB, Rathburn M, Mandel NS. Calcium pyrophosphate crystal deposition: A kinetic study using a type I collagen gel model. Scanning Microscopy 1990;4:175-180

56. Mandel NS, Mandel GS. Nucleation and growth of CPPD crystals and related species in vitro. In: *Calcium in Biological Systems*. RP Reuben, G Weiss, JW Putney, eds. New York: Plenium Press, 1984:711-717.

57. Wheatley DN, Agutter PS. Historical aspects of the origin of diffusion theory in 19th century mechanistic materialism. Perspectives in Biology and Medicine 1996;40:139-156

58. Agutter PS, Malone PC, Wheatley DN. Intracellular transport mechanisms: a critique of diffusion theory. Journal of Theoretical Biology 1995;176:261-272

59. Pommert JF, Klaebe A, Perie J, Lebugle A, Puech J. Observation and analysis of crystalline phases in processed cheese. Journal of Food Science 1988;53:1367-1370

60. Ryan LM, McCarty DJ. Understanding inorganic pyrophosphate metabolism: toward prevention of calcium pyrophosphate dihydrate crystal deposition. Annals of the Rheumatic Diseases 1995;54:939-941

61. Urban JPG. Solute transport in articular cartilage and the intervertebral disc. In: *Connective Tissue Matrix, Part 2*. DWL Hukins, ed. Boca Raton: CRC Press Inc. 1990:44-65.

62. Hearn PR, Russell RGG. Formation of calcium pyrophosphate crystals in vitro: implications for calcium pyrophosphate crystal deposition disease (pseudogout). Annals of the Rheumatic Diseases 1980;39:222-227

63. Cheng P-T, Pritzker KPH. Ferrous [Fe^{2+}] but not Ferric [Fe^{3+}] ions inhibit *de novo* formation of calcium pyrophosphate dihydrate crystals: Possible relationships to chondrocalcinosis and hemochromatosis. Journal of Rheumatology 1988;15:321-324

64. Cheng P-T, Pritzker KPH. Inhibition of calcium pyrophosphate dihydrate crystal formation: Effects of carboxylate ions. Calcified Tissue International 1988;42:46-52

65. Dijkgraaf LC, Liem RSB, De Bont LGM, Boering G. Calcium pyrophosphate dihydrate crystal deposition disease: a review of the literature and a light and electron microscopic study of a case of the temporomandibular joint with numerous intracellular crystals in the chondrocytes. Osteoarthritis and Cartilage 1995;3:35-45

66. Taylor MG, Simkiss K, Greaves GN, Harries J. Corrosion of intracellular granules and cell death. Proc R Soc Lond 1988;B234:463-476

67. Rosenberg H. The isolation and identification of "volutin" granules from tetrahymena. Experimental Cell Research 1966;41:397-410

68. Munk N, Rosenberg H. On the Deposition and Utilization of Inorganic Pyrophosphate in Tetrahymena Pyriformis. Biochim Biophys Acta 1969;177:629-640

69. Rosenberg H, Munk N. Transport Phenomena Associated with the Deposition and Disappearance of Pyrophosphate Granules in Tetrahymena Pyriformis. Biochim Biophys Acta 1969;184:191-197

70. Taylor M, Simkiss K, Greaves GN. Amorphous structure of intracellular mineral granules. Biochemical Society Transactions 1986;14:549-552

71. Taylor MG, Simkiss K. Structural and analytical studies on metal ion-containing granules. In: *Biomineralization.* S Mann, J Webb, RJP Williams, eds. Chemical and Biochemical Perspectives. New York: VCH, 1989:427-460.

72. Nielsen AE, Christoffersen J. Mechanisms of crystal growth and dissolution. In: *Biological Mineralization and Demineralization.* GH Nancollas, ed. Berlin: Springer Verlag, 1982:37-77.

73. Rogers LB, Reynolds CA. Interaction of pyrophosphate ion with certain multivalent cations in aqueous solutions. J Amer Chem Soc 1949;71:2081-2085

74. Cheng P-T, Pritzker KPH, Kandel RA, Reid A. Analytical scanning and transmission electron microscopy and x-ray microdiffractometry of calcium pyrophosphate dihydrate crystal deposits in tissues. Scanning Electron Microscopy 1983;369-377

75. McCarty DJ. Identification of calcium pyrophosphate dihydrate crystals. Clinical and Expermental Rheumatology 1992;10:555-556

76. Omar SA, Cheng P-T, Nyburg SC, Pritzker KPH. Application of scanning and transmission electron microscopy, x-ray energy spectroscopy, and x-ray diffraction to calcium pyrophosphate crystal formation in vitro. Scanning Electron Microscopy 1979;11:745-749

77. McCarty DJ. Synovial fluid. In: *Arthritis and Allied Conditions*. 12th Ed. McCarty DJ, Koopman WJ, eds. Philadelphia: Lea & Febiger, 1993:71-76.

78. Blair JM, Sorensen LB, Arnsdorf MF, Lal R. The application of atomic force microscopy for the detection of microcrystals in synovial fluid from patients with recurrent synovitis. Seminars in Arthritis and Rheumatism 1995;24:359-369

79. Shinozaki T, Pritzker KPH. Polyamines enhance calcium pyrophoshate dihydrate crystal dissolution. Journal of Rheumatology 1995;22:1907-1912

80. Holmyard DP, Shinozaki T, Pritzker KPH. FE-SEM examination of calcium pyrophosphate dihydrate crystal dissolution. Proceedings of Microscopical Society of Canada 1997;24:86-87.(Abstract)

81. Osterheld RK. Kinetics of the aqueous reversion of pyrophosphate. J Phys Chem 1958;62:1133-1135

82. Van Wazer JR, Griffith EJ, McCullough JF. Structure and properties of the condensed phosphates. VII. Hydrolytic degradation of pyro- and tripolyphosphate. J Amer Chem Soc 1955;77:287-291

83. Van Wazer JR. *Phosphorus and its compounds: Volume 1*. New York: Interscience, 1958.

84. Flodgaard H, Fleron P. Thermodynamic parameters for the hydrolysis of inorganic pyrophosphate at pH 7.4 as a function of [Mg^{2+}], [K^+], and ionic strength determined from equilibrium studies of the reaction. Journal of Biological Chemistry 1974;249:3465-3473

85. Shimabayashi S, Kusada O, Nakagaki M. Hydrolysis of diphosphate ion to orthophosphate ion in the presence of calcium ion. Fundamental study of the development of the mixed crystal deposition disease. Chem Pharm Bull 1987;35:2163-2170

86. Colvin ME, Evleth E, Akacem Y. Quantum chemical studies of pyrophosphate hydrolysis. J Amer Chem Soc 1995;117:4357-4362

87. Egan EP, Jr., Wakefield ZT. Thermodynamic properties of calcium pyrophosphate, 10 to 1700°K. Journal of American Chemical Society 1957;79:558-561

88. Grynpas MD, Cheng P-T, Pritzker KPH. Dissolution of calcium pyrophosphate crystals by free radicals generated by neutron irradiation. Journal of Rheumatology 1987;14:1073-1074

89. McCarty DJ, Palmer DW, Halverson PB. Clearance of calcium pyrophosphate dihydrate crystals in vivo. I. Studies using ^{169}Yb labeled triclinic crystals. Arthritis and Rheumatism 1979;22:718-727

90. McCarty DJ, Palmer DW, James C. Clearance of calcium pyrophoshpate dihydrate crystals in vivo. Arthritis and Rheumatism 1979;22:1122-1131

91. McCarty DJ, Palmer DW, Garancis JC. Clearance of calcium pyrophosphate dihydrate crystals in vivo. Arthritis and Rheumatism 1981;24:706-710

92. Sin YM, Sedgwick AD, Moore A, Willoughby DA. Studies on the clearance of calcium pyrophosphate crystals from facsimile synovium. Annals of the Rheumatic Diseases 1984;43:487-492

93. Xu Y, Cruz TF, Cheng P-T, Pritzker KPH. Effects of Pyrophosphatase on Dissolution of Calcium Pyrophosphate Dihydrate Crystals. Journal of Rheumatology 1991;18(1):66-71

94. Xu Y, Cruz TF, Pritzker KPH. Alkaline phosphatase dissolves calcium pyrophosphatase dihydrate crystals. Journal of Rheumatology 1991;18:1606-1610

95. Xu Y, Pritzker KPH, Cruz TF. Characterization of chondrocyte alkaline phosphatase as a mediator in the dissolution of calcium pyrophosphate dihydrate crystals. Journal of Rheumatology 1994;21:912-919

96. Shinozaki T, Xu Y, Cruz TF, Pritzker KPH. Calcium pyrophosphate dihydrate (CPPD) crystal dissolution by alkaline phosphatase: interaction of alkaline phosphatase on CPPD crystals. Journal of Rheumatology 1995;22:117-123

97. Shinozaki T, Pritzker KPH. Regulation of alkaline phosphatase: Implications for calcium pyrophosphate dihydrate crystal dissolution and other alkaline phosphatase functions. Journal of Rheumatology 1996;23:677-683

98. Hench LL. Bioceramics and the origin of life. Journal of Biomedical Materials Research 1989;23:685-703

99. Kulaev IS. Inorganic polyphosphates in chemical and biological evolution. In: *The Biochemistry of Inorganic Polyphosphates*. Chichester: Wiley-Interscience, 1979:194-205.

100. Weber AL. Formation of pyrophosphate on hydroxyapatite with thioesters as condensing agents. BioSystems 1982;15:183-189

101. Jung A, Bisaz S, Fleisch H. The binding of pyrophosphate and two diphosphonates to hydroxyapatite crystals. Calcified Tissue Research 1973;11:269-280

102. McGaughey C. Binding of polyphosphates and phosphonates to hydroxyapatite, subsequent hydrolysis, phosphate exchange and effects on demineralization, mineralization and microcrystal aggregation. Caries Res. 1983;17:229-241

103. Meyer JL. Can biological calcification occur in the presence of pyrophosphate? Archives of Biochemistry and Biophysics 1984;231:1-8

104. Amjad Z. The influence of polyphosphates, phosphates, and polycarboxylic acids on the crystal growth of hydroxyapatite. Langmuir 1987;3:1063-1069

105. Vasudevan TV, Somasundaran P. Interaction of pyrophosphate with calcium phosphates. Langmuir 1994;10:320-325

106. Kim EE, Wyckoff HW. Reaction mechanism of alkaline phosphatase based on crystal structure. J Mol Biol 1991;218:449-464

107. Mathew M, Schroeder LW. The crystal structure of calcium ammonium hydrogenpyrophosphate $CaNH_4HP_2O_7$. Acta Crystallographica 1977;B33:3025-3028

108. Takagi S, Mathew M, Brown WE. The structure of diammonium tricalcium bis (pyrophosphate) hexahydrate. Acta Crystallographica 1980;B36:2526-2529

109. Calvo C. The crystal structure of α-$Mg_2P_2O_7$. Acta Crystallographica 1967;23:289-295

110. McDonald WS, Cruickshank DWJ. A reinvestigation of the structure of $Na_4P_2O_7{\bullet}10H_2O$. Acta Crystallographica 1967;22:43-48

111. Schneider S, Collin RL. Crystal structure of manganese pyrophosphate dihydrate, $Mn_2P_2O_7{\bullet}2H_2O$. Inorganic Chemistry 1973;12:2136-2139

112. Dumas Y, Galigne JL. Structure cristalline du pyrophosphate tetrapotassique trihydrate, $K_4P_2O_7{\bullet}3H_20$. Acta Crystallographica 1974;B30:390-395

113. Middlemiss N, Calvo C. Crystal structure of tetraammonium diphosphate, $(NH_4)_4P_2O_7$ at 22°C. Canadian Journal of Chemistry 1976;54:2025-2028

114. Calvo C, Au PKL. Crystal structure of $Cd_2P_2O_7$[1]. Canadian Journal of Chemistry 1969;47:3409-3416

115. Calvo C. The crystal structure and phase transitions of β-$Zn_2P_2O_7$. Canadian Journal of Chemistry 1965;43:1147-1153

116. Ng HN, Calvo C. The crystal structure of $KAlP_2O_7$. Canadian Journal of Chemistry 1973;51:2613-2620

117. Badraoui AE, Pivan JY, Maunaye M, Louer M, Louer D. Hydrothermal synthesis, structure refinement and properties of the new vanadium (III) diphosphate $H_3OVP_2O_7$. Journal of Alloys and Compounds 1996;245:47-53

13

CALCIUM PHOSPHATE IN PHARMACEUTICAL PRODUCT DEVELOPMENT

Mansoor A. Khan and Indra K. Reddy

College of Pharmacy and Health Sciences,
Northeast Louisiana University, Monroe, LA-71209 USA

INTRODUCTION

Calcium phosphate is a widely used pharmaceutical excipient as binder and filler in solid oral dosage forms which include compressed tablets and hard gelatin capsules. A review of the marketed tablets and capsules indicated that several of them contain calcium phosphate. Some of the commonly used dosage forms are Benadryl®, Sudafed®, Decadron®, Norvasc®, Ex-Lax®, and Gas-X®. The wide acceptance of calcium phosphate is primarily due to its excellent flowability and compressibility characteristics. The relatively low cost and its elemental calcium value adds to its popularity. It is mainly used as anhydrous and dihydrate forms of dibasic calcium phosphate, and tribasic calcium phosphate. While the calcium phosphate dihydrate is commercially available from different sources, Emcompress® from Mendell and Di-Tab® from Stauffer are very popular. The present chapter focuses on the physical and chemical properties of these calcium phosphates, flow property measurements, compressional behavior and the forces involved, tablet compression studies, and certain excipient incompatibilities. Calcium phosphates have also been used as anti-caking or flocculating agents in suspensions, and adsorbent or thickening agents in ointments and creams. However, the present chapter is focused towards its use in solid dosage forms. Since the present book is intended for a multidisciplinary readership, an introduction is provided for the method of preparation of tablets, type of tablets, function of tablet excipients, and the evaluation of compressed tablets.

METHODS OF TABLET PREPARATION

Tablets are oral solid unit dosage forms in which one usual dose of the drug has been accurately placed. When compared to liquid oral dosage forms, tablets are less expensive to ship, easy to store, least likely to undergo microbial attack, and offer better shelf-life stability. Further, tablets offer greatest dose precision and the least content variability. Tablet technology lends itself to the design of a variety of therapeutic products with a broad range of drug release characteristics, i.e., immediate, controlled-release, delayed-release, release-in stages, etc.

In addition to the active or therapeutic ingredient(s), tablets contain a variety of other inert ingredients (or additives) with specialized functions. These ingredients contribute to the physical form, stability, taste, appearance and overall effectiveness of the tablet dosage form. Therefore, the excipients may be included in tablet formulations to improve physical appearance, enhance stability and facilitate disintegration process following administration. The most common tablet excipients include diluents (or fillers), binders, anti-adherents, glidants, lubricants, colorants, and disintegrating agents. Although these ingredients, in general, are non-therapeutic, some of them have shown to influence the release of the drug substances which in turn affects the clinical outcome. The production methods employed may also affect the physiological availability of the drug from tablets. In light of these reasons, the personnel involved in large scale production of tablets should not only consider the physiological availability of tablet formulation but also the non-therapeutic excipients, processes, and equipment. Therefore care must be exerted in the selection as well as evaluation of excipients to ensure that the therapeutic efficacy of the active ingredient(s) will not be affected.

Compressed tablets are prepared by three methods including: A. dry granulation, B. wet granulation, and C. direct compression. Of the three methods, wet granulation is the most widely used method for the production of tablets.

Wet Granulation

Substances having crystalline structure may be directly compressed into a tablet without additional treatment. However, most substances require pretreatment to impart free flowability in the tablet machine and subsequent compression into tablets. The pretreatment in which fine powders are converted to free flowing granules is referred to as *granulation*. Granulation ensures uniformity in filling of the die cavity, formation of a tablet under compressional forces, and easy ejection of the tablet after compression. In wet granulation, a solution of a binder is used which is more effective than the same amount of dry binder.

The process of wet granulation may be described in the following steps:

A. The powdered active ingredient(s) and inactive tablet excipients are weighed and thoroughly mixed in a motor driven blender. After thorough blending, powders are passed through a sifter or screen to eliminate clumps or compacts.
B. The blended ingredients are moistened by adding appropriate quantities of granulating solution. The granulating solution is added gradually with kneading to obtain wet mass. The addition of optimal amount of binder is critical in the preparation of an effective tablet. While insufficient binder results in poor adhesion leading to capped or chipped tablets, excessive amounts of granulating solution may lead to a hard tablet with longer disintegration time.
C. The wet mass is passed through a set of sieves (generally through # 6 or 8 size mesh in a sifter) to obtain granules or pellets.
D. The wet granules are dried in an oven at a suitable temperature. The granules are typically spread on shallow trays. In drying granules, it is desirable to maintain a residual amount of moisture to maintain the granulation ingredients in a hydrated state.
E. The dried granules are passed through another set sieves with appropriate mesh sizes, generally through #16 or 20.
F. A lubricating agent (lubricant) is added to the granules to ensure free flowability in tablet machine. The lubricated granules are blended and compressed into tablets under compressional forces. Generally, a compressional force of 3,000 to 4,000 lbs is applied in making a tablet. Single and multi- rotary press machines are available to compress the tablets commercially.

Dry Granulation

For tablet ingredients that are sensitive to moisture or elevated temperatures used during the drying cycles, dry granulation serves as an alternative method to wet granulation. For dry granulation, tablet ingredients which have sufficient inherent binding or cohesive properties are suitable. In this process, all formulation ingredients are weighed, mixed together and made into large masses (slugs or pellets of approximately one inch diameter) by compaction. The large masses are subsequently broken up by a mill, passed through a suitable size mesh to obtain granulation, lubricated, and finally compressed into tablets in a tablet machine.

The process of dry granulation may be described as follows:

A. Weighing of the active ingredient(s) and tablet excipients
B. Blending of the mixture
C. Preparing large masses of the mixture by compaction
D. Crushing/sizing the large masses into smaller granules
E. Adding the lubricant(s) and blending
F. Compressing tablets

Direct Compression

Certain substances possessing free flowing as well as cohesive characteristics can be compressed directly in a tablet machine without the need of any granulation process. While theoretically they do not need the granulation process, a closer look at the material would reveal that they are microgranulations themselves.[1] They are small crystallite randomly embedded in a matrix of some amorphous material which readily imparts the desired characteristics of consolidation including plastically deformation to relieve internal stress and strong bonding surfaces.[2] It is necessary that the drug possess good flow and compressible characteristics. When the amount of drug is 25% or less of total tablet weight, it is customary to use vehicles such as dicalcium phosphate dihydrate, tricalcium phosphate, calcium sulfate, spray dried lactose and microcrystalline cellulose. Following are some examples of the directly compressible excipients:

Diluents:	Nu-tab, Avicel, anhydrous lactose, spray-dried lactose, dicalcium phosphate dihydrate and granular mannitol
Binders:	Starch, carboxymethyl cellulose and cross linked polyvinyl pyrrolidone
Lubricants:	Magnesium stearate, stearic acid, Cabosil®, polyethylene glycols, and waxes with high melting points
Glidants:	Fumed silicon dioxide, talc

TYPE OF TABLETS

There are many types of tablets and the classification of which is based on A. types of excipients used, B. method of preparation, C. site of absorption, and D. route of administration. Of many types, compressed tablets are the most commonly used. Compressed tablets, which are prepared by single compression of a formulation containing the drug and excipients, generally referred to as uncoated tablets, but some tablets (e.g., Aspirin, Tylenol®, Valium®) may be coated after the compression.

Characteristics and qualities of compressed tablets include A. elegant in appearance, B. the drug in tablet must be bioavailable (*bioavailability* refers to the rate and extent of the drug that is available to systemic circulation from the administered dosage form, and it is checked by disintegration and dissolution tests), C. uniformity in weight, thickness and drug contents, D. sufficiently strong and resistant to withstand pressures during manufacture, packaging, shipping, shelf-life and use, and E. retain all the functional attributes throughout its shelf-life, including drug stability and efficacy

TABLET EXCIPIENTS

A tablet consists of two components which are A. one or more drug(s), and B. one or more non-therapeutic agents (also known as tablet additives or excipients). A drug can be absorbable for intended systemic effect(s) or non-absorbable (i.e., antacids and adsorbents for local effects within the gastrointestinal tract). Non-therapeutic agents include diluents, binders, disintegrants, lubricants, glidants, colorants, flavorants, sweeteners and adsorbents.[3] The additives have some kind of effect on the physical, chemical and biopharmaceutical properties of finished tablets. Selection of these agents is the biggest challenge to a formulator. Following is a brief discussion on the ingredients used in tablet formulations.

Diluents

These are substances which make up the major portion of the tablet and hence are also known as *bulking agents* or *fillers*. Moisture sensitive drugs need additional precautions in the selection of diluent materials. Generally, diluents with bound water are preferred over un-bound water. Each diluent material behaves differently under an applied compressional force. Such variations strongly influence the hardness which in turn affects the disintegration, dissolution and biopharmaceutical profile of the tablet. It is very important to determine the exact compressional force required to produce a tablet of desired hardness. Under similar compressional forces, tablets made of a same active ingredient but with different diluents have different hardness. It is imperative that different hardness results in different dissolution which in turn gives rise to different absorption. Some commercially available diluents include lactose, starch, mannitol, sorbitol, microcrystalline cellulose, calcium phosphate-trihydrate and calcium sulfate-dihydrate.

Binders

Binders are the materials added to tablet formulation to add cohesiveness to powders, thereby providing the necessary bonding to form granules, which under compaction form tablets. Binders are selected on the basis of A. compatibility and/or B. power of cohesiveness.

Disintegrating Agents

Disintegrating agents are materials which facilitate the breakup of a tablet after its administration. These agents are added either prior to or after the granulation. Disintegration depends upon the sorption of water by the binder when in aqueous environment leading to swelling and finally rupture of tablet.

Lubricants

The main function of the tablet lubricant is to reduce the friction arising at the interface of tablet and die wall during compression and ejection. The lubricant ingredient may possess anti-adherent and glidant properties. Most lubricants are hydrophobic in nature. Amongst the lubricants available, magnesium stearate is the most commonly employed excipient. The usual range of lubricants used in tablet dosage forms ranges from 1-3%.

Glidants

Glidants are the materials used to improve the flow characteristics of granules from hopper to feed tabletting machine. The content uniformity and weight of tablets are very much dependant on the flow of granules. Glidants should always be added in the dry state prior to compression.

Adsorbents.

Adsorbents are substances capable of holding quantities of fluids in an apparently dry state and generally used with ethereal oils, fluid extracts and eutectic melts.

EVALUATION OF COMPRESSED TABLETS

When the tablet is administered by the oral route, it passes into the gastrointestinal tract where it disintegrates and/or dissolves, and the dissolved drug is then absorbed and elicits its therapeutic effect. Tablets, in general, are designed in such a way as to release their drug to the body rapidly and completely so that the therapeutic effects will be realized quickly. On the other hand, tablets may be designed to release the drug slowly over a longer (12 to 24 hours) interval, producing a prolonged or sustained therapeutic effect. The latter type of tablets are commonly referred to as *controlled-release* or *sustained-release* tablets. Tablets, once prepared, must meet a set of quality control specifications which include tablet weight, content uniformity, tablet thickness, tablet hardness, tablet disintegration, and for drug dissolution. These characteristics must be controlled within the production of a batch of tablets as well as between the batches in order to ensure the therapeutic efficacy and outcome. Following is a brief discussion on the routine quality tests.

Content Uniformity

Content uniformity is determined by assaying the drug contents in tablets by grinding a large sample of tablets and analyzing an aliquot. Generally a sample of 30 tablets are selected. Out of 30, 10 tablets are assayed individually for their drug content. At least 9 must assay within ± 15% of the declared content (label claim) and none should exceed ± 25%. If more than 2 exceed the 15% limit, the other 20 tablets are also assayed. None of these should fall outside the limits.

Tablet Weight Variation

The United States Pharmacopoeia (USP) has a weight variation standard to which the official tablet must conform.[4] Periodic checks are made to ensure that the tablet weight is within the USP specified range.

Tablet Thickness

Measurement of thickness or diameter is performed by Micrometer Calipers, Thickness Gauge, or by Ames Dial Comparator. Generally, the average thickness of 10 tablets is determined and the average thickness of each tablet should be ± 5%.

Tablet Hardness

Measured in the units of kgs. or lbs. by Strong-Cobb Hardness Tester. Typically, a compressional force of 3,000 - 4,000 lbs force is used. However, certain types of tablets (such as lozenges, buccal and chewable tablets) are made hard to dissolve slowly, and some are made soft (e.g., hypodermic, compressed tablet triturates and sublingual tablets).

Hardness depends on density, porosity particle size, applied pressure and also on the nature of drug and excipients. Generally, a force of 4 kg to break a tablet is considered to be the minimum for a satisfactory tablet. The hardness range for tablets should be between 4-10 kg.

Friability

Another means of determination of hardness of a tablet is friability. Friability is the tendency of a tablet product to crumble. Friabilator is the instrument (a plexiglass drum having a baffle inside) used to determine friability of a given tablet by allowing the tablet to roll and fall within a rotating tumbling apparatus (fall six inches on each turn for 4 minutes or 100 rotations). The tablets (usually, a sample size of 25) are weighed before and after a specified number of rotations and the loss in weight, expressed in %, is a measure of tablet's friability. Generally, a weight loss must be less than 0.8% for a given product to pass the test. Resistance to loss of weight indicates the tablet's ability to withstand abrasion and fracture in handling, packaging, and shipment.

Tablet Disintegration

Disintegration refers to penetration of the tablet by an aqueous liquid, disruption of internal bonds and the subsequent breakdown of the tablet. The disintegration is followed by dissolution which is a prerequisite for drug absorption. The scheme of events that lead to the absorption of drugs from solid dosage forms can be represented as follows:

Drug in solid dosage forms

↓ DISINTEGRATION

Granules or aggregates

↓ DISSOLUTION

Drug in solution

↓ ABSORPTION

Therapeutic effect

Therefore, the first step in the overall scheme of events leading to therapeutic effect is to disintegrate and discharge the drug to the gastrointestinal fluids for dissolution.

All USP official tablets must pass the official test for disintegration. *In Vitro* disintegration testing apparatus consists of a basket-rack assembly containing 6 open-ended glass tubes held vertically upon a 10-mesh stainless steel wire screen. During testing, a tablet is placed in each of the six tubes of the basket and the basket is placed in the immersion fluid. The wire screen always being maintained below the level of the fluid (generally water at 37° C). The basket is raised and lowered by a mechanical device (29-32 cycles/min.)

For these tests, complete disintegration of the tablet (or capsule) is defined as *that state in which any residue of the unit, except fragments of insoluble coating (or capsule shell), remaining on the screen of the test apparatus is a soft mass having no palpably firm core.* For compressed uncoated and sublingual tablets, the USP end point of disintegration test should be within 30 minutes (sublingual disintegrate within 2 minutes). For buccal and some chewable tablets, USP end point is less than 4 hours. For plain coated tablets, initial soaking in water at room temperature for 5 minutes is allowed to remove any water-soluble external coating followed by disintegration tests in simulated gastric and intestinal fluids at 37° C. Enteric coated tablets are similarly tested, except that the tablets are permitted to be tested in the simulated gastric fluid for an hour after which no sign of disintegration must be seen. The time of testing for enteric-coated tablets, generally, is 2 hours but may vary according to the individual monograph. In all the cases, if 1 of 6 tablets fail to disintegrate completely within the specified time, tests are repeated on 12 additional tablets, and not less than 16 of the total of 18 tablets tested must disintegrate completely to meet the official USP standard.

Dissolution Rate

Dissolution rate is the time required for a drug to dissolve in the fluids at the absorption site (or the rate at which a drug is available at the absorption site). Dissolution rate when combined with other physico-chemical data including solubility and pKa, enables one to predict the potential *in vivo* absorption of a new drug substance. It is understood that a drug must first dissolve in gastrointestinal tract fluids in order to be absorbed. Many times, dissolution rate is the rate limiting step in the absorption process. For example, increasing the dissolution rate of digoxin would increase its biological activity.

Dissolution rate affects onset, intensity, duration of response, and overall bioavailability. Dissolution rate is routinely determined in water, 0.1 N HCl, and pH 7.4 buffer. Dissolution rate can be increased by either decreasing the particle size or by formation of soluble salts of the substance.

The dissolution test is carried out in a special dissolution apparatus. This test is carried out where a tablet is placed in a small wire mesh basket which is immersed in the dissolution media contained in a 1000 mL flask. The drug absorption and bioavailability largely dependent upon having the drug in the dissolved state.

PROPERTIES OF CALCIUM PHOSPHATES

Dibasic calcium phosphate, anhydrous, has a molecular weight of 136.06, while the dihydrate and trihydrate forms have molecular weights of 172.09 and 502.32 respectively.[5] Their empirical formulae are $CaHPO_4$, $CaHPO_4.\ 2H_2O$, and $Ca_3(PO_4)_2Ca(OH)_2$. They are white, odorless, and tasteless powders which are relatively insoluble in water and alcohol. They are soluble in dilute mineral acids. The dihydrate forms degrade at temperatures below 100° C, and the trihydrate form has a melting point of 1670° C. The reported tapped bulk densities are; anhydrous form of dicalcium phosphate (0.76 g/cc), dihydrate form of dicalcium phosphate (0.94 g/cc), and the tricalcium phosphate form (0.96 g/cc). The reported untapped bulk densities are 0.70 g/cc, 0.83 g/cc, and 0.88 g/cc in the same order.

The reported crystal systems are Triclinic, Monoclinic, and Triclinic for the anhydrous dicalcium phosphate, dihydrate dicalcium phosphate and the tricalcium phosphate. The dicalcium phosphates are aggregates of small crystallites and undergo brittle fracture for consolidation upon the application of high compression pressures.[6] The solubility product, K_{SO}, have been reported to be 2.5×10^{-7} and 4.7×10^{-59} at 25° C for dicalcium phosphate and tricalcium phosphate respectively.[5] At 37° C, these values were found to be 1.9×10^{-7} and 2.3×10^{-59}. The same study also reported a pH value of 7.0-7.4 for dicalcium phosphate and 6.6-7.0 for tricalcium phosphate. All the three calcium phosphates are reported to be non-hygroscopic, which is presumably due to the metastable nature of the compound.

METHOD OF MANUFACTURE

Dibasic calcium phosphate is prepared by dissolving apatite or some other phosphate mineral in sulfuric acid and then filtering this solution.[7] Addition of lime water precipitates the calcium phosphate which is then dried to contain 0.1 to 0.2% moisture. Tribasic calcium phosphate is obtained similarly by using oxydapatit or other rocks instead of apatite. The equilibrium moisture content is ≤ 2.5%.

FLOWABILITY CHARACTERISTICS

The flowability of tablets and granules is an important parameter to be considered especially when they have to be compressed. The flow properties may change because of various factors such as temperature, chemical changes, moisture, particle size, shape, porosity, density and surface texture.[8] Further, there are forces of friction and cohesion between the particles. These forces will depend, to a large extent, upon the geometry of the particles and their size distribution. Spherical or near-spherical particles that are uniform have the best flow characteristics. Fine particles hinder flow by filling up the voids between larger particles. Emcompress® and Di-Tab® particles are fairly uniform with near-spherical shape, and therefore, provide good flow properties.[9-10] Powders can be non-cohesive or cohesive and accordingly may have a steady or gushing flow. Flowability is a measure for the steady flow where the flow is controlled as in dry sand. Floodability is a measure of gushing flow where the entire bulk tries to move as a solid mass. There are many methods to determine the flow properties including the angle of repose,[11] angle of spatula, and flow rates through funnel. Carr[12] has described a method in which a number of parameters related to flow such as angle of repose, angle of spatula, compressibility and uniformity coefficient are measured and scored according to a weighting system. The sum of all the points is described as the flowability index value. The angle of repose is the angle formed by a heap of granules when allowed to fall from a funnel on a flat surface. Angle of spatula is the angle formed when a spatula is raised from a flat surface out of the bulk pile. Angle of repose values between 30 and 40 are considered good but the values of less than 30 are excellent. Percent compressibility is a numerical value obtained from the measurements of tapped (T) and untapped (U) bulk densities. It is given by ((T-U)/T)*100. Percent compressibility values of 5-15 are considered excellent, and this value for Emcompress® has been reported to be 15.[13] Uniformity Coefficient is determined by sieving the powder and determining the ratio of the sieve through which 60% of the material passed to the sieve through which 10% of the material passed. A ratio of 2 or less is considered excellent.

For floodability, Carr proposed extra parameters to be considered in addition to the flowability parameters. These are angle of fall, angle of difference, and dispersibility. By the summation of points of all these parameters, floodability index values are obtained. The angle of fall is the new angle of repose when the granules are mechanically disturbed. Angle of difference is the numerical value obtained by the difference of the angle of repose and the angle of fall. Dispersibility of granules is its ability to become fluidized. A commercial testing machine from Hosokawa Iron Works, Osaka, Japan is now available for measuring flowability and floodability values.

Carr's flowability index values are much more reliable than measuring a single parameter such as angle of repose because the flow properties are also related to packing density and mechanical arrangements.[14] In an optimization study, Carr's Flowability index has been used as a response variable for the evaluation of flow properties of some novel ibuprofen-Eudragit coprecipitates.[15] These flow property measurements, however, have not been reported for calcium phosphates. From the available information on its particle shape which is near-spherical,[10] angle of repose studies,[1] compressibility characteristics[5] and non-hygroscopic nature,[16] the dicalcium phosphates should have a high flowability index value. Further, in a study involving common excipients such as pregelatinized starch, microcrystalline cellulose (Avicel pH-102), powdered cellulose, hydrous lactate, a compressible sugar, and dibasic calcium phosphate, the latter showed the best flowability properties.[6] In the same study, the bulk density of calcium phosphate was found to be the highest for dicalcium phosphate (Figure 1).

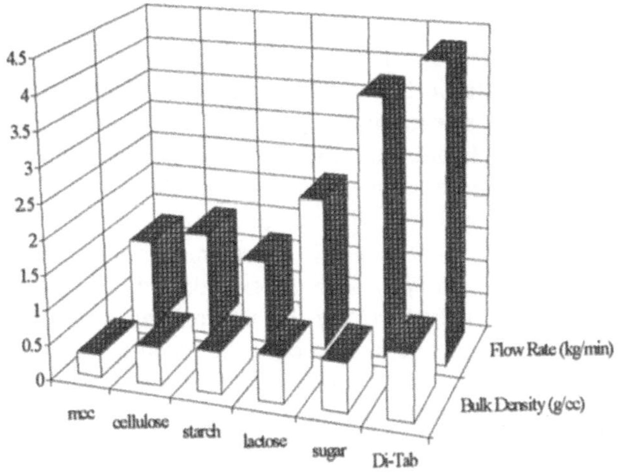

Figure 1. Bulk densities and flow rates of some common excipients.[6]

COMPRESSIONAL BEHAVIOR AND THE FORCES INVLOVED

The recent advancements in tabletting technology have led to development of high speed tabletting presses which can compress up to 10,000 tablets per minute. Similar developments in the capsule fillings have yielded machines which can fill up to 2500 capsules per minute.[6] These developments are certainly a direct result of better understanding of the excipient properties including their compressional behavior. Because of this understanding, several manufactures have now provided improved/modified excipients such as the Emcompress® and Di-Tab® brands of dicalcium phosphates that are amenable to high speed production and yet provide the desired quality.

When the powdered ingredients or granules are compressed in a die by the upper and lower punch, numerous forces act on them. The behavior of the tablet depends upon the chemical composition and physical form of additives,[17-18] moisture content,[19] particle size and morphology,[20-21] flow,[22-23] preparation methodology,[24] and the nature of excipient deformation when pressures are applied.[25-26] Since calcium phosphates, particularly dicalcium phosphate, are used primarily for their excellent compression behavior, the present section will be covered in greater detail.

The process of compression refers to the reduction in volume of the powder when it is compressed between hard surfaces while the entrapped air escapes. Consolidation refers to the sequence of events taking place in the material upon compression that yields hard compacts. When compression is applied, powdered material comes closer with a reduction in porosity. When the compression is increased, the particles deform elastically or plastically depending upon the nature of the material.[1] The deformation is considered elastic when the material recovers reversibly upon the release of pressure, and plastic when it is irreversible. The tablet formation with microcrystalline cellulose predominantly involves the elastic deformation. The elasticity is dependent on the equilibrium of attractive and repulsive inter- and intramolecular forces.[27] It has also been shown to be dependent on the different mechanical and thermophysical interactions inside the heterogeneous, disperse powder-air compact.[28-30] The plastic deformation takes place when the magnitude of compression pressure is greater than certain 'yield value' beyond which there is no reversible recovery. With this compression, the particles try to fill the remaining voids and the applied load is still less than the breaking strength. When the compression pressure increases further, the shear strength or breaking strength of the material exceeds and begins to fracture. This fracture is called the 'brittle fracture' which is the predominant mechanism in hard and brittle substances. Therefore, the overall compaction includes bulk reduction and rearrangement, elastic and/or plastic deformation, brittle fracture and mechanical interlocking. For the consolidation process, several mechanisms have been proposed which include cold welding due to surface free energies for particles in very close contact,[1] fusion of particles due to melting by frictional heat,[31-32] brittle fracture which happens to be the predominant mechanism for consolidation in dicalcium phosphates,[33] and mechanical interlocking or rebonding.[22-23]

In the tablet press, when the powder or granules are compressed between the punches, the downward motion of the upper punch forces the material down with a considerable friction along its boundary. This results in a significant loss of energy to the die wall and therefore, the force in the lower punch is much lesser than the applied force of the upper punch. Further, as the upper punch moves up, the lower punch has to force out the tablet by an ejection force. The compaction also squeezes the material into close contact with the die wall which provides a great radial force and forms a hard 'skin' of distorted strongly bonded particles. This force can be evaluated by diametrically crushing the tablets and measuring the force just needed to rupture them. This force is known as crushing strength or tablet hardness. The extent of compaction can be evaluated by measuring the relative volumes, density and porosity of the tablets. The forces of the upper and lower punches can be monitored by strain gauges. By monitoring the forces involved, the crushing strength, tablet characteristics at various preselected upper punch force, and the compaction behavior can be studied. The crushing force and tablet disintegration times were found to increase with increasing upper punch forces for sulfathiazole compacts.[11] The porosity of the tablets gradually decreased but the surface area increased to about 1600 kg/sq. cm and subsequently decreased. This was found to be due to fragmentation of granules initially, which was followed by bonding of the freshly formed surfaces at higher pressures.

In a classical study, Train[34] discovered that magnesium carbonate compacts had even thickness when a lubricant was used, but had dense compacts at the peripheral ring near upper punch and in a lower central region. This finding helps us understand why some core of hard compact remains during the disintegration tests. Train also suggested that fragmentation and rebonding are competitive processes and the rebonding prevails during the final stages of compression.

Solid excipients for tabletting at different stages of compression can be characterized by a number of methods. At the pretabletting stage, Hiestand and Smith[35] have used indices of bonding, Heckel[36-37] and Hersey and Rees[38] used yield values, Kerridge and Newton[39] and Roberts et al.[40] used the modulus of elasticity, and Tsardaka and Rees[41] used the plastic compliance and apparent viscosity. At the tabletting stage, DeBlay and Polderman[42] have used dynamic compression values with changing stress for the determination of work done on the lower punch during initial and repeated loading, Long,[43] Leigh et al.[44] and Obiorah[45] have used the residual die wall pressures and the Poisson's ratio from radial and axial pressure cycles. At the compacted state, while still under the compression load, materials have been characterized by stress relaxation methods at constant strain by David and Augsburger,[46] Rees and Rue,[47] and constant stress by Tsardaka and Rees.[48] Further, Heckel Plots[1] have been used to provide yield values, and strain-rate sensitivity on elastic recovery. For characterization at the post-tabletting stage, Radebaugh et al.[49] used non-destructive methods to determine the viscoelastic properties of consolidated particulate materials, and Patel and Staniforth[50] and Rees and Rue[51] examined the response of finished tablets to applied loads. Konkel and Mieleck,[52-53] and Mitrevej and Augsburger[54] have used tablet instrumentation methods to characterize the compression behavior of tablets.

Prior to the compression of granules into tablets, the granules are dried in special drying cabinets or ovens that are thermostatically controlled. In some instances, the granules are dried by being suspended and agitated by a stream of warm air in fluid bed dryers. In drying granulations, a residual amount of moisture in the granulation is critical for effective production of tablets. The small amounts of moisture is necessary to maintain certain tablet excipients such as gums in a hydrated state. For some binders, the effectiveness is dependent upon the presence of minute amounts of moisture. Further, the residual moisture contributes to the reduction of the static electric charges on the particles. However, an excessive amount of moisture in a granulation can cause rupturing of coatings later placed on the compressed tablets. The stability of the products containing moisture-sensitive active ingredients may also be adversely affected by the excessive moisture content of the products. In the drying process, therefore, care must be exerted to maintain an optimal and uniform moisture content. Chowhan[55] investigated the effect of low and high humidity aging on the hardness, disintegration time, and dissolution rate of naproxen tablets prepared with dicalcium phosphate. It was found that on aging in a high humidity atmosphere, the tablet hardness decreased and so were the disintegration and dissolution rates. The loss of disintegration function was attributed to a partial loss of starch binder. The dissolution rate decrease was a consequence of disintegration rate decrease where the surface area exposed to the dissolution medium decreased. The drug dissolution rate was also decreased at low humidity conditions. This was apparently due to the limited dissolution and recrystallization of the calcium phosphate in the available water in the tablet. Infrared spectroscopy and x-ray diffraction studies did not indicate any appreciable physical changes.

Under given tableting conditions, the intergranule bond strength is dependant to a large extent on the nature of the excipients. The tablet hardness, for example, is determined by the extent to which granules bond together and the strength of the intergranule bonds relative to the strength of the granules. Bonding of the granules in the plane parallel to the direction of compression appear to be sponger than bonding in the plane normal to the direction of compression. Because of the flattening or distortion of the granules, the tablets are harder in the parallel plane. When intergranule bonds are strong, they offer resistance to the propagation of faults across the grain boundary. In such cases, the strength of the compact formed at a given compaction pressure should be smaller with coarse granules than with fines. This was in agreement with hexamine and sodium chloride compacts.[56] When hexamine granules were coated with stearic acid, the intergranule bonds were weakened, thereby decreasing the hardness of the compacts. Further, the particle-size effect was not observed. For substances that do not involve fault propagation, granule size had little or no effect on tablet hardness.

TABLET COMPRESSION STUDIES WITH CALCIUM PHOSPHATES

Several studies have indicated that Emcompress® exhibited pronounced fragmentation tendency.[57] A high particulate surface area was observed regardless of the particle size of the starting material. In another study, Alderborn and Nystrom[58] reported that the initial particle size influenced the compact strength only slightly. These workers have interpreted the dominating bonding mechanism to be the consolidation with intermolecular forces.

A compression profile also evaluates the breaking strength of tablets at a certain pressure. The effect of magnesium stearate on the compression strength or breaking strength helps in the evaluation of bonding mechanism of calcium phosphates. Logically, for plastic deformations, the well known lubricant magnesium stearate should decrease the tablet hardness and for brittle fracture with bonding, it should not have a considerable effect on the tablet hardness. Therefore, a drop in breaking strength with magnesium stearate or the length of mixing with magnesium stearate indicates plastic deformation. Cartensen[5] has indicated that the tablets prepared with anhydrous dicalcium phosphate and Di-tab® did not change the hardness of tablets with an increase in magnesium stearate amounts, even when the compaction force increased from 0 to 40 KN. However, for tricalcium phosphate the hardness reduced considerably with an increase in magnesium stearate from 0.5 to 1% in the compaction pressure range from 10 to 15 KN. At still higher pressures, the tablets were capping. Further, the increase of magnesium stearate concentration to 2% also yielded tablets of poor quality. This property appears to be the reason why tricalcium phosphate is not as widely used as dicalcium phosphate as a filler in tabletting. Rudnic and Kottke[59] have provided a comparative evaluation of some directly compressible fillers on grading scale from 1 (poor) to 5 (good/high) with 0 as none. This is shown in Figure 2.

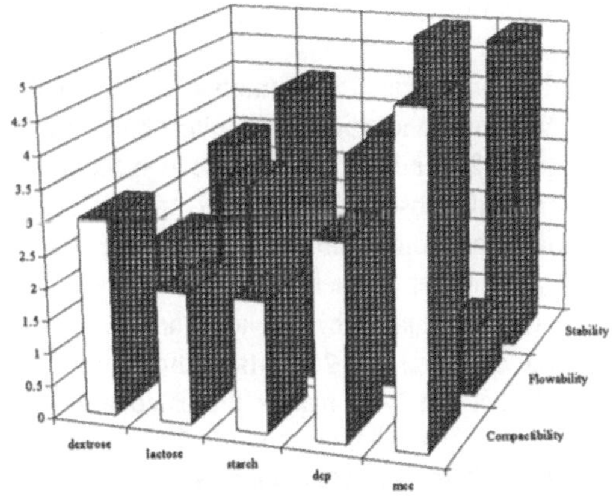

Figure 2. Comparative evaluation of some commonly used tablet excipients.[59] Dcp and mcc are dicalcium phosphate and microcrystalline cellulose, respectively.

When a formulator is experiencing hardness problems with lactose or starch as the filler material, it is worth exploring dicalcium phosphate or microcrystalline cellulose. But if there are flow problems as well, then dicalcium phosphate should be a good choice. As shown in Figure 2, even for stability reasons, dicalcium phosphate appears to be at least as good as microcrystalline cellulose. The differences between excipients are due to several reasons highlighted earlier, but an interesting study shows the compactibility behavior of dicalcium phosphate and microcrystalline cellulose. Castillo and Villafuerte[60] have evaluated the compactibility and hardness of tablets containing mixtures of microcrystalline cellulose and dicalcium phosphate with varying amounts of a lubricant/surfactant sodium lauryl sulfate. It was clearly shown that the effect of sodium lauryl sulfate on the compactibility of tablets with a continuous phase of dicalcium phosphate was negligible at low concentrations of microcrystalline cellulose. However, when the microcrystalline cellulose concentrations were increased, the tablet hardness was greater at low sodium lauryl sulfate concentrations. Conversely, the tablets with a continuous phase of microcrystalline cellulose showed a decrease in their tablet hardness with decreasing concentrations of dicalcium phosphate regardless of the sodium lauryl sulfate concentration. The conclusions that could be drawn from the studies are that the intensity of the microcrystalline cellulose effect to increase the hardness of tablets of dicalcium phosphate is dependent on the sodium lauryl sulfate proportion and the intensity of the dicalcium phosphate effect to increase the hardness of microcrystalline cellulose is not dependent on the sodium lauryl sulfate concentration.

Hunter and Ganderton[61] have also indicated that the porosity of granules obtained with calcium phosphate are independent of the moisture content and the granule strength is directly related to the degree of mechanical interlocking. Even the type of mixer used had very less influence on the porosity of dicalcium phosphate granules at equivalent moisture content. As compared to calcium phosphate, lactose granules were harder with the addition of moisture. Even the intragranular porosity was reduced with the addition of moisture.

By a holographic evaluation process, Silvennoinen et al.[62] evaluated the compacts of microcrystalline cellulose and dicalcium phosphate dihydrate and showed compressive modulii of 7.27 and 105.30 for the respective compacts with zero porosity. The tensile modulii observed were 7.21 and 102.91 in the same order. Microcrystalline cellulose and dicalcium phosphate have also been evaluated for crushing strengths of slugs either on their own or in various combinations.[63-64] The mean yield pressures were evaluated by Heckel Analysis. A blend containing 75 parts of microcrystalline cellulose and 25 parts of dicalcium phosphate yielded the hardest tablets. The Heckel Analysis results indicated a shift away from plastic deformation when dicalcium phosphate was gradually increased. Further the brittle nature of dicalcium phosphate was confirmed with a very high mean pressure value. The authors confirmed that the fragmentation of dicalcium phosphate led to enhanced bonding of compaction, and therefore hardness, when mixed with microcrystalline cellulose.

INCOMPATIBILITIES

Certain incompatibilities have been reported with calcium phosphates. Therefore, proper care is required when these chemicals are used as tablet additives. Tricalcium phosphate has been reported to be incompatible with tocopherol acetate at higher temperatures. This incompatibility is due to the large number of hydroxyl groups in the compound and its large surface area. There is a good possibility of ion substitution with barium, magnesium, and iron ions. The dicalcium phosphate has been found to delay the absorption and bioavailability of tetracyclines. The dihydrate excipients are generally good for water-sensitive drugs because they do not release the bound water easily. However, the dicalcium phosphate releases moisture at temperatures between 40 to 45° C. Therefore, if this temperature is attained in any of the unit operations in tabletting for a moisture-sensitive drug, dicalcium phosphate should not be used. One more difficulty with dicalcium phosphate has been reported to be its high density. This is because, when the tablets containing dicalcium phosphate are subjected to dissolution, the powder tends to settle down and cause erroneous results. Dicalcium phosphate is not recommended for use with strong acid salts of weak bases. The tricalcium phosphate is not recommended for use with water soluble vitamin B salts and esters of vitamin E.

CONCLUSIONS

Calcium phosphate has a variety of applications in pharmaceutical oral dosage forms including tablets and capsules. The wide use of calcium phosphate is mainly attributed to its excellent flowability and compressibility, and owing to these characteristics, it is commonly included in the product formulations to be compressed directly in a tablet machine without the need of any granulation process. Various factors affect flow properties of a chemical which include temperature, moisture, particle size and shape, porosity, density, and surface texture. Many routine methods are available to determine the flow properties. Advances in compressional technology and the underlying mechanisms including forces involved led to the development of commercial brands of dicalcium phosphates which lend themselves to the modern day high speed tablet production.

REFERENCES

1. Lachman L, Lieberman HA, Kanig JL. *Theory and Practice of Industrial Pharmacy*, Lea & Febiger, Philadelphia, 1986.

2. Shangraw RS. Raw materials of natural origin in solid dosage forms Part 1. Drug and Cosm Ind. 1978;122:68

3. Ansel HC, Popovich, NG. Pharmaceutical Dosage Forms and Delivery Systems, 5th ed. Lea and Febiger, Philadelphia, 1990.

4. *United States Pharmacopoeia*, XX1 Revision. US Pharmacopoeial Convention, Inc. 1985.

5. Carstensen JT, Ertell C. Physical and chemical properties of calcium phosphates for solid state pharmaceutical formulations. Drug Dev Ind Pharm 1990;16:1121- 33

6. Wallace JW, Capozzi JT, Shangraw RF. Performance of pharmaceutical filler/binders as related to methods of powder characterization. Pharm Technol 1983;94:96-104

7. *Handbook of Pharmaceutical Excipients*, American Pharmaceutical Association, 1994.

8. Thomson F. "Flow properties used in analysis" In *Handbook of Powder Science and Technology*, Fayed, Otten, ed. Van Nostrand Reinhold, England, 1991.

9. Mendell. Product identification on Emcompress®. Sept, 1996.

10. Stauffer. Product information on Di-Tab®, 1976.

11. Rawlins EA. *Bentley's Textbook of Pharmaceutics*, ELBS Edition, London, 1975.

12. Carr RL. Evaluating flow properties of solids Chem. Eng 1965;163-68

13. Jones TM. The influence of physical characteristics of excipients on the design and preparation of tablets and capsules. Pharm ind 1977;39:469-76

14. Martin A, Swarbrick J Cammarata A. *Physical Pharmacy*, Lea & Febiger, Philadelphia,

15. Khan MA, Bolton S, Kislalioglu MS. Optimization of process variables for the preparation of ibuprofen coprecipitates with Eudragit S100. Int J Pharm 1994;102:185-92

16. Schwartz, JB. Oral Solid Dosage Forms. In Remington's Pharmaceutical Sciences, 18th ed. Mack Publishing Company, Pennsylvania, 1990.

17. Marshall PV, York P. The compaction properties of nitrofurantoin samples crystallized from different solvents. Int J Pharm 1991;67:59-65

18. Shaikh NA, Abidi SE, Block LH. Evaluation of ethylcellulose as a matrix for prolonged release formulations. I. Water soluble drugs: Acetaminophen and Theophylline. Drug Dev Ind Pharm 1987;13:1345-69

19. Chowhan ZT, Palagyi L. Hardness increase induced by partial moisture loss in compressed tablets and its effect on In Vitro dissolution. J Pharm Sci 1978;67:1385-89

20. Marks AM, Sciarra JJ. Effect of size on other physical properties of granules and their corresponding tablets. J Pharm Sci 1968;57:497-504

21. Pitkin C and Carstensen JT. Moisture content of granulation. J. Pharm Sci 1973;62:1215

22. York P. Crystal engineering and particle design for the powder compaction process. Drug Dev Ind Pharm 1992;18:677- 721

23. Wiay PF, The physics of tablet compaction revisited. Drug Dev Ind Pharm 1992;18:627-58

24. Khan MA, Karnachi AA, Singh SK, Sastry SV, Kislalioglu SM Bolton S. Controlled Release Coprecipitates: Formulation Considerations. J Control Rel 1995;37:131-41

25. Karnachi AA, DeHon RA, Khan MA. Compression of Indomethacin Coprecipitates with Polymer Mixtures; Effect of Preparation Methodology. Drug Dev Ind Pharm 1995;

26. Karnachi AA, DeHon RA, Khan MA. Plackett-Burman Screening of Coprecipitates with Polymer Mixtures for Controlled Drug Delivery. Die Pharmazie 1995;50:550-55

27. Roberts RJ, Rowe RC, York P. The relationship between young's modulus of elasticity of organic solids and their molecular structure. Powder Technol 1991;65:136-46

28. Bassam F, York P, Rowe RC, Roberts RJ. Effect of particle size and source on variability of Young's modulus of microcrystalline cellulose powders J Pharm Pharmacol 1988;40S:68-73

29. Bassam F, York P, Rowe RC, Roberts RJ. Young's modulus of powders used as pharmaceutical excipients. Int J Pharm 1990;64:55-60

30. Roberts RJ, Rowe RC, York P. The measurement of critical stress intensity factor (K_{IC}) using three point single edge notched beam (SNEB) testing. Int J Pharm 1993;91:173-82

31. Karehill PG, Glazer M, Nystrom C. Studies on direct compression of tablets. XXIII. The importance of surface roughness for the compactibility of some directly compressible materials with different bonding and volume reduction properties. Int J Pharm 1990;64:35-43

32. Karehill PG, Nystrom C. Studies on direct compression of tablets. XXI. Investigation of bonding mechanisms of some directly compressed materials by strength characterization in media of different dielectric constant (relative permittivity) Int J Pharm 1990; 61:251-60

33. Alderborn G, Pasanen K, Nystrom C. Studies on direct compression of tablets. XI.Characterization of particle fragmentation during compaction by permeametry measurements of tablets. Int J pharm 1985;23:79-86

34. Train D, An investigation into compaction of powders. J Pharm Pharmacol 1956; 8:745- 61

35. Hiestand EN, Smith DP. Indices of tableting performance. Powder Technol 1984;38:145-59

36. Heckel RW. Density-pressure relationship in powder compaction. Trans Met Soc, AIME, 1961a;221:671-75

37. Heckel RW. An analysis of powder compaction phenomena. Trans Met Soc, AIME, 1961b;221:1001-08

38. Hersey JA, Rees JE. The effect of particle size on the consolidation of powders during compaction. Proc. Particle Size Analysis Conference, Bradford, UK,1970.

39. Kerridge JC, Newton JM. The determination of the compressive Young's modulus of pharmaceutical materials. J Pharm Pharmacol 1986;38:79P

40. Roberts RJ, Rowe RC, York P. A dynamic mechanical method for the determination of the Young's modulus of powdered drugs and excipients. J Pharm Pharmacol 1989;41:30P

41. Tsardaka KD, Rees JE. Plastic deformation and retarded elastic deformation of particulate solids using creep experiments. J Pharm Pharmacol 1989;41:28P

42. DeBlaey, CJ, Polderman J. Compression of pharmaceuticals. Pharm Biweekly 1971;

43. Long WM. Radial pressure in powder compaction. Powder Metall 1960;6:73-86

44. Leigh S, Carless JE and Burt BW, Compression Characteristics of some pharmaceutical materials. J Pharm Sci 1967;56:888-92

45. Obiorah BA. Possible prediction of compression characteristics from pressure cycle plots. Int J Pharm 1978;1:249-55

46. David ST, Augsburger LL. Plastic flow during compression of directly compressible fillers and its effects on tablet strength. J Pharm Sci 1977;66:155-59

47. Rees JE, Rue PJ. Work required to cause failure of tablets in diametral compression. Drug Dev Ind Pharm 1978a;4:131-56

48. Tsardaka KD, Rees JE. Effect of porosity on the apparent viscosity of a particulate solid, derived using creep analysis. Proc. 6th International Conference on Pharmaceutical Technology, APGI, Paris, IV 169-78, 1992.

49. Radebaugh GW, Babu SR, Bondi JN. Characterization of viscoelastic properties of compacted pharmaceutical powders by a novel non destructive technique. Int J Pharm 1989;57:95-105

50. Patel CI, Staniforth JN. Determination of apparent failure viscosity of tablets. J Pharm Pharmacol 1987;39:647-50

51. Rees JE, Rue PJ. Time-dependent deformation of some direct compression excipients. J Pharm Phramacol 1978b;30:601-07

52. Konkel P, Mielck JB. A compaction study of directly compressible vitamin preparation for the development of a chewable tablet. Part I. Pharm Tech 1992;16:138-46

53. Konkel P, Mielck JB. A compaction study of directly compressible vitamin preparation for the development of a chewable tablet. Part II. Pharm Tech 1992;16:42-50

54. Mitrevej A, Augsburger LL. Adhesion of tablets in a rotary tablet press. Part I Instrumentation and preliminary studies of variables affecting adhesion. Drug Dev. Ind. Pharm 1980;6:331-37

55. Chowhan ZT. The effect of low- and high- humidity aging on the hardness, disintegration time and dissolution rate of dibasic calcium phosphate-based tablets. J Pharm Pharmacol 1980;32:10-14

56. Shotton E, Ganderton D. The strength of compressed tablets. The relation of particle size, binding and capping in tablets of sodium chloride, aspirin and hexamine. J Pharm Pharmacol 1961;13:44T-56T

57. Alderborn G, Borjesson E, Glazer M, Nystrom C. Studies on direct compression of tablets.-XIX The effect of particle size and shape on the mechanical strength of sodium bicarbonate tablets. Acta Pharm Suec 1988;25:31-40

58. Alderborn G, Nystrom C. Studies on direct compression of tablets. IV. The effect of particle size on the mechanical strength of tablets. Acta Pharm Suec 1982a;19:381-90

59. Rudnic EM and Kottke M. "Tablet Dosage Forms" In *Modern Pharmaceutics*, Gilbert S. Banker, Christopher T Rhodes, ed. New York and Basel, Marcel Dekker Inc, 1996.

60. Castillo S, Villafuerte L. Compactibility of ternary mixtures of pharmaceutical powders. Pharm Acta Helv 1995;70:329-37

61. Hunter BM, Ganderton D. The influence on pharmaceutical granulation of the type and capacity of mixers. J Pharm Pharmac 1973;25S:71-78

62. Silvennoinen R, Raatikainen P, Ketolainen J, Ketolainen P, Paronen P. Holographic evaluation of bending and integrity of pharmaceutical powder beams. Int J Pharm 1996;131:209-17

63. Kochar SK, Rubinstein MH, Barnes D. The effects of slugging and recompression on pharmaceutical excipients. Int J Pharm1995;115:35-43

64. Kochar SK, Rubinstein MH. Slugging and recompression on characteristics of some blends of pharmaceutical excipients. Int J Pharm 1994;112:241-47

14

RECENT ADVANCES IN DENTIFRICE TECHNOLOGIES

Abdul Gaffar and John Afflitto

Colgate Palmolive Technology Center
Piscataway, New Jersey 08855 USA

INTRODUCTION: BACKGROUND ON EVOLUTION OF TECHNOLOGIES IN DENTIFRICES

Dental diseases including cavities (caries), tartar (calculus), sore gums (gingivitis), and periodontitis (loss of tissue supporting the teeth) are worldwide problems. The annual cost of all dental care in the U.S. exceeds $37 billion,[1] out of which roughly $6 billion is spent to repair the ravages of decay. However, the cost of dental disease cannot simply be measured in monetary terms. Other factors also need to be considered; for example, the loss of teeth leading to impaired chewing ability, speech problems, and changes in facial aesthetics which can cause embarrassment. The well-being of a person may also be compromised due to the associated dental pain, inability to chew properly, and potential of the infection spreading from the mouth to other parts of the body.

Currently, a tremendous amount of time is spent by dentists and hygienists to clean the teeth and associated structures to prevent dental disease. Alternative methods to prevent dental diseases which can be used by the general population are being developed to reduce the amount of time spent with the dental professional.

Historically, dentifrices or toothpastes were developed for keeping teeth clean and free of stains. The essential ingredients of a toothpaste are: a thickening agent, an abrasive cleaning agent, a surfactant, a humectant, flavor, and active therapeutic agent.[2] One of the first dentifrices contained an abrasive (precipitated calcium carbonate) and a small amount of powdered soap. This toothpaste was irritating to the tissues of the mouth because the pH was relatively high due to its soap content. After the Second World War, many companies undertook scientific research to develop dentifrices which were milder, gentler and also had therapeutic properties. Instead of soap, a synthetic detergent, sodium lauryl sarcosinate was introduced in the toothpaste. Besides preventing irritation, the synthetic detergent improved the taste and was also shown to control plaque acids which cause cavities.[3] Figure 1 provides an overview of the evolution of technologies in toothpaste. The category is driven by scientific advances and consumer benefits which have been broadly classified as a good smile.

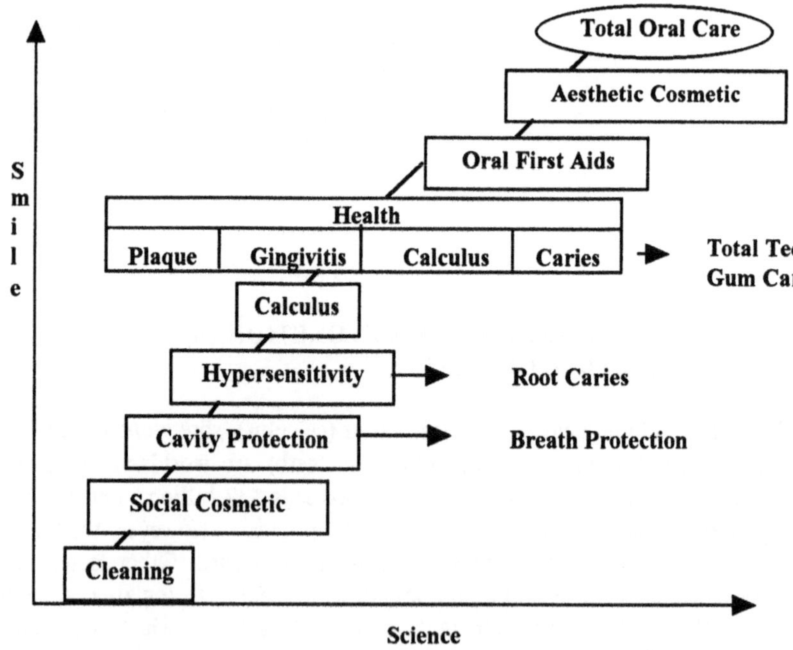

Figure 1. Evolution of technologies in toothpaste.

CLEANING, POLISHING AND FRESH BREATH

Stains on Teeth

Extrinsic discoloration is the stain found within the acquired pellicle or the superficial layer of the teeth.[4] This type of stain may originate from chromogenic foods and drinks or oral therapeutic agents such as chlorhexidine and stannous fluoride.[5] It is well-known that the color, composition, mechanism of formation, and tenacity of stains vary according to the causative agent. Several investigators have studied mechanisms of stain formation and developed methods to remove dental stain.[6]

Methods to Remove Stains

The most commonly accepted procedure to remove discoloration from tooth surfaces is by using abrasives such as prophy pastes or a combination of abrasives and surface-active agents found in toothpastes. In the former method, superficial stains are removed by an abrading action, while in the latter method, the surface-active agents solubilize or loosen the stain and the abrasive removes the stain. These methods prevent stain accumulation and, to a certain extent, remove extrinsic stain.[7] Several laboratory tests have been developed to assess relative effectiveness of different abrasive systems, but none of these tests correlate with the *in vivo* stain removal on teeth. Recently, an *in vivo* method has been developed to measure stain quantitatively.[8] The method uses reflectance spectroscopy to measure color objectively. This instrument has a pulsed xenon arc lamp which provides illumination through 30 optical fibers. The chromameter is also equipped with six highly sensitive photo cells--three photo cells to monitor the illumination output and three cells to measure the light reflected from the tooth surface being examined. The meter then converts all colors within the range of human perception into a digital mode (Minolta Corp., 1988) which is recorded in the tristimulus L*a*b* color space, as established by the Commission Internationale de L'Eclairage (CIE).[9-11] To obtain accurate and reproducible color readings *in vivo*, a custom chromameter measuring jig was fabricated for each tooth under investigation. The purpose of the custom measuring jig is to ensure that tooth color measurements are recorded from the same location on the tooth surface, preventing loss of reflected light and interference from external light sources. The tooth color is then calculated by averaging the L*a*b* color parameters for each tooth and calculating the color differences between the initial measurements and those measurements taken at each phase of the stain removal process, using the equation:

$$\Delta E = \{(\Delta L^*)^2 + (\Delta a^*)^2 + (\Delta b^*)^2\}^{1/2} \quad (1)$$

where ΔE is the difference in color (the more positive the value, the whiter the color); ΔL^* is the change in lightness (the greater the ΔL^*, the whiter the teeth); and Δa^* and Δb^* are chromacity values; i.e., the amount of redness and the amount of yellowness.[12-14]

Using this method, one can assess cleaning ability *in vitro*. Typical results of one such study are shown in Figure 2 which compared traditional fluoride in silicon dioxide abrasive (SiO_2) vs. fluoride/SiO_2 with sodium bicarbonate (SBC)/peroxide mixture. As can be seen in Figure 2, the addition of peroxide enhanced stain removal *in vitro*.

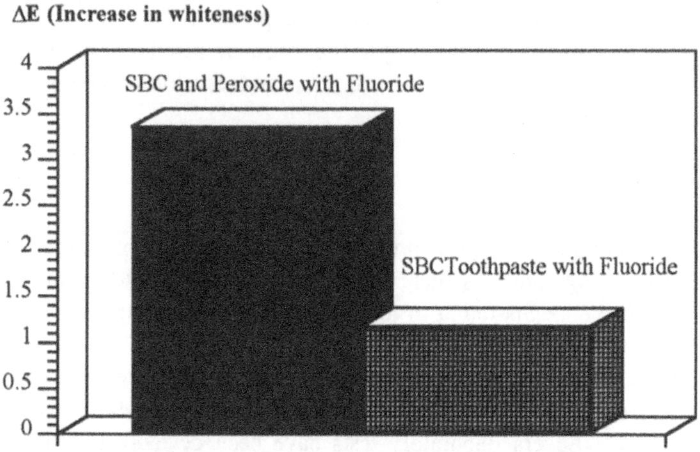

Figure 2. Cleaning efficacy of SBC/peroxide/tartar toothpaste *In Vitro*.

To confirm the laboratory data of the toothpaste an independent clinical study was performed. In this study, the cleaning ability of the new Baking Soda and Peroxide was tested versus a leading baking soda toothpaste. Results (Figure 3) of this study confirmed the cleansing ability of the product. Statistical analysis of CIE L*a*b* parameters again showed significant ($p<0.05$) cleaning can be attained after one month usage of the sodium bicarbonate/peroxide/anti-tartar toothpaste, and the bicarbonate and peroxide product was significantly better versus the leading bicarbonate toothpaste.[14]

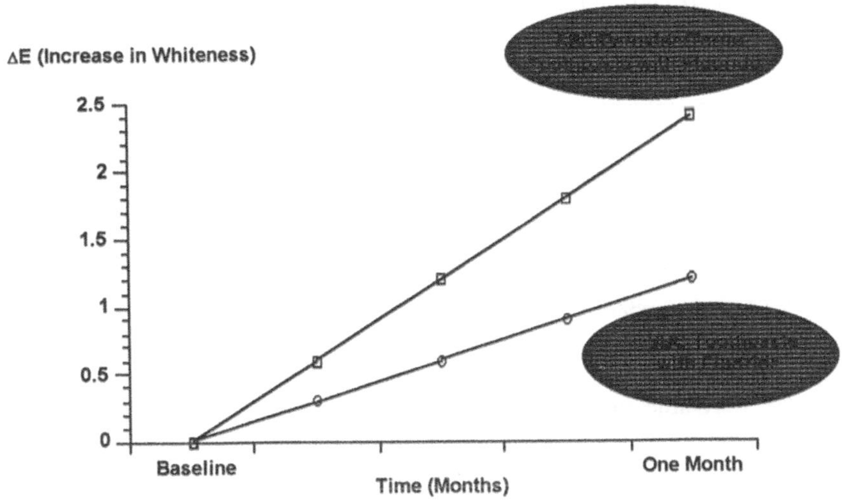

Figure 3. *In Vivo* cleaning effects of SBC peroxide toothpaste.

Dentifrices to Reduce Offensive Bad Breath

Local mouth odor is caused by oral bacteria reacting with salivary proteins to form volatile sulfur compounds (VSC). Tonzetich[15] has shown that hydrogen sulfide, methyl mercaptan and dimethyl sulfide (H_2S, CH_3SH and $(CH_3)_2S$, respectively) are the major components of mouth odor. A gas chromatographic method was developed to objectively measure VSC directly from mouth air as an alternative to the organoleptic/sensory method. This instrumental method has in turn permitted investigators to carry out studies in a number of areas relevant to human oral malodor.[16]

There are two methods currently available to assess the magnitude of oral malodor. The first is the organoleptic or sensory rating approach and the second is the GC instrumental method.[17,18] A study was conducted to determine the correlation between these two methods in a controlled clinical study. An excellent correlation ($r = 0.78$) has been established between the instrumental method and sensory evaluation.[19] Using the analytical technique, the effect of dentifrices on mouth odor have been evaluated in a variety of clinicals. Typical data from such a study are summarized on Table 1, which shows a significant reduction in mouth odor (VSC) as measured by the instrument. Typically, these studies are run double-blind cross-over. A baseline reading is taken in the morning. The subjects then brushed with a placebo or an active dentifrice and then readings are taken three or 12 hours post treatment to assess the effects. Table 1 shows the effects of sodium bicarbonate dentifrice on offensive mouth odor. A significant reduction in the odor was evident for three hours post-brushing.[19]

Table 1. Three hour effect of sodium bicarbonate dentifrices on VSC as compared to fluoride dentifrice.

Treatment	N	Initial[a] (VSC)	Final[a] (VSC)	% Inhibition
Na Bicarbonate	16	14.86 (±2.44)	8.35[b] (±3.19)	43.8
Fluoride	16	14.82 (±2.21)	10.26[b] (±4.18)	30.77

[a] = Mean ± standard deviation
[b] p = 0.01 between final treatment means; significance between dentifrices: p = 0.01
VSC = Volatile Sulfur Compounds

Bicarbonate ions react with sulfur volatiles to convert them to non-volatile forms. Both H_2S and CH_3SH are gases at 37°C, and both will dissolve in water to some degree. In the dissolved state, a fraction of these gases will dissociate, converting from volatile acids to non-volatile ionic forms in the presence of $NaHCO_3$.

Dentifrice could even provide a more pronounced long-lasting effect on odor if it contains a special antibacterial agent. Such a study was run in humans and summarized in Table 2, which shows the effect in a cross-over study of triclosan/copolymer/NaF dentifrice (TCF) vs. a placebo. The results demonstrated a significant long-lasting overnight effect against VSC formation when compared to the baseline and the placebo.[20] Thus, today the technologies are available in dentifrices which provide a sustained and long-lasting effect on mouth odor. We anticipate more product technologies in the future which may provide sustained protection against mouth odor.

Table 2. Triclosan/copolymer/fluoride (TCF) vs. a matching placebo and a negative control dentifrice extended use / overnight effect.

Treatment Dentifrice	N	Initial Mean	Final Mean	% Inhibition	p Value
TCF	14	14.57 (±2.71)	7.92 (±1.35)	45.64	0.0001
Matching Placebo	14	12.53 (±2.79)	11.01 (±3.67)	12.13	0.14
Negative Placebo Control	14	14.44 (±2.89)	11.83 (±3.43)	18.67	< 0.01

THERAPEUTIC DENTIFRICES

In the early 1950's several epidemiological studies indicated that fluoride had a protective effect against dental decay.[21] As a result, a number of investigators evaluated the effect of fluoridated toothpaste. In 1955, stannous fluoride was incorporated in dentifrice and was shown to be clinically effective against caries. Based on some of these studies, the ADA (American Dental Association) granted a seal of approval in 1960. Subsequently, sodium monofluorophosphate (MFP) was substituted for stannous fluoride and probably remains the most tested fluoride compound around the world.[22] Currently, other fluorinated compounds such as sodium fluoride and amine fluoride are used to prevent caries formation. Several mechanisms have been proposed to explain the effectiveness of fluoride against caries. These include the uptake of fluoride in the teeth, the effect of fluoride in neutralizing plaque acids, and the effect of fluoride in increasing the remineralization of sub-surface enamel lesions. A caries lesion forms when local demineralization occurs in tooth enamel and possibly in underlying dentin. Demineralization, or dissolving of the teeth mineral, results when organic acids penetrate the surface of enamel. These acids are released by specific oral bacteria due to metabolism of sugars and fermentable carbohydrates. Since bacteria form an adhering film on teeth surfaces called "dental plaque," the released organic acids are held in intimate contact with the tooth surface for extended period of time. The mechanism of demineralization can be shown by the following equation:

$$Ca_{10}(PO_4)_6(OH)_2 + 8H^+ \longrightarrow 10Ca^{2+} + 6HPO_4^{2-} + 2H_2O \qquad (2)$$

Under normal conditions, the tooth mineral, known as hydroxyapatite (HAP), is relatively insoluble. However, under acidic conditions hydroxyapatite dissolves and calcium and phosphate ions diffuse out of tooth surface and are washed away by saliva. Even though the demineralizing acids are found in plaque coating the tooth surface, current scientific evidence shows that tooth decay begins below the surface of enamel as shown in Figure 4.

This phenomenon can be explained by the fact that fluoride from external sources such as drinking water, food, and toothpaste reacts with hydroxyapatite to form fluorohydroxyapatite (FHA). This compound is found in highest concentration in the surface of enamel (outer 60 microns contain about 10 times more fluoride than that of inner enamel layer). Solubility studies have shown that fluoridated apatite is much more resistant to an acid attack than hydroxyapatite. Hence, the bacterial acids are thought to enter the enamel through tiny pores, resulting in demineralization of the subsurface structure which contains less fluoroapatite. Clinically, demineralized surface lesions can be visualized as opaque white spots or radioluscent areas in a radiograph as shown in Figure 4. As indicated above, the fluoride reacts with hydroxyapatite to form fluorohydroxyapatite according to the following equation:

$$Ca_{10}(PO_4)_6(OH)_2 + 2F^- \longrightarrow Ca_{10}(PO_4)_6F_2 + 2OH^- \qquad (3)$$

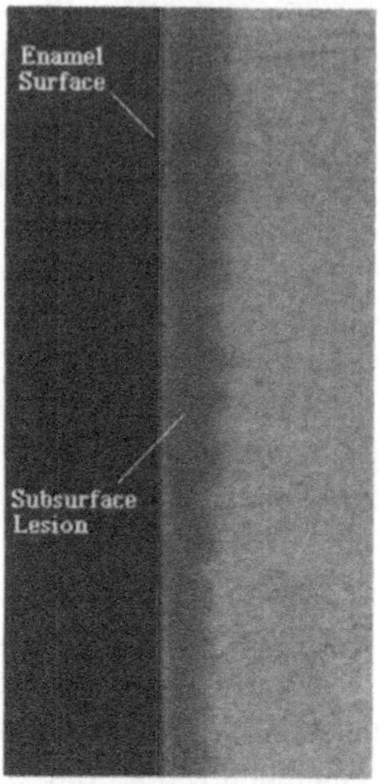

Figure 4. Demineralized subsurface lesion. Clinical observation shows that the fluoridated dentifrices are capable of remineralizing white spot lesions to FHA as shown in Figure 5.

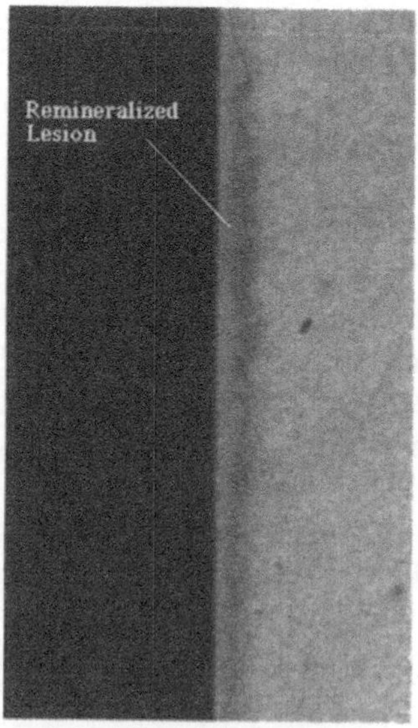

Figure 5. Remineralization *in situ* white spot lesions.

The remineralization of demineralized areas can be measured by using the model of intra-oral remineralization. A comparison was made between different fluoridating agents such as sodium monofluorophosphate (MFP) vs. sodium fluoride with the same abrasive system, silicone dioxide, and the remineralization of the lesions was evaluated both on dentine and enamel *in vivo*.[23] The results of enamel and dentin remineralization/demineralization behavior of artificial caries lesions are compared in the absence or presence of fluoride. In the presence of fluoride (Figure 6), the mean percent mineral gain was 26% for dentine and 16% for enamel, respectively. In the absence of fluoride, the mineral loss was 34% and 30%, respectively. Both fluoride species gave similar type results. The study demonstrated that the dentifrices containing fluoride preparations were effective in fortifying the early caries lesions, *in vivo*. Fluoride also reduces tooth decay, partly because of this property of enhanced remineralization. There are, however, other mechanisms by which fluoride works, including its ability to inhibit the production of acids by oral bacteria.[24]

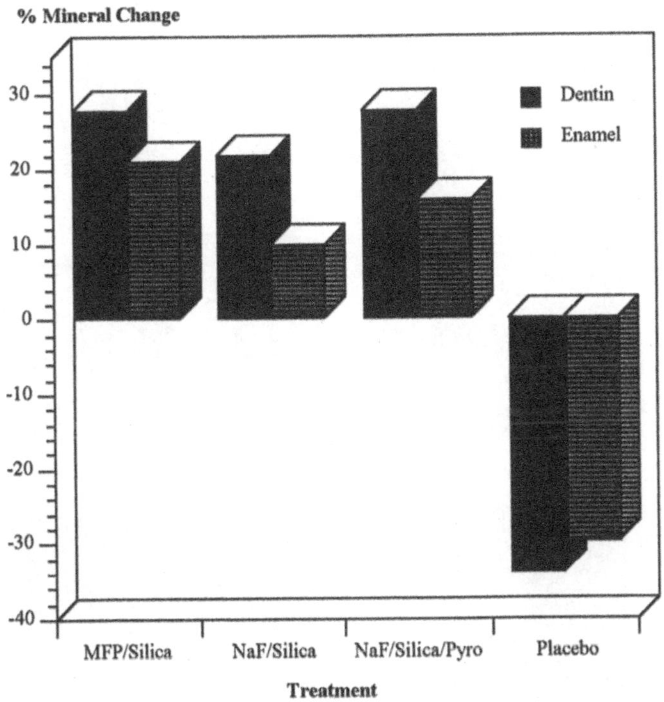

Figure 6. Mean percent mineral gain *in situ* following brushing with fluoridated toothpaste.[23]

Overall, recent data shows that people in developed countries are keeping their teeth longer due to the long-term effects of fluoride. Unfortunately, as the population ages, gum tissues recede, resulting in exposure of root surfaces and giving rise to root surface caries (Figure 7). A topical lesion can be arrested by the combined treatment of fluoride-containing toothpaste followed by professionally applied gel as a dental office treatment (Table 3).[25]

Table 3*. Percent root surface lesions arrested at 4, 8 and 12 months.

Group	4 Mths.	8 Mths.	12 Mths.
MFP/Dical 1000 ppm F	18	28	40
MFP/Dical** 1000 ppm F and 5000 ppm F Professionally applied gel weekly	39	75	91

* De Paola[25]

** Dicalcium phosphate dihydrate

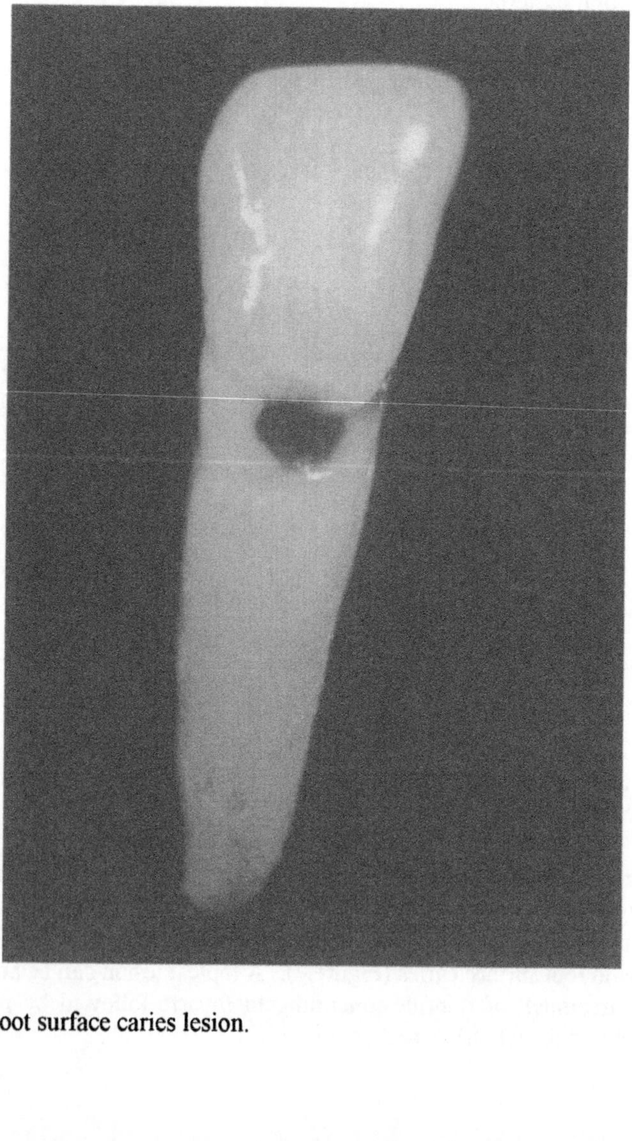

Figure 7. Root surface caries lesion.

Technologies to Enhance Effectiveness of Fluoride

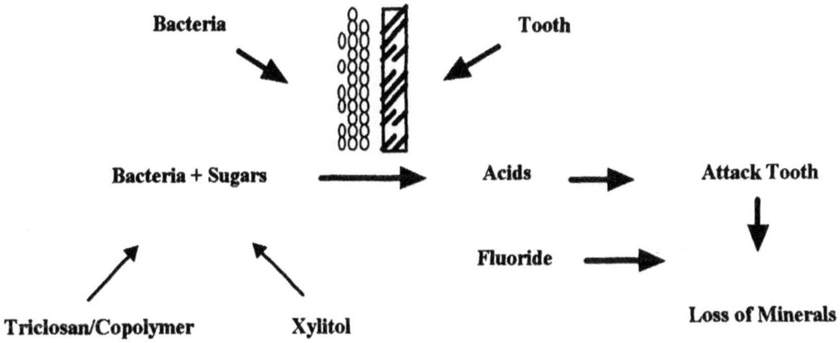

Figure 8. A schematic to depict approaches to enhance fluoride effectiveness.

Figure 8 describes in a cartoon form the different approaches by which fluoride efficacy can be enhanced. One approach is to reduce the sugar challenge in the plaque. For this purpose, xylitol has been proven to be effective.[26] Xylitol is a five-carbon sugar which is not metabolized by plaque bacteria to produce lactic acid. Xylitol is known to reduce plaque acids, but its synergistic effect with fluoride has not been demonstrated. In recent studies, it has been shown that xylitol from toothpaste is retained in the plaque for up to one hour post-brushing, and it reduced plaque acids *in situ* when given in combination with the fluoride. In the *in situ* remineralization studies, xylitol-containing fluoride dentifrice was significantly ($P = 0.05$) more effective in the remineralization of artificial caries lesions than fluoride by itself, as shown on Figure 9. This has now been confirmed in a three-year caries clinical study showing that addition of 10% xylitol with fluoride in dentifrice provided 12% incremental reduction when compared to fluoride toothpaste by itself.[27,28] Dentifrices containing 10% xylitol + NaF are sold in Europe for an improved cariostatic effect. Mechanistically, xylitol reduces the acid challenge and permits fluoride to exert its effects under a less cariogenic challenge, with improved remineralization of teeth and a consequent reduction in caries.

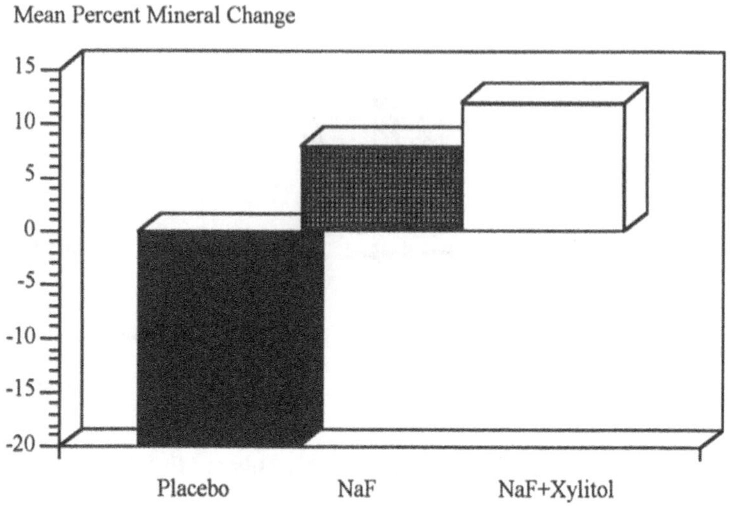

Figure 9. Remineralization *in vivo* of aritificial caries lesions byXylitol/Fluoride, placebo and NaF dentifrices. Differences between NaF/Xylitol is > NaF (P = 0.05).

Reducing Bacterial Challenge (Antibacterial)

Another approach to improve the cariostatic effect of fluoride would be to reduce bacteria in dental plaque by using antibacterial agents. Recently, an antibacterial agent with a special polymer delivery system has been introduced in toothpaste. It contains the antibacterial agent, triclosan, plus a copolymer delivery system. The data published in the literature indicates that the combination of antibacterial and fluoride was more effective in remineralization of early caries lesions than fluoride by itself (Figure 10). Long-term clinical data from field trials are needed to confirm these short-term studies.[29]

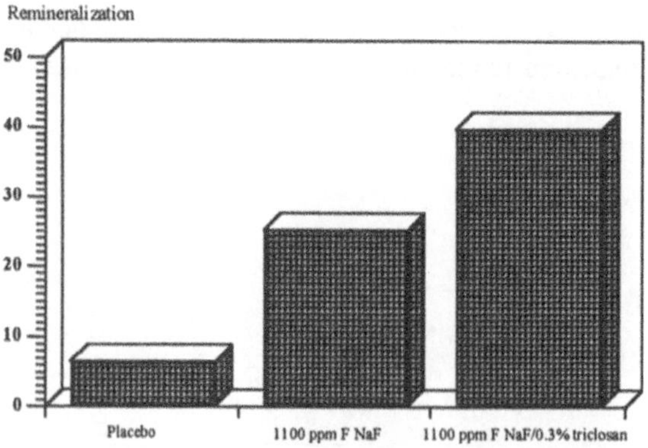

Figure 10. Anti-caries efficacy of 1100 ppm F NaF/0.3% Triclosan/2% Copolymer vs. 1100 ppm F NaF in short-term intra-oral model.

Fortifying the Tooth Surface

Another way of increasing the anti-caries effect of fluoride is fortifying tooth surfaces. In this case, AHP, azacycloheptane diphosphonate, which has been shown to prevent enamel dissolution, was tested in long-term clinical studies against caries. Animal and *in vitro* studies indicated that the combination is more effective than fluoride against caries. This has been confirmed in a three-year clinical done in Iceland.[30] Thus, a whole gamut of technologies are available for enhancing the effectiveness of fluoride and all of these approaches have been successful in clinical trials. In the future, it is expected that new anti-caries agents will be developed for dentifrices based on materials derived from biotechnology sources. These materials from natural sources are likely to provide a cariostatic effect equal to fluoride.[31]

ANTI-CALCULUS (TARTAR) + ANTI-CAVITIES TECHNOLOGIES

Incidence of Calculus Formation

The success of fluoridated toothpaste in preventing caries has resulted in a shift of focus to other problems of the oral dentitions; i.e., calculus or tartar formation. A comprehensive oral survey by the National Institute of Dental Research in the U.S. showed that 34% of school-age children and 20-30% of adults have tartar above the gumline, while 60-65% have deposit below the gumline.[32] Currently, the only method to remove calculus is by mechanical means. This procedure is labor-intensive and may cause discomfort to the patients. Therefore, there was a need to develop preventive agents for incorporating in the toothpaste to reduce tartar build-up around the teeth and gumline.

Chemically, dental calculus is composed of both inorganic and organic components, and the organic material is a combination of epithelial cells, leukocytes, microorganisms, and polysaccharides. A combination of analytical techniques (x-ray diffraction, electron microprobe infrared absorption spectroscopy) established that components of calculus consist of one of the major calcium phosphate compounds which include: carbonated hydroxyapatite (CHA), dicalcium phosphate dihydrate (DCPD), octacalcium phosphate (OCP), and magnesium-substituted beta tricalcium phosphate (beta-TCMP) (Table 4). Relative abundance of these calcium and phosphates differ from specimen to specimen and even between outer/inner layers of same specimen. Recent x-ray diffraction, infrared analysis of human dental calculi deposited on denture and other surfaces show that the inorganic component of calculus obtained from denture surface is principally carbonated hydroxyapatite (CHA), while material from tooth surfaces have mixed calcium phosphate phases.[33]

Table 4. Mineral Composition of Human Calculus

Dicalcium Phosphate Dihydrate (DCPD), $CaHPO_4 \cdot 2H_2O$
Octacalcium Phosphate (OCP), $Ca_8H_2(PO_4)_6 \cdot 5H_2O$
Whitlockite (Mg Substituted Tricalcium Phosphate), βTCMP
Carbonated Apatite

Although the general method of removing calculus is by mechanical means, several agents have been developed to inhibit the transformation of amorphous calcium phosphate salts to a crystalline phase, thus reducing calculus formation. Among the most effective inhibitors are pyrophosphate, pyrophosphate plus copolymers, and zinc salts. In general, two types of tests are used to evaluate the inhibitors. One test follows the spontaneous formation of HAP in a supersaturated (with calcium and phosphate) environment which is similar to plaque fluid (Figure 11). The second test, known as the seeded crystal growth of hydroxyapatite, has a driving force for precipitation equivalent to the saliva environment (Figure 12). These tests were used to investigate the relative efficacy of these inhibitors. The results are summarized in Table 5, which shows that the most effective inhibitor is pyrophosphate. Inhibition is defined as 50% inhibition of HAP formation (in minutes) when compared to HAP formation in the absence of an inhibitor. Also, the combination of pyrophosphate and a copolymer (PVMA/MA) provides enhanced efficacy when compared with pyrophosphate by itself. Zinc salts, on the other hand, require a higher concentration for effectiveness. The relative clinical efficacy of these agents in various dentifrices are summarized in Table 6. Available data from several independent clinical studies indicate that calculus inhibition ranged from 26 to 30% for the pyrophosphate/sodium fluoride combination to 50% for the copolymer/pyrophosphate combination. Zinc salts require a higher concentration for efficacy.[34] With lower concentrations of zinc, efficacy against supragingival calculus is not proven.

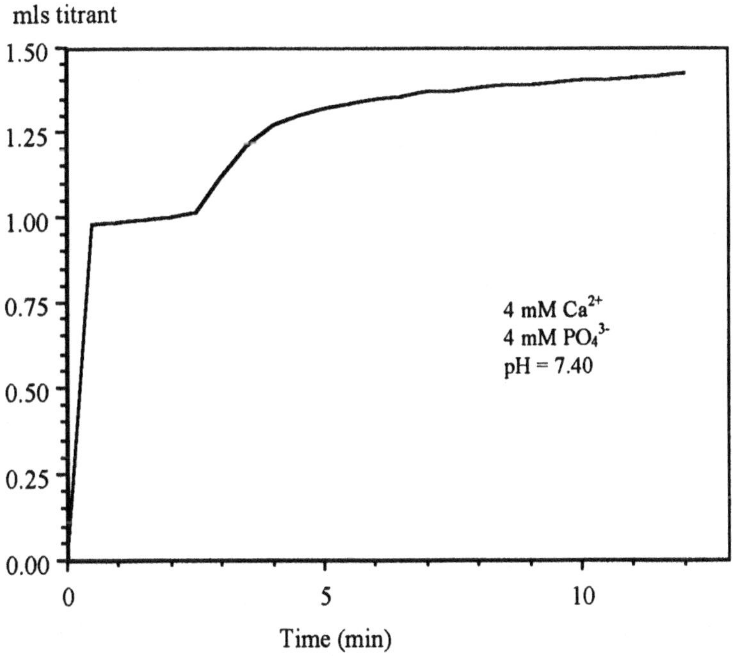

Figure 11. HAP Formation.

$$5Ca^{2+} + 3HPO_4^{2-} + H_2O \longrightarrow Ca_5(PO_4)_3\,OH + 4H^+ \qquad (5)$$

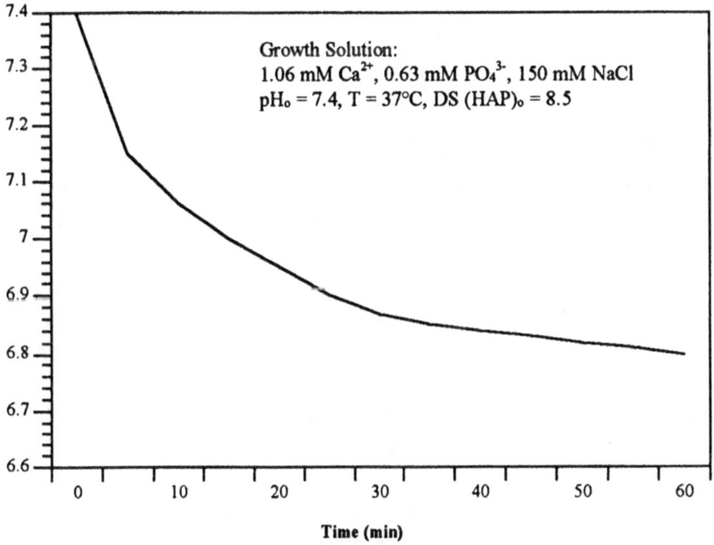

Figure 12. Seeded crystal growth of HAP.

Table 5. Calculus control technologies.

Compound	Relative Efficacy Inhibition * (ppm)
Pyrophosphate	4.0
Pyrophosphate + Copolymer	3.0
Zinc	60.0

* = 50% inhibition of HAP formation vs. a control

The mechanism by which inhibition of calculus formation occurs has been elucidated.[35] The calcium and phosphate ions from saliva or the plaque fluid precipitate to form an amorphous phase which matures to crystalline calcium phosphate in the absence of inhibitor. In the presence of inhibitor, the amorphous phase is stabilized and conversion of its crystalline phase is delayed or inhibited.

Table 6. Clinical efficacy of toothpastes in humans (from Volpe et al.[34]).

Toothpaste	Mean Reduction in Calculus vs. Placebo
3.3% Pyrophosphate + NaF	26%
3.3% Pyrophosphate + 1% PVM/MA/NaF	50%
1.3% Pyro + 1.5 PVM/MA/NaF	47%
0.5% Zinc Citrate + MFP	14%
2% Zinc + Sodium Fluoride	38-50%

PVM/MA: copolymer polyvinymethylether and maleic acid

Most of the current formulations used to inhibit calculus formation also contain fluoride. When the application of an inhibitor of calcium and phosphate crystal growth coexists with fluoride, a crystal growth promoter, one needs to understand how they work together. Fluoride ion accelerates the formation of HAP. How do the two agents coexist in the same system and still be effective? Early data indicated that the crystal growth inhibitor worked on the tooth surface, while the fluoride ion works within the teeth.[36] The effect could be explained by the fact that calculus formation occurs on the tooth surface whereas the demineralization occurs in the subsurface region of enamel. The presence of pellicle on the tooth allows selective transfer. The fluoride can diffuse into the enamel, but the inhibitor is excluded. The mechanism has been further elucidated by the studies conducted on salivary proteins, natural inhibitors of calculus formation. Crystal growth inhibition was explained by the fact that the addition of fluoride did not activate secondary growth sites of crystals. Crystal growth, on the other hand, was explained on the basis of an increasing driving force of precipitation caused by fluoride and incomplete blockage of crystal growth sites by the inhibitor, on the basis of a steric effect.[37] In addition, studies of fluoride/cation interactions in the formation and properties of apatite showed that the fluoride effect was the more dominant effect.[38] The dominant effect of fluoride on remineralization has been confirmed in *in vivo* studies. Clinical trials have shown that the effectiveness of fluoride in presence of an anti-calculus agent was not compromised.[39]

Relationship between Calculus Formation and Gingival Recession

Recession of the gingival margin is common and an important sign of periodontal disease. Gingival recession, which is widespread in adults, can occur on all tooth surfaces and is usually related to the severity of attachment loss in periodontal diseas.[40] Several other factors have been identified including age, frequency of brushing, malocclusion, and presence of tooth deposits. A long-term study in animals has shown the relationship between accumulation of supragingival calculus and gingival recession.[41] The study also showed that a reduction in supragingival calculus formation resulted in the concomitant improvement in gingival recession. This has now been confirmed in a series of long-term clinical studies in humans.[42] Collectively, the results of these studies has shown the relationship between gingival recession and calculus accumulation in the selected population. Thus, calculus reducing agents provide both aesthetic and therapeutic benefits, reducing supragingival calculus and mitigating gingival recession.

MULTI-BENEFIT TECHNOLOGIES FOR DENTIFRICES

The next evolution in dentifrice technology was to incorporate antibacterial agents with fluoride and tartar-reducing agents. In 1683, Anthony von Leeuwenhoek was first to demonstrate the presence of bacteria on teeth.[43] Oral bacteria attach to tooth surfaces in the form of dental plaque. The film shows a highly organized structure of aggregated bacteria containing 10^{12} bacteria per gram of film, in addition to other material including polysaccharides, lipids and proteins. Although the bacteria do not normally invade the gum tissues, the products of bacteria can cause tissue pathology. As early as 1890, W.D. Miller, considered the father of dental microbiology, in a classic book, recommended the ways by which to counter or remedy the ravages of dental disease--when he said, "We may endeavor by hygienic measures to secure the best possible development of teeth; by repeated thorough systemic cleansing of oral cavity and to teeth to so far reduce the amount of fermentable substances, as to mutually diminish the production of the acid as well as to rob bacteria of organic matter necessary to their rapid development; by prohibiting or limiting consumption of such food or luxuries which may verily undergo acid fermentation so as to remove the cheap source of fermentable product injurious to teeth, and by proper intelligent use of antiseptic to destroy bacteria or at least limit their number and activity."[44] For a number of years attempts have been made to develop something to address Miller's last point. Approaches used have included the use of antibiotics, antiseptics and anti-inflammatory agents. These agents suffer from a variety of deficiencies including low efficacy, development of resistant organisms, disturbance of the oral microbial ecology, and undesirable side effects such as staining of the teeth.[3]

Our current understanding of the pharmacology of antibacterial, antiplaque agents indicates that antibacterial activity is not the only factor in achieving a sustained effect on plaque formation. If the teeth are brushed twice a day and 90% of the bacteria are killed, plaque formation will be inhibited only for six hours.[45] Because the mouth is an open system and plaque formation is a continuous process, chances of continuous reinfection is always present. For antibacterial agents to be effective, they should have:

- Broad spectrum activity
- Retention on the surface of teeth and oral surfaces
- Good taste
- Low toxicity
- Must not disturb the oral microbial ecology

Among the most widely-used agents are cationic antibacterials, such as chlorhexidine, which is based on bis-biguanide chemistry. These agents are effective in the mouth because of their substantivity (good retention to oral surfaces and a release for up to 12 hours after use of the mouthwash). However, these agents suffer from major deficiencies; i.e., they disturb the taste perception, cause discoloration of teeth, and increase tartar formation.[46] Therefore, there was an urgent need to develop another class of compounds that could fulfill the above criteria, which would not cause these side effects and could be used daily. A non-cationic antibacterial agent, triclosan, fulfilled this need. Triclosan is a highly effective broad spectrum antibacterial agent. The relative efficacy of triclosan and chlorhexidine are summarized in Table 7. Unlike cationic antibacterials, triclosan does not induce staining of the teeth or disturb taste perception.

Table 7. Intrinsic antibacterial activity (mcg/mℓ).

Agents	Strep. mutans	A. viscosus	F. Nucleatum
Chlorhexidine	0.19	0.5	3.1
Triclosan	0.78	0.78	0.78

Although triclosan is widely used in personal care products, its substantivity and retentivity in the mouth was poor. In a short-term clinical trial, a triclosan-containing rinse and other antibacterials (Table 8) were compared for effectiveness against plaque. The results showed that triclosan required a special delivery system to improve its retention in the mouth.[47] These include either the copolymer (PVM/MA) or polyvinylphosphonic acid (PVPA). These polymers increase the retention of triclosan in the mouth for more than 12 hours and the agent is released slowly into the plaque. Dentifrices containing triclosan, copolymer, and fluoride have now been widely tested in long-term clinical trials against plaque, gingivitis, tartar and cavity formation. These are summarized in Tables 9, 10, and 11. Today, the multi-benefit dentifrices which provide protection against dental diseases are the state-of-the-art in dentifrice technology worldwide. These will have a major impact in preventing plaque-associated diseases such as dental cavities, tartar, and gingivitis formation.[48]

Table 8. Comparative *in vivo* studies on de novo plaque formation.

Treatment	Mean P on All Surfaces + S.D.	SNK Group
Placebo	1.46 ± 0.12	A
0.12% CH	0.53 ± 0.17	B
SnF_2 (Rinse)	1.10 ± 0.16	C
0.06 Triclosan	1.00 ± 0.14	C
0.06 Triclosan + Gantrez	0.72 ± 0.17	B
0.06 Triclosan + PVPA	0.67 ± 0.16	B

Gantrez = PVM/MA
PVPA = Polyvinylphosphonic acid
SNK = Student Newman Keuls test for significance at P=0.05

Table 9. Plaque and gingivitis efficacy Triclosan/copolymer dentifrice long-term clinical studies (0.3% Triclosan/2.0% copolymer in a sodium fluoride/silica base.

Investigators	Location	Duration	Plaque Efficacy vs. Placebo * (Q-H Index)	Gingivitis Efficacy vs. Placebo* (GS Index)
Garcia-Godoy and co-workers, 1990	Dominican Republic	7 months	-58.9%	-87.5%
Cubells and co-workers, 1991	Spain	6 months	-24.9%	-57.5%
Deasy and co-workers, 1991	United States	6 months	-32.3%	-57.1%
Tiratana and co-workers, 1993	Thailand	6 months	-32.9%	-38.3%

Lindhe and co-workers, 1993	Sweden	6 months	-31.2%	Significantly less bleeding sites**

* Plaque and gingivitis efficacy results pertain to data obtained at the final clinical examination. All percentages relating to plaque and gingivitis efficacy of the triclosan/copolymer dentifrice were calculated relative to the placebo dentifrice and were statistically significant at the 99% level of confidence. Q-H index refers to the Quigley-Hein (Turesky Modification) Plaque Index; GS Index refers to the Gingivitis Severity Index of Palomo and co-workers.

**At the conclusion of the study, the triclosan/copolymer dentifrice group had significantly less bleeding sites (and significantly more gingivitis-free sites) than the placebo dentifrice group.[34]

Table 10. Calculus inhibition by dentifrices containing Triclosan / copolymer/NaF.

Study	Duration	% Reduction vs. Placebo
Schiff U.S.A.	3 Months	-23.1
Lobene U.S.A.	6 Months	-36.2
Volpe U.S.A.	3 Months	-35.5

Table 11. Anti-caries efficacy of dentifrices with Triclosan/copolymer/NaF.

Study	Duration	Fluoride Dentifrice		Total	
		DFS	DFT	DFS	DFT
Feller/Krieger U.S.A.	26 months	1.64	0.53	1.55	0.48
Mann, Israel	26 months	3.84	0.95	3.72	0.90

TECHNOLOGIES TO REDUCE TOOTH SENSITIVITY

Multi-Benefit Anti-Hypersensitive Products

The next evolution of toothpaste chemistry was developed as a means to prevent pain caused by sensitive teeth; i.e., hypersensitivity. Dentinal hypersensitivity is defined as an acute, localized tooth pain in response to thermal, tactile, or air blast stimulation to exposed dentine surfaces.[49] Normally, the roots of teeth are covered by the gingival or gum tissue but when the gum recedes, the underlying tooth surface is exposed. Once exposed, with time, abrasion and erosion will remove the thin layer of cementum, thus exposing underlying porous dentine. Exposure of the dentine surface to dietary or bacterial acids can expose the dentine pores or tubules at the surface. It is well known that exposure and the presence of open tubules on the surface is associated with increased dentinal hypersensitivity as shown in Figure 13. The dentine tubes contain fluid.[50]

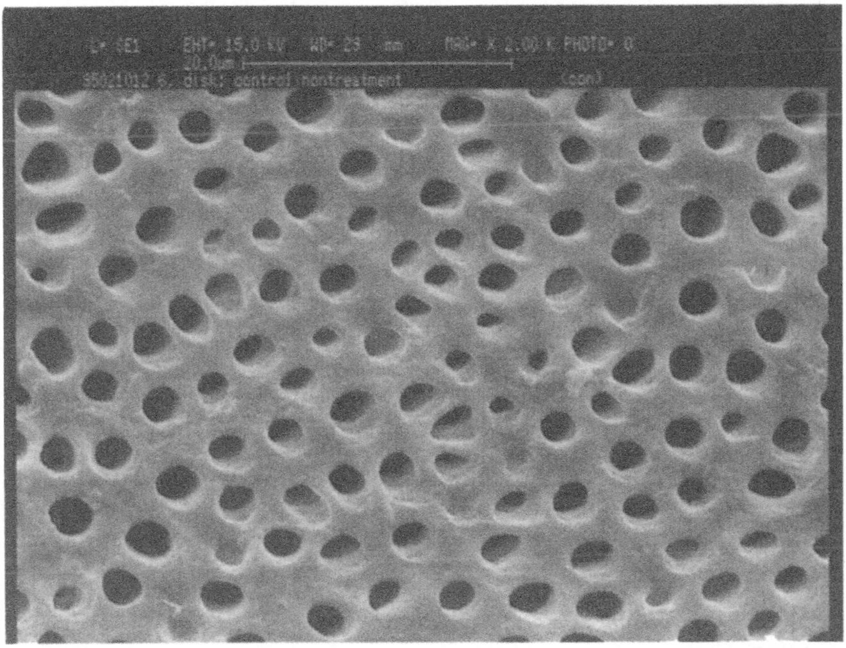

Figure 13. Open dentinal tubules in hypersensitive teeth.

Mechanistically, hot or cold stimuli can cause this fluid to expand or shrink, stimulating underlying pulpal nerve resulting in pain. Currently, salts of potassium and strontium are available as preventive therapies in OTC toothpaste. Strontium salts control pain by occluding exposed, open dentine tubules. Various other agents such as potassium nitrate are believed to cause reduction in nerve activity by altering the threshold of pulpal nerve excitation. These approaches have been combined in a single toothpaste containing potassium nitrate and copolymer which adhere to tooth surfaces.[51] Figure 14 shows occlusion which can result from in vitro treatment of dentine with such a toothpaste. Unfortunately, this therapy requires two to three weeks treatment before a reduction in sensitivity is observed. Therefore, there is currently a strong need for a rapid and fast reactive material in toothpaste which could rapidly reduce dentinal hypersensitivity.

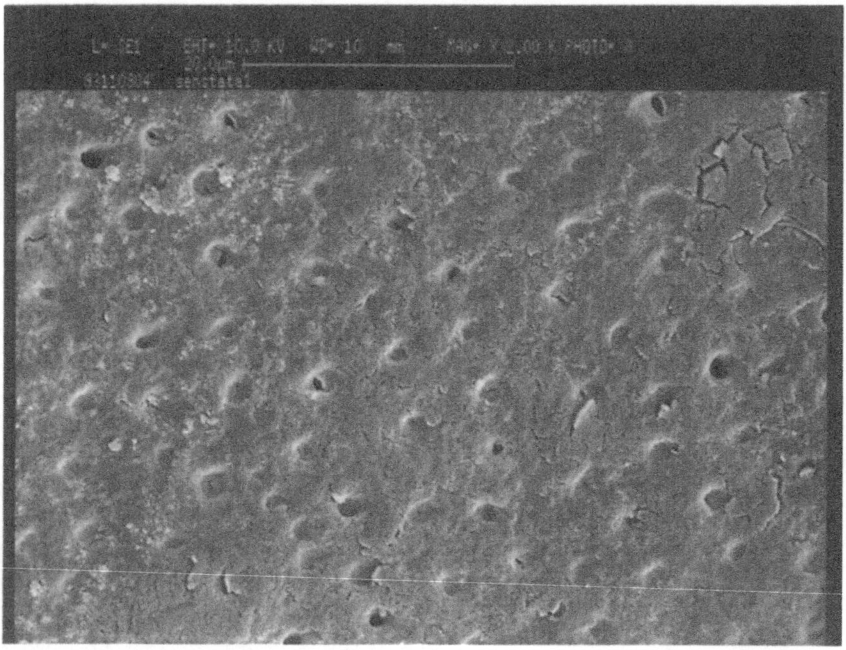

Figure 14. Occlusion of open tubules by the yrophosphate/Copolymer/KNO_3/NaF dentifrice.

TECHNOLOGIES FOR WHITENING OF TEETH

Intrinsic Stain

Unlike extrinsic discolorations, intrinsic discoloration is due to the presence of chromogenic material within enamel or dentin. This type of staining is related to the penetration of discoloring agents during development, illness or trauma.[50] Developmental tooth staining can occur either pre- or post-eruptively. The most common type of pre-eruptive staining is endemic fluorosis occurring as a result of excessive fluoride ingestion during tooth development. Fluorosis is considered a form of enamel hypoplasia because chronic intake of fluoride causes an alteration in ameloblast metabolism which results in a hypomineralized, porous enamel subsurface. Generally, staining due to fluoride manifests itself in three different ways: simple fluorosis, opaque fluorosis and fluorosis with pitting. Simple fluorosis appears as a brown pigmentation on a smooth enamel surface. Opaque fluorosis appears as gray or white specks on the tooth surface while fluorosis with pitting has defects in the enamel and the color appears to be darker. Certain inherited defects such as *dentinogenesis imperfecta* and *amelogenesis imperfecta* also cause pre-eruptive dental stain. In *dentinogenesis imperfecta* the pulp chamber is obliterated or reduced in size and the teeth appear reddish-brown to gray. *Amelogenesis imperfecta* is a similar disorder that affects the enamel and the teeth may appear yellowish or brownish, pitted or smooth. Hematological disorders such as *erythroblastosis fetalis, thalassemia* and sickle cell anemia may also cause pre-eruptive stain. Tooth color may range from greenish-blue to brown as a result of accumulation of blood pigments within enamel and dentin.[51] Tetracycline staining is another form of discoloration which develops during odontogenesis. Tooth staining occurs when tetracycline binds to calcium and becomes incorporated within the hydroxyapatite crystal as the tooth is mineralizing. Therefore, both dentin and enamel become discolored. However, a majority of stain resides within dentin because of a greater surface area available to tetracycline for chelation. The severity of stain depends on the duration, time (age) and the specific type of tetracycline administered. Typically, teeth are susceptible to tetracycline stain with chlortetracycline; yellow stain with tetracycline dimethylchlortetracycline and oxytetracycline. The color of stain has also been shown to change with age. Teeth that have a yellow pigmentation may darken to brown or gray-brown due to photo-degradation of photo-oxidation of tetracycline.[52]

Teeth may also require post-eruptive intrinsic stains. A common example of this is staining due to trauma where a hemorrhage occurs in the pulp chamber and blood gets trapped within the dentinal tubules. Post-eruptive dental stain may also arise when secondary dentin, tertiary dentin and pulp stones are deposited as a result of the normal aging process. Dental procedures may also lead to tooth discoloration, such as release of metals from amalgam restorations or incomplete obturation of the pulp chamber during endodontic treatment.[58]

Methods to Remove Stain

The most commonly accepted procedure to remove discoloration from tooth surfaces is by using abrasives such as prophy pastes or a combination of abrasives and surface active agents found in toothpastes (see section under "Cleaning, Polishing and Fresh Breath"). In the former method, superficial stains are removed by an abrading action while in the latter method, the surface active agents solubilize or loosen the stain and the abrasive removes the stain. These methods prevent stain accumulation and to certain extent remove extrinsic stain but generally do not fulfill the needs of a patient with gross discolorations because all stain is difficult to remove. Total removal of all stain requires agents that decolorize the intrinsic stain. The two most common agents that have the ability to remove both types of stain are hydrochloric acid and hydrogen peroxide. Hydrochloric acid is generally used to remove stains associated with fluorosis.

Hydrogen peroxide has a broader spectrum of action and 30-35% solutions have been effectively used to remove the majority of extrinsic and intrinsic dental discolorations, including certain types of tetracycline stains which reside primarily in dentin. Hydrogen peroxide is also less destructive to the tooth structure compared to hydrochloric acid.[53]

Currently, two types of methods are available to chemically remove internal and external dental stain using hydrogen peroxide.[54] The oldest is an in-office procedure that uses a 30-35% hydrogen peroxide solution in combination with heat and light to accelerate stain removal.[55] A more recent method involves fabricating a mouth guard like appliance where a peroxide preparation is placed while the appliance is worn over the teeth during stain removal. This method uses 10% carbamide peroxide or 3% hydrogen peroxide as the active ingredient and is sometimes referred to as "nightguard vital bleaching," or "dentist-prescribed home-applied" technique.[55] When compared to the in-office procedure, "nightguard vital bleaching" tends to be slower acting and sometimes not very effective against certain types of stain because lower concentrations of peroxide are used.[55]

To assess the stain removal capabilities of oxygenating agents, an *in vivo* technique to quantitatively measure stain was developed. It was necessary to develop a method which gives a consistent and accurate evaluation of tooth color because traditional methods rely on standard shade guides which require subjective visual grading. The basic problem with this is that the subjective component of color is highly variable and is dependent on external factors such as ambient lighting and other factors such as interpretation of individual assessors.[56] To overcome these difficulties and provide an accurate evaluation of stain removal by various agents, tooth color was measured by reflectance spectroscopy using the Minolta chroma meter with 45°/0° illumination/observation geometry (see earlier section).

In a majority of the current at home use products, carbamide peroxide is used as a vehicle to deliver hydrogen peroxide. Upon contact with water, carbamide peroxide breaks down into urea and hydrogen peroxide.[57] The action of urease in the oral cavity enables urea to break down into ammonia and carbon dioxide which is thought to have a cleansing effect.[58] Undegraded urea may, under certain circumstances; e.g., in presence of abrasives and surfactants, also help in solubilization of surface proteins and stains by denaturing proteins and disrupting non covalent interactions such as hydrogen bonding on the tooth surfaces.[58] The active bleaching agent is hydrogen peroxide which acts on the chromophore. The mechanism by which this occurs is not well understood; however, studies with tea and wine suggest that bleaching is a result of direct oxidation of the stain by hydrogen peroxide.[58] This has been confirmed by finding that the bleaching reaction involves the nucleophilic perhydroxyl anion which reacts with the stain.[59] Some authors have proposed a free radical bleaching mechanism, but this is not likely to occur because the kinetics do not favor this type of a reaction. Even if free radicals were produced, the presence of the salivary peroxidase system[60] and catalase in the mouth would rapidly eliminate these molecules. Other authors have proposed that hydrogen peroxide acts as a mechanical cleansing agent due to its oxygen releasing action.

The kinetics of stain removal by the "nightguard vital bleaching" method can be enhanced in several ways; e.g., raising the pH would tend to favor the production of perhydroxyl anions which are responsible for the oxidation of the chromogenic material. Peroxide activators would also tend to favor the production of the reactive anions, thus increasing the stain removing power. Raising the temperature would also tend to increase the rate of stain removal. Unfortunately, these methods are not practical due to the chemical and physical limitations; e.g., high pH aqueous hydrogen peroxide preparations tend to be chemically unstable. The safety and the clinical efficacy of peroxide activators remains unknown and finally, it may not be possible to increase intra-oral temperature during nightguard bleaching. To overcome these limitations, an alternative method or the broad spectrum approach to enhance the stain removal process was developed. In this method, simultaneous extrinsic and intrinsic stain removal is thought to occur by combining dentifrice technology with the bleaching technology. Basically, surfactants have been included in the peroxide preparation to reduce the surface energy and solubilize and loosen extrinsic stain. Abrasives have been added to aid in mechanical cleaning and remove denatured protein which may accumulate stain.[61] To prevent breakdown of peroxide and maintain a constant whitening effect, chelating agents such as pyrophosphate have also been added. The pyrophosphate has an additional advantage because it is known to prevent re-deposition of stain due to crystal growth inhibition or anti-calculus properties.

The clinical efficacy of a whitening product designed on the broad spectrum approach has been reported.[61,62] In one such study, the clinical stain removal efficacy was examined using the "dentist-prescribed home-applied" tooth whitening paste. In this study, three types of products were examined. The first product contained 10% carbamide peroxide in an anhydrous glycerin base. The second product contained a broad spectrum product (10% carbamide peroxide formulated in a toothpaste-like base which has the capability to remove extrinsic dental stain) and the third was a matching placebo of the second product and did not contain peroxide. The results of this study are shown in Figure 15. As can be seen from Figure 16a and Figure 16b, neither the oxygenating agent alone nor the placebo, which is capable of removing extrinsic stain removed all the stain. The greatest stain removal was obtained using a broad spectrum formulation.

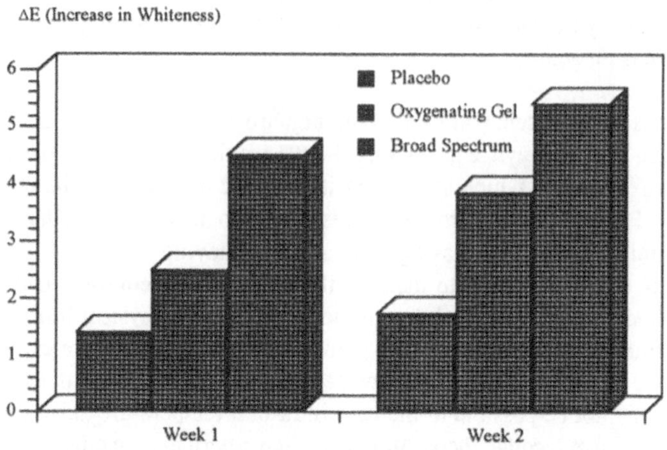

Figure 15. Clinical evaluation of a stabilized peroxide system on whitening.

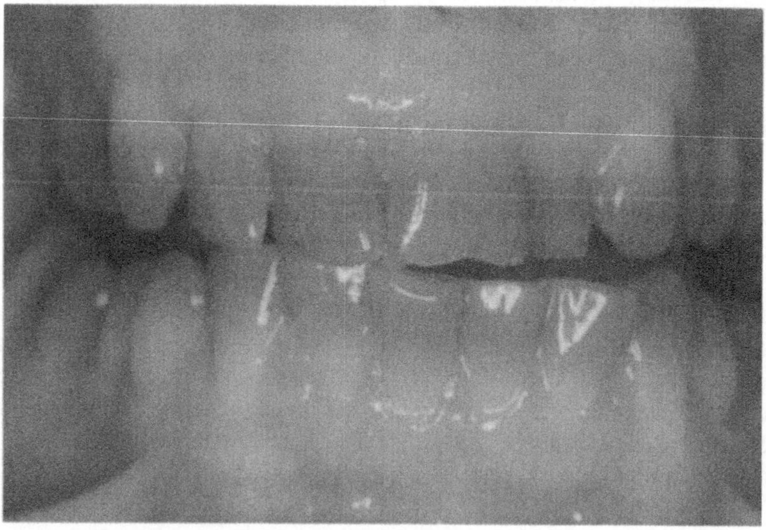

Figure 16a. Tooth staining prior to use of peroxide paste.

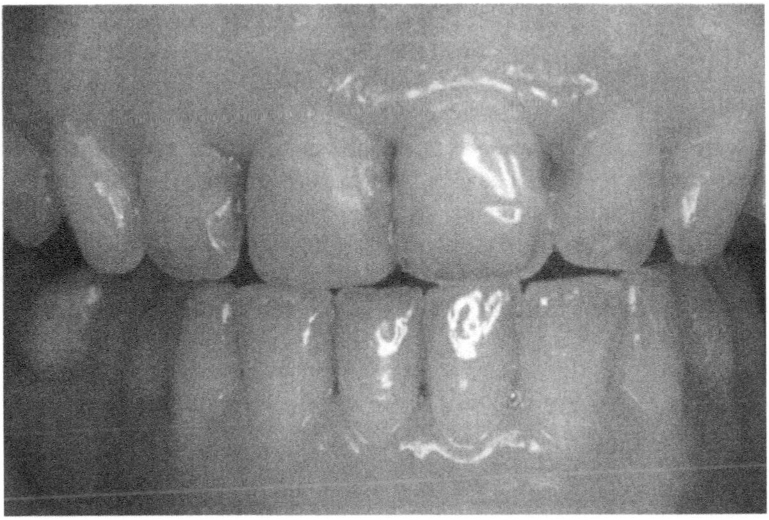

Figure 16b. Tooth staining two weeks post-peroxide paste use.

For a procedure to be practical, i.e., used in the dental practice or by the consumer, it must be safe. To investigate the safety of our broad spectrum approach, several *in vitro* studies wre done. Microhardness measurements of treated and untreated teeth (Table 11) indicated no statistically significant changes in the Knoop Hardness Numbers (KHNs) of enamel and dentin. Interestingly, a directional increase in hardness of treated teeth was observed which may be a result of the remineralizing properties of dicalcium phosphate dihydrate found within the broad spectrum whitening product.[57] Scanning electron micrographs (SEM) of teeth treated with the broad spectrum product also showed no changes in the surface morphology of enamel when compared to treatment with enamel etching agents.[57] This finding was recently confirmed by Mehra *et al.*[58] who also showed the broad spectrum approach was not damaging to tooth structures. Similarly, the surface atomic percentages of carbon, nitrogen, oxygen, calcium and phosphorus of treated, placebo treated and untreated teeth were not significantly different (Table 12). Thus, dentifrices to reduce stain or whiten teeth are now in vogue in developed markets.

Table 12. Microhardness of enamel and dentin.

Treatment	Mean Microhardness (KHN)(+/-SD)	
	Enamel	Dentin
Saliva	332.83 (25.52)	63.62 (11.18)
Placebo	341.50 (47.77)	66.53 (2.82)
Broad Spectrum	359.63 (8.86)	64.72 (1.97)

Thus, we started dentifrice technology with a basic cleaning function, but it has now evolved toward a total oral care goal (Figure 1). As we look to the future, we can visualize the next generation of chemistry in toothpaste related to the control of the infections in the mouth. These will include controlling the overgrowth of bacterial and mycotic infections in immune compromised patients; i.e., AIDS. New technology will be available in toothpaste to rapidly control the pain caused by tooth sensitivity. This could provide quick and sustained relief against tooth sensitivity. With the use of therapeutic agents, we are reducing the edentulous population. We can eventually visualize that toothpaste become a major vehicle for providing total oral care for the conditions in the mouth, and technologies are evolving towards that goal.

ACKNOWLEDGMENT

We sincerely thank Theresa Aguilar for skillfully typing the manuscript and making importation suggestions in format and contents.

REFERENCES

1. Dental Spending Hits $37 Billion. Am Dent Assoc News, Jan. 6, 1992

2. Hein, J. The industrial contribution to safe and effective dentifrices. Community Dent Oral Epidemiol 1980;8:230-6

3. Fischman, SL. "Hare's teeth to fluorides, historical aspects of dentifrice use." In *Clinical and Biological Aspects of Dentifrice*, G. Emery and G. Rolla, ed. Oxford University Press, 1972;1-7.

4. Harris, NO. "Dentifrices, Mouthrinses and Oral Irrigators." In *Primary Preventive Dentistry*, N.O. Harris and A.G. Christien, ed. Norwalk, CT: Appleton and Lange, 1991;141-147.

5. Addy M, Moran J, Griffiths AA, Wallis-Wood NJ. Extrinsic tooth discoloration by metals and chlorhexidine. 1. Surface protein denaturation of dietary precipitation? Br Dent J 1985;189:281-5

6. Eriksen HM, Gijermo P. Chemical plaque control and prevention of extrinsic staining. Acta Odont Scand 1983;41,87-91

7. Feinman RA, Goldstein RE, Garber DA. *Bleaching Teeth.* Chicago, Il: Quintessence Publishing, 1987,18-33.

8. Kowitz GM, Rustogi KN, Wong R, Curtis JP. Use of chroma meter to assess changes in tooth color (Abs). J Dent Res 70 (Sp Iss):787

9. Commission Internationale de L'Eclairage. Recommendations on uniform color spaces. Color differences equations. Suppl 2 to CIE Publications 15(E13.1). Paris, France: Bureau Central de la Cie, 1971

10. Woziak, WT. *Proposed guidelines for the acceptance program for dental shade guides.* Chicago, Il: Am Den Assoc, 1987.

11. Gegaulf AG, Rosenstiel SF, Langhorst KJ, Johnson, WM. Evaluating tooth color change from carbamide peroxide gel. J Am Dent Assoc 1993;124:65-72

12. Sproull, RC. Color matching in dentistry. Part II. Practical application of the organization of color. J Prosthet Dent 1973;29:556-66

13. O'Brien WJ, Groh CL, Boenke KM. A new, small color difference--equation for dental shades. J Dent Res 1990;69:1762-4

14. Kowitz GM, Nathoo SA, Rustogi KN. Clinical comparison of Colgate Platinum tooth whitening system and Rembrandt Gel Plus. Compend Contin Dent Edu 1994;Suppl 17:S646-51

15. Tonzetich, J. Production and origin of oral malodor. A review of mechanisms and method of analysis. J Periodont 1977; 48:13-20

16. Solis-Gaffar M, Niles HP. Instrumental evaluation of mouth odor in a human clinical study. J Dent Res 1975;54:351-7

17. Richter VJ, Tonzetich J. The application of instrumental technique for the evaluation of odoriferous volatiles from saliva and breath. Arch Oral Biol 1984;9:49-53

18. Massler M, Emslie RD, Bolden TE. Fetor ex ore. Oral Surg 1951;4:110-25

19. Niles HP, Gaffar A. Relationship between sensory and instrumental evaluation of mouth odor. J Soc Cos Chem 1993;44:101-7

20. Hiles HP and Gaffar A. "Advances in mouth odor research." In *Bad Breath Research Perspective*, Mel Rosenberg, ed. Tel Aviv, Israel: Ramot Publishing--Tel Aviv University, 1995;53-61.

21. Griffiths, M. Dentifrices--an item of historical interest. New Zealand Dent J 1966;62:296-301

22. Blinkhorn AS, Clarkson JE. The international scientific assembly on the comparative anticaries efficacy of sodium fluoride and MFP dentifrices. Am J Dent 1993;6(Sp Iss):1-6

23. Sullivan RJ, Fletcher R, Bachiman R, Penungonda B, Legeros RZ. Intra-oral comparison and evaluation of the ability of fluoride dentifrices to promote the remineralization of caries-like lesions in dentin and enamel. J Clin Dent 1995;6:135-8

24. Fluoride conference report. Mechanism of action and recommendation for use. J Dent Res 1990;69:505-676

25. DePaola, PF. "Caries in aging population: what are we learning." In *Cariology for the Nineties*, W.H. Bowers and L.A. Tabak, ed. Rochester University Press, 1993;27-35.

26. Bar, A. Caries prevention with xylitol: a review of scientific evidence. World Rev Nutr Diet 1988;55(1):1-27

27. Gaffar A, Blake-Haskins J, Sullivan RJ, Simone AJ. Cariostatic effect of xylitol/fluoride in dentifrice. Int Dent J 1997 (In press)

28. Sintes J, Escalante G, Stewart B, McCool J, Garcia L, Volpe AR. Enhanced anticaries efficacy of a 0.243% sodium fluoride/10 xylitol/silica dentifrice: 3 year results. Am J Dent 95:8;231-5

29. Zhang YP, Sullivan RJ, Din C, Miller S, Schimd R, Gaffar A. Anticaries efficacy of a triclosan/copolymer/NaF dentifrice *in vivo*. ADA/FDI Congress 1996; Abs 136:157

30. Koch G, Bergmann-Arnadottir AI, Bjornason, Finnbogason S, Hoskuldsson O, Karlsson R. Caries preventive effect of fluoride dentifrices with and without anticalculus agents. A 3 year controlled clinical trials. Caries Res 1990;24:72-9

31. Mandel, ID. "Dental Caries: Another Extinct Disease." In *Cariology for the Nineties*, W.H. Bowen and L.A. Tabak, ed. University Rochester Press, 1983;1-5.

32. Mandel, ID. Rinses for the control of supragingival calculus formation. Int Dent J 1992;42:270-5

33. Gaffar A, Legeros RZ, Gambogi RJ, Afflitto J. Inhibition of formation of calcium phosphate deposits on teeth and dental materials: Recent Advances. Adv Dent Res 1995;9(4):419-26

34. Volpe AR, Petrone M, Davies RM. A review of calculus clinical efficacy studies. J Clin Dent 1993;4:71-81

35. Gaffar A, Volpe AR, Lindhe J. "Recent advances in plaque/gingivitis control." In *Chemical Biological Aspects of Dentifrices*, G. Emery and G. Rolla, ed. Oxford University Press, 1992;229-47.

36. Gaffar A, Aoba T, Afflitto J, Esposito AO, Moreno EC. "Structure-activity relationshp between *in vitro* inhibition of HA crystal growth and *in vivo* anticalculus effect." In *Recent Advances in the Study of Dental Calculus*, Ten Cate Jon, ed. IRI Press, 1989;155-171.

37. Margolis H, Varughese K, Moreno EC. Effect of fluoride on crystal growth of calcium apatite in the presence of salivary inhibition. Calcif Tissue 1982;39:33-40

38. Legeros RZ, Kojkowsaka R, Tiu Cot, Legeros JP. Fluoride/cation interaction in the formation and the stability of apatites. J Fluor Chem 1988;44:53-64

39. Ripa LW, Leske GS, Triol CW. Clinical study of anticaries efficacy of three fluoride dentifrices containing anticalculus agent: three year results. J Clin Dent 1990;2:29-33

40. Baelum, V. Pattern of periodontal breakdown in adult Tanzanians. Scand J Dent Res 1987;95:221-8

41. Gaffar A, Coleman EJ, Afflitto J. Influence of an anticalculus mouthrinse treatment on the dentition and periodontium in beagle dogs. J Clin Dent 1991;3 (Supp):B12-B18

42. Triratana J, Rustogi KN, Volpe AR. The effect of anticalculus dentifrice on supragingival calculus formation and gingival recession in Thai adults. A one year study. J Clin Dent 1991;2:22-6

43. Dobell, C. *Anthony van Leuwenhock and His Little Animals.* New York: Harcourt Brace and Company, 1932.

44. Miller, WD. *The Microorganisms of the Human Mouth*, K.E. Koenig, ed. Krager Basel, 1973.

45. Stralfors, A. Disinfection of dental plaque in man. Odont Tidska 1962;70:154-61

46. Löe H, Schiott CR. The effect of mouthrinses and topical application of chlorhexidine on the development of plaque and gingivitis in man. J Periodont Res 1970;5:79-83

47. Gaffar A, Nabi N, Kashuba B, Williams M, Herles S, Afflitto J. Antiplaque effects of dentifrices containing triclosan/ copolymer/NaF system vs. dentifrices without the copolymer. Am J Dent 1990;3:S7-S14

48. Gaffar A, Afflitto J, Nabi N, Kruger I, Olsen S. Recent advances in plaque, gingivitis, tartar and caries prevention technology. Int Dent J 1994;44:63-70

49. Miller S, Gaffar A, Sullivan RJ, Truong T, Stranick M. Evaluation of a new dentifrice for treatment of sensitive teeth. J Clin Dent 1994;5 (Sp Iss):71-9

50. Shafer WJ, Hine WK, Levy BM. *A Textbook of Oral Pathology*. Philadelphia: W.B. Saunders Co., 1983;41-51.

51. Fletcher M, Farina M, Smith M, Woodbury S. Bleaching of discolored teeth. PA Dent J 1992;9:29

52. McEvoy, SA. Chemical agents for removing intrinsic stains from vital teeth. II. Current techniques and their clinical applications. Quintessence Int 1989b;20:279-84

53. McEvoy, SA. Chemical agents for removing intrinsic stain from vital teeth. I. Technique development. Quintessence Int 1989a;20:323-29

54. Haywood VB, Heymann HO. Nightguard vital bleaching. Quintessence Int 1989;20:173-6

55. Haywood VB, Heymann HO. Nightguard vital bleaching. Quintessence Int 1991;22:515-23

56. Kowitz GM, Rustogi KN, Wong R. Use of chroma meter to assess changes in tooth coloration. J Dent Res 1991;70 (Sp Iss):78

57. Osol A, Pratt R, Altschule MD, ed. *Hydrogen Peroxide Solution: U.S. dispensatory and Physicians Pharmacology*. Philadelphia: Lippincott, 1967;853.

58. James AP, MacKirdy IS. Chemistry of peroxygen bleaching. Chemistry & Industry 1990;Oct:641-5

59. Evans DF, Upton WM. Studies on singlet oxygen in solution. J Chem Soc Dalton 1985;12:2525-29

60. Tenovuo J, Pruitt KM. Relationship of the human salivary peroxidase system to oral health. J Oral Pathol 1984;13:573-84

15

PREPARATION AND APPLICATION OF CALCIUM PHOSPHATE FIBERS

Koji Ioku

Department of Advanced Materials Science and Engineering, Yamaguchi University, 2557 Tokiwadai, Ube 755, JAPAN

INTRODUCTION

Inorganic fibers and whiskers have been used for many years, finding applications mostly as thermal insulation and reinforcement in composites based on polymers, metals and ceramics. Various fibrous materials such as glass, carbon, SiC, Si_3N_4, Al_2O_3, ZrO_2, $K_2Ti_6O_{13}$ and many others[1] have been prepared for such purposes. Asbestos fibers have been widely used as natural fibers due to their high tensile strength, high thermal and chemical stability, and low cost. Unfortunately, an application of the fibrous materials may be connected with a health risk. Asbestos is reported to be a cause of many serious diseases including lung cancer.[2] The main factors to be considered when addressing carcinogenicity are believed to be the morphology and the chemical composition of the materials.

Stanton et al.[3] and Pott[4] reported that the carcinogenic effect of the fibrous materials was restricted to long and thin fibers with the diameter smaller than about 0.25 mm and with the length longer than about 8 mm. From this point of view on morphology, many commercially available fibers and whiskers may be safe, but their chemical compositions must also be taken into account. For example, alumina, zirconia, titania, carbon, silicon carbide, and silicon nitride are known as bioinert materials.[5] These compounds are non-toxic, but they are not biocompatible. There is a real need for preparation of safe fibrous materials, which may replace existing biohazardous whiskers and fibers in many applications.

CALCIUM PHOSPHATE FIBERS

Calcium phosphates contains only non-toxic species, such as Ca, P, (H_2O, OH^-, CO_3^{2-} etc.). In contrast with the fibers above mentioned, bioactive calcium phosphates exhibit excellent biocompatibility due to their chemical and crystallographic similarities with the mineral constituents of bones and teeth. Fibrous calcium phosphates can be used as insulating agents, packing media for column chromatography and so on. It is a very promising material for preparation of composites, because it should act both as a bioactive and biocompatible phase and as a reinforcement for composites. In comparison with the other fibers and whiskers above mentioned, fibrous calcium phosphates should not be health hazardous because of their excellent biocompatibility. Even if some of the crystals have a dangerous shape, they should dissolve in a human body without causing any health problems.

β-Calcium Metaphosphate Fibers

Glass-related materials based on the calcium phosphate system have been investigated for biomedical applications. When the CaO-P_2O_5 glasses are reheated under a temperature gradient around the glass transition temperature (T_g), unidirectionally crystallized glass-ceramics having a fiber-reinforced composite structure are successfully prepared.[6] These glass-ceramics composed of crystalline β-calcium metaphosphate (β-$Ca(PO_3)_2$) fibers.

It has been reported that $Ca(PO_3)_2$ glass is nontoxic.[7] Fibrous β-$Ca(PO_3)_2$ was prepared by extraction from the crystallized glass in the CaO-P_2O_5 system.[8] The mixture with CaO/ P_2O_5 = 0.85 in molar ratio prepared using raw materials such as $Ca(H_2PO_4)_2{\cdot}H_2O$ and H_3PO_4 was melted at 1250 °C for 2 hr. The melt was cooled to a room temperature, and then the resultant glass was heated at 600 °C for 48 hr for the crystallization. The crystallized products containing fibrous β-$Ca(PO_3)_2$ crystals was effectively leached with 0.1N NaOH aqueous solutions at 60-70 °C. The fibrous β- $Ca(PO_3)_2$ crystals were filtered from the solution and then dried in air.

The fibers had aspect ratio from 30 to 120 with diameters of 1-5 mm. In other conditions, the longer fibers could be prepared as shown in Figure 1. The fibrous products consisted of β-$Ca(PO_3)_2$ crystals, with a small amount of $Ca_2P_6O_{17}$ crystals. It is expected that the fibers can be used as fillers for high-strength composite-biomaterials and for biomedical cushions such as a periodontal substitute.

Figure 1. SEM image of β-$Ca(PO_3)_2$-based fibers extracted from crystallized product of calcium phosphate glass by leaching with water. The photograph was reproduced by courtesy of Dr. Toshihiro KASUGA.

As one of the application, porous ceramics with a skeleton composed of crystalline β- $Ca(PO_3)_2$ -fibers were prepared by heating the fiber-compact at 800 °C for 5 hr.[9]

Porous ceramic biomaterials are important in applications as artificial bones, especially as the packing media for bone defects in orthopedic practice. The porous material of β-$Ca(PO_3)_2$ has a large porosity of >60% with large-size spaces. The porous materials with porosities of 65 to 73% had the compressive strengths of 2-8 MPa by using the long fibers of about 100 mm.[10] Experiments for evaluations of compatibility in a living body are in preparation. While the dense composite materials with a relative density of >95% consisted of β-$Ca(PO_3)_2$, fibrous crystals with CaO-P_2O_5 glass were prepared by hot-pressing.[11] These ceramics showed a high bending strength of 150-220 MPa and a low Young's modulus of 30-60 GPa. It is found that this material has mechanical properties close to those of natural cortical bones.

Hydroxyapatite Fibers and Whiskers

Hydroxyapatite ($Ca_{10}(PO_4)_6(OH)_2$;HA) is the main mineral constituent of hard tissues (bones and teeth) in human body, and thus it is one of the most biocompatible materials.[12] It has been used in medicine for many years in the form of small unloaded implants, powders, and coatings.

Preparation techniques for HA fibers and whiskers are roughly divided into four groups as shown in Table 1. A. the pyrolysis method of precursor fibers[13,14], B. the hydration method for calcium phosphates[15], C. the homogeneous precipitation method using urea[16,17] and D. the decomposition method using chelating agents.[18-20]

Table 1. Preparation techniques for fibers and whiskers of hydroxyapatite ($Ca_{10}(PO_4)_6(OH)_2$).

Techniques	Character
Pyrolysis Method[13,14]	Fibrous polycrystal
Hydration Method[15]	Rod-shaped crystal 1.51≤Ca/P≤1.63 Trace of carbonate ions
Homogeneous Precipitation Method[16,17]	Fibrous polycrystal Rod-shaped crystal 1.56 ≤ Ca/P ≤ 1.66 High levels of carbonate ions
Decomposition Method using Chelating Agents[18-20]	Whisker, Single crystal 1.59 ≤ Ca/P ≤ 1.67 Low levels of carbonate ions

In the first method, alginic fibers containing Ca and P was used as precursors for HA fibers. The precursor fibers were heated at 900 °C to obtain HA fibers. In the second method, calcium phosphates with lower Ca/P molar ratio than HA (Ca/P of stoichiometric HA is 1.67) were hydrated by vapor exposure treatment under hydrothermal conditions to obtain HA fibrous materials. The third method utilized a continuous increase of pH in the solution containing calcium and phosphate ions at the temperatures from 80 °C to the temperatures under hydrothermal conditions. In the case of the fourth method, chelating agents like ethylenediaminetetraacetic acid, lactic acid, or citric acid were used. During the heat treatment, which was usually carried out under hydrothermal conditions, Ca-complexes with chelating agents decompose, followed by the precipitation of HA whiskers.

Pyrolysis Method

The Fibrous HA was prepared by using sodium alginate (Na-Alg.).[13,14] The starting materials were prepared by dissolving $Na_4P_2O_7$ into 5% Na-Alg. solution, while the mixed solution of 0.15 mol $Ca(CH_3COO)_2$ and 0.1 mol $CaCl_2$ was used as the spinning solution with the pH of 6.0 to 7.5. The starting materials were spun through a nozzle of 0.1 mm ϕ into the spinning solution in a vessel of 16 cm^3 capacity. Gelatinous fibers obtained were aged, dried and calcinated. β-$Ca_2P_2O_7$,

β-$Ca_3(PO_4)_2$ and HA ($Ca_{10}(PO_4)_6(OH)_2$) were formed by heating at temperatures above 600 °C in air. HA without other calcium phosphates was formed at 900 °C. The optimum condition for preparing fibrous HA was found to be as follows; Na-Alg./$Na_4P_2O_7$ = 3.0, pH = 6.5. It was noted that, for the obtained fibrous HA heated in air a porous surface was obtained.

In order to remove pores appearing during the heating process, the dried fibers formed under the optimum condition were heat-treated under reduced pressure (at 133 Pa and 900 °C for 3h). The resulting fibrous HA was polycrystalline material of about 30-40 mm in diameter. It was also confirmed that the HA phase in the fibers was maintained even after heat-treated at 1300 °C for 1h in air.

Hydration Method

The rod shaped particles of HA were prepared by hydration method from a-tricalcium phosphate ($Ca_3(PO_4)_2$) powders by vapor exposure treatment under hydrothermal conditions at the temperatures from 105 to 250 °C for 20 hr.[15] The materials obtained at a temperature of 105 °C had aspect ratio less than 40 with the length from 10 to 40 mm and with the diameter of about 0.5 mm (Figure 2).

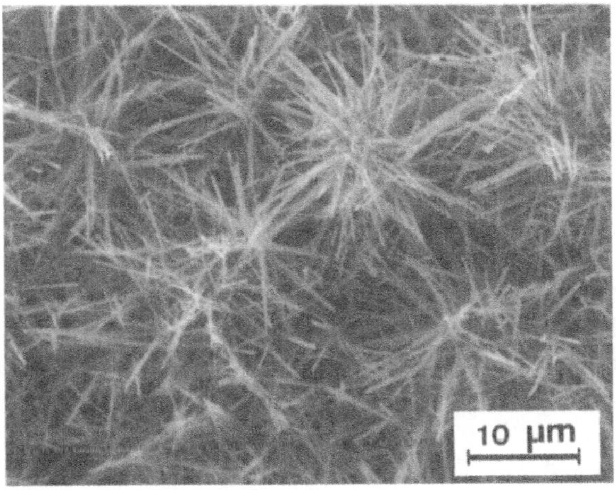

Figure 2. SEM image of the hydroxyapatite rod-shaped crystals prepared from α-tricalcium phosphate by vapor exposure treatment at 105 °C for 20 hr.

According to the chemical analysis, the apatite prepared by this method was calcium deficient hydroxyapatite. The rod shaped HA obtained at 105 °C had the Ca/P molar ratio of 1.58. It was lower than that of stoichiometric HA of 1.67. The Ca/P molar ratio of HA increased with an increase in temperature (Figure 3).

When powders of α-TCP was formed into cylindrical shape, cylindrical porous ceramics of apatite of about 50% porosity could be prepared directly.[15] The compressive strength of the porous ceramics was about 20 MPa. This value was comparable to that of spongy bone. The porous ceramics of calcium deficient HA prepared by the hydration method had the inter-connecting pore structure. Therefore, newly formed bone must be introduced into this ceramics.

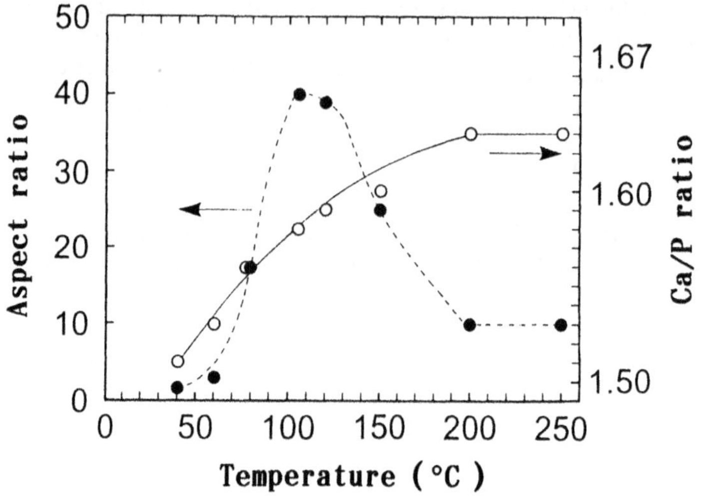

Figure 3. The Ca/P molar ratio and the aspect ratio of hydroxyapatite rod-shaped crystals prepared from α-tricalcium phosphate by vapor exposure treatment.

Homogeneous Precipitation Method

Fibrous and whisker-like materials of HA containing large quantities of carbonate ions were obtained by homogeneous precipitation method using urea.[16] The mixed aqueous solution with Ca/P=1.67 in molar ratio of hydroxyapatite stoichiometry was prepared from $Ca(NO_3)_2$, $(NH_4)_2HPO_4$ and urea ($(NH_2)_2CO$). Then it was refluxed at the temperatures from 80 to 100 °C for 17-170h, or it was treated hydrothermally at the temperatures from 120 to 150 °C for 0-6 hr. The fibrous materials obtained at 80-100 °C had aspect ratio less than 200 with the length from 20 to 150 mm and with the diameter of about 1-5 mm. The fibrous materials prepared hydrothermally had the length of 200-300 mm. They have not been verified yet to be single crystals. The flexible sheets and porous compacts of HA containing carbonate ions were obtained from the materials. Needle-like HA crystals of 0.5 mm in diameter and 10mm in length[17] were prepared by the hydrolysis of monetite ($CaHPO_4$) or brushite ($CaHPO_4 \cdot 2H_2O$), using decomposition of urea at high temperature to accelerate the reaction.

Decomposition Method using Chelating Agents

HA whiskers were obtained by the hydrothermal treatment of aqueous solutions containing β-$Ca_3(PO_4)_2$ and citric acid at a temperature of 200 °C under saturated vapor pressure for 1 hr.[18] Before the hydrothermal treatment, the solution was clear and colorless without any precipitation, but it turned a faint yellow after the hydrothermal treatment, suggesting the decomposition of the Ca chelate. Diameter, length and aspect ratio of the HA whiskers prepared hydrothermally are in the range of 0.1-1 mm, 20-30 mm, and 20-30 mm, respectively. It is of particular importance that most of the whiskers have the same morphology and size. The HA whiskers have a slightly lower Ca/P molar ratio (1.63) than in stoichiometric HA of 1.67, and have a trace of carbonate ions in their structure.

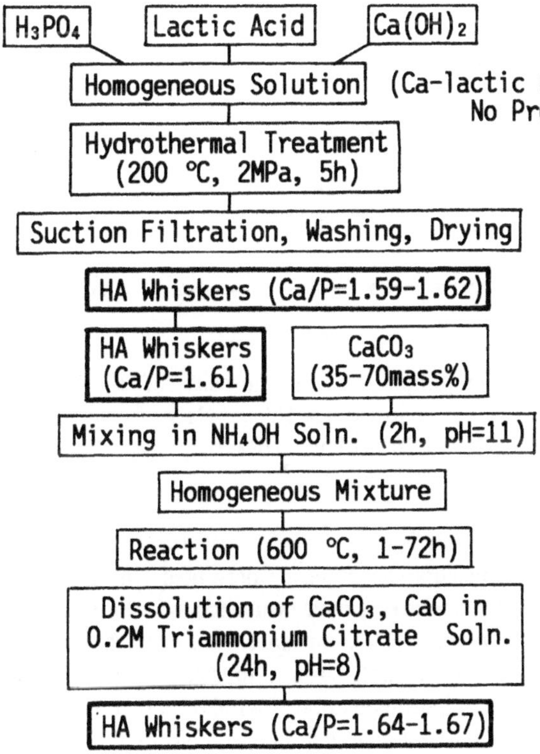

Figure 4 Experimental procedure for preparation of hydroxyapatite whiskers and for the control of Ca/P ratio of hydroxyapatite whiskers.[19]

With many other starting chemicals such as H_3PO_4, $Ca(OH)_2$ and lactic acid etc., various hydrothermal conditions was tried to control the shape of whiskers of HA.[19,20] The Ca/P ratio of HA whiskers is in the range of 1.59-1.62, which makes them thermally unstable. The stoichiometry of HA crystals prepared by the decomposition of chelating agents under hydrothermal conditions seems to be almost independent of the pH and Ca/P ratio in the starting solution. Reaction with calcite at 600 °C allows the control of Ca/P ratio of the whiskers up to the stoichiometric value of 1.67 (Figure 4).

The morphology of the crystals remains almost unchanged after this treatment. The diffusion of calcium from calcite into the bulk of HA seems to be responsible for the increase of the Ca/P ratio of the whiskers.[19]

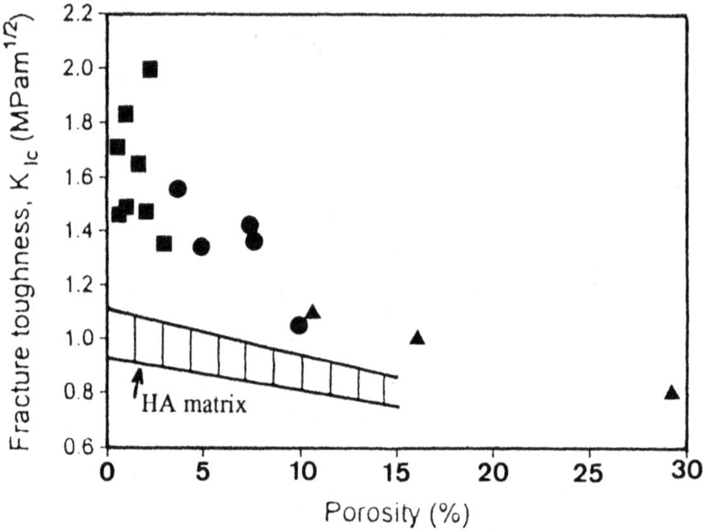

Figure 5. Fracture toughness K_{IC} vs. porosity of the hydroxyapatite / hydroxyapatite whiskers composites (▲:HA/10%HA(w) sintered, ■:HA/20-30%HA(w) HIP-ed, ●:HA/10-30%HA(w) HP-ed). The Ca/P ratio of the hydroxyapatite whiskers was 1.66 in all case.[22]

Applications of HA whiskers have been presented in the literatures.[21,22] Hydrothermally prepared HA whiskers have been used as reinforcements of the HA ceramics. HA whisker-reinforced HA composites have been fabricated by normal-sintering, hot pressing (HP) and hot isostatic pressing (HIP) at 1000-1100 °C. HA/0-30% HA-whiskers composites exhibit 40-100% higher toughness than the non-reinforced HA matrix without degradation of the biocompatibility, because the HA whiskers act both as a reinforcement and as a biocompatible phase. The fracture toughness of the dense HA/HA-whiskers composites with relative densities of 97.0-99.5% was 1.4-2.0 MPa·m$^{1/2}$ as shown in Figure 5.[22]

Mechanical Properties of Hydroxyapatite Single Crystals

The single crystals give us the basic information for mechanical properties of the material. In order to use hydroxyapatite fibers and whiskers for wide applications, we must know the mechanical properties of hydroxyapatite single crystals.

Because of the insufficiency of mechanical strength[23-26], hydroxyapatite ceramics have been applied to unloaded implants such as the porous material for bony defects and the dense material for a percutaneous device.

The mechanical strength of ceramics highly depends on crystal structural disorder at grain boundaries, as well as potential flaws such as pores, inclusions and coarse grains. The crystal structural disorder makes residual strain and the flaws become stress concentration sites. These factors induce the fracture of ceramics at a stress level from 1/10 to 1/100 times less than the theoretical strength. In the case of hydroxyapatite ceramics, the random orientation of crystal grains could highly increase the residual strain after a sintering process because of a large discrepancy in thermal expansion coefficient in a- and c-directions in the temperature range of 1000 to 1150 °C.[27]

Single crystals of hydroxyapatite (HA) and OH-carbonated hydroxyapatite (CHA) were prepared hydrothermally by the hydrolysis of $CaHPO_4$ without and/or with supplementation of carbon dioxide in the form of dry ice.[28-30] $CaHPO_4$ was prepared from 99.99% pure grade calcium carbonate, reagent grade orthophosphoric acid solution and ultra-pure water to eliminate magnesium which inhibits the crystal growth of hydroxyapatite. Crystal growth runs were performed by slowly hydrolyzing $CaHPO_4$ with a temperature gradient application by a gradually heating vessel (Figure 6). The obtained single crystals were colorless and hexagonal prismatic and were bounded by α-faces.

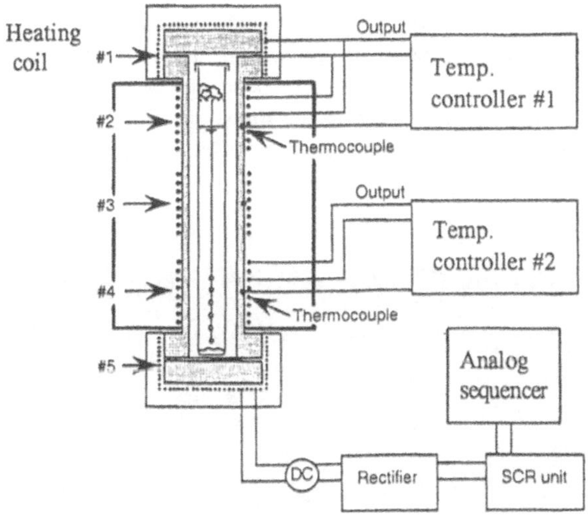

Figure 6. Schematic illustration of apparatus for hydrothermal crystal growth for hydroxyapatite and carbonate-containing hydroxyapatite.[28]

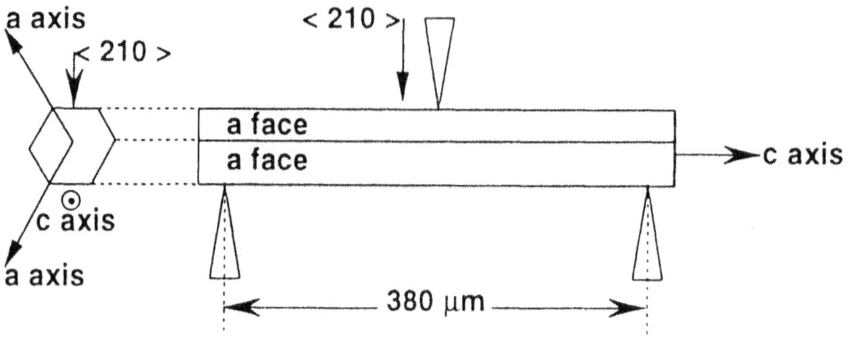

Figure 7. Schematic illustration of three-point bending test.[30]

The bending strength of the crystals was obtained in air and in water. The bending strength of the single crystals was determined in three-point bending tests using a modified ultra micro hardness tester with three knife blades 50 mm long and 380 mm thick for an indenter and two supporting points (Figures 7 and 8). The application of 50-mm-long knife blades to the indenter and the supporting points guarantees the degree of parallelization between the indenter and the supporting points to be less than 0.04°. The crystals, 15-70 mm in thickness and 500-1000 mm in length, selected for the three-point bending tests have neither cracks nor fluid inclusions microscopically and show complete extinction at their extinction positions between crossed poles in polarizing microscopic observation. The bending strength in water was measured by bending the single crystals in a drop of ultrapure water filled in the space between the supporting points. The three-point bending strength of HA single crystals and CHA single crystals were shown in Table 2.[29,30] The single crystals were well deflected as shown in Figure 8.

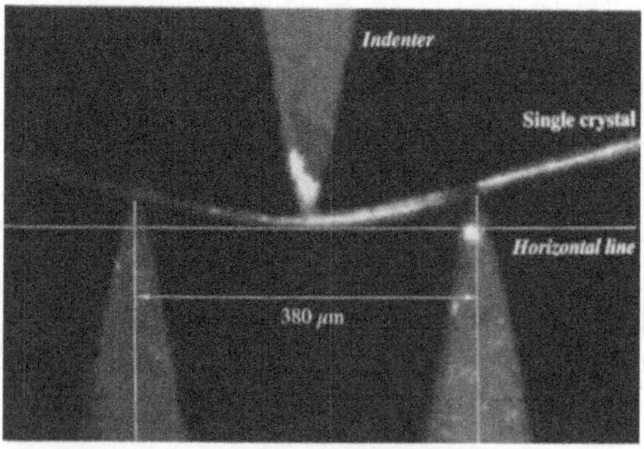

Figure 8. Expanded view of three-point bending test for the hydroxyapatite single crystal. The photograph was reproduced by courtesy of Dr. Atsuo ITO.

The bending strength of HA single crystals was 500 ± 184 MPa in air.[29] This value was about 2.5 times higher than that of dense HA ceramics.[23-26] The strength was comparable to that of dense alumina ceramics[31], and about three times higher than that of cortical bone (161.56 ± 46.86 MPa).[32] The average strength in water was 9% lower than that in air due to stress corrosion.

This decrease in strength, however, was not statistically significant. The bending strength of CHA single crystals containing carbonate of 0.09 CO_2 mass % was 468 ± 205 MPa and 361 ± 199 MPa in air and in water, respectively.[30] These comparison show that the insufficient bending strength of HA ceramics is caused not by the mechanical properties of HA itself but by other factors including the mechanical properties of the grain boundary, size and distribution of flaws, and residual strain. Therefore, the strength of hydroxyapatite single crystals must be enough for applications as reinforcement in composites.

Table 2 Bending strength of single crystals of hydroxyapatite (HA) and OH-carbonated hydroxyapatite (CHA) in comparison with dense hydroxyapatite ceramics, dense alumina ceramics and cortical bone.

Material	Bending Strength (MPa)
HA Single crystal in air	500 ± 184 [29]
HA Single crystal in water	454 ± 203 [29]
CHA Single crystal in air	468 ± 205 [30]
CHA Single crystal in water	361 ± 199 [30]
Dense HA ceramics	113~196 [23-26]
Dense alumina ceramics	450~595 [31]
Cortical bone	162 ± 47 [32]

REFERENCES

1. Cook TF Inorganic fibers-A literature review. J Am Ceram Soc 1991;74:2959-78

2. Mossman BT, Bignon J, Corn M, Seaton A, Gee JBL. Asbestos: Scientific developments and implications for public policy. Science 1990;247:294-301

3. Stanton MF, Layard M, Tegeris A, Miller F, May M, Morgan E, Smith A. Relation of particle dimension to carcinogenicity in amphibole asbestoses and other fibrous minerals. J Natl Cancer Inst Inst 1981;67:965

4. Pott F. Some aspects of the dosimetry of the carcinogenic potency of asbestos and other fibrous dust. Staub-Reinh. Luft 1978;38:486

5. Hench LL. Bioceramics: From concept to clinic. J Am Ceram Soc 1991;74:1487-1510

6. Abe Y, Kasuga T, Hosono H. Preparation of high-strength calcium phosphate glass-ceramics by unidirectional crystallization. J Am Ceram Soc 1984;67:C142-44

7. Nagase M, Abe Y, Chigira M, Udagawa E. Toxicity of silica-containing glasses demonstrated in mice. Biomaterials 1992;13:172-75

8. Kasuga T, Abe Y. Preparation of calcium phosphate fibers. Phosphorus Res Bull 1992;2:17-20

9. Kasuga T, Inoue T, Tsuji K, Ota Y, Abe Y. Preparation of porous ceranics with calcium metaphosphate fiber skeleton for biomedical use. J Am Ceram Soc 1995;78:245-47

10. Kasuga T, Ota Y, Tsuji K, Abe Y. Preparation of high-strength calcium phosphate ceramics with low modulus of elasticity containing β-$Ca(PO_3)_2$ fibers. J Am Ceram Soc 1996;79:1821-24

11. Ota Y, Kasuga T, Abe Y. Preparation and compressive strength behavior of porous ceramics with b-$Ca(PO_3)_2$ fiber skeletons. J Am Ceram Soc 1997;80:225-31

12. Aoki H. *Medical Applications of Hydroxyapatite*, Tokyo, St.Louis: Ishiyaku EuroAmerica, 1994.

13. Iwasaki H, Kaneko Y. Preparation of fibrous hydroxyapatite by use of sodium alginate. J Soc Mat Sci Jpn 1988;37:60-64

14. Iwasaki H, Fibrous hydroxyapatite. Ceram Jpn 1989;24:295-300

15. Ioku K, Eguchi Y, Goto S. Hydrothermal preparation of porous hydroxyapatite. Proc. Int. Conf. High Pressure Science and Technology (Joint AIRAPT-16 & HPCJ-38, Aug. 25-29, 1997, Kyoto) in press.

16. Kinoshita M., Kishioka A, Hayashi H, Itatani K. Preparation of fibrous calcium phosphates by homogenious precipitation method and their thermal changes. Gypsum & Lime 1989;219:79-87

17. Kinoshita M, Itatani K, Nakamura S, Kishioka A. Preparation and morphology of carbonate-containing hydroxyapatite by homogeneous precipitation and hydrothermal methods. Gypsum & Lime 1990;227:207-215

18. Yoshimura M, Suda H, Okamoto K, Ioku K. Hydrothermal synthesis of biocompatible whiskers. J Mater Sci 1994;29:3399-3402

19. Suchanek W, Suda H, Yashima M, Kakihana M, Yoshimura M. Biocompatible whiskers with controlled morphology and stoichiometry. J Mater Res 1995;10:521-29

20. Fujishiro Y, Fujimoto A, Sato T, Okuwaki A. Coating of hydroxyapatite on titanium plates using thermal dissociation of calcium-EDTA chelate complex in phosphate solutions under hydrothermal conditions. J. Colloid and Interface Sci 1995;173:119-127

21. Suchanek W, Yashima M, Kakihana M, Yoshimura M. Processing and mechanical properties of hydroxyapatite reinforced with hydroxyapatite whiskers. Biomaterials 1996;17:1715-23

22. Suchanek W, Yashima M, Kakihana M, Yoshimura M. Hydroxyapatite/Hydroxyapatite (Whiskers) Composites without Sintering Aids-Mecha nical Properties and Microstructural Evolution. *Bioceramics* Vol.9 (Proc. 9th Int. Sympo. Ceram. Med., 24-27 Nov. 1996, Otsu, Jpn.). T. Kokubo, T. Nakamura, F. Miyaji Eds., Japan: Pergamon/Elsevier, 1996:153-56.

23. Jarcho M, Boren CH, Thomas MB, Bobick J, Kay JF, Doremus RH. Hydroxyapatite synthesis and characterization in dense polycrystalline form. J Mater Sci 1976;11:2027-35

24. Thomas M.B., Doremus R.H., Jarcho M., Salsbury R.L. Dense hydroxyapatite: Fatigue and fracture strength after various treatments, from diametral tests. J Mater Sci 1980;15:891-94

25. Akao M, Aoki H, Kato K Mechanical properties of sintered hydroxyapatite for prosthetic applications. J Mater Sci 1981;16:809-12

26. De With G, Van Dijk HJA, Hattu N, Prijs K. Preparation, microstructure and mechanical properties of dense polycrystalline hydroxyapatite. J Mater Sci 1981;1592-98

27. Nakamura S, Otsuka R, Aoki H, Akao M, Miura N. Thermal expansion of hydroxyapatite β-tricalcium phosphate ceramics. Thermochimica Acta 1990;165:57-72

28. Ito A, Nakamura S, Aoki H, Akao M, Teraoka K, Tsutsumi S, Onuma K, Tateishi T. Hydrothermal Growth of Carbonate-Containing Hydroxyapatite Single Crystals. J Crystal Growth 1996;163:311-17

29. Ito A, Teraoka K, Tsutsumi S, Tateishi T. Single Crystal Hydroxyapatite: Preparation, Composition and Mechanical Properties. In *Bioceramics* Vol.9 (Proc. 9th Int. Sympo. Ceram. Med., 24-27 Nov. 1996, Otsu, Jpn.), T. Kokubo, T. Nakamura, F. Miyaji Eds, Japna: Pergamon/Elsevier, 1996:189-92.

30. Teraoka K, Ito A, Onuma K, Tateishi T, Tsutsumi S. Bending strength of synthetic OH-carbonated hydroxyapatite single crystals. J Biomed Mater Res 1997;34:269-72 31.

31. Hulbert SF. *Introduction to Bioceramics*. L.L Hench., J. Wilson Ed., Singapore:World Scientific 1993:25-40.

32. Evans FG. *Mechanical Properties of Bone*. A.R. Burdi Ed., Illinois:Charles C Thomas 1973:97.

16

DEVELOPMENT OF CALCIUM PHOSPHATE INHIBITNG POLYMERS FOR COOLING WATER APPLICATIONS

Zahid Amjad

Advanced Technology Group, BFGoodrich Company
Brecksville, Ohio 44141 USA

INTRODUCTION

The prccipitation and deposition of mineral scales on heat exchangers resulting from dissolved ions is a common problem in the use of water for industrial processes such as coooling, boiler, desalination, geothermal, secondary oil recovery utilizing water flooding techniques, and clothing washing machines. The scales consist primarily of phosphates, sulfates, carbonates, and silicates of alkaline earth metal, particularly calcium and magnesium. The insulating nature of scales on heat exchanger surfaces results in lost heat transfer efficiency and tube failures.[1] The major cause of RO system failures is membrane fouling by mineral scales and suspended matter, which results from the accumulation of foulants on the surfaces of semipermeable membranes. The effects of fouling in RO system can lead to membrane replacement, loss in production, poor quality, and reduced quantity of produced water.[2]

In addition to scaling problems, cooling water systems and boilers constructed of carbon steel also experience corrosion problems caused by the presence of dissolved oxygen. In RO system, the presence of dissolved oxygen poses serious problem as it oxidizes hydrogen sulfide present in the feed stream resulting in deposition of elemental sulfur. Once formed, removal of sulfur is difficult and expensive to remove, and in many cases results in membrane replacement.

Over the years, different inhibitors have been incorporated in water treatment formulations to provide corrosion protection. For many years, chromate-based corrosion inhibitors were considered as the performance standard in the cooling water treatment industry. Unfortunately, chromates are toxic and due to environmental regulations their use has declined considerably in recent years. Accordingly, there has been a trend away from chromate-based corrosion inhibitors. Among the more environmentally acceptable mild steel corrosion inhibitors which have been displacing chromates since the 1960s are molybdates, silicates, polyphosphates/phosphate, phosphonates, and zinc salts. From the perspective of their versatility, cost, and performance, phosphates have become a new performance standard.[3] Additionally, phosphate-based cooling water treatment (CWT) programs require more careful system parameter control than do chromate programs, mainly because they rely upon relatively high phosphate levels for corrosion protection. These high phosphate levels, and especially cooling systems operating at alkaline pH and high calcium hardness water, can lead to calcium phosphate precipitation and fouling of heat exchangers. The key to the succesful use of phosphate based CWT program lies in the proper selection of polymeric calcium phosphate inhibitor. The polymer serves a dual role. It controls the thickness of the calcium phosphate film on the metal surface and also prevents precipitation of the calcium phosphate in the recirculating water.

It is well established that the presence of soluble metal ions (i.e., iron, manganese, zinc, aluminum, etc.) in make-up water can lead to heat exchanger fouling in cooling water systems.[4,5,6] It is generally agreed that iron(II) and Mn(II) do not cause significant fouling problem as long as they remain in soluble form. At typical cooling water system pH values, and especially in the presence of oxidizing agents (i.e., oxygen, chlorine), a mixture of metal hydroxides can precipitate. It is well documented that iron(III), when present at low concentrations, can significantly reduce the performance of a polymeric calcium phosphate inhibitor.[7,8]

In cooling systems, water temperature plays a significant role both on the kinetics and thermodynamics of calcium phosphate precipitation, as well as on corrosion rate. Many scale forming salts show inverse solubility relationship with temperature (i.e., solubility decreases with increasing solution temperature). Since temperature on heat exchanger surface is typically high (e.g., 130°F), there is high potential for calcium phosphate to precipitate and deposit on the heat exchangers. Temperature also influences corrosion process. A general rule is that the corrosion rate of mild steel will double for every 25-50°F increase, up to 160°F.[9]

Additionally, temperature can also effect the performance of polymeric inhibitors especially if they exhibit poor thermal stability.

The influence of solution pH on corosion rate as well scaling tendency has been investigated. It is well known that alkaline pH and high alkalinity generally produce reduced corrosion rate. However, an increase in solution pH especially in phosphate based CWT program can increase the calcium phosphate scaling potential, thus requiring incorporation of high performance calcium phosphate inhibiting polymer in the formulations. Additionally, if the polymeric inhibitors are not hydrolytically stable (i.e., hydrolyze in alkaline pH), their use in a CWT formulation will be limited (compromised).

Considering the factors discussed above, the role of polymers in phosphate-based CWT programs is to control the precipitation of calcium phosphate and allow the use of the phosphate levels necessary to provide effective corrosion protection under a variety of operating conditions. The following technical criteria[4,7] has been suggested in selecting a polymer for phosphate-based treatment program:

- Calcium phosphate inhibition
- Retention of activity in the presence of soluble iron
- Compatibility with organic cationic flocculant
- Dispersion of particulate matter
- Calcium ion tolerance
- Compatibility with chlorine
- Hydrolytic and thermal stability
  - as raw material
  - as formulation component
  - when applied in end-use application
- Environmental acceptability

During the last two decades, a variety of polymeric inhibitors containing different functional groups (i.e., -COOH, -COOR, -OH, $-SO_3H$, etc.) have been developed and are currently being used in phosphate-based CWT programs. This chapter examines the development of calcium phosphate inhibitors for cooling water applications. The performance of these inhibitors especially in view of process variables (i.e., iron, calcium, phosphonate, cationic flocculant, etc.) will be compared. Also, the ability of these polymers to disperse particulates will be discussed. In addition, a polymer structure-performance correlation will also be presented.

EXPERIMENTAL

Reagent grade chemicals and grade A glassware were used. Stock solutions of calcium chloride and disodium hydrogen phosphate were prepared using distilled water, filtered through 0.22 micron filter paper (Millipore Corp.) and analyzed for calcium and phospate as described previously.[7] The polymeric materials used were selected from commercial and experimental materials. All inhibitors solutions were prepared on a dry weight basis. The desired concentrations were obtained by dilution.

Subsaturated calcium phosphate solutions were prepared in a double-walled glass cell, maintained at constant temperature, by adding stock solutions of calcium chloride and disodium hydrogen phosphate to the distilled water. Spontaneous precipitation was induced by by raising the pH of this acidic solution with 0.10M NaOH. The pH of the solution was maintained at a constant pH value (± 0.01) using a Metrohm pH stat (Model 600 series, Brinkmann Instruments, Westbury, N.Y.). Hydrogen ion measurements were made with a combination electrode equilibrated at the required temperature. The electrode was calibrated before each experiment using standard buffer solutions. The calcium phosphate supersaturated solutions were continuously stirred with a Teflon-coated stirring bar at ~ 400 revolutions per minute. Experiments involving inhibitors were performed by adding the inhibitor to the subsaturated calcium phosphate solution before inducing spontaneous precipitation. Precipitation of calcium phosphate in this subsaturated solution was monitored by analyzing aliquots of the filtered (0.22 micron filter paper) solution for phosphate by the spectrophotometric method.[7] The pH-stat apparatus used is illustrated in Figure 1.

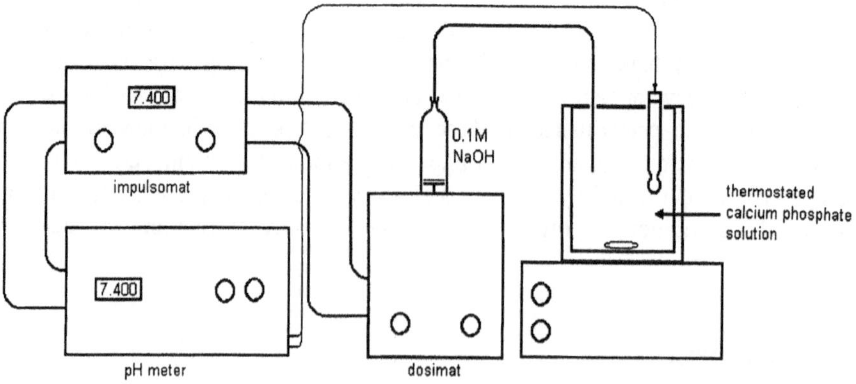

Figure 1. pH-stat apparatus used in calcium phosphate precipitation experiments.

The polymer efficiency as a calcium phosphate inhibitor was calculated using the following equation:

$$\% \text{ Inhibition (I)} = [(PO_4)_{exp} - (PO_4)_{final}] / [(PO_4)_{initial} - (PO_4)_{final}] \times 100\% \quad (1)$$

where:

$(PO_4)_{exp}$ = concentration of the phosphate in the filtrate in the presence of the inhibitor at known time.

$(PO_4)_{final}$ = concnetration of phosphate in filtrate in absence of inhibitor at 20 hour.

$(PO_4)_{initial}$ = concentration of phosphate at the beginning of the experiment.

Table 1 lists the polymers investigated in this study.

Inhibitor	Structure	Acronym
poly(acrylic acid)	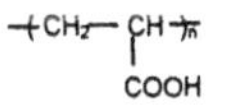	PAA
poly(maleic acid)	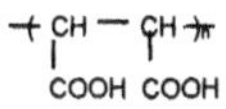	PMA
poly(methacrylic acid)	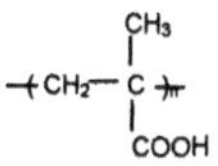	PMAA
poly(acrylamide)	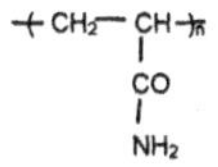	PAM
poly(2-acrylamido-2-methyl propane sulfonic acid)	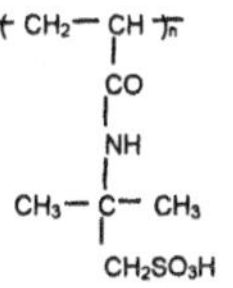	PSA
poly(acrylic acid:hydroxyl propyl acrylate)	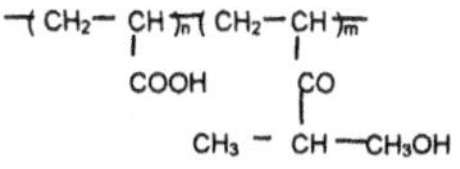	AA:HPA
poly(acrylic acid:vinyl acetate)	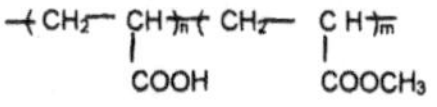	AA:VAc
poly(acrylic acid:diacetone acrylamide)	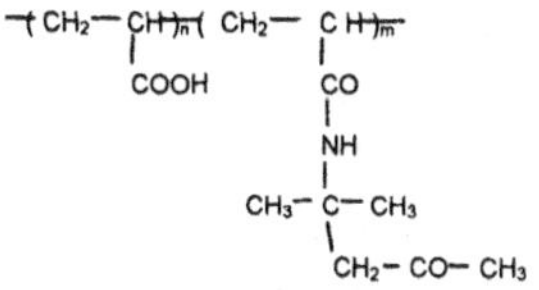	AA:DACAM
poly(acrylic acid:2-acrylamido-2-methyl propane sulfonic acid:sulfonated styrene)	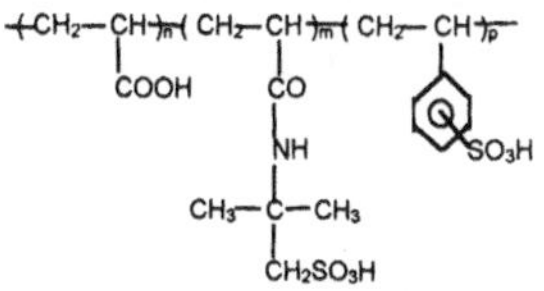	AA:SA:SS
poly(acrylic acid:dimethyl itaconate)	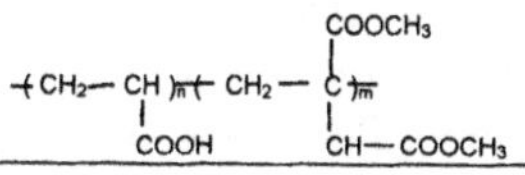	AA:DMI

RESULTS AND DISCUSSION

Using the pH-stat technique, a large number of experiments were conducted in which the depletion of phosphate from solution was monitored spectrophotometrically. The experiments reported here have good reproducibility (± 6% or better). Initiation of precipitation by foreign particles, which are always present, does not significantly influence the precipitation, since the results were reproducible. In addition, the experiments were repeated after having filtered the solutions prior to mixing through 0.1 micron membrane filters without any change in the observed results. However, the possibility of heterogeneous nucleation on the walls of the glass cells used cannot be precluded, although experiments were conducted in two different experimental set-ups.

Calcium Phosphate Inhibiting Polymers

During the last two decades a variety of copolymers have been developed and are currently being used as component of phosphate based formulations. The role of calcium phosphate inhibiting polymers is to prevent calcium phosphate precipitation and deposition on heat exchanger, disperse suspended matter, and also provide protection against corrosion while maintaining soluble phosphate in circulating water. Over the last several years the polymers which have been developed and/or used as calcium phosphate inhibitors fall into the following two categories:

Homopolymers

- poly(acrylic acid), PAA
- poly(methacrylic acid), PMAA
- poly(maleic acid), PMA
- poly(acrylamide), PAM

Copolymers

- acrylic acid: hydroxyl propyl acrylate, AA:HPA
- maleic acid: sulfonated styrene, MA:SS
- acrylic acid: methacrylic acid: acrylamide: carbitol acrylate AA:MAA:AM:CA
- acrylic acid: 2-acrylamido-2-methylpropane sulfonic acid , AA:SA
- acrylic acid: vinyl acetate, AA:VAc
- acrylic acid:diacetone acrylamide, AA:DACAM
- acrylic acid:dimethylitaconate, AA:DMI
- acrylic acid: 2-acrylamido-2-methylpropane sulfonic acid: sulfonated styrene, AA:SA:SS

In the following section we will discuss the performance of the above mentioned polymers in terms of their efficacy for calcium phosphate inhibition, especially in the presence of iron (III), cationic flocculant, high calcium hardness, their ability to disperse suspended matter, and also their tolerance to high calcium waters. Calcium phosphate inhibition data will also be presented to demonstrate the role of polymer molecular weight as well as polymer stability under alkaline pH conditions as it relates to polymer performance.

Calcium Phosphate Inhibition

Effect of polymer concentration

By conducting a series of precipitation experiments in the presence of varying concentrations of polymer, it is possible to describe the concentration-performance relationship for a given inhibitor as illustrated in Figure 2. As seen in the Figure (curve B), the addition of relatively small amount (i.e., 5 ppm) of PAA has a negligible influence on the precipitation reaction. However, at 15 ppm, performance of PAA is substantially improved (curve-D) and at 50 ppm maximum inhibition (>80%) is obtained (Curve F) for at least 20 hr. It should be noted that in the absence of polymer (curve A) calcium phosphate precipitation is essentially over in ~ 1 hr, while the addition of 50 ppm of PAA maintains >80% phosphate in solution for upto 20 hr.

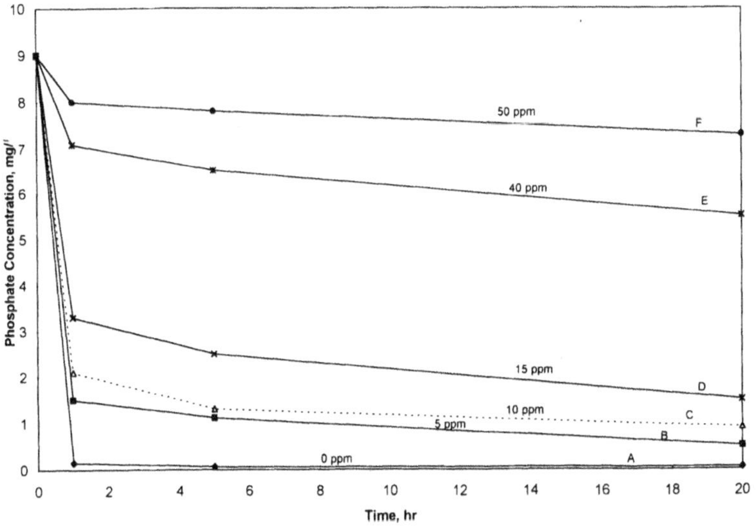

Figure 2. Plots of phosphate concentration as a function of time in the presence of varying concentrations of poly(acrylic acid), PAA, (140 mg/ℓ Ca, 9 mg/ℓ PO_4^{3-}, pH 8.50, 50°C).

Figure 3 compares the performance of PAA and a terpolymer in which carboxyl group has been partly replaced by two sulfonic acid containing monomers, namely 2-acrylamido 2-methylpropane sulfonic acid and sulfonated styrene, AA:SA:SS. As illustrated in Figure 3, the performance of polymer strongly depends on the concentration of the polymer present in the calcium phosphate supersaturated solution. It is worth noting that at 5 ppm the polymer concentration the calcium phosphate inhibition obtained for PAA is ~ 5% compared to >90% for the terpolymer. From the practical point of view, the inhibition data presented in Figure 3 clearly shows that PAA will not deliver the performance expected by the polymer used in the phosphate-based water treatment formulation.

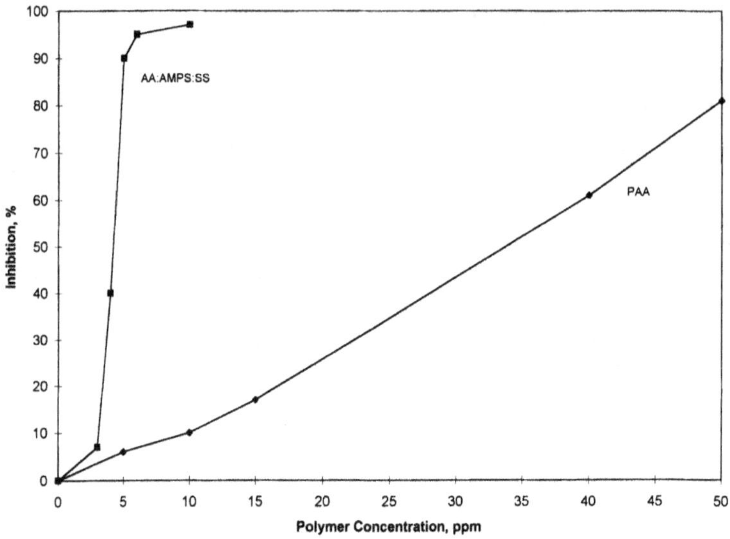

Figure 3. Calcium phosphate precipitation in the presence of polymers. Plots of % inhibition as a function of polymer concentration (140 mg/ℓ Ca, 9 mg/ℓ PO_4^{3-}, pH 8.50, 50°C, 20 hr).

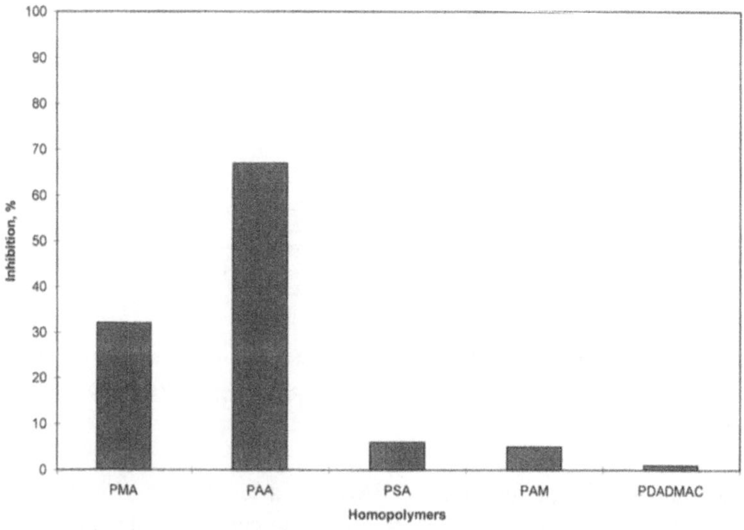

Figure 4. Performance of homopolymers as calcium phosphate inhibitors. Plots of % inhibition for various homopolymers (140 mg/ℓ Ca, 9 mg/ℓ PO_4^{3-}, pH 8.50, 50°C, 20 hr, 10 ppm polymer).

The inhibition data for a number of homopolymers (i.e., PAA, PMA, PMAA, PSA, PAM, etc.) collected under similar experimental conditions (Figure 4) clearly shows that polymers which are devoid of -C00H groups (i.e., PAM, PSA, etc.) are ineffective calcium phosphate inhibitors compared to polymers which do contain COOH groups. For example, inhibition values obtained for PAA and, PMA at 40 ppm concentration are 61 and 33% respectively, compared to 5 and 4% obtained for the other polymers. The data presented in Figure 4 highlights the importance of the interaction of the -COOH group of the polymer with calcium on the calcium phosphate nuclei in the overall performance of the inhibitor. This observation is consistent with previous reports regarding the performance of PAA, PAM, and PSA polymers as dicalcium phosphate dihydrate[10] and hydroxyapatite[11] crystal growth inhibitors. The poor performance exhibited by PMA, although has two carboxyl group per monomer unit compared to only one carboxyl group in the acrylic acid monomer unit, may be due to the precipitation of PMA with calcium ion, thus reducing the effective concentration of the PMA and/or Ca-PMA salt providing the surface for calcium phosphate precipitation. It is worth noting that carboxyl containing homopolymers have also been shown to be effective inhibitors in seeded growth studies involving other sparingly soluble salts, including CaF_2,[12] $CaSO_4 \bullet 2H_2O$,[13] and $BaSO_4$.[14]

Effect of polymer molecular weight

In recent years, the influence of molecular weight (MW) on the performance of polymers as scale inhibitors has been the subject of numerous investigations. Amjad,[10-12,14-16] in studies on the evaluation of PAA in controlling the precipitation of $BaSO_4$, $CaSO_4.2H_2O$, $CaHPO_4.2H_2O$, $Ca_5(PO_4)_3OH$, and calcium phosphonate from aqueous solutions, reported that the MW of the polymer plays an important role in the inhibition of various scale forming salts. Results of these studies indicate that especially for PAA, the optimum performance is obtained with ~2,000 MW.

Excellent examples of the molecular weight effects are provided by AA:HPA and AA:VAc copolymers. The calcium phosphate inhibition data collected at 20 hr for two copolymers illustrated in Figure 5 cleary shows that the higher molecular weight version of AA:HPA and AA:VAc polymers exhibit enhanced performance when tested at 10 ppm concentration, compared to the low molecular weight version. The optimum performance obtained with copolymer in the molecular weight range 10 - 15,000 for calcium phosphate is consistent with the results of earlier investigations on calcium phosphates and calcium phosphonates.[17,18] William and Ruehrwein[19] have shown that one of the factors determining the effectiveness of a polymer is its molecular weight. They suggested that MW influences the kinetics of adsorption of the polymer onto growing crystals. It was postulated that there should be an optimum MW for a given polymer at which a maximum effect would be obtained. Wilson[20] in his study on the effect of phosphinocarboxylic acids of varying MW (1,000 to 7,500) as scale inhibitors reported that whereas 2,200 MW polymer showed the best performance in inhibiting the precipitation of $CaSO_4.2H2O$ and $CaCO_3$, in the case of $BaSO_4$ the polymer exhibiting optimum performance had a MW of 3,500.

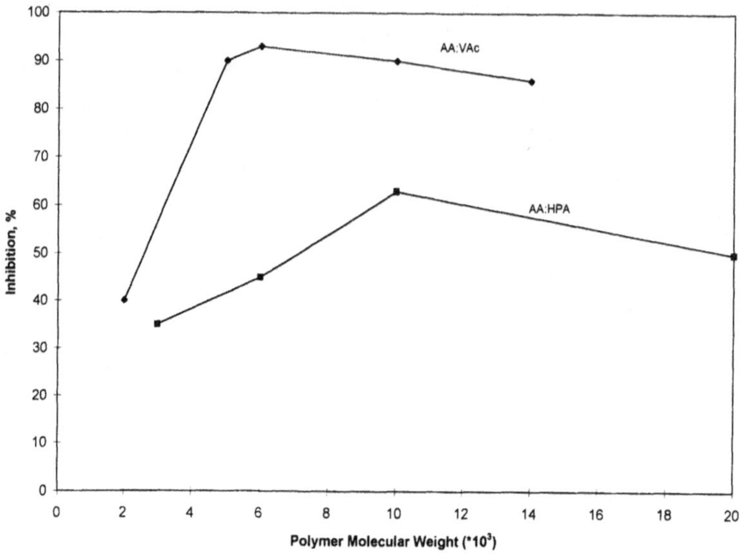

Figure 5. Calcium phosphate precipitation in the presence of AA:HPA and AA:VAc copolymers. Plots of % inhibition as a function of polymer molecular weight (140 mg/ℓ Ca, 9 mg/ℓ PO_4^{3-}, pH 8.50, 50°C, 20 hr, 10 ppm polymer).

Howie-Meyers et al.,[21] in their study on the crystal growth inhibition of hydroxyapatite (HAP) by polycarboxylates, reported that the optimum MW range for crystal growth inhibition of HAP appears to be about 1,000 to 2,000 for acrylate/maleate copolymers and slightly higher for poly(acrylates). The reason for the existence of an optimum MW range is attributed to the maximum adsorption on HAP of polymers in that MW range.

It is interesting to note that whereas ~2,000 is an optimum molecular weight for PAA, copolymers where COOH groups have been partly replaced with other bulkier groups showed optimum performance at much higher molecular weight. This suggests that both the interaction between calcium on the calcium phosphate nuclei and -COOH group as well as the ionic charge of the polymer play important roles in the overall inhibitory activity of the polymer.

Effect of polymer composition

The effect of polymer composition on calcium phosphate precipitation was investigated by a series of precipitation experiments conducted under similar experimental conditions in the presence of copolymers of varying compositions. Excellent examples of substitution of COOH group with other monomers containing different functional groups are shown in Figure 6, in which the percent inhibition is plotted as a function of polymer composition for AA:VAc and AA:DACAM polymers. As illustrated in Figure 6, poly(acrylic acid), PAA, shows very poor inhibitory activity (~10 % inhibition). However, as AA content is replaced with VAc, the inhibitory power of the polymer increases, reaching an optimum between 90:10 to 80:20 polymer compositions and decreasing therafter. A similar performance-composition profile for a polymer containing a different functional group (diacetone acrylamide) is also observed (Figure 6).

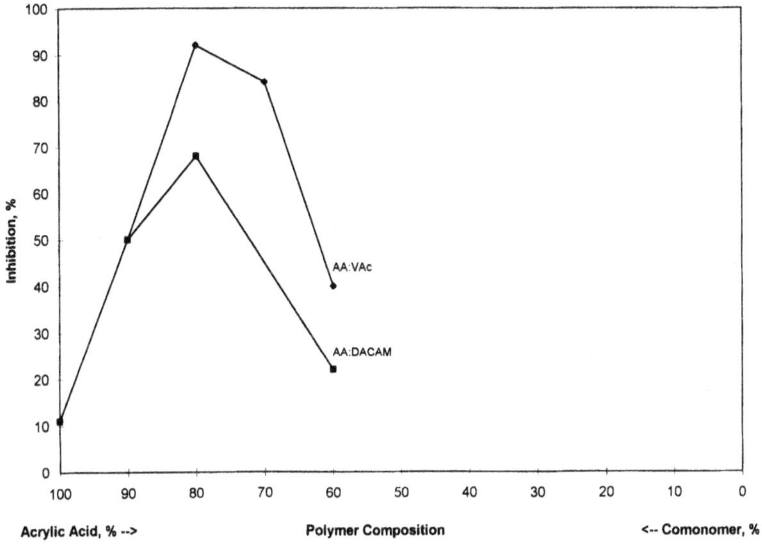

Figure 6. Calcium phosphate precipitation in the presence of AA:VAc and AA:DACAM copolymers. Plots of % inhibition as a function of polymer composition (140 mg/ℓ Ca, 9 mg/ℓ PO_4^{3-}, pH 8.50, 50°C, 20 hr, 10 ppm polymer).

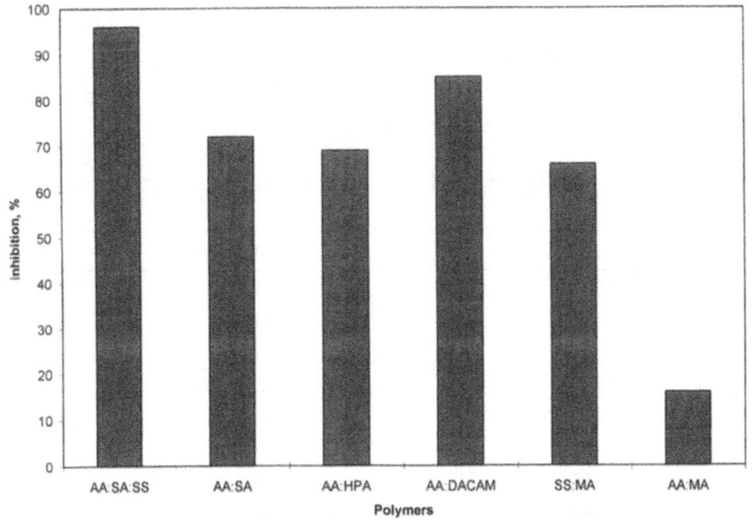

Figure 7. Calcium phosphate inhibition data for various copolymers (140 mg/ℓ Ca, 9 mg/ℓ PO_4^{3-}, pH 8.50, 50°C, 20 hr, 10 ppm polymer).

A comparison of calcium phosphate inhibition provided by a variety of copolymers is shown in Figure 7. Clearly, the best inhibition is provided by AA:SA:SS, a terpolymer of acrylic acid, sulfonic acid, and sulfonated styrene, and the least is provided by AA:MA. Other copolymers show moderate activity at equal concentrations. The effectiveness of AA:SA:SS as a calcium phosphate inhibitor is surprizingly better than AA:SA polymer, suggesting that both the ionic charge as well as the chain length or the bulkiness of the comonomers exhibit positive influence in improving the overall inhibitory power of the polymer. From the data presented in Figure 7, the ranking of the polymers (in terms of decreasing effectiveness) is:

AA:SA:SS>AA:DACAM>AA:SA>AA:HPA>SS>MA>>AA:MA

Hann and Robertson[22] have described the importance of polymer structure to effectiveness of dispersion of particulates. For a polymer to effectively disperse particulates, the polymer must adsorb on the particle surface and provide steric or electrostatic particle-particle interaction so that agglomeration does not occur. The authors suggested that a terpolymer containing AA, SA, and nonionic groups as being ideally suited for dispersion of iron. The weakly acidic acrylic acid units adsorbs onto particulates, while the the strongly acidic and ionized sulfonic acid groups provide chain extention into bulk water and enhance electrostatic repulsion. The nonionic groups provide steric repulsion by preventing the polymer molecule from being adsorbed flat against the surface of the particle. This in turn may force charged groups further away from the particle surface and increase particle-particle repulsion.

The importance of the steric factors can be seen in simple model copolymers of acrylic acid and acrylate esters. By varying the nature of the acrylate ester, one can change the way the copolymer adsorbs onto calcium phosphate forming micronuclei and can influence the efficiency of the polymer. Table 2 shows the effectiveness of a series of polymers as calcium phosphate inhibitors under a standard test conditions.

Table 2. Calcium phosphate inhibition vs. polymer structure for acrylic acid-acrylate ester copolymers.

polymer	composition	mol.wt.	% Inh.
PAA	100	5,100	16
AA:MeA	80:20	6,000	78
AA:Ea	80:20	6,000	77
AA:nPrA	80:20	6,000	50
AA:iPrA	80:20	6,000	52
AA:nBuA	80:20	6,000	30
AA:iBuA	80:20	6,000	30
AA:sBuA	80:20	6,000	38
AA:tBuA	80:20	6,000	51

MeA, methyl acrylate; Ea, ethyl acrylate; nPrA, n-propyl acrylate; iPrA, i-propyl acrylate; iBuA, i-butyl acrylate; sBuA, s-butyl acrylate; tBuA, t-butyl acrylate

From Table 2, it is apparent that the acrylic acid homopolymer is not particularly effective as a calcium phosphate inhibitor. However, when acrylic acid is copolymerized with lower alkyl acrylate (or methacrylate) esters, performance is dramatically enhanced. The performance is directly related to the steric bulk and hydrophobic nature of the comonomer. The methyl and ethyl groups provide the optimum steric bulk in this series. Particularly telling is the butyl series where the more sterically bulky, but relatively less hydrophobic, t-butyl acrylic gives substantially higher performance than the less sterically demanding, but more hydrophobic n-butyl and isobutyl acrylates. Sec-butylacrylate is intermediate in performance.

Iron Sensitivity

The presence of iron in cooling water systems, whether originating from the feed water or as a result of iron-based metal corrosion in the system or the carry-over from the clarifier operating on iron-based flocculating agents, can have a profound effect effect on the performance of polymers and the CWT program. It should be noted that in addition to inhibiting calcium phosphate precipitation and dispersing iron oxide, the ability of the polymer to resist the negative effect of iron plays an important role in the overall selection of the polymer for an effective CWT program.

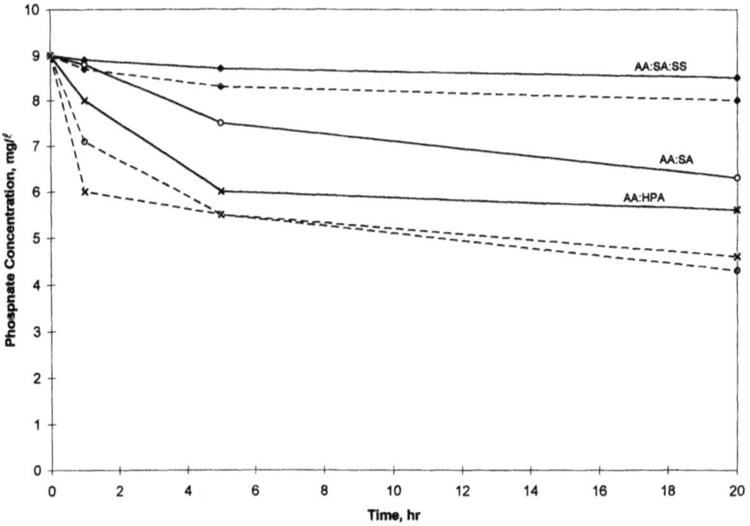

Figure 8. Effect of 1.0 ppm iron (III) in calcium phosphate precipitation. Plots of phosphate concentration as a function of time for ter- and copolymers (140 mg/ℓ Ca, 9 mg/ℓ PO_4^{3-}, pH 8.50, 50°C, 10 ppm polymer).

The iron sensitivity test for several calcium phosphate inhibiting polymers were performed in the presence of 0 and 1.0 ppm of iron (III). The inhibition data for these polymers illustrated in Figure 8 clearly demonstrates that terpolymer (i.e., AA:SA:SS) exhibits better tolerance to iron than the copolymers of AA:HPA and AA:SA. For example, residual phosphate concentrations obtained at 10 ppm polymer concentration in the presence of 0 ppm of iron (III) and 10 ppm polymer concentrations are 8.8, 6.7, and 5.8 mg/ℓ for AA:SA:SS, AA:SA and AA:HPA, respectively, compared to 8.3, 4.4, and 4.6 mg/ℓ phosphate concentrations obtained in the presence of 1 ppm of iron. In other words, loss in phosphate concentration in the presence of 1 ppm iron for terpolymer was 5% compared to 33% and 21% observed for copolymers of AA:SA and AA:HPA. From a practical point of view, if iron is encountered in cooling water systems, a polymer which provides better overall performance in reducing the calcium phosphate scaling potential, especially in the presence of soluble iron should be considered in developing the phosphate based CWT programs.

Hardness Sensitivity

Recently, from a water conservation point of view, there has been an increasing trend in operating cooling water systems at high cycles of concentrations resulting in greater driving force for the precipitation of scale forming slats. To combat this increasing scale potential, a high performance polymer is required in the CWT formulations in order to prevent the deposition of mineral scales on the heat exchanger sufaces. The tendency for calcium phosphate precipitation increases with increasing hardness (i.e., calcium ion) in the recirculating water. Another measure of the performance of a calcium phosphate polymer is its efficacy in high hardness water.

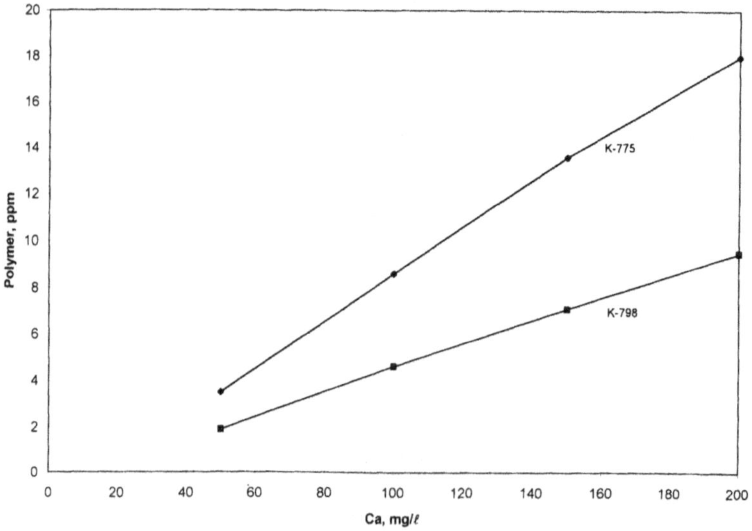

Figure 9. Effect of polymer concentration on calcium phosphate concentration. Plots of ppm polymer needed to achieve >90% inhibition as a function of calcium concentration (9 mg/ℓ PO_4^{3-}, pH 8.50, 50°C, 20 hr).

In Figure 9, the influence of calcium hardness is demonstrated in low hardness (50 mg/ℓ), medium hardness (100 mg/ℓ) and high hardness (200 mg/ℓ). As illustrated in Figure 9 amount of polymers (i.e., AA:SA and AA:SA:SS) required to achieve >90% inhibition increases with increasing calcium ion concentrations. For example, increasing the calcium ion concentration from 50 mg/l to 200 mg/l results in ~ five fold increase in AA:SA polymer concentration to maintain > 90% phosphate in solution. Similar trend was also observed in the case of AA:SA:SS terpolymer. The data shown in Figure 9 also indicates that in the calcium ion concentration studied terpolymer performance was always better than the copolymer. Thus, in cooling water applications where potential for high calcium hardness exists, the incorporation of a polymer which can perform under stressed conditions into CWT formulation will ensure better overall performance of the system.

Cationic Flocculant Sensitivity

Iron- and aluminum- based flocculating/coagulating agents (e.g., ferric sulfate, ferric chloride, alum, etc.) have been used for years to clarify influent water. These compounds neutralize the charge of the colloidal particles in the water, and they hydrolyze to form insoluble precipitates that, in turn, entrap additional particles. In most cases these large particles (flocks) are removed via settling in a clarifier and are recollected as sludge. Occasionally, clarifier upsets cause the metal-ion-containing particles to carryover. In such instances, iron- or aluminum-based deposits can form on the heat exchangers. Although aluminum- and iron-based compounds show good properties in terms of clarifying the water, the optimum performance of these compounds is very sensitive to water pH and alkalinity. The development of organic polymers has overcome the problems associated with the application of inorganic flocculant/coagulant. Among the cationic organic polymers, DMAC (diallydimethylammonium chloride) is the most commonly used polymer for water clarification. It is known that low levels of cationic polymer, if present in the recirculating water, can interact with the highly anionic charged calcium phosphate inhibiting polymer to form an insoluble polymer salt.

The results of calcium phosphate precipitation experiments carried out in the presence of 1.5 ppm DMAC concentration and under similar test conditions (240 mg/ℓ Ca, 15 mg/ℓ PO_4, 20 ppm polymer , pH 9.0, 50°C) are shown in Figure 10 . As it can be seen, 1.5 ppm DMAC has a pronounced effect on the performance of AA:SA and AA:SA:SS polymers. For example, residual phosphate values obtained in the presence of 0 and 1.5 ppm DMA are 14.4 and 13.3 mg/ℓ (Figure 10 curves A,C) compared to 12 and 1.5 ppm obtained in the presence of 1.5 ppm DMAC. The calcium phosphate inhibition data on co- and terpolymers presented in Figure 10 clearly shows that terpolymer (curves A, B) is more tolerant to DMAC than the copolymer of AA:SA (curves C, D).

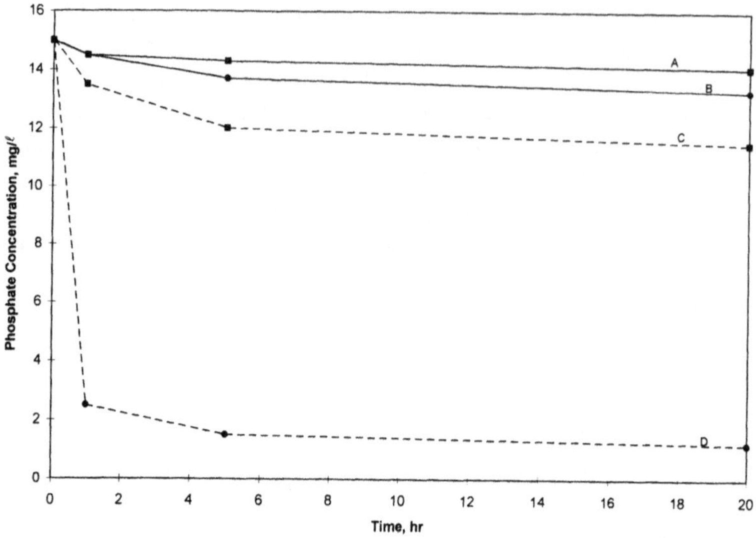

Figure 10. Effect of cationic polymer (DMAC) on the performance of AA:SA and AA:SA:SS polymers. Plots of phosphate concentration as a function of time in the presence of 1.5 ppm DMAC (240 mg/ℓ Ca, 15 mg/ℓ PO_4^{3-}, pH 9.00, 50°C, 20 hr).

Phosphonate Sensitivity

Phosphonates are a family of compounds widely used in many industrial applications. These phosphonates differ structurally from polyphosphates in that they have a P-C bond rather than P-O bond. The structural differences account for their superior stability under extremes of pH and temperatures. Phosphonates exhibit excellent performance in preventing the precipitation of calcium carbonate and many other sparingly soluble salts. In addition, phosphonates are also effective in controlling mild steel corrosion. In many CWT programs, especially all-organic, phosphonates are key components and play an important role in providing protection against corrosion and calcium carbonate scale formation on heat exchangers. It is well known that under stressed conditions (i.e., high temperature, high hardness, high pH, etc.), phosphonates can react stoichiometrically with the calcium ion leading to calcium phosphonate precipitation.[23-25] In addition, solution phosphonate concentration can be depleted due to calcium phosphonate precipitation to the extent that severe corrosion and calcium carbonate scaling can occur.

The results of the experiments to evaluate the effectiveness of polymers as Ca-HEDP inhibitors are illustrated in Figure 11. As can be seen (Figure 11) that homopolymers (i.e., PAM, PSA, etc.) which do not contain -C00H group show poor performance(< % I) compared to C00H containing homopolymer (i.e., PAA, PMA). Figure 11 also shows that, compared to homopolymers, the copolymers (i.e., AA:HPA, AA:DACAM, AA:SA, AA:DMI, AA:SA:SS, etc.) show better performance as Ca-HEDP inhibitors. The overall ranking of the polymers in terms of decreasing effectiveness is:

AA:SA:SS>AA:DACAM>AA:DMI>AA>SA>AA:HPA>>PAA

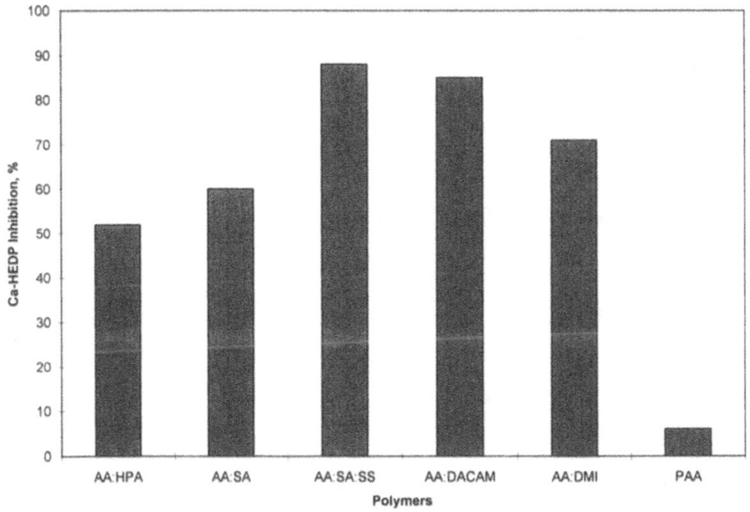

Figure 11. Ca-HEDP inhibition data for various polymers (150 mg/ℓ Ca, 15 mg/ℓ HEDP, pH 8.50, 50°C, 20 hr, 10 ppm polymer).

Hydrolytic Stability

Hydrolytic stability of water treatment polymers can be an important consideration in the choice of a water treatment polymer. A properly chosen polymer will be stable both under conditions of storage and under use conditions. In polymers used for water treatment, carbon-carbon bonds are generally quite stable. Carbon-hetroatoms bonds, however, are not stable, particularly hetroatoms attached to carbonyl. For example, poly(acrylic acid) is considered hydrolytically stable. Poly(acrylamide) is not, and, for example, under alkaline conditions, degrades to ammonia and salt of poly(acrylic acid). Acrylate esters, too, are subject to hydrolysis under both acidic and basic conditions. Yet, acrylamide polymers and copolymers containing acrylate esters are used successfully in many cooling water treatment applications if formulations and use conditions are kept within the limits required by each polymer system. Particularly desireable are polymers which are stable to alkaline conditions because these conditions allow for neutral solubility of several ingredients into one-drum formulation.

Table 3 gives some generic hydrolytic stability guidelines.

Table 3. Hydrolytic Stability of Various Polymerized Monomers

Polymerized Monomer	Structure	Stable Under Alk. Conds	Comments
Acrylic Acid	$CH_2 = CH$ COOH	Yes	No hydrolytically labile bonds
Maleic Acid	$CH_2 = CH$ COOH COOH	Yes	No hydrolytically labile bonds
Styrene Sulfonic Acid	$CH_2 = CH$ 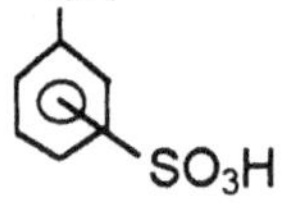	Yes	No hydolytically labile bonds
Acrylate Ester	$CH_2 = CH$ COOR	No	Ester group hydrolyzes to alcohol (ROH) and acrylate salt
Acrylamide	$CH_2 = CH$ $CONH_2$	No	Amide group hydrolyzes to ammonia and acrylate salt
Substituted Acrylamide	$CH_2 = CH$ CONHR	Maybe	Some acrylamide are resistant to hydrolysis

Of course, hydrolysis may or may not affect polymer performance. A case in point is the copolymer of acrylic acid with vinyl acetate. The inhibition data for several acrylic acid and vinyl acetate copolymers and acrylic acid homopolymer under standard suggest that optimum performance is obtained with 80:20 coplymer. This polymer was studied for hydrolytical stability at 22 ^{o}C and pH = 2.5. As shown in Figure 12 the performance of the polymer degrades with storage time. For example, hydrolyzed polymer lost ~ 80% of its inhibitory activity when tested for calcium phosphate inhibition under standard test conditions. It interesting to note that the performance of hydrolyzed AA:VAc was only slightly better than PAA indicating that most of the vinyl acetate activity was lost due to hydrolysis. The hydrolysis of the AA:VAC copolymer to a copolymer of acrylic acid and vinyl alcohol was confirmed by instrumental method.

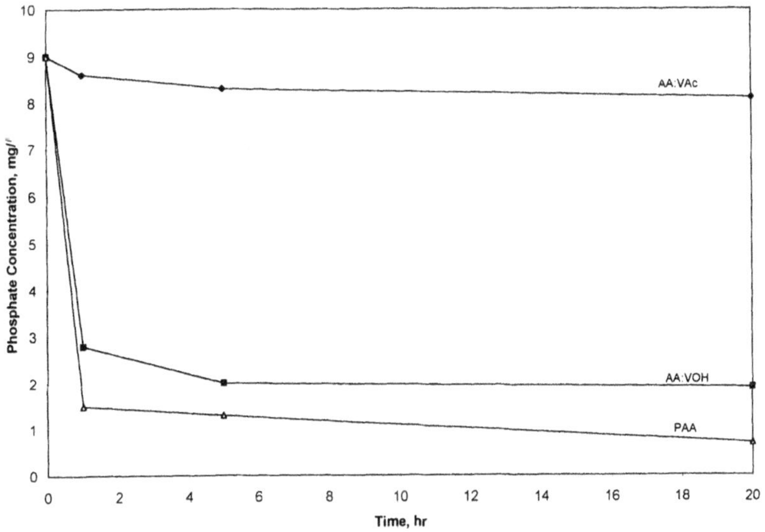

Figure 12. Effect of hydrolysis on the performance of AA:VAc as calcium phosphate inhibitor. Plots of phosphate concentration as a function of time for AA:VAc, AA:VOH, and PAA (140 mg/ℓ Ca, 9 mg/ℓ PO_4^{3-}, pH 8.50, 50°C, 10 ppm polymer).

Calcium Ion Tolerance

Many polymers form insoluble salts under conditions frequently encountered in cooling water systems. The deposition of these insoluble calcium salts can impede heat transfer and form the basis of fouling, corrosion, and biological growth. The trend toward the operation of cooling water systems under increasingly severe conditions (e.g., high hardness, high alkalinity, and higher pH and temperature) have favored the formation of insoluble calcium-polymer salts. For this reason, the calcium ion tolerance or the ability of a polymer to remain soluble in the presence of high hardness ions is of increasing importance. When polymer forms an insoluble salt with calcium ion, not only does it give additional scaling problems but it will also will decrease the effective concentration of polymer, thus leading to calcium phosphate precipitation and corrosion problems.

Figure 13 presents calcium ion tolerance data for several polymers commonly used in water treatment formulations. As illustrated in Figure 13, PAA and PMA exhibit poor calcium ion tolerance. The data in Figure 14 further indicates that copolymers of AA-SA show moderate tolerance toward calcium and some polymers (i.e., AA:HPA, AA:SA:SS, and AA:SA:NI) have excellent calcium ion tolerance. The overall ranking (from best to worst) of the calcium ion tolerance of commonly used polymers is:

AA:SA:SS = AA:SA:NI = AA:HPA > AA:SA >> PAA > PMA

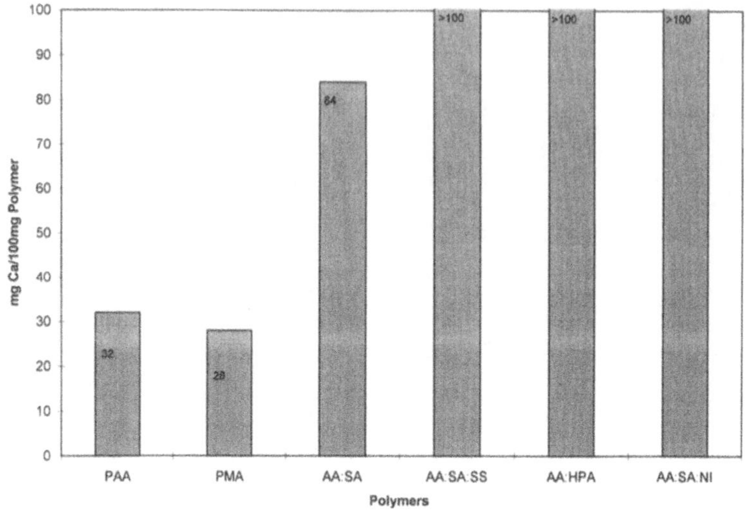

Figure 13. Calcium ion tolerance for various polymers (pH 8.0, 22°C).

From a practical viewpoint, this implies that terpolymer is better suited in hardness waters and will provide greater protection in the event of system upsets, such as overdosing.

Particulate Matter Dispersion

Fouling, as opposed to scaling, is the deposition of solids normally in suspension in the recirculating water. The largest number of suspended solids are in the form of colloids. Fouling can be of two types: natural and artificial. Natural fouling involves materials that exists in the system as a result of factors external to the system. These could be silt, mud, clay, organics, sand, iron, and debris. Artificial fouling is caused more by the system characteristics, such as corrosion products, clarifier carry-over, aluminum and iron phosphate, and microbiological growth.

The control of suspended solids can be done mechanically, chemically, or both. Chemicall control is best achieved by using dispersant. A calcium phosphate inhibiting polymer should suspend inorganic particulate matter and silt and prevent their deposition on heat exchanger surfaces. Because iron is generally considered to be one of the cooling water treatment industry's most troublesome problems, the iron oxide dispersion is normally used to compare the performance of calcium phosphate inhibiting polymers. Generally, anionic polymers are known to exhibit good dispersancy property and are believed to function by increasing the negative charge on solid particles. In addition, various factors such as pH, temperature, settling time, particle size, and water chemistry are known to affect the dispersancy power of the polymer.

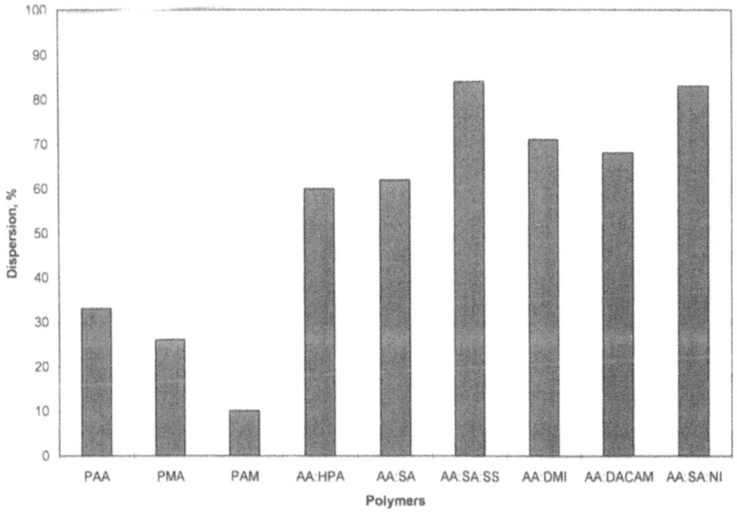

Figure 14. Performance of various polymers as iron oxide dispersants (200 ppm iron oxide, 1 ppm polymer, 3 hr, pH 7.7-8.0, 100 ppm Ca, 30 ppm Mg, 671 ppm SO_4, 60 ppm bicarbonate, 23°C).

The iron oxide dispersancy was evaluated for a number of homo and copolymers commonly used in cooling water applications. Figure 14 details the performance of several polymers under standard test conditions. As shown in Figure 14, homopolymers of AA, MA, and AM provide relatively poor dispersion of iron oxide. However, both AA:HPA and MA:SS showed very good iron oxide dispersion. It is intersting to note the best performing iron oxide dispersants are terpolymers containing AA, SA, SSA, and NI groups. The dispersancy data of these polymers suggest that the best performing dispersants are those polymers which impart more negative charge as well stabilize the iron oxide particles. Based on the dispersancy data, the effectiveness of the polymers as iron oxide dispersants follows the following order (from best to worst):

AA:SA:SS = AA:SA:NI > AA:DACAM > AA:SA > AA:HPA > PAA > PMA >> PAM

CONCLUSIONS

- The results presented in this chapter support the following conclusions regarding the performance of polymers of varying composition, molecular weight, and functional groups under a variety of test conditions commonly encountered in cooling water applications.

- Homopolymers such as poly(acrylamide), poly(acrylic acid), poly(maleic acid), and poly(2-acrylamido 2-methyl propane sulfonic acid) exhibit poor performance as calcium phosphate inhibitors

- Incorporation of bulky comonomers containing anionic and/or nonionic groups in the polymer significantly increases the inhibitory activity of the polymer. Examples of comonomers include: A. Anionic: sulfonated styrene, 2-acrylamido-2-methyl propane sulfonic acid, etc. B. Nonionic: vinyl actetate, hydroxylpropyl acrylate, dimethyl itaconate, substituted acrylamides, etc.

- Polymer concentration is critical to the performance of calcium phosphate inhibiting polymer. An increase in co- or terpolymer concentration results in a significant increase in the inhibition of calcium phosphate precipitation. The ability of polymer to perform at low concentration is an important selection criteria.

- The efficacy of a polymer strongly depends upon molecular weight, with optimum performance achieved in the 10,000 to 15,000 molecular weight range.

- Calcium ion concentration significantly impacts the ability of a polymer to inhibit calcium phosphate precipitation. The higher the calcium ion concentration in the calcium phosphate supersaturated solutions, the greater the driving force and the poorer the polymer performance.

- The presence of iron (III) decreases the performance of calcium phosphate inhibiting polymers. Co- and terpolymers containing sulfonic and/or substituted acrylamides and ester groups exhibit varying degrees of tolerance to the presence of iron.

- The presence of cationic flocculating polymer (i.e, diallyldimethy ammonium chloride) adversely impacts the inhibitory performance of copolymers. The incorporation of sulfonated styrene in a terpolymer greatly improves its tolerance to cationic polymer.

- Co- and terpolymers, compared to homopolymers, exhibit good to excellent dispersancy power for iron oxide particles.

REFERENCES

1. *Water Formed Scale Deposits*, Cowen, J. C. and Weintritt (Eds.), Gulf Publishing Company, Houston, Texas, 1976.

2. *Reverse Osmosis: Membrane Technology, Water Chemistry, and Industrial Applications*, Amjad, Z. (Ed.), Van Nostrand Reinhold , New York, NY, 1992.

3. Freedman L. Phosphate-Based Corrosion Inhibitors, Hercules Incorporated, Technical Report, Wilmington, Delaware, 1982

4. Zuhl R, Amjad Z, Masler W. A novel polymeric material for use in minimizing calcium phosphate fouling in industrial water systems, Cooling Tower Institute, 1987 Annual Meeting, Paper No. TP-87, Houston, Texas, 1987

5. Hoots JE and Crucil, GA. Role of polymers in the mechanisms and performance of alkaline cooling water programs, CORROSION/86, Paper No. 13, NACE, Houston, Texas, 1986

6. Fivizzani KP, Dubin L, Fair BE, Hoots JE. Managanese Stabilization by Polymers for Cooling Water Systems, CORROSION/89 Paper No. 433, NACE, Houston, Texas, 1989

7. Amjad Z, Pugh J, Zibrida J, Zuhl R. Polymer performance in cooling water: the influence of process variables, Materials Performance 1997;36:32-38

8. Smyk EB, Hoots JE, Fivizzani KP, Fulks KE. The design and application of polymers in cooling water programs, CORROSION/88, Paper No. 14, NACE, Houston, Texas, 1988

9. Fontana MG and Greene, NR Corrosion Engineering, McGraw Hill, New York, 1978

10. Amjad Z. Constant composition study of dicalcium phosphate dihydrate crystal growth in the presence of poly(acrylic acid), Langmuir, 1989;5:1222-1225

11. Amjad Z. Performance of polymeric additives as hydroxyapatite crystal growth inhibitors, Phosphorous Research Bulletin, 1995;5:1-12

12. Amjad Z. Inhibition of Calcium Fluoride Crystal Growth by polyelectrolytes, Langmuir, 1991;7:2405-2408

13. Amjad Z. Kinetics of crystal growth of calcium sulfate dihydrate: the influence of polymer composition, molecular weight, and solution pH, Can. J. Chem. 1988;66:1529

14. Amjad, Z. Inhibition of barium sulfate precipitation: effects of additives, solution pH, and supersaturation, Water Treatment, 1994;9:47-56

15. Amjad Z and Masler W. The inhibition of calcium sulfate dihydrate crystal growth by polyacrylates and the Influence of Molecular Weight, CORROSION/85, Paper No. 357, NACE, Houston, TX 1985

16. Amjad Z. Calcium sulfate dihydrate (gypsum) scale formation on heat exchanger surfaces: the influence of scale inhibitors, J. Colloid Interface Sci. 1988;123:523-536

17. Amjad Z. Effect of precipitation inhibitors on calcium phosphate scale formation, Can J of Chem 1988;67:850-856

18. Amjad Z. Performance of polymers as precipitation inhibitors for calcium phosphonate, Tenside, 1997;2:102-107

19. Williams FV and Ruehrwein RA. Effect of polyelectrolytes on the precipitation of calcium carbonate, J Amer Chem Soc 1957;79:4898-4900

20. Wilson D. Influence of Molecular Weight on Selection and Application of Polymeric Scale Inhibitors, CORROSION/94, Paper No. 48, NACE International, Houston, TX 1994

21. Howie-Meyers CL, Yu K, Elliott D, Vasudevan T, Aronson MP, Ananthapadamanabhan KP, Somasundaran P. "Crystal Growth Inhibition of Hydroxyapatite by Polycarboxylates: Role of Calcium and Polymer Molecular Weight". In: *Mineral Scale Formation and Inhibition,* Amjad, Z. Ed., Plenum Publishing Corporation, New York, NY 1995.

22. Hahn WM and Robertson ST. Control of iron and silica with polymeric dispersants, Paper No. IWC-90-29, Fifty-First Annual Meeting, International Water Conference, Pittsburgh, PA 1990

23. Hogue RD. Sensitivity of the Dequest^{R} phosphonates to elevated levels of calcium, Report No. 8433, Monsanto Company, September 13, 1976

24. Browning FH and Fogler AS. Effect of precipitation conditions on the formation of calcium-HEDP precipitates, Langmuir 1996;12:5231-5238

25. Masler WF and Amjad Z. Advances in the control of calcium phosphonate with a novel polymeric inhibitor, CORROSION/88, Paper No. 11, NACE, Houston, TX1988

17

MECHANISM OF CALCIUM PHOSPHATE SCALE FORMATION AND INHIBITION IN COOLING SYSTEMS

Libardo A. Perez

BetzDearborn Water Management Group
Trevose, Pennsylvania 19053 USA

INTRODUCTION

Cooling systems are an integral part of an industrial plant overall process. The cooling system removes heat from heat transfer surfaces and expels it into the air, allowing the process to run smoothly. There are three types of cooling water systems: once-through, open recirculating and closed recirculating systems.

Once through systems use cooling water on a one-time basis prior to discharge. These systems uses large amounts of water to remove heat from the process streams. They are mainly used in plants that are near to rivers, lakes and other inexpensive source of surface water. Once trough systems have the advantage that evaporation does not take place and the amount of dissolved solids remains essentially the same as the supplied water. Any potential for scale formation results from the increase in temperature of the cooling water. On the other hand, corrosion in these systems is primarily the result of relatively low pH values and dissolved oxygen and the presence of corrosive contaminants that may be present in the water. Figure 1 presents the schematic of a once through system.

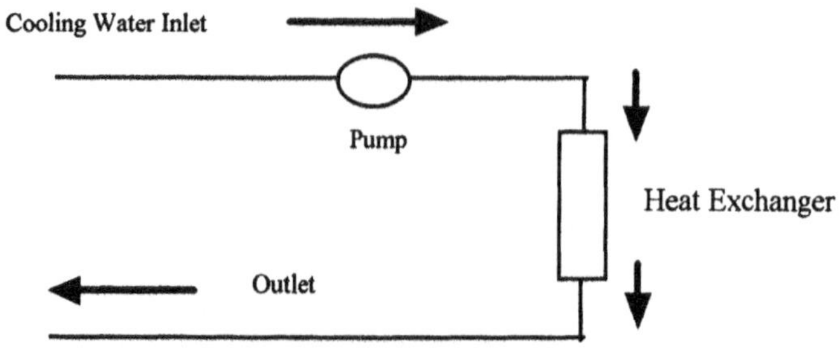

Figure 1. Representation of a Once-Through System.

Once-through systems are extensively used in the power generation industry. However, due to environmental regulations and the desired to control thermal pollution, these type of systems are becoming less common.

Open recirculating systems reuse water by recycling it across a cooling tower. In the cooling tower, conduction and evaporation remove heat from the cooling water so that the water can return to the system to repeat the process. With evaporation comes the need to replenish the water removed from the system (makeup water). Because of the evaporation that takes place, the concentration of dissolved solids in the recirculating water increases. This creates a numbers of potential problems. These problems are generally related to corrosion, scale or fouling which can occur within the cooling system. Makeup water is also added to replace water that is removed from the system (Blowdown) either in order to keep a desired level of concentration of total dissolved solids in the system and/or any loss of the water in pump glands or drift. A representation of an open recirculating system is given in Figure 2.

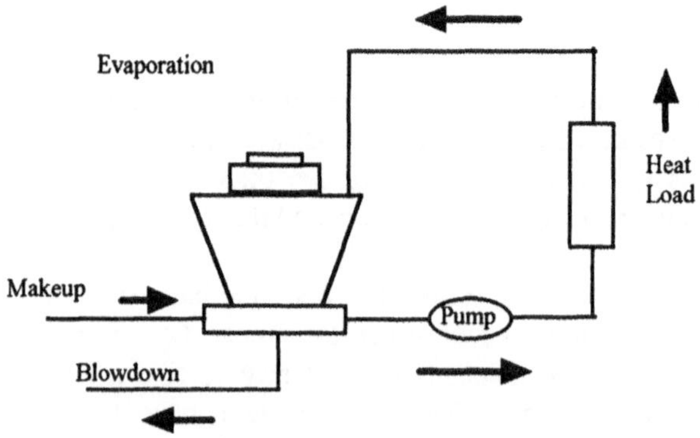

Figure 2. Representation of an open recirculating system.

Closed recirculating systems (Figure 3) recirculates a fixed volume of water in a closed loop. The heat removed from the heat exchanger surface is absorbed by the cooling water. The resulting higher temperature water is then cooled down by circulating the water back through another exchanger, which is cooled by other methods. The only makeup to a closed water system is to replace the amount loss via leaks. Due to the lack of evaporation, the potential for scaling in this type of system is very low unless very high hardness water is used as makeup. Corrosion, however, is the main problem. In addition, corrosion by-products can cause fouling on the heat exchanger surfaces.

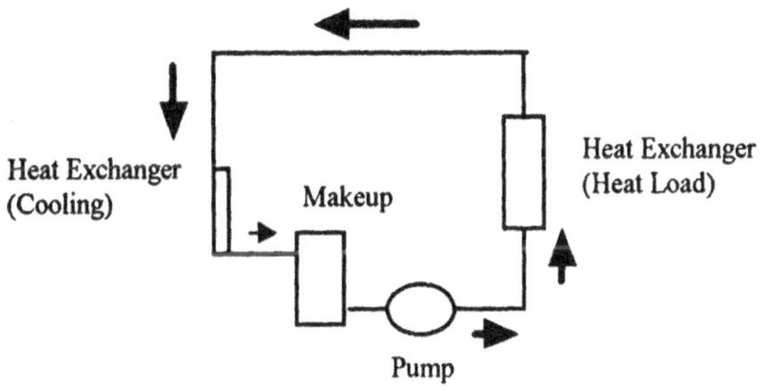

Figure 3. Schematic of a closed recirculating system.

The water circulating through a cooling system could become supersaturated with respect to several scale forming salts as, for example, calcium carbonate, calcium phosphate, calcium phosphonate, and calcium sulfate. This supersaturation state with respect to some of these salts could be the result of several parameter such as pH, temperature, or evaporation. Supersaturated waters have a tremendous potential for precipitation and subsequent scale formation due either to heterogeneous nucleation on the cooling system surfaces or to the adherence of bulk precipitated crystals to the system surfaces. The problem increases dramatically due to the fact that most scale forming salts, such as calcium carbonate and calcium phosphate, have a decreased solubility at elevated pH and/or temperature.

Calcium phosphate scale results from the presence of orthophosphate in the circulating water. The main source for phosphate in the circulating water is its use as a chemical treatment to prevent corrosion on low carbon steel water pipes and heat exchanger equipment. The mechanism for corrosion inhibition by orthophosphate is not very well understood. Some authors claim that orthophosphate act primarily as an anodic inhibitor.[1,2] Others believe that orthophosphate exhibits both anodic and cathodic protection.[3] It has been suggested that orthophosphate precipitates with calcium and, to some extent, with ferric ions to form a barrier film on the surface. Because this is a precipitated barrier film, the potential for calcium phosphate scale exits. If not properly controlled, excessive calcium phosphate precipitation can result. To prevent calcium phosphate scale formation, the use of calcium phosphate inhibitor is required when using this type of corrosion control.

Other water sources also contribute to increased levels of phosphate in cooling systems. Surface waters used as makeup could have high content of phosphate coming from agricultural runoff or for sewage plants. In addition, ortho-phosphate can be generated in cooling systems treated with phosphonate as scale or corrosion inhibitors. Phosphonates revert to orthophosphate in system treated with oxidizing biocides.

MECHANISMS OF SCALE FORMATION IN COOLING SYSTEMS

Deposition is one of the major problems encountered in cooling-water systems. Deposition on the heat transfer surfaces, piping and cooling tower fill surfaces can cause under deposit corrosion, increased pressure drop on water side across heat exchangers and pipes, and loss of heat transfer efficiency. Deposition problems are divided in two types, scale formation and fouling. Scaling is the precipitation from solution of sparingly soluble salts on the surfaces of the cooling water system. Fouling in cooling water systems is the deposition on the surfaces of suspended solids present in the circulating water, other than sparingly soluble salts. Fouling can include, among other things, corrosion products, biological growths, dirt, dust, and water-borne sediments. The two terms, fouling and scaling, are often used interchangeably.

A necessary condition for precipitation to occur is the attainment of a supersaturation state.[4] Normally, the solubility of a salt in water increases with increasing temperature (i.e. sodium chloride), however, the solubilities of most scale forming salts in cooling water systems decrease with increasing temperature (inverse solubility) and/or pH (i.e. calcium carbonates and calcium phosphates). In cooling systems, a supersaturation state can be reached by just an increase in the temperature of the water with no evaporation. This is the case of once-through systems in which the water temperature is increased by adsorbing heat from the heat exchangers and the water is immediately discharged. In open recirculating systems, supersaturation is reached by the combined effects of increased temperature from heat exchange and increased concentration of dissolved salts from evaporation in the cooling tower.

Attainment of a supersaturation alone, however, is not sufficient for a system to begin precipitation. It has been observed that some supersaturated solutions may remain for long periods of time without forming a new phase. Such solutions are termed metastable and the corresponding region the metastable zone.[5,6] The life of the metastable solution and the width of the metastable zone are inversely proportional to the supersaturation and are also dependent on other factors such as degree of agitation, ionic impurities, and ultrasonic radiation. At supersaturation levels higher than the metastable region, the solution becomes unstable and a new phase is formed. The stability of this new metastable phase depends upon two consecutive steps: A. overcoming the free energy that exists, which stabilizes the phase, and B. changing to an state of lower energy and consequently of larger stability. The first step is the process of nucleation and the second step is the process of crystal growth.

Nucleation

During nucleation the smallest stable molecular aggregate (embryos) of a crystalline phase is formed. This process controls the size, number, structure and morphology of the precipitated crystals and as such it also plays an important role in determining a number of their chemical and physical properties.

The free energy changes in the nucleation process are such as that the spontaneous growth of the molecular aggregates is only allowed if they have achieved certain critical size. The overall free energy goes through a maximum at some critical size r*. Embryos in solution will tend to decrease their overall free energy by changing their size. Those smaller than the critical size achieve this by redissolving, and those larger than r* will continue to grow.[7,8] Therefore, as it is shown in Figure 4, the minimum size of a stable nucleus is the critical cluster size, and this is the smallest cluster that can decrease its free energy by growing.

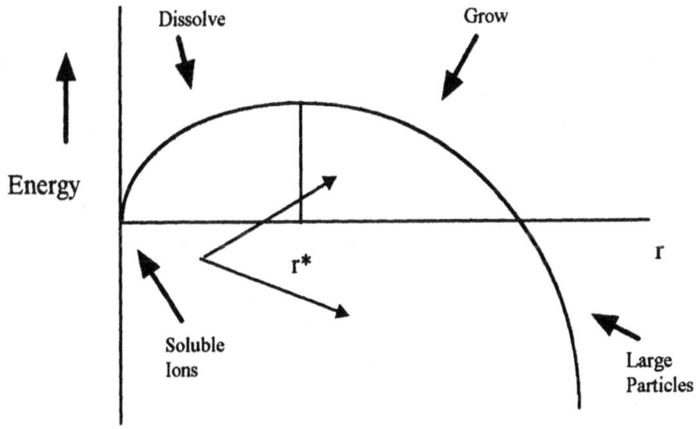

Figure 4. Profile of the minimum size for the formation of a stable nucleus.

If nucleation takes place in perfectly clean homogeneous solutions in the absence of a surface or foreign particles (ions, dust, wall surfaces, suspended solids, etc.), it is referred to as homogeneous nucleation. When foreign bodies are present so as to assist the system in overcoming the activation barrier imposed by surface free energy considerations, the process is termed heterogeneous nucleation. Primary nucleation is the term given to the two processes to distinguish them from the secondary nucleation process in which nuclei are often generated in the vicinity of parent crystals in a supersaturated solution. An schematic of the nucleation process is given in Figure 5.

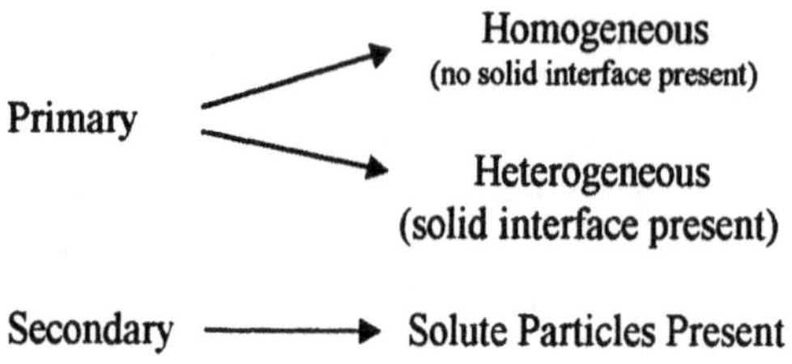

Figure 5. The nucleation process.

Homogeneous nucleation is practically impossible to occur in real systems since foreign bodies are always present. Heterogeneous nucleation, which is initiated by the catalytic effect of the extraneous particle surfaces, walls, grain boundaries, pores, etc., is the common mechanism in real systems. The essential conditions for heterogeneous nucleation is that the surface must be wetted by the phase formed in the presence of the mother phase. Due to the catalytic action of the foreign surfaces, heterogeneous nucleation occurs at much lower supersaturation than homogeneous nucleation , since the molecules in the substrate affect the ordering process and lower the energy barrier for nucleation. In the case of cooling systems, the walls of a pipe, the surface of heat exchangers, and the surfaces of the other components of the cooling system (i.e., tower fill) can serve as nucleating site similar to other foreign substances. As illustrated in Figure 6, corroding surfaces are ideal for scale to take place since they offer more nucleation sites. In addition, insoluble corrosion products also act as nucleation site for scale formation. Active nucleation sites can also be offered by welds and other internal stress present on the metal surfaces as well as suspended matter, such as clays, silt and biological matter.

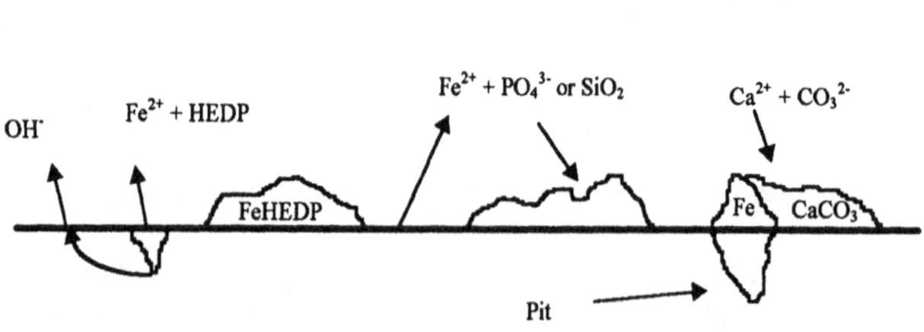

Cathodic Reaction: $2e^- + 2H_2O + O_2 \longrightarrow 4\,OH^-$

Figure 6. Deposition of salts on corroded surfaces.

Supersaturation has an enormous effect on nucleation. As the supersaturation of a solution increases, the size of the critical nucleus, the number of molecules necessary to form the cluster of critical size r*, and the free energy barrier all decrease. Below a critical value of the supersaturation, S*, the nucleation rate is negligible; once this supersaturation is reached, the rate of nucleation increases very rapidly. Below S* crystals will grow but nucleation rate is basically zero. In the absence of crystals the solution will be metastable. The nucleation rate is determined by the rate at which nuclei overcome the maximum in the free energy barrier. This free energy plays a similar role to the activation energy in chemical reactions, the difference being that the free energy barrier for nucleation is not constant and it is extremely dependent on supersaturation.

Nucleation rates are very difficult to determine. The most common method to investigate nucleation rate in solution is by measuring the induction time. This is defined as the time that elapses before the detection of crystals in solution. The problem with this method is that it is dependent on the detection system that is employed. Caution should be taken on the induction time values reported in the literature. The values of the induction time are enormously influenced by the presence of impurities in solution, by the degree of agitation of the solution and by other mechanical stimuli. In addition, the induction time also depends on the time taken to form stable nuclei and on the time necessary to grow to a detectable size. [9]

Crystal Growth

As soon the energy barrier for the formation of a stable nucleus has been surmounted, nuclei larger than the critical size will spontaneously adjust to a state of greater stability. This process is achieved by the deposition of the excess of solute from the solution to the surface of the crystals. As a result of this process, an increase of crystal size and a decrease of the concentration of the crystal lattice ions in solution are obtained. The driving force of crystallization is the free energy difference ΔG, between the solid and the supersaturated solution. [10,11,12]
This can be expressed by

$$-\Delta G_{min} = RT \ln S \tag{1}$$

where S is the saturation ratio defined by

$$S = a/a_s \tag{2}$$

where a and a_s are the mean ionic activities of the crystallizing electrolyte in supersaturated solution, and in a saturated solution, respectively. For non-equivalent solutions, S may be defined by

$$S = (\Pi a_i^{\nu} / K_s)^{1/\nu} \qquad (3)$$

where Πa_i^{ν} is the ionic product, ν_i, the stoichiometric number of ion number i in the formula for the electrolyte, $\nu = \Sigma\nu_i$, and K_s the solubility product.

The relation between the ion activity (a) and ion concentration [c] in solution is given by the expression;

$$(a) = \gamma_i [c] \qquad (4)$$

where γ_i is the activity coefficient of the ion i, and is a function of the ionic strength of the solution.

A very important factor to consider in the calculation of solubilities and saturation ratio in a mixed ionic environment is the interaction between specific ions, which can result in the formation of a soluble ion-pair or complex. The ion-pair formation reduces considerably the "free" or "ionized" concentrations of the ions forming the complex. In cooling systems, a large number of such interactions is possible, and the resulting free ion concentration or activities are impossible to measure directly. In this case, it is necessary to utilize computer programs to calculate, ab initio, the free ion concentration of all relevant ions species in solution from the knowledge of other parameters such as pH, total ion concentrations, stability constants of the formed complexes, and temperature. The calculation of the free ion concentrations also allows the calculation of the ionic strength and the various activity coefficients. By using this technique, it is possible to calculate the activity product of the ions of a given salt in solution. By comparing the value of the ionic product with the thermodynamic solubility product of the salt, an estimate of the degree of supersaturation of the solution with respect to the salt can be calculated. The degree at which a solution is supersaturated is a good measurement of the potential for scaling to occur.

An important step in the growth process is the interfacial reaction. The reaction may take place in a number of steps, but regardless of the details of these, the geometry of the growing interface will be of considerable importance. Growth units are molecules or ions. These units can occupy energetically favorable positions on the surface. On a crystal surface, terraces are separated by monatomic or monomolecular steps. Within steps, kink sites may form. Based upon nearest neighbors considerations, the most favorable position for the addition of a new growth unit is at a kink in a step, because bonds may be formed with three nearest neighbors. The next most favorable position is at the front of a step and the least favorable is a lone growth unit on the surface.

Crystal growth involve several stages, one or more of which can establish the overall rate. The nature of these steps will depend upon the system from which growth takes place. The process may include: A. diffusion of the dissolved atoms or molecules (Growth Unit) from the solution to the growth interface. B. Adsorption of the atoms or molecules on a flat terrace. C. Surface diffusion of the growth unit from the terrace to a step. D. Incorporation of the growth unit into the crystal lattice at a kink site. The overall reaction must involve these steps in addition to dehydration of the cations in solution and perhaps the penetration through the hydration layer of the crystal.

Many models have been proposed in order to describe crystal growth, but a completely successful theory has not been developed. A suitable model should be able to describe or predict the absolute value of the rate of crystal growth and its dependence on chemical and physical conditions of the system, such as fluid dynamic, area of solid surface, degree of supersaturation, and temperature. Some modern theories of crystal growth are able to correlate most of these parameters to the rate of growth, although some of the data required in calculation must be considered as adjustable parameters. In the case of sparingly soluble salts more studies suggest that these crystals grow by a surface controlled mechanism.

In cooling systems, the number of crystalline phases that can precipitate is relatively small, being limited, for the most part, to solids such as the calcium carbonates, calcium phosphates, calcium sulfates, calcium and magnesium silicates and perhaps zinc and iron phosphates. These salts, however, may precipitate in a number of different phases and polymorphs which makes the elucidation of the mechanism of their precipitation more difficult. In many cases, mixed phases may form by the growth of one phase upon another in metastable supersaturated solutions. In some cases, an epitaxial formation may take place; this involve the regularly oriented growth of a crystalline substance on another surface. It should be noted, however, that nucleation of one phase by another does not imply a true epitaxial growth.

Salts that precipitate in a number of different phases, may undergo phase transformation. In these cases, the formation of the most stable phase is usually preceded by the precipitation of one or more precursor phases. According to the Ostwald's rule of stages, the least stable phase of highest solubility is formed preferentially during a sequential precipitation.

Adhesion

Precipitation alone does not lead to scale formation. It has been suggested that scaling is the result of precipitation followed by the adhering of precipitates to surrounding surfaces.[13] This would lead to the development of nucleation sites on surfaces for subsequent crystal growth leading to scale buildup. In some instances, precipitates can be tolerated as long they do not adhere to the surfaces of the system. A particle adhere to a solid surface if mutual attraction prevails. The adhesion is caused mainly by the dispersion forces or the van der Waals forces. In addition to these forces, there are specific bonds between the atoms or ions located on two adhering surfaces, such as calcium bridging or hydrogen bonding.

The adhesion process is influenced by the interfacial properties, the composition of the solution, and by the hydrodynamic and electric forces. Perhaps the most important parameter is the sign of the surface charge; if the particle and the substrate have opposite charge, the electrostatic attraction will cause fast deposition and strong adherence. If the interacting surfaces have charge of the same sign, deposition is not expected to occur and spontaneous detachment of adhered particles is expected to occur. The surface charge is related to the activity of the potential determining ions.

Adherence of deposits tend to be more tenacious at the gas/liquid/solid interfaces like those that exists in cooling towers or in heat exchangers in which bubble formation occurs. The degree of adhesion is also related to the nature of the surface. Corroding surfaces are more likely to have extremely adherent deposits.

Flocculation

Flocculation, the aggregation or agglomeration of small suspended particles in a liquid to form large clusters, is a widespread phenomenon, occurring in a diverse range of applications including cooling water systems where precipitated minerals or other suspended materials can undergo flocculation and adhere to the system surfaces to form deposits. In general, flocculation is the reduction in particle number, whatever the mechanism involved.[14] Coagulation, is a term often used to represent this process, and in the literature it is usually used either in place of, (or together with) the term flocculation. Scientifically, coagulation implies the formation of compact clusters of particles, leading to the macroscopic separation of a coagulum. Flocculation implies the formation of a loose or open network (floc) which may or may not separate macroscopically.

There are several ways by which the number of particles in a suspension may decrease. One is a consequence of the large interfacial free energy associated with small crystals, which would tend to decrease by crystallization to a value for which the surface area is minimal. Sedimentation under gravity would also lead to a reduction in number with time, but this is only important for large dense particles since thermal agitation is normally sufficient to keep small colloidal particles dispersed. For dispersion of solid particles in liquid media the primary cause of instability is flocculation in which particles under the influence of random Brownian motion come into close contact and form clusters (flocculates) of aggregates and/or primary particles (perikinetic flocculation). In flowing systems or stirred suspensions, fluid motion can bring about particle collisions and the process is then termed as orthokinetic flocculation.

At least three major types of interactions are involved in the approach of colloidal particles, namely A. the London-van der Waals force of attraction B. The coulombic force (repulsion or attractive) associated with charged particles. C. The repulsive force arising from solvation, adsorbed layers, etc. The electrostatic charge at the surface of most particles arises either from the preferential adsorption of certain ions from solution or from the ionization of groups at the particle surface. Electrostatic forces, due to the interaction of the electrical double layer surrounding the particles, always lead to repulsion between particles if they are of the same chemical nature and have surface charge and surface potentials of the same sign and magnitude. The addition of electrolytes (higher ionic strength) reduces the energy barrier due to electrostatic repulsion between surfaces of like charges and results in a faster flocculation.

In cooling water systems flocculation must be controlled to avoid deposition on the surfaces of the cooling system. Typically this is achieved by the addition of dispersant agents to prevent particle agglomeration and adherence to metals and other surfaces present in the system. The dispersant protects the surface by affecting particle surface attractive forces through surface charge modification.

Operating Conditions Leading to Deposition

As it has been discussed, the potential for scale typically found in cooling systems increases by increasing temperature and pH. Operating systems at higher pH, however, is a practice that is increasing not only because it provides better corrosion control but also for safety considerations (acid feed elimination), and water conservation (higher cycles). Keeping corrosion under control is perhaps the most important factor in a cooling systems because it reduces the possibiliy of equipment failure and also prevents induced corrosion scale from occuring.

The presence of high concentration of suspended matter is also a potential for increasing scale formation since, as already discussed, the suspended matter may serve as substrate for nucleation and subsequent growth of scale forming salts.

METHODS USED TO PREVENT SCALE FORMATION

Scale control in cooling water systems has become more important in the last decade because of the tendency to operate cooling systems at higher cycles to conserve water, at higher pH to improve safety, and at higher temperature and heat transfer rates as a mean to maximize system efficacy.[15] There are mainly two ways to control scaling in cooling water systems: mechanically and chemically.

Mechanically, scale control can be achieved by making changes in the process, for example, operating the system at lower cycles of concentration, which lead to higher operation costs, or by decreasing the heat load. In addition, the scale forming ions can be removed from the feedwater.

Chemically, chelating agents, scale inhibitors, dispersants, corrosion and biocides are used as components of programs designed to prevent scale and corrosion control. Acids are used to keep pH low as a mean of lowering the supersaturation of the potential scale forming salts. Acids are also used to remove scale present in the system. Figure 7 summarize the methods used for scale control.

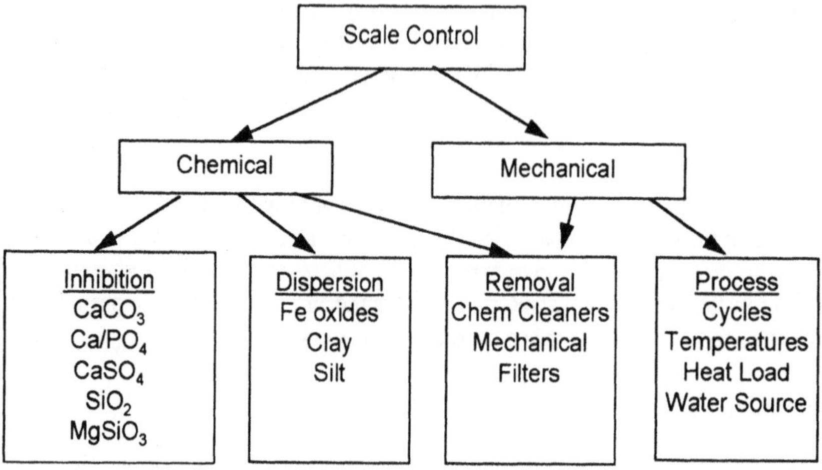

Figure 7. Scale control methods.

Chemical inhibitors to control undesirable scale in cooling water systems has been extensively investigated by numerous researchers who have developed many of them aiming to reduce or even eliminate the possibility of scale formation on all the surfaces present in the cooling system.

The use of threshold concentrations of these inhibitors has a strong effect on the rate of nucleation and subsequent growth of the crystal. Inhibitors decrease the growth rate by adsorbing on the surface of the growing crystal. They can affect the relative interfacial energies of the individual faces and block active growth sites. In addition to reducing the growth rate, inhibitors can also lead to a modification of crystal habit by affecting the growth rate of each face to a different extent. Figure 8 gives a profile of the role of the inhibitor on preventing scale formation.

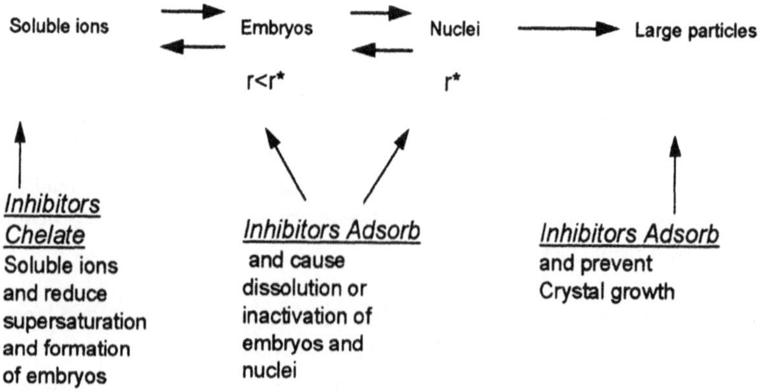

Figure 8. Inhibitor role on scale prevention.

Most scale inhibitors are anionic in nature and they range from simple chemical compounds (e.g., phosphonates) to complex natural products (starch and lignosulfonantes) and to a variety of synthetic homo and heteropolymers. In selecting a scale inhibitor, attention must be given to some parameters that may affect its performance. The effect of the use of chlorine and other oxidizing biocides on the stability of the inhibitors should be determined. The hydrolytic stability of the inhibitor should be known as well as the calcium and iron tolerance of the inhibitor and the effect of suspended solids. The compatibility of the inhibitor with the other component of the water treatment program should also be determined.

CALCIUM PHOSPHATE SCALE FORMATION

Because of the use of orthophosphate as a corrosion inhibitor and the tendency to run cooling systems at higher pH (higher alkalinity) to provide corrosion protection, calcium phosphate scale has become more common in cooling water systems. Make-up waters containing high phosphate concentrations can increase the potential for calcium phosphate deposits on the heat exchanger surfaces when the water is cycled up. Phosphate formed from the reversion of phosphonates and other organic phosphorus containing compounds also increase the potential for scale formation. Cooling waters can become supersaturated with respect to at least four calcium phosphate phases. It is often assumed that the thermodynamically most stable hydroxyapatite ($Ca_5(PO_4)_3OH$, HAP) is a suitable prototype for scale prediction. However, it is generally accepted that other phases such as amorphous calcium phosphate ($Ca_9(PO_4)_6$, ACP), dicalcium phosphate dihydrate ($CaHPO_4{\bullet}2H_2O$, DCPD) and octacalcium phosphate ($Ca_8H_2(PO_4)_6{\bullet}5H_2O$, OCP) as well as defect apatite may participate. Many kinetics studies of calcium phosphate precipitation suggest that the formation of apatite is usually preceded by one or more precursor phases (Figure 9).[16,17,18] It seems that during the precipitation of calcium phosphate, ACP is formed first followed by the nucleation of OCP, which serve as a template for HAP growth. Due to the presence of other ions in cooling waters, defect apatite can readily form. Defect apatites can be expressed in the structural formula

$$Ca_{10-w}\,M_w(PO_4)_{6-x}A_x(OH)_2 \qquad (5)$$

where M = Mn, Mg, Zn, Fe, Cu and A = SiO_4, CO_3, AlO_4, FeO_4.

Dicalcium phosphate dihydrate	$CaHPO_4{\bullet}H_2O$, DCPD
Octacalcium phosphate	$Ca_8H_2(PO_4)_6{\bullet}5H_2O$. OCP
Hydroxyapatite	$Ca_5(PO_4)_3OH$, HAP

Amorphous calcium phosphate (ACP)

ACP ⟶ Precursor ⟶ Stable Phase

Figure 9. Calcium phosphate phases.

The solubility of calcium phosphate(s) is strongly dependent on solution pH. The scaling tendency of calcium phosphate is a complex function of pH, supersaturation, calcium and phosphate ions activity, ionic strength, and temperature.

Several studies have demonstrated that the formation of the most thermodynamically stable phase, hydroxyapatite, is invariably preceded by the formation and subsequent transformation of other phases.[19,20,21] These precursors, however, can be kinetically stabilized by the presence of other species present in the solution. The transformation of one phase into another is also markedly influenced by the presence of ions that may retard or in some cases accelerate the transformation rate.[22]

PREVENTION OF CALCIUM PHOSPHATE SCALE FORMATION

The use of phosphate to control corrosion is a widely used practice, the mechanism of the process, however, is not well understood. The technique is based on a well controlled metal phosphate precipitation from solution. Inhibitors are used to prevent an extensive precipitation from solution but at the same time to allow the formation of a metal phosphate barrier film strong enough to prevent corrosion.[23,24,25]

Several studies have been conducted to understand the formation and inhibition of the most stable calcium phosphate phase (hydroxyapatite).[26-30] Many water soluble polymers with different polymer composition have been developed and studied as calcium phosphate scale inhibitors. These polymers are mainly co- or ter-polymers having acrylic acid as one of the monomers and other monomer containing sulfonic or acrylamide groups.[31-37] In addition to act as calcium phosphate scale inhibitors, some of these polymers act as dispersant of particles suspended in solution. The ability of some of these polymers to act as both scale control agents and dispersants depend markedly not only on the ratio of monomer units but also on the molecular weight of the polymer. Polymers having same mole ratio of the monomers but different molecular weight can have significantly different scale inhibitory properties.[38,39]

Development of a New Calcium Phosphate Scale Inhibitor

As discussed before, the trend in the cooling industry is to operate cooling systems at higher cycles of concentrations, higher temperature, and alkaline pH. Under this conditions, the potential for scale formation increases dramatically due to the increased supersaturation under such stressed conditions. Studies have demonstrated that terpolymers of acrylic acid/allyl hydroxy propyl sulfonate ether/polyethylene glycol allyl ether (AA/AHPSE/PEGAE), and terpolymers of acrylic acid/allyl hydroxy propyl sulfonate ether/polyethylene glycol monomethacrylate (AA/AHPSE/HEM) are effective in controlling calcium phosphate deposition at water temperature higher than those founded in typical cooling systems and at very high cycle of concentrations as those expected under zero blowdown conditions. In addition to scale control, blends of these inhibitors with orthophosphate and phosphonic acids gave corrosion protection to mild steel surface under the above mentioned stressful conditions.

The polymers that were synthesized and tested are tabulated in Table 1. They were prepared from aqueous solutions and characterized by carbon-13 NMR.[40]

Table 1. Polymer compositions and characteristics

Polymer No.	Monomers	Ratios	Viscosity	pH
1	AA/AHPSE/PEGAE$_{n=10}$	6/1/1	11.7	4.4
2	AA/AHPSE/PEGAE$_{n=4}$	6/1/1	10.4	4.3
3	AA/AHPSE/EP1404	6/1/1	11.8	4.5
4	AA/AHPSE/PEGAE$_{n=10}$	6/2/1	10.0	4.5
5	AA/AHPSE/PEGAE$_{n=4}$	6/2/1	9.7	4.5
6	AA/AHPSE/EP1404	6/2/1	11.6	4.6
7	AA/AHPSE/PEGAE$_{n=4}$	10/1/1	10.0	4.4
8	AA/AHPSE/PEGAE$_{n=15}$	10/1/1	12.5	4.5
9	AA/AHPSE/PEGAE$_{n=4}$	3/1/1	7.8	4.6
10	AA/AHPSE/PEGAE$_{n=10}$	3/1/1	9.3	4.6
11	AA/AHPSE/HEM$_{n=5}$	3/1/1	11.9	4.1
12	AA/AHPSE/HEM$_{n=10}$	3/1/1	12.7	4.3
13	AA/AHPSE/HEM$_{n=5}$	6/2/1	8.8	4.0
14	AA/AHPSE/HEM$_{n=10}$	6/1/1	10.1	4.0
15	AA/AHPSE/HEM$_{n=10}$	6/2/1	10.1	4.1
16	AA/AHPSE/HEM$_{n=5}$	6/1/1	9.1	4.0
17	AA/AHPSE/HEM$_{n=5}$	10/1/1	12.9	4.1
18	AA/AHPSE/HEM$_{n=10}$	10/1/1	14.1	4.1

EP1404= polyethylene/propylene glycol allyl ether 14=EO units ; 04=PO units
n = average number of PEGAE or HEM units in the polymer

Two types of tests were used to evaluate the effectiveness of these polymers. A static beaker test was used to screen them as inhibitors for calcium phosphate and calcium phosphonate precipitation. A laboratory open recirculating cooling system unit was used to study the effectiveness of the polymers to prevent scale and corrosion.

Table 2 summarizes static calcium phosphate inhibition results for these polymers at varying treatment levels. The tests were conducted by adding the treatment to a solution containing calcium and magnesium ions at the described conditions. Another solution containing phosphate and carbonate ions was added and the mixture incubated at 70 °C (158 °F) for 18 hours. The pH of the solution at 70 °C was 8.2. After 18 hours, a measured portion of the hot solution was filtered through a 0.22 μm membrane and analyzed for phosphorus determination by using inductively coupled plasma atomic emission (ICP). From the analysis results the percent inhibition was calculated by using the following relationship:

$$\% \text{ Inhibition} = \frac{\text{ppm PO}_4\text{ (treated)} - \text{ppm PO}_4\text{ (control)}}{\text{ppm PO}_4\text{ (stock)} - \text{ppm PO}_4\text{ (control)}} \times 100\ \%$$

The test conditions were: 400 pm Ca^{2+} as $CaCO_3$, 100 ppm Mg^{2+} as $CaCO_3$, 50 ppm Alkalinity as $CaCO_3$, 10 ppm PO_4^{3-}, pH 8.2 at 70 °C., 18 hours test duration.

Table 2. Beaker Test Calcium Phosphate Inhibition

Polymer No.	Treatment	% Inhibition* at ppm Active		
		10.0	15.0	20.0
1	AA/AHPSE/PEGAE$_{n=10}$	4 (4)	80 (4)	95 (4)
2	AA/AHPSE/PEGAE$_{n=4}$	2 (3)	48 (4)	98 (4)
3	AA/AHPSE/EP1404	20 (4)	18 (2)	88 (2)
4	AA/AHPSE/PEGAE$_{n=10}$	8 (4)	82 (4)	75 (2)
5	AA/AHPSE/PEGAE$_{n=4}$	3 (4)	83 (4)	92 (2)
6	AA/AHPSE/EP1404	19 (4)	38 (2)	90 (2)
7	AA/AHPSE/PEGAE$_{n=4}$	2 (2)	9 (2)	90 (2)
8	AA/AHPSE/PEGAE$_{n=15}$	1 (2)	3 (2)	80 (2)
9	AA/AHPSE/PEGAE$_{n=4}$	15 (4)	96 (4)	96 (4)
10	AA/AHPSE/PEGAE$_{n=10}$	19 (4)	94 (4)	95 (4)
11	AA/AHPSE/HEM$_{n=5}$	2 (2)	98 (2)	100 (2)
12	AA/AHPSE/HEM$_{n=10}$	7 (2)	97 (2)	99 (2)
13	AA/AHPSE/HEM$_{n=5}$	1 (2)	95 (2)	100 (2)
14	AA/AHPSE/HEM$_{n=10}$	0 (2)	12 (2)	94 (2)
15	AA/AHPSE/HEM$_{n=10}$	2 (2)	78 (2)	95 (2)
16	AA/AHPSE/HEM$_{n=5}$	5 (2)	63 (3)	96 (2)
17	AA/AHPSE/HEM$_{n=5}$	0 (2)	0 (2)	2 (2)
18	AA/AHPSE/HEM$_{n=10}$	0 (2)	0 (2)	2 (2)

* Average % Inhibition with number of tests in parenthesis

Recirculating test were conducted by using a laboratory cooling system (Figure 10). These recirculator units are designed to provide in the laboratory a realistic measure of the ability of a treatment to provide corrosion and scale formation under heat transfer conditions. The treated water is circulated through heat exchanger metal tubes contained in the auto-steam generation condenser units, and through a corrosion coupon by pass rack into which corrosion coupons are inserted. After passing through the heat exchangers, the water is cascaded through a cooling tower containing tower film fill. The water velocity through the exchanger can be controlled in the range of 0 to 8 ft/sec. The pH, temperature and conductivity of the system are automatically controlled. The evaporation rate can be fixed by maintaining constant the flow of air through the cooling tower.

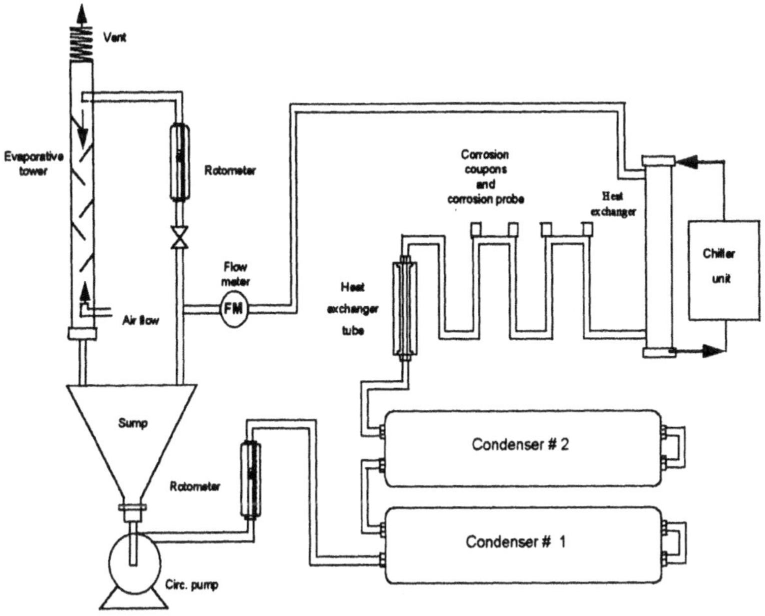

Figure 10. Schematic of a the laboratory cooling system.

Polymers number 9 and 10 (Table 1) were tested in the recirculating unit. The cycled water composition for these tests was: 600 ppm Ca^{2+} as $CaCO_3$, 300 ppm Mg^{2+} as $CaCO_3$, 50 ppm Alkalinity as $CaCO_3$, 12 ppm PO_4^{3-}. The water velocity through the condenser heat exchangers was 2.8 ft/sec and the pH was controlled to 7.2. Both terpolymers were able to prevent scale and improve the capability of orthophosphate to control corrosion on mild steel (LCS) and admiralty heat transfer surfaces in seven day test duration. The coupons corrosion rates in both cases were in the 0.6 to 0.8 mpy range.

In another test, polymer 9, at 10 ppm (active), was able to prevent scale on the LCS heat transfer tubes at 140 °F water temperature and 160 °F skin temperature in a seven day test. Corrosion rates were excellent at this relatively high temperatures, 0.83 mpy for the LCS coupons and 0.14 for the admiralty coupons.

Diffuse reflectance FTIR spectroscopy analysis conducted on the LCS heat exchanger surface (Figure 11) showed the presence of phosphate on the blue colored film that formed on the surface. The spectrum shows a band at 1042 cm^{-1} with shoulder at approximately 1100 cm^{-1} and at 950 cm^{-1}. This is consistent with those reported in the literature for bonding of orthophosphate at the iron oxide aqueous interface.[41] The spectrum also shows stretching characteristic of C-O stretching frequencies. X-ray Photo Electron Spectroscopy analysis (Figure 12) showed that the species present on the surface corresponded to PO_4^{3-}, C-(C,H), C-O, O-C=O, Ca^{2+}, FeO, and $R\text{-}SO_3^-$. These results indicate that in addition to the expected component of the film, iron, phosphate and calcium ion, the polymer is also adsorbed on the film and that may also play a role on the corrosion prevention.

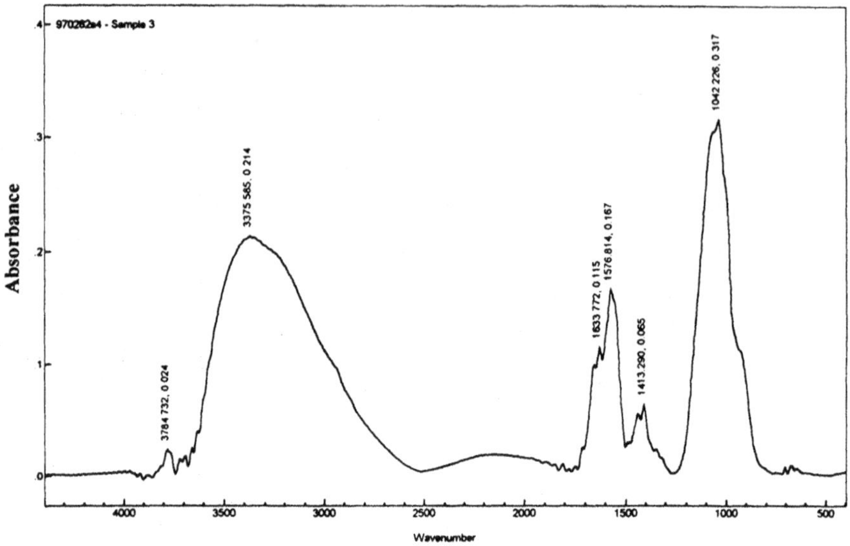

Figure 11. Diffuse Reflectance FTIR of the LCS Surface

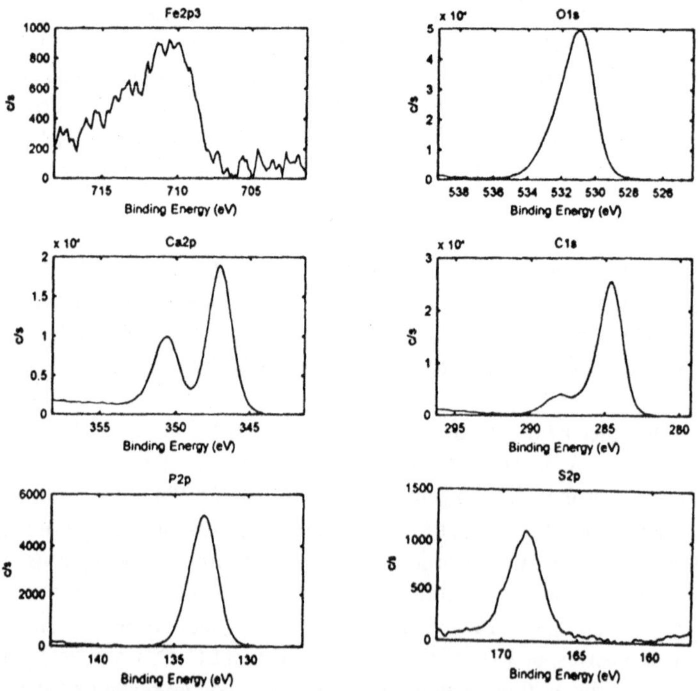

Figure 12. X-ray Photoelectron Spectroscopy Analysis of LCS Surface.

REFERENCES

1. Sullivan PJ and Hebburn BJ. Cooling towers changes in cooling water treatment programs. Industrial Water Treatment, November/December 1993;53-58

2. Several Contributors. Cooling water systems. Power 1987;103-137

3. Crucil GA, MacDonald JR, and Smyk EB. Role of polymers in the mechanism and performance of phosphate-based cooling water programs. IWC 1987;373-380

4. Garside J. *Nucleation in Biological Mineralization and Demineralization*, GH Nancollas ed., 1982.

5. De Jong EJ. Nucleation: A Review, in *Industrial Crystallization* 78, EJ Jong and SJ Jancic eds., 1979.

6. Nyvlt J, Sohnel O, Matuchova M, and Broul M. *The Kinetics of Industrial Crystallization*, New York; Elsevier 1985.

7. Dirksen JA and Ring TA. Fundamentals of crystallization: Kinetic effects on particle size distributions and morphology, Chemical Engineering Science 1991;36:2389-2427

8. Nielsen AE. *Kinetics of Precipitation*, Oxford: Pergamon Press, 1964.

9. Mullin JW. *Crystallization*, 2nd ed. London: Butterworth, 1972.

10. Nielsen AE. Electrolyte crystal growth mechanisms. J of Crsytal Growth 1984;67:289-310

11. Furedi-Milhofer H. Spontaneous precipitation from electrolytic solutions. Pure and Applied Chemistry 1981;53:2041-2055

12. Zhang JW and Nancollas GH. Mechanism of Growth and Dissolution of Sparingly Soluble Salts. *Review in Mineralogy*, vol. 23, MF Hochelle Jr. and AF White eds., Chelssea, Michigan, 1990.

13. Junghahn H and Werkstoffe U. *Korrosion*, vol 13, 1962.

14. Parfitt GD. Fundamental Asects of Dispersion. in *Dispersion of Powders in Liquids*, GD Parfitt ed., Applied Science Publications, 1981.

15. Perez LA and Freese DT. Scale prevention at high LSI, high cycles, and high pH without the need for acid feed. Corrosion 97, Paper No. 174, New Orleans, 1997

16. Nancollas GH, LoRe M, Perez L, Richardson C, and Zawacki SJ. Mineral phase of calcium phosphate. The Anatomical Records 1989;224:234-241

17. Boistelle R and Lopez-Valero I. Growth units and nucleation: The case of calcium phosphates. J Crystal Growth 1990;102:609-617

18. Nelson DGA, Salimi MH, and Nancollas GH. Octacalcium phosphate and apatite overgrowths: A crystallographic and kinetic study. J Colloid Int. Science 1986;110:32-39

19. Perez LA, Shyu LJ, and Nancollas GH. The phase transformation of calcium phosphate dihydrate into octacalcium phosphate in aqueous suspensions. Colloids and Surfaces 1989;38:295-304

20. Wong ATC and Czernuszka JT. Transformation behaviour of calcium phosphate 1. Theory and modelling, Colloids and Surfaces A: Physicochemical and Engineering Aspects 1993;78:245-53

21. Nancollas GH. Phase Transformation During Precipitation of Calcium Salts. in *Biological mineralization and Demineralization*, GH Nancollas ed., Springer-Verlag, New York, 1982.

22. Ebrahimpour A. Kinetics of dissolutionn, growth and transformation of calcium salts and the development of spectroscopy and dual constant composition methods. Ph. D. Thesis, Sate University of New York at Buffalo, 1990

23. Nancollas GH. phosphate precipitation in corrosion protection: Reaction mechanism. Corrosion 1983;38:77-82

24. Tippet RJA and Kuhn VR. On-line monitoring of extended phosphate programs in cooling water systems. IWC, paper No. 63, 1994

25. May RC, Geiger GE, and Bauer DA. A new nonchromate cooling water treatment utilizes high orthophosphate levels without calcium phosphate fouling. Material Performance 1981;20:17-25

26. Salimi MH, Heuhebaert JC, and Nancollas GH. Crystal growth of calcium phosphates in the presence of magnesium ions. Langmuir 1985;1:119-122

27. Amjad Z. The influence of polyphosphates, phosphonates, and poly(carboxylic acids) on the crystal growth of hydroxyapatite. Langmuir 1987;3:1063-1069

28. Abbona F, and Franchini-Angela M. Crystallization of calcium and magnesium phosphates from solution of low concentration. J of Crystal growth 1990;104:661-671

29. Zawacki SJ, Heughebaert JC, and Nancollas GH. The growth of non-stoichiometric apatite from aqueous solution at 37 °C. II: Effect of pH upon the precipitated phase. J Colloid Interface Sci 1990;135:33-44

30. Heughebaert JC, Zawacki SJ, and Nancollas GH. The growth of non-stoichiometric apatite from aqueous solution at 37 °C. I. methodology and growth at pH = 7.4. J Colloid Interface Sci 1990;135:20-32

31. Imai T, Uchida T, Ano S, and Tsuneki T. A newly developed polymer to inhibit scale in cooling water systems. Material Performance 1989;28:41-44

32. Boffardi BP. Corrosion and deposit control in alkaline cooling water systems. 6th European Symposium on Corrosion Inhibitors, Ferrara, Italy, Publ. Universita Degli Studi di Ferrara, 195; 1007-1039

33. Amjad Z and Masler WF. Scale control with terpolymers containing styrene sulfonic acid; with methacrylic acid or slats and 2-methaacrylamido-2-methylpropanesulfonic acids; effective with or without soluble metal compounds. US Patent 4952327; August 28, 1990

34. Chen F. Water treatment polymers and methods of use thereof; antideposit, antiscale. US Patent 4944885, July 31, 1990

35. Chen F. N-(hydroxyalkyl) acrylamide copolymers for corrosion control. Proceeding of the ACS Division of Polymeric Materials: Science and Engineering 1985;53:843-848

36. Chen F. calcium phosphate scale control methods; acrylic acid-monohydroxyalkyl allyl ether copolymers. US Patent 4929362, May 29, 1990

37. Knudsen JG, Al-Saud A, Breske TC and Chenoweth JM. Fouling Characteristics of Cooling Water Containing Phosphate Corrosion Inhibitors and the Inhibiting Effect of Copolymer Additives. *HTD*, vol. 108, Publ. by American Soc of Mechanical engineers (ASME), New York, 1989; 297-304

38. Amjad Z. Effect of precipitation inhibitors on calcium phosphate scale formation. Can. J Polym Chem 1989;67:850-856

39. Hann WM, Natoli J. Acrylic acid polymers and copolymers as deposit control agents in alkaline cooling water systems. Corrosion 84, Paper No. 315, New Orleans, 1984

40. Freese DT, Perez LA, Bair KT, and Chen F. Method of inhibiting scale and controlling corrosion in cooling water systems. US Patent 5575920, November 19, 1996

41. Person P, Nilsson N, and Sjoberg S. Structure and bonding of orthophosphate ions at the iron oxide-aqueous interface. J of Colloid and Interface Sci 1996;177:263-275

18

THE ROLE OF CALCIUM PHOSPHATE IN INTERNAL BOILER WATER TREATMENT

Jasbir S. Gill

Calgon Corporation
Pittsburgh, Pennsylvsania 15230, U.S.A.

INTRODUCTION

This chapter is devoted to boiler water scale/deposit control in general, with a detailed description of various treatments based on phosphate chemistry. Depending on the boiler steam pressure, scale control is achieved by a combination of internal treatment and improvement in the quality of feed water to the boiler. The pretreatment of the boiler feed water consists of using multimedia filters to reduce the suspended solids and colloidal material in feed water, simple lime softening(hot or cold), soda ash or zeolite softening. A total demineralization and removal of organic compounds from the softened water are achieved by using ion exchange, reverse osmosis membranes, and electrodialysis. Any of these techniques are used alone or in combination with others. The feed water is further conditioned to remove dissolved oxygen for corrosion control. However, all boilers rely on some internal scale/deposit control treatment.

BOILER

In simple terms, a boiler is a container in which water is continuously evaporated to form steam by applying a heat source. There are many different types of boilers, however, these can be divided into two main categories, namely, shell boiler and water tube boilers. The boiler engineers determined very early that heating a single large vessel of water was extremely inefficient, that increasing the water/heat interface for improving thermal efficiency was necessary. This lead to the fire tube design which not only increased the heated surface exposed to the water, but also tends to distribute steam formation more uniformly throughout the mass of the water. Many different configurations of the tubes to maximize the water/heat interaction resulted in many different types of designs in practice.[1]

Watertube designs have relatively small drums, with different configurations of tubes carrying water-steam mixtures and are heated from the outside. Tubes carrying higher temperature water/steam mixture to the steam drum are called risers and relatively lower temperature water carrying tubes are called downcomers.

Typical firetube boilers are low pressure heating (15 psig) units producing steam up to 15,000 lbs/hr. Some power boilers may go up to 250 psig. On the other hand, typical Watertube boilers are generally operated at higher pressures (150-5000 psig), producing steam on an average 35000 lbs/hr and up to 200,000 lbs/hr maximum. Watertube boilers are more efficient than firetube boilers.[2]

Water treatment of boilers has the following three broad aims:

A. To prevent the formation of scale and deposit on heating surfaces.
B. To prevent corrosion of the boiler, steam line, and condensate system.
C. To maintain a high level of steam purity.

These objectives can be achieved by improving the feed water quality for the boiler using several options as a part of the pre-treatment. The water is further conditioned by the addition of conditioning chemicals. Alternatively, the water may be treated entirely by adding various chemicals directly into the boiler drum by the method known as internal treatment. For low pressure boilers, sometimes the boiler feed water is used as a raw water without pre-treatment and only internal treatment is used for controlling scale, deposit, and corrosion. However, the high pressure boilers must use a complete pre-treatment, chemical conditioning of the feed water, and internal boiler water treatment to offset deficiency of the pre-treated feed water to the boiler.

Following are the broad components of the boiler water chemical treatment:

A. Scale inhibitors are used to prevent the precipitation of hard minerals as scale on boiler heat transfer surfaces and boiler feed water system including economizers, pumps, etc.

B. Deposit control additives are used to condition or disperse sludge to make it non-adherent to boiler metals so that it can be readily removed by blowdown.

C. Corrosion control is achieved by adjusting the water pH, alkalinity, eliminating dissolved oxygen from water, and deposit control additives to inhibit under deposit corrosion.

D. Corrosion control of condensate systems is also important to the boiler water treatment. This is generally achieved by adjusting the pH in the condensate using neutralizing and filming amines.

E. Preserve the steam purity so that it does not cause corrosion or deposit of the turbine. Untreated silica in the boiler water, due to its volatility, can be transported with the steam and then ultimately form deposit on the turbine blades.

SCALE CONTROL

Scale/deposit in feed water systems and boilers are a mixture of many minerals. Although, sometimes the deposit may be dominated by one component as much as 95% of it. Table 1 gives the components of the boiler deposit with some salient features of each component.

Table 1. Boiler system Scale/Deposit Components

Mineral	Chemical Name	Chemical Formula	Deposit Characteristics
Calcite, Araganite	Calcium Carbonate	$CaCO_3$	Deposit of variable hardness besides sludge
Anhydrite	Calcium Sulfate	$CaSO_4$	Hard adherent deposit
Brucite	Magnesium Hydroxide	$Mg(OH)_2$	Very rare, but soft material
Hydroxyapatite	Basic Calcium Phosphate	$Ca_{10} (PO_4)_6(OH)_2$	Soft sludge
Althausite or Holtedahlite	Magnesium Hydroxyphosphate	$Mg_3(PO_4)_2.Mg(OH)_2$	Thick sticky deposit
Haematite, goethite, etc.	Anhyrated and hydrated iron oxides	Fe_2O_3 $FeO.OH$	Act as binder for other deposit
Acmite Serpentine Xonotlite	Mixed silicates with Ca, Mg, and other tervalent ions	$Na_2O.Fe_2O_3.4SiO_2$ $3MgO.2SiO_2.2H_2O$ $5CaO.5SiO_2.H_2O$	Usually soft sludge but can be hard deposit

Some deposits consist of a number of layers that may differ from one another in composition and in texture.[3-5] A close examination of these layers can help elucidate the history, mechanism, and kinetic of deposit formation. Figure 1 shows a picture of a fouled tube with scanning electron micrograph and energy dispersive X-ray.

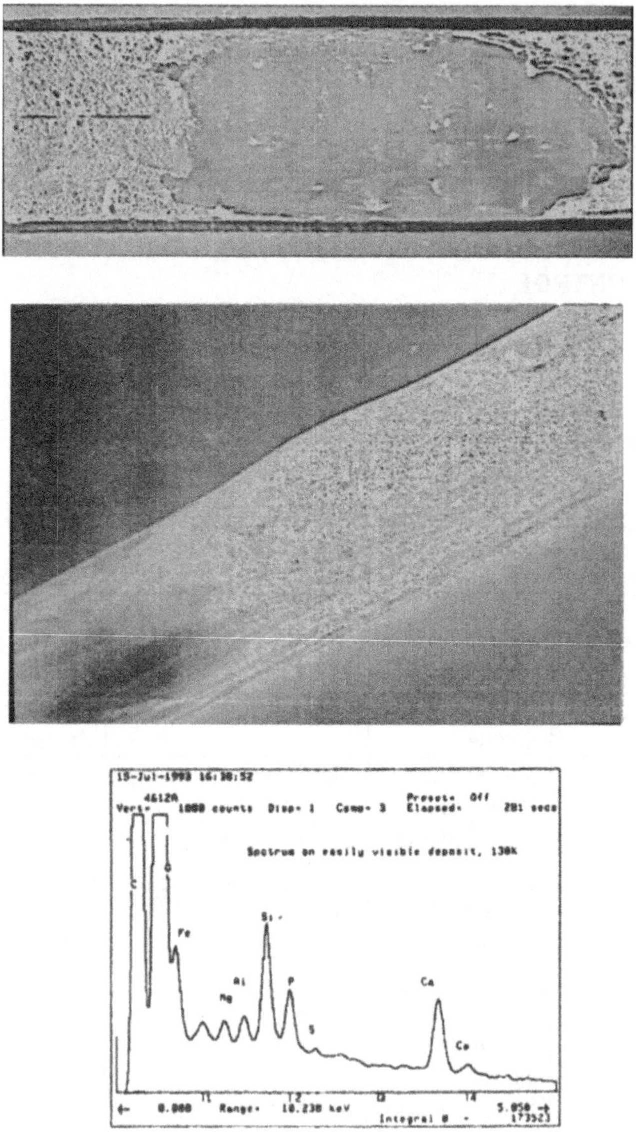

Figure 1. The top picture shows a section of the tube with deposit, the middle picture shows the tube bulge due to overheating and ready to burst, and the bottom picture shows the EDX of the deposit on the tube.

Scale/deposit causes many serious problems in the operation of a boiler including:

A. There is a loss of boiler efficiency due to the insulation caused by the scale/deposit. The extent of the loss of efficiency is shown in Table 2. The direct effect of scale in reducing the thermal efficiency of a boiler in terms of fuel consumption may be small, but the indirect effect as for turbine efficiency loss may be significant.[6]

B. Deposit formation, due to its insulating property, results in over heating of the boiler tube and ultimately to tube failure. As shown in Figure 2, this results in overheating the tube to heat the water to achieve the same temperature as without the tube deposit. This could have serious and very costly consequences on the boiler operation.

C. Under deposit corrosion of the tube also results in costly tube replacement.

D. Loose deposit can accumulate in such positions in the boiler as to impair the circulation, thus starving some of the tubes of water, leading to tube failure due to over heating.

Table 2. Loss of Boiler Efficiency Due to Scale Deposit

Scale Deposit Thickness (Inch)	Loss of Efficiency (%)
1/64	4
1/16	11
1/8	18
1/4	28
1/2	60

From the point of view of scale control, a boiler system can be divided into three parts:

A. The pre-boiler equipment such as economizers, feed heaters, and feed water pumps.

B. The boiler itself.

C. The superheater, turbine, and the condensate system.

There is a great deal of similarity of the formation of scale/deposit in the first two parts, therefore, these are considered together in this chapter. However, the deposit in the superheater, turbines, and condensate system is a result of impurities in the steam. Except for the corrosion treatment of the condensate system, usually no scale deposit treatment is required in these parts. The deposit is prevented by improving the quality of the steam coming from the steam drum. The discussion in this chapter will be limited to scale deposit control in pre-boiler and the boiler itself.

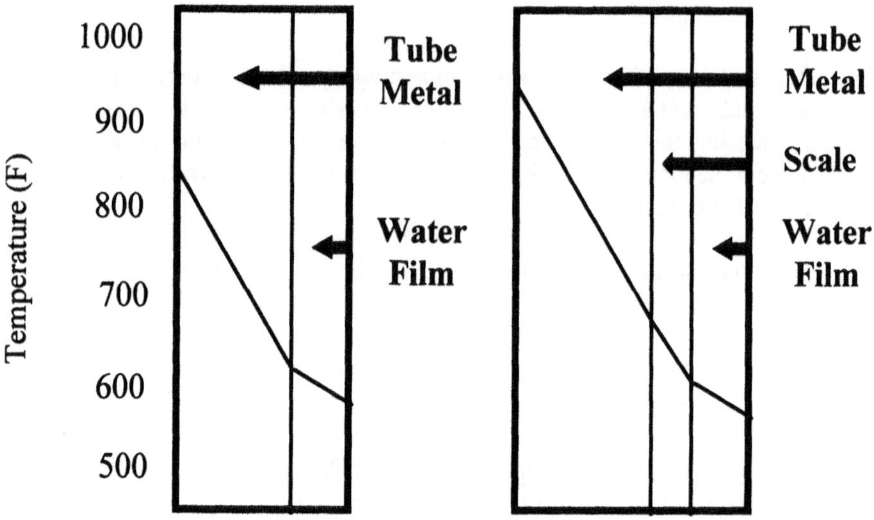

Figure 2. The effect of fouling on heat transfer causing an increase in tube temperature.

There are three main sources of deposit in the boiler:

1. The minerals exceeding their solubility limits at higher temperatures.

2. Suspended solids and corrosion products.

3. Deposit caused by a phenomenon known as 'hideout'. This occurs in boilers with a high rate of heat input per unit area of heating surface. It is often suggested that the deposit is caused due to an increased concentration of dissolved substances, such as sodium phosphate, sodium ferrous phosphate, sodium sulfate, and alkalinity in the film of the boiler water next to the area of high heat input.[7,8] Hideout could also occur due to the formation of a very large stationary steam-filled spaces known as steam blankets, and boiler water salts are concentrated at the edges of the blanketed area. The high concentration in the localized area raises the boiling point, making it stationary and limiting boiling mainly to the outside of this film.[2] Eventually some of the least soluble materials, and especially those with negative solubility with temperature are actually deposited on the metal surfaces. Although this process is reversed when the boiler load (input heat) is reduced and once again the concentrations of lost chemicals start to increase.

However, the damage may already have been done by deposit forming insulating layer and thus causing metal temperatures to increase. Depending on the chemistry of the hideout material, there is also a risk of failure from hydrogen embrittlement and dissolution of corrosion protecting magnetite film. The prevention of hideout will be discussed under various phosphate treatments.

Typical scales in the boiler or pre-boiler equipment consist of:

Calcium carbonate
Calcium sulfate
Calcium Phosphate
Magnesium Hydroxide
Magnesium Silicate
Iron oxide

There are several treatment strategies one can use depending on the boiler pressure and the pre-treatment of the boiler feed water. Boiler feed water consists of a large amount of condensate return and the make-up water. Generally, the condensate return is of excellent quality water. The make-up water also goes through a number of pre-treatment steps to remove suspended and dissolved materials from the water. For very high pressure boilers, very high quality feed water is desired. Limits may be based on solids content, alkalinity, hardness, silica, etc., which are keyed to boiler operating pressure. For boilers operating at 50-300 psig, for example, the American Boiler Manufacturers Association recommends a maximum of 700-3500 ppm total dissolved solids (TDS) in boiler water. The limit range drops with rising pressure, e.g., 100 ppm for boilers exceeding 1000 psig.[9] The deposit is also controlled by using continuous blow down which is usually controlled by conductivity of the boiler water in a specified range. This can assure the boiler operator that the boiler is not exceeding solubility of scale forming minerals.

DEPOSIT CONTROL

Deposit control is achieved by the use of crystal modifiers/dispersants, so that the precipitated mineral is not adhering to the boiler surface. Sludge conditioners are also used to precipitate non-adherent minerals instead of the minerals that can cause hard deposits. Dispersants can also help control the deposit. A good corrosion control eliminates the iron oxide deposition. Usually the boiler feed water is deaerated and chemically treated for maximum oxygen removal. The risk of corrosion is further reduced by using neutralizing or filming amines. Antifoam agents also play an important role in the overall cleanliness and operation of the boilers.

Scale/Deposit Inhibitors

In this section scale and deposit control will be treated together. There are several different treatments which are employed for deposit control based on the boiler operation.

A. Chelant treatment
B. Precipitation Methods such as Calcium Carbonate and Calcium Phosphate
C. Combination of Chelant/Phosphate treatment
D. Polymeric Dispersion treatment
E. All volatile treatment

Chelant treatment program

This program consists of feeding directly to the boiler one or more relatively strong chelating agents such as sodium salts of ethylene diamine tetracetic acid (EDTA) or nitrilo triacetic acid (NTA). These chelating agents selectively form stoichiometric water soluble complexes with calcium, magnesium, and other chelatable substances. In the alkaline environment of the boiler, chelating agents such as EDTA forms selectively stronger complex with calcium or magnesium rather than the iron metal surface of the boiler (Figure 3). The degree of success in forming selective water soluble complexes depends upon specific boiler conditions such as temperature, pH, heat flux, residence time, concentration, stability constant for the complex, and oxidation state of the metal ion in the feed water.[10,11]

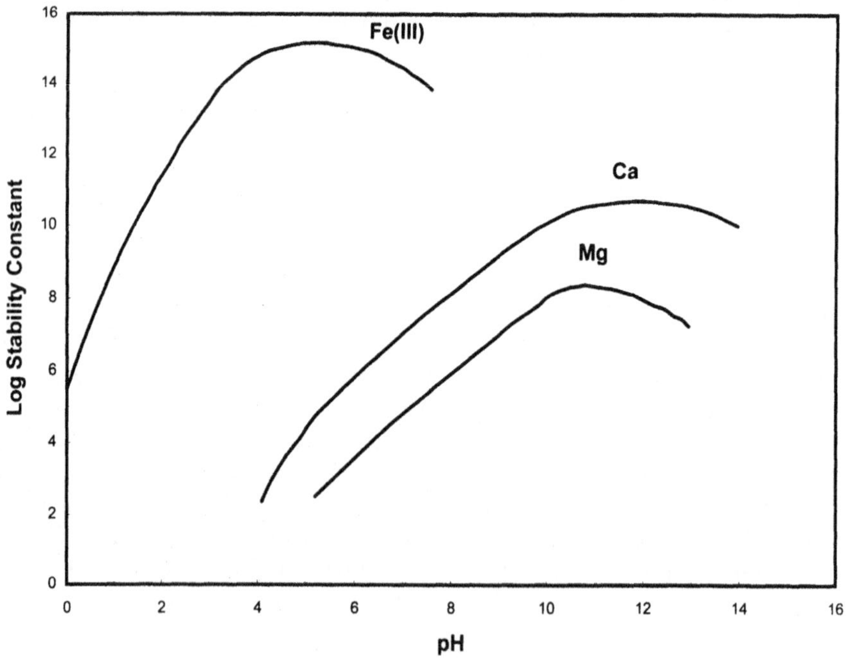

Figure 3. EDTA complexes with Ca, Mg, and Fe (III).

The chelants are quite effective at pressure up to 1200 psig with EDTA and practical limit for NTA is about 850 psig.[7] However, the direct analysis for the chelant becomes progressively more questionable due to the thermal degradation of the chelant to an intermediate, that while still an effective chelant is not apparently found in water with field test. This could lead to over feeding the chelant which may react with protective boiler metal oxide surface. The chelant should be fed to properly deaerated feed water at a point well down stream of all nonferrous alloys. Dissolved oxygen may deactivate the chelant as well as accelerate corrosion rates. The chelant program is very effective and economical when used in systems where pre-treatment systems operate reliably to remove all but trace quantities of calcium to less than 2 ppm total hardness.[10]

Another group of chelating, known as organic phosphonates, is also used in low pressure boilers below 400 psig.[9] HEDP (1-hydroxy ethylidene 1-1-diphosphonic acid), AMP (amino tri(methylene phosphonic acid), and other similar phosphontes and phosphinochemicals are commonly used, in water treatment industry, at substoichiometric dosage to keep calcium in solution. These chemicals have advantage over the traditional chelating agents that they can be used economically at feed water hardness levels of 50 ppm and even higher. They do not pose as great a threat as traditional chelants, since they do not form as strong complexes as EDTA or NTA. Phosphonates can be fed directly to the boiler or the feed water. HEDP or AMP will gradually decompose to orthophosphate but up to 50% is generally available to solubilize hardness entering the boiler water.[9]

Precipitation Methods

Since different minerals have different tendencies to adhere on to the boiler tubes, precipitation of sludge forming or soft deposit producing minerals is preferred over the minerals which form hard tenacious deposits. Calcium sulfate forms very hard deposit while calcium carbonate is relatively softer and calcium phosphate is formed only as a soft sludge. There are two distinct programs which are widely used for boiler water deposit control.

I. Carbonate cycle.
II. Phosphate Treatment.

Carbonate Cycle:

Since calcium carbonate is less adherent onto the boiler surface and less soluble than calcium sulfate, sodium carbonate (soda ash) is added to the boiler water to promote the formation of calcium carbonate instead of calcium sulfate. The chemistry of this concept where converting sulfate to carbonate was first elucidated by R.E.Hall in 1927 by maintaining a ratio of CO_3^{2-}/SO_4^{2-} in excess of a similar ratio derived from the equilibrium solubility of calcium carbonate and calcium sulfate.[12] However, this treatment is applicable to low pressure boilers, generally less than 200 psig since above this temperature sodium carbonate hydrolyzes easily and a sufficiently excess of carbonate alkalinity can not be maintained. Another short coming of this program is that with temperature the degradation of the carbonate into CO_2 can result in lower boiler water pH, leading to boiler corrosion.

Hardness is never completely precipitated with alkalinity in the carbonate cycle, therefore, there is always some hardness left in the boiler water which can form hard scale/deposit with excess silica or sulfate in the water.

Phosphate treatment:

This treatment is most commonly applied for sludge conditioning so that precipitated minerals are not adherent to the boiler tubes. Phosphate treatment provides much better reduction of Ca hardness in boiler water than the carbonate cycle, due to its extremely low solubility.[13] The precipitated material can be easily collected in boiler mud drum for removal by blow down. Sometimes the phosphate treatment program is a common term used for very specific objectives based on the boiler operating conditions. All different phosphate programs employ one or more of the following phosphate sources:[2]

Orthophosphate
Trisodium phosphate
Disodium phosphate
Monsodium phosphate
Polyphosphates
Sodium hexametaphosphate
Tetrapotassium pyrophosphate
Stable phosphonates, such as HEDP or PBTC

If hideout is a problem, potassium salts of these phosphates can sometimes eliminate the problem because of the higher solubility of potassium salts.

There are four types of phosphate program which are commonly employed.[2]

I. Conventional phosphate precipitation program
II. Coordinated phosphate program
III. Congruent phosphate program
IV. Equilibrium phosphate program

Conventional Phosphate Precipitation Program:

The idea of this treatment is again to prevent the precipitation of calcium sulfate in boiler waters which contains some calcium. Any of the above mentioned sources of phosphates, most preferably trisodium phosphate, along with caustic is added to the boiler water to precipitate calcium hydroxyapatite as per the following chemical reaction.[2,14]

$$10Ca^{2+} + 6PO_4^{3-} + 2OH^- \rightarrow 3Ca_3(PO_4)_2 \bullet Ca(OH)_2 \quad \text{or } Ca_{10}(PO_4)_6(OH)_2 \qquad (1)$$

This reaction along with the formation of $Mg(OH)_2$ by the reaction of magnesium and hydroxide, provide internal softening of the boiler water. If the water contains small amounts of silica, it can react with magnesium hydroxide to form serpentine which can be removed along with calcium phosphate and magnesium hydroxide sludge. This treatment is suitable for boilers up to 1000 psig which has low hardness (<50-60 ppm). An adequate amount of OH^- alkalinity (20-200 ppm, depending on the pressure, generally lower at higher pressures). Additional OH^- is required in proportion to the amount of silica level in the water. This can be calculated based on the rule of thumb in the industry; 1.7 ppm OH^- expressed as $CaCO_3$ for 1 ppm of silica as SiO_2. However, a large excess of 'free' hydroxide may result in a phenomenon known as "gouging" where free caustic is concentrated beneath the boiler deposits which may have caused by corrosion products. 'Free' caustic can also contribute to embrittlement and hydrogen damage. High phosphate level (>50 ppm) can also cause problem as such an environment would favor the formation of sticky magnesium hydoxyphosphate rather than sludge forming magnesium hydroxide.

Coordinated Phosphate Treatment:

As pointed out in the discussion on the conventional phosphate treatment, there can be serious consequences of excess caustic and phosphate. The coordinated phosphate treatment limits free hydroxide to only formed by some hydrolysis of phosphate in the boiler water.[15]

$$PO_4^{3-} + H_2O \rightarrow HPO_4^{2-} + OH^- \qquad (2)$$

Thus, this program provides the removal of hardness as a manageable soft sludge without adding OH^- alkalinity. Coordinated phosphate was also proposed by R.E. Hall, first in 1942. Since then there has been numerous refinements to the original concept. The program is applicable in 1000 to 1500 psig. It is applied by controlling the PO_4 residual and pH of the boiler water so as to always operate on or below the theoretical curve of pH versus trisodium phosphate. When pH exceeds the theoretical curve (Figure 2), sodium dihydrogenphosphate or even phosphoric acid is added to lower the pH to titrate the 'free' OH^-. For the sake of ease of application the program uses two phosphate solutions one for pH suppression and one for pH elevation. $Na:PO_4$ is 2.2:1 for pH suppression component and 2.8:1 for the pH elevation component. However, control limits for the coordinated phosphate/pH treatment program are defined primarily by the operating pressure of the boiler and the steam purity requirements; higher pressures require lower phosphate concentrations. This control can be guided by following Figure 4. Sodium phosphate buffers the water and that is the reason why systems with too little phosphate are quite sensitive to feed water pH changes. In phosphate treated systems, pH of the boiler water is directly related to the phosphate concentration and sodium to phosphate ratio. Systems operating with too high ratio of $Na:PO_4$ leads to 'free' OH^- and suffer caustic corrosion, cracking, and other drawbacks of the free caustic. On the other hand, systems operating with $Na:PO_4$ too low lead to acid attack and other failures due to excess phosphate as described under conventional phosphate treatment. Sodium to phosphate ratio of 2.85:1 to 3.0: with 5-15 ppm PO_4 in softened or demineralized make a very successful boiler water treatment.

Congruent Phosphate Treatment:

This program is further refinement of the coordinated phosphate treatment. It is most suitable for boilers operating at high pressures in the range of 1500 to 2500 psig. The goal of this treatment is to achieve "zero" free hydroxide to prevent caustic corrosion.[15] It require very high quality demineralized feed water and relatively low level of PO_4, 2-5 ppm. As shown in Figure 4, Na:PO_4 ratio is controlled between 2.3:1 and 2.6:1. A separate chemical feed system for each component of this phosphate treatment allows independent adjustment of the chemicals to achieve the required PO_4 level and Na:PO_4 ratio. The operator achieves the system control envelop by using various control vectors as shown in Figure 5.

A. If the pH is low (to move from X to D) add caustic.
B. If the pH is high (to move from X to B) decrease the alkaline feed mix or add sodium dihydrogenphosphate.
C. If the PO_4 level is too high (to move from X to A), increase the blowdown rate.
D. If the PO_4 level is too low (to move from A to X), decrease the blowdown rate.
E. If the PO_4 level is high and the pH is low (to move from B to X), decrease acidic phosphate feed rate.
F. If the PO_4 level and pH are low (to move from X to C), increase alkalinephosphate feed rate.
G. If the PO_4 level and pH are high (to move from C to X), decrease alkaline phosphate feed rate.

This program is prone to phosphate hideout, especially in dirty boilers. In this case sometime the use of potassium salts of phosphate instead of the sodium salts may eliminate the problem.

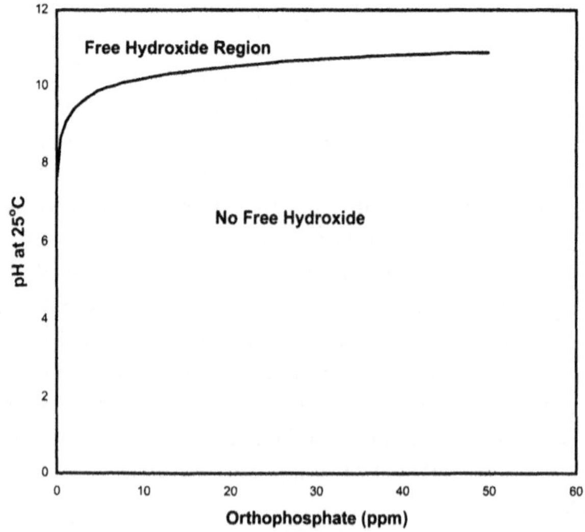

Figure 4. pH-orthophosphate relationship for conventional phosphate treatment.

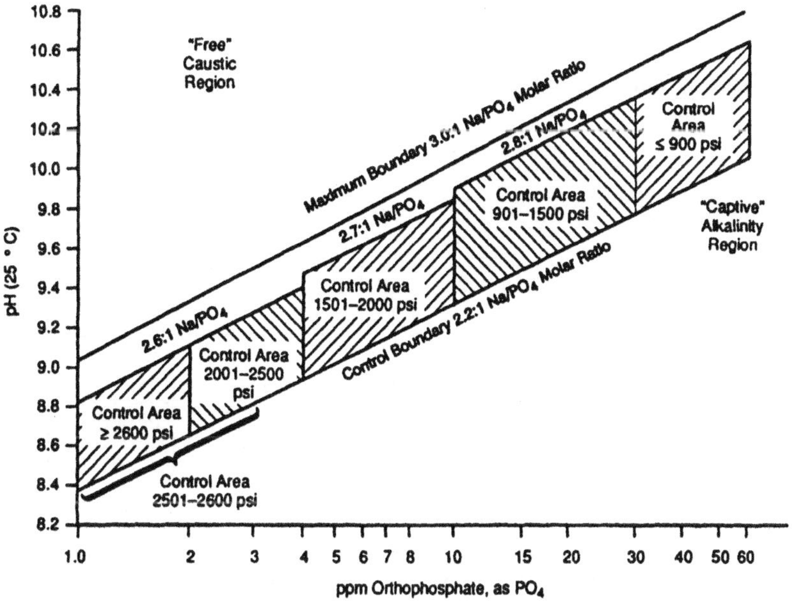

Figure 5. Phosphate Control Strategy for coordinated and congruent phosphate treatment.

Equilibrium Phosphate Treatment:

This treatment program uses very low level of phosphate, less than 2.5 ppm.[16] Unlike the congruent phosphate treatment, equilibrium phosphate tolerates some 'free' hydroxide alkalinity, less than 1.0 ppm. There is no guidelines on the Na:PO_4 ratio but the pH is controlled in a very narrow range of 9.3-9.6 as shown in Figure 6.

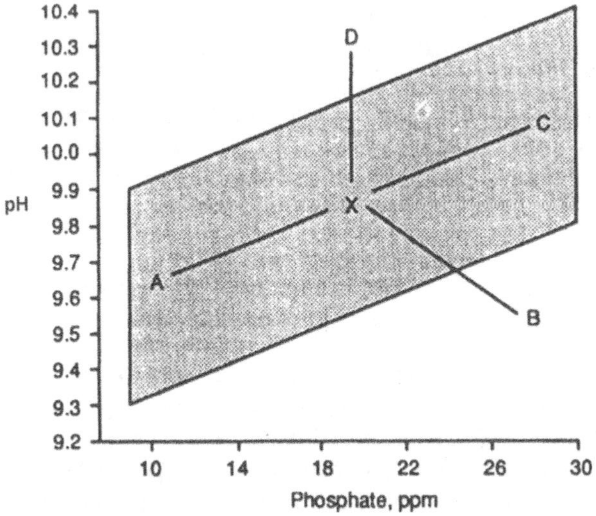

Figure 6. System control envelope for internal phosphate boiler treatment.

Table 3 provides the summary of all the four phosphate treatments.[2]

Table 3. Summary of the Phosphate Treatment Programs.

Treatment	PO_4 ppm	OH^- ppm	$Na:PO_4$	pH	Pressure Range (psig)	Water Quality µmhos
Conventional	20-40	20-250	N/A	11-12	<1000	<150 to 5000
Coordinated	5-20	Trace	2.85:1 to 3:1	9-11	1000-1500	<150
Congruent	2-5	Zero	2.3:1 to 2.6:1	9-9.5	1500-2500	<100
Equilibrium	<2.5	<1.0	N/A	9.3-9.5	>2000	<100

Although silica is either taken out of the water during pre-treatment of the boiler feed water or removed with magnesium hydroxide in the internal boiler treatment, but deliberate addition of soluble silica to phosphated treated boilers has been claimed to be helpful in certain cases.[17] In waters containing high magnesium hardness, silica prevents the formation of sticky, gooey magnesium hydroxyphosphate by forming serpentine. Magnesium silicate as serpentine forms a manageable sludge along with calcium hydroxyapatite which leave the boiler in blowdown rather than a hard boiler deposit.

The examination of the deposit analysis can reveal many aspects of the treatment program, its failures, successes, and corrective actions.

A. The presence of calcium sulfate in the deposit is very undesirable and it indicates that the proper phosphate residuals had not been maintained in the boiler water treatment. Calcium sulfate scale is formed when either hardness is very high for phosphate or chelant or sulfate level is very high in the water.

B. Calcium carbonate can not be formed in the boiler but it can form in the boiler feed system. Calcium carbonate under unusual conditions is formed in the boiler only when phosphate or chelant residues have not been properly maintained.

C. Formation of magnesium phosphate indicates the presence of extremely high phosphate residuals.

D. Iron oxide deposits predominately occur because of corrosion in the condensate system and the corrosion products are being returned to the boiler. Sometimes improperly pretreated well-water can also cause iron oxide problem. The solution to this problem is the proper application of filming or neutralizing amines to provide corrosion protection in the condensate system. Some mechanical adjustments in the pretreatment system may eliminate the severity of the iron problem in the make-up water.

E. The presence of calcium phosphate deposit in boiler scale is usually an indication of inadequate feed of an organic sludge dispersant or poor blow down control.

F. Similarly, the formation of calcium or magnesium silica scale are also related to either poor blow down control or insufficient organic dispersant. Calcium silicate in the deposit could also indicate operational periods of phosphate residuals and low alkalinities.

G. Copper oxide deposits are mostly associated with high pressure boilers having a very high percentage of return condensate in the boiler feed water. This is again due to corrosion in the steam and condensate system which can be corrected by the addition of neutralizing amines.

H. Sodium-alumino-silicate, analcite, is usually associated with aluminum carry over from the clarifier where the primary coagulant is alum. This can be corrected by feeding chemicals in the clarifier to maintain the alumina content in the clarified water at less than 0.1 ppm. Application of chelating agent such as EDTA can also help in controlling analcite deposit.

Combination of Chelant/Phosphate Treatment:

The previously described two treatments namely, the use of chelant and precipitation methods can be applied as free standing treatments for boiler deposit control. However, in reality the two treatments are often combined to obtain the most optimal results. In this treatment program, some chelant is used to make soluble complexes of the hardness and other tervalent metal ions which are not precipitated as sludge, either with alkalinity or phosphate. Theoretically, the treatment strategy completely eliminates the dissolved hardness and other deposit forming inorganic ion from the boiler water. Formation of calcium hydroxyapatite and calcium-chelant soluble complex are in competition, and thermodynamically precipitation of calcium hydroxyapatite is favored over the soluble complex, provided sufficient amount of phosphate is available. Therefore, chelating agent combines only with the calcium left over from the calcium hydroxyapatite equilibrium.

Polymeric Dispersion Treatment:

These chemicals have long been used and known as sludge conditioners. As pointed out earlier under the section "Precipitation Methods" in which the boiler water deposit is controlled by converting undesirable minerals to desirable mineral sludge. The organic dispersants help by further conditioning the sludge to make it more fluidized for easy removal. These organic dispersants are adsorbed on to crystals faces during their formation, resulting in crystals distortion (Figure 7). Thus the precipitated salts are readily dispersed and do not adhere to the boiler surface rather easily transported with the blowdown.

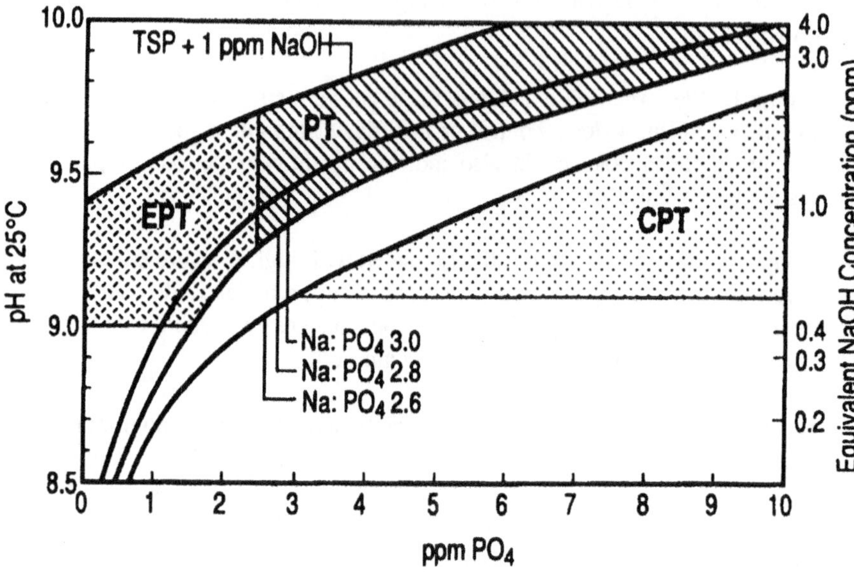

Figure 7. Boiler water equilibrium phosphate operating ranges.

Many of the earlier organic dispersants/sludge conditioners have their origin in naturally occurring organic compound in everyday life used materials such as coffee grounds (tannins), potato skins (starch), wood sawdust (lignins, cellulose), manure (lignins, starch, etc.), sugars, and seaweeds. These materials have some chelating properties as well as crystal modification. Lignins are a major non-cellulose component in wood and it is usually converted to complex lignosulfonates by digesting wood with sulfite for pulp manufacturing. This material is a byproduct of paper making and is stable up to 2200 °F. Lignosulfonate has, however, been replaced by more potent synthetic polymers. Tannin is also a complex organic material, extracted from wood bark, coffee ground, etc., and is no longer in common use. Initially these natural materials were replaced with modified natural materials, such as, lignin sulfonate (modified wood extract), bisulfited tannins (modified wood extract), carboxymethyl cellulose, CMC (modified wood extract). These products are still in some use, particularly in boilers operating at 100 psig or less. Natural materials are effective for dispersing calcium carbonate scale but is not very effective for phosphate based deposits. Thus as the phosphate treatment gained more acceptance, the need for new synthetic polymers was realized and most of the natural and modified natural organic materials have been replaced with more effective synthetic dispersants.

Synthetic dispersants/ sludge conditioners:

These are water soluble synthetic polymers which are more stable and have wider application than the previously employed natural materials. They can be generally classified as synthetic carboxylated and /or sulfonated homopolymer or copolymers. Table 4 lists of some of the commonly used dispersants.

Table 4.

Polyacrylates or modified polyacryaltes such as polymethacrylates
Polymaleates
Polyacrylamides
Sulfonated polymers
Acrylic/acrylamide copolymers
Acrylic/maleic copolymers
Phosphino/carboxylic copolymers
Sulfonated styrene/carboxylic copolymers and terpolymers
Acrylic/2-acrylamido-2-methylpropylsulfonic acid copolymers or terpolymers with other monomers
Acrylic/2-hydroxypropylacrylic copolymer
Acrylic/sulfophenomethallyl ether/ maleic acid terpolymers

Each polymer has unique characteristic depending on the composition, molecular weight, and the end groups. Their use is guided by the boiler operating conditions, such as pressure, hardness, iron, silica, etc. Most of these polymers are excellent crystal growth inhibitors in addition to crystal modifiers. Figure 8 shows calcium carbonate crystals as modifies by a common scale inhibitor.

Figure 8. Calcium carbonate crystal distortion with deposit control additive.

Polyacrylate is commonly used dispersant at 5-20 ppm dosage, for boilers operating up to 1000 psig. It is very effective for calcium and magnesium phosphate scales but is not very effective for iron transport. Polymethacrylate, polyacrylamide, and sulfonated styrene based copolymers are excellent for iron transport. These polymers are very effecive under high pressure conditions and at relatively very low dosage of 5-10 ppm and sometimes even as low as 2-3 ppm. Acrylic/2-acrylamido-2-methylpropylsulfonic acid copolymers or terpolymers with other monomers are very effective dispersants and iron transport agents for boilers operating at lower pressures.

All Volatile Treatment (AVT):

AVT is commonly used in electric utility high pressure once through boilers operating at in excess of 2200 psig and nuclear systems.[2,9,18] Once through boilers have no tolerance for solids, as all feed water entering boiler, exits as steam. Such boilers require full-flow condensate polishing, turbine protection from any deposit caused by the volatiles carried with steam, and use AVT. In this treatment program, chemicals are added only to control pH and oxygen for corrosion prevention. Thus, in practical terms AVT is a condensate/feed water treatment and not boiler water treatment. Amines in the form of ammonia, morpholine, monoehanolamine, etc. are used for pH control which depends on the system metallurgy, e.g., pH 8.8 to 9.1 with copper in the system and 9.1 to 9.5 for ferrous metal. Oxygen is controlled by the addition of reducing agents such as hydrazine and its analogues and other substitutes. Since the feed water used with this treatment has a cation conductivity of <0.2 μS/cm, it provides no buffering against salt intrusion caused by condenser leakage. AVT thus requires very close monitoring of feed water impurities caused by any source of pretreatment or other operational failures.

Reducing environment is excellent for copper corrosion control; however, such an environment may be harmful for ferrous alloys such as carbon and mild steel. The studies in 1970's has shown that carbon and mild steel undergo minimal corrosion in pure water having 50-200 ppb (parts per billion) dissolved oxygen and the corrosion rate increases as dissolved oxygen drops below this range.[9] This lead to a new treatment program called oxygenated treatment for corrosion control which represents a significant departure from earlier boiler water treatment philosophy. In this treatment, either hydrogen peroxide or oxygen is added to boiler feed water and the dissolved oxygen is held at about 200 ppb and cation conductivity below 0.15 μS/cm: those two perimeters determine the iron content resulting from corrosion. This treatment strategy protects metal surfaces and provides a buffer against occasional increases in acid forming impurities. Dissolved oxygen controlled to 50-200 ppb forms stable oxide film to protect metal surfaces.

CONCLUSIONS

The aim of the deposit control treatment in boilers is to prevent the formation of scale and deposit in the boiler water feed system, the boiler itself, steam line, condensate system, and the turbines. These objectives can be achieved by improving the feed water quality using several options as pretreatment. These pretreatment options consist of multimedia filters, lime softening, soda ash or zeolite softening, and a complete demineralization using ion exchange, reverse osmosis, and electrodialysis. The corrosion control is achieved by removing dissolved oxygen and operating the boiler and condensate system in a highly alkaline region. The pretreatment of the feed water and the internal boiler treatment are generally used in combination, depending on the boiler pressure. The feed water for the boilers operating at lower pressure uses less pretreatment and rely more on the internal boiler treatment such as sludge conditioners and dispersants. However, boilers operating at high pressures uses extremely high purity feed water and use internal treatment such as chelating agents, dispersants, and all volatile treatment. The electric utility boilers operating at pressures greater than 2200 psig and nuclear systems use water which has cation conductivity of <0.15 mS/cm. These once through boiler systems use minimum internal treatment which is limited to dissolved oxygen control and pH adjustment.

REFERENCES

1. Marks RH, A Power Special Report on Steam Generation, June 1964; Power Magazine.

2. Harfst WH, Selby KA, Quigly JT, Basic Boiler Water Treatment Course #6973, Department of Engineering Professional Development, College of Engineering, University of Wisconsin-Madison, 1977.

3. Clark LM, Bunn CW, The Scaling of Boiler. Part IV. Identification of Phases in Calcium Silicate Scalles. F. Soc. Chem. Ind., Lond. 1940;59:155

4. Clark LM, Bunn CW, Symposium on the identification of Water-Formed Deposits, Scales and Corrosion products by Physico-Chemical Methods. Proc. Amer. Soc. Test Mat. 1943;43:1269.

5. Clark LM, Identification of Minerals in Boiler Deposits: Examples of Hydro-Thermal Synthesis in Boilers. Miner. Mag. 1948;28:539

6. Brandvold DC, *Water Treatment, Industrial, Commercial & Municipal.* Jacksonville, Fl: DCB Interprises, Inc., 3rd Edition 1994.

7. Hall RE, water Conditioning for High Pressure Steam Generation, Report of Steam Station Engineering Committee of National District Heating Association. 1936.

8. Straub FG, Hideout of Sodium phosphate in High Pressure Boilers. Trans. Amer. Soc. Mech. Engrs. 1950;72:479

9. *Boiler protection*, Power Magazine. 1993;33.

10. Cowen JC, Weintritt DJ, *Water-Formed Scale Deposits.* Houston,Texas: Gulf Publishing Company, 1975.

11. Metcalf JR, Boiler Chelant Treatment: An update. Ind. Water Eng. 1971;Jan:1971.

12. Hall RE, A Physico-Chemical Study of Scale Formation and Boiler Water Conditioning. Bull. Carneg. Inst. Tech. 1927:14;1

13. Smith RM, Martell AE, *Critical Stability Constants.* New York &london: Plenum Press, 1976:4.

14. Hamer P, Jackson J, Thurson EF, *Industrial Water Treatment Practice.* London: Butterworth& CO. (Publishers) Ltd 1961.

15. Steele CW, Winters MA, Willis JA, SPE-Enhanced Co-Ordinated PO_4/pH control Improves Bolier Operating reliability. Proceedings of International Water Conference, Pittsburgh, PA. 1992: IWC-92-409.

16. Carter PB, Selby AK, Coversion of a high Pressure Drum Boiler from AVT to Equilibrium Phosphate. Proceedings of International Water Conference, Pittsburgh, PA. 1992: IWC-92-56;472.

17. Welsh JW, Reducing Boiler Sludge with a Silica Equilbrium. Pwr. Generat. 1949: 53(3); 76

18. Doodley B, Paterson S, Phosphate Treatment: Boiler Tube Failures Lead to Optimum Treatment. Proceedings of International Water Conference, Pittsburgh, PA. 1994: IWC 94-50; 420.

19

EQUIPMENT FOULING IN THE DAIRY APPLICATION: PROBLEM AND PRETREATMENT

Georges Daufin

National Institute for Agriculture, INRA, Dairy Technology Research Laboratory, LRTL, F - 35042 RENNES Cedex FRANCE

Jean-Pierre Labbé

National School of Chemistry of Paris F - 75231 PARIS Cedex 05 FRANCE

INTRODUCTION TO FOULING

The dairy industry processes huge volumes of milk every day worldwide. Factories processing more than 1 million liters a day are not rare. Milk undergoes a variety of transformations so as to propose consumers various products such as drinking milk, yogurts, butter, cheeses, etc. Among the corresponding processes, heat treatments aim at giving milk and milk products either satisfactory hygienic or physico-chemical properties. Besides, membrane processes have emerged for more than 25 years. They are developping more and more for the concentration and the selective separation of milk components. In both kinds of processes milk and whey, co-products of cheese preparation, are the major fluids which are processed.

During heating of milk in a heat exchanger, an undesirable deposit is formed on the surface of the stainless steel wall. The process of deposit formation is called fouling. The deposit mainly consists of proteins and inorganic compounds, the relative contribution of which depends on heating temperature (60 % proteins in the range 70 - 90 °C ; 80 % inorganic compounds between 110 and 140 °C). It is well known that a direct link exists between fouling and the heat-denaturation of serum proteins when dairy liquids are processed at temperatures above 70 °C. Another major contribution to fouling is an inorganic precipitation, as calcium phosphates are easier to precipitate on increasing temperature.

Fouling of heat exchangers by dairy liquids is a major problem in the dairy industry, since it causes the following undesirable effects : decrease in heat-transfer coefficient, increase in pressure drop, hence decreasing pumping efficiency, contamination of the product processed by loosened deposits.

To remove a deposit it is common practice to clean the heat-treatment equipment for dairy liquids at least once a day. The extra processing costs as a result of fouling can be divided into:[1]

- increased capital costs for oversizing the heat-transfer equipment, and its early replacement because of a shorter technical life-span ;
- increased energy (fuel) costs for operation, due to higher heat-transfer coefficient and lower pressure ;
- costs for product losses, due to substances remaining on the wall of the heat exchanger and to plant shut-down for cleaning and start-up procedures ;
- high maintenance costs, due to the need for a cleaning-in-place installation and equipment for effluent treatment ;
- costs for energy, water and detergents needed for cleaning ;
- costs of cleaning time, during which no production can take place.

To summarize, fouling is a costly affair. Hence, it would be a great benefit to solve or at least to reduce the problems it causes. Fouling has been studied for many years; recent reviews are by Lalande et al.[2] and Fryer et al.[3] It is clear from these that both processes of serum protein denaturation and precipitation of calcium phosphates play a key role in fouling. Several researchers have studied the link between denaturation of serum proteins and fouling, but there is no agreement on which kinetic data for the denaturation should be taken. Fouling is not simply a mass-transfer-controlled process, and it is not yet clear whether it is controlled by a surface and/or by a bulk reaction. The mechanisms by which proteins and inorganic compounds are deposited on a stainless steel surface are not precisely understood. There is no agreement either on which component adsorbs first, organic or inorganic.

The processing of dairy fluids by reverse osmosis (RO), nanofiltration (NF), ultrafiltration (UF) or microfiltration (MF) is characterized by a progressive decline in flux with time (at constant pressure). A number of phenomena acting simultaneously reduce the permeate flux. Chronologically, at least for UF, it is possible to identify three separate phases in flux decline.[4,5] In the first few minutes, the rapid initial drop in flux is due primarily to concentration polarization, which consists in developing a concentration gradient of the components retained near the membrane. Concentration polarization builds up rapidly and is a function of the hydrodynamic conditions in the membrane system. It is independent of the physical properties of the membrane (for totally retentive membranes) and the corresponding loss of flux is reversible: flushing the membrane with water restores the portion of flux loss due to concentration polarization. After these few minutes, the flux keeps decreasing rapidly and membrane fouling is observed from now on, the corresponding loss of flux being then irreversible: flushing the membrane with water will not restore the portion of flux loss due to fouling anymore but after the membrane will be chemically cleaned. A quasi-steady-state period follows where the slow decline in flux may be due to further deposition of particles or to consolidation of the fouling layer.

When water is filtered through a membrane, the permeate flux (J) is proportional to the transmembrane pressure ΔP as described by Darcy's law:

$$J = \frac{\Delta P}{R_m m_p} \qquad (1)$$

where R_m is the membrane hydraulic resistance and μ_p the viscosity of the permeate.

The effect of fouling is generally considered as an addition of a second resistance (R_f) resulting from a deposit on the membrane surface (including the membrane pores). However, there are considerable discrepancies on the mechanism through which concentration polarization reduces the permeate flux. A first possibility lies in adding a third resistance to R_m and R_f in equation (1). The second view is that the osmotic pressure (ΔΠ) induced by the accumulation of macromolecular solutes at the membrane surface actually makes a contribution to transmembrane pressure decrease and hence to lower permeate flux. For macrosolutes of less than 100 $kg.mol^{-1}$. in size, such as most of the whey proteins, the osmotic pressure model is considered to be more appropriate.

From these conclusions, equation (1) can be modified to account for membrane fouling and concentration polarization, as:

$$J = \frac{\Delta P - \Delta \Pi}{\mu_p.(R_m + R_f)} \qquad (2)$$

Concentration polarization and fouling are serious problems which have limited the development of membrane technology, in particular MF, in dairy applications. Fouling with milk or whey is a complex phenomenon because of the large number of dairy components known to contribute to membrane fouling. The compositions of milk and whey are shown in Table 1. Contributions to fouling come from : fat or lipid material, proteins, peptides and inorganic compounds, especially calcium phosphates. Bacteria and particulate matter such as casein fines may also add their contribution.

Table 1. Composition of typical liquid dairy streams (% w/v)

Component	Whole milk	Skim milk	Whey	
			Sweet	Acid
Total solids	12-13	9-10	6.0-7.0	6.0-6.5
Fat	3.5-4.0	0.1-0.2	0.1-0.5	0.1
Proteins (total nitrogen x 6.38)	3.2-3.3	3.2-3.4	0.7-0.9	0.7-0.8
Lactose	4.5-5.0	4.6-5.0	4.2-5.0	4.2-4.9
Ash	0.6-0.8	0.6-0.8	0.5-0.6	0.5-0.7
Calcium	0.12	0.12	0.047	0.103
Potassium	0.15	0.17	0.161	0.143
Sodium	0.05	0.05	0.054	0.048
Magnesium	0.013	0.011	0.008	0.01
pH	6.6-6.7	6.6-6.7	5.6-6.3	4.4-4.7

Membrane fouling severely alters process performance for permeability and selectivity of separations, makes it necessary to shut-down the plant for cleaning and start-up procedures. Hence, membrane fouling has been studied for many years and a recent review has been published by the International Dairy Federation.[6]

The present review will consider the physico-chemical properties of milk components in conditions encountered in heat exchanger and membrane processes. Then, an overview of the phenomena which rule fouling by proteins and inorganic compounds will be presented while describing heat treatment of milk products and cross-flow filtration of milk and whey.

MILK CHEMISTRY; STABILITY OF MILK COMPONENTS

The knowledge of milk chemistry is important for understanding how physical variations (for example shifts in pH or temperature), impact on membrane fouling.[7,8] In this section, the properties of the major milk proteins linked to the inorganic system will be briefly summarized.

Caseins

Bovine milk contains approximately 3.5 % proteins: caseins (80 %) are usually defined as those proteins which get insoluble around pH 4.6 at 20 °C and whey proteins (20 %). The main caseins are α_{s1}-, α_{s2}-, β- and κ- in the approximate ratios 40:10:35:12 respectively and several minor proteins including g-caseins.[9] Caseins are water soluble except at their iso-electric points (IEPs) (typically 4.1 - 5.4) at moderate temperatures (> 20 °C).

In standard milk, approximately 95 % of caseins exist as micelles, with molecular masses of about 10^8 g.mol^{-1} and mean diameters in the 50-500 nm range. Micelles are made up from casein submicelles and micellar calcium phosphate (MCP) i.e., MCP is not a purely inorganic material ; it consists of colloidal calcium phosphate linked to caseins through phosphorylated side chains. It is generally accepted that k-caseins are found at micelle surface and are partly responsible for micelle stability.[10] Upon proteolysis of κ-caseins by rennet and removal of the casein-macropeptide, a micelle is destabilized : caseins aggregate and gel in the presence of calcium.[9] Micelle stability is also dependent upon calcium phosphate bonds and some additional inner interactions, most probably hydrophobic interactions.[10]

Micelle stability is affected by temperature, pH and calcium content. By decreasing temperature the solubility of MCP increases and the micelle shrinks. A proportion of β-casein also migrates from within the micelle to the bulk solution. Increasing the temperature reverses these effects and also increases the probability of calcium phosphate precipitation.

As pH is lowered, calcium and phosphate dissolve and are displaced from the micelle ; below pH 4.9, MCP is totally soluble. Caseins are released from the micelle and remain soluble at low temperature.

Whey proteins

Whey proteins, strongly associated with fouling, represent about 20 % of milk proteins; some of their properties are shown in Table 2. b-Lactoglobulin, the major whey protein, is a typical globular protein and exists as a dimer at room temperature between pH 5.2 and 7.0. Above 40 °C or below pH 3.5, the dimer begins to dissociate to monomers. Octamers may form between pH 5.2 and 3.5, with a maximum number at pH 4.4 - 4.7 and 0°C. a-Lactalbumin, less abundant, is the smallest and has a spherical shape. It is known to bind calcium.

Table 2. Whey protein characteristics

Protein	% of whey proteins	molecular mass g mol^{-1}	IEP
b-lactoglobulin	55 - 65	18 400	5.35 - 5.49
a-lactalbumin	15 - 25	14 200	4.2 - 4.5
Immunoglobulins	10 - 15	80 000 - 900 000	5.5 - 8.3
Bovine serum albumin	5 - 6	66 300	5.1
Proteose-peptone	10 - 20	4 000 - 80 000	5.1 -6.0
b-caseins	1 - 2	24 000	4.7
minor proteins	< 0.5	30 000 - 100 000	-

The reader should be cautious about results of membrane fouling studies with individual proteins, especially bovine serum albumin (BSA). Most of these experiments were performed with modified ionic environments which are completely different from that of the milk and whey system ; the results, therefore, are not necessarily indicative of the actual behaviour of the protein in its natural environment.

Inorganic compounds

For many years milk salts have been recognized as playing a major part in the heat stability of milk. Our understanding has been more and more focused on their interaction with caseins, particularly calcium and phosphate ions. Increasingly, their role in influencing stability has been interpreted as being mediated by their ability to maintain or destroy the structure of the casein micelle. Bovine milk contains cations of calcium, magnesium, sodium and potassium, together with chloride and the anions of citrate and phosphate, all typically in the 5 - 40 mM concentration range (Table 3).[11]

Table 3. Typical salt composition of bovine milk[a]

Constituent	Concentration (mM)
Phosphate (total)[b]	20.9
Phosphate (colloidal)[b]	9.7
Calcium (total)	30.1
Calcium (colloidal)	20.6
Citrate	9.8
Na^+	25.5
K^+	36.8
Cl^-	30.3

a After Holt[11] Table 1, recalculated from the original results of White & Davies[12]

b Concentrations are inorganic phosphate only.

The concentrations of calcium and inorganic phosphates in milk are relatively high, approximately 30 and 20 mmol/l respectively.[15] The behaviour of these ions dominates the inorganic equilibria in milk. Approximately two-thirds of the calcium and one half of the inorganic phosphate are colloidal. It is presumed that all colloidal inorganic phosphate is present as micellar calcium phosphate (MCP), whereas colloidal calcium is partly incorporated in the MCP and partly bound, first to caseins through their phosphoserin ends, and to a lesser extent to whey proteins.[12] Calcium may also interact with lipid material. The structure of MCP involves a multicomponent biphasic system strongly dependent on pH with three acid-base pairs linked to PO_4^{3-}. Citrate, magnesium and active groups of casein appear to play a role in MCP, a complex system with unique properties, quite different from those of any simpler form of calcium phosphate.[11]

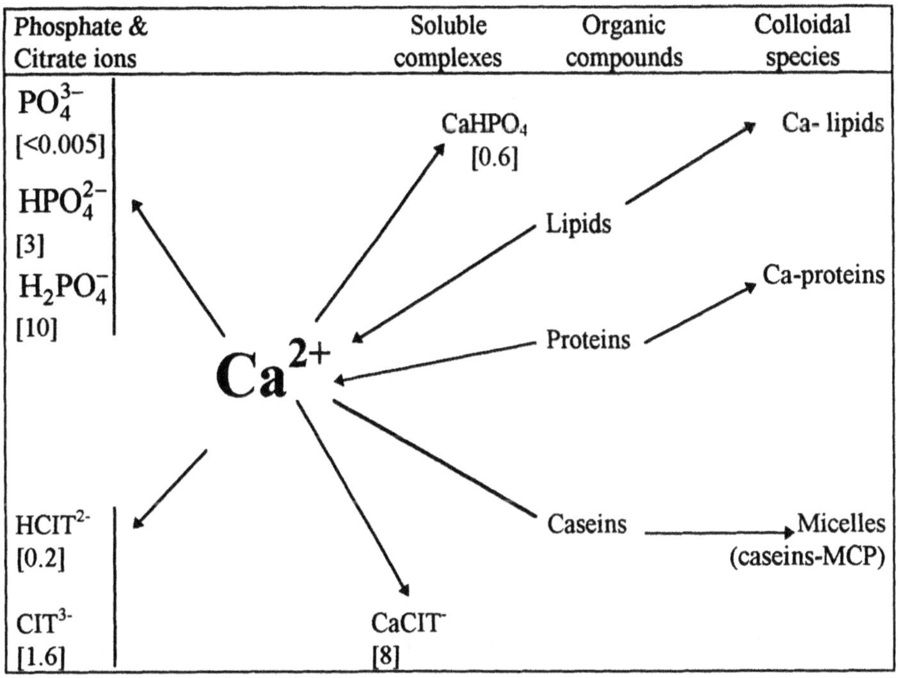

Figure 1. Calcium and inorganic phosphate equilibria in milk with the equilibria associated with calcium highlighted. The values in square brackets are concentrations of the specific ions (mmol/ℓ) calculated by Brulé.[13]

Among the variety of soluble calcium and inorganic phosphate species in the aqueous phase of milk, $CaCit^-$ and $H_2PO_4^-$ are in the highest concentrations (Figure 1). They pass readily through an UF membrane whereas the colloidal material is retained.

Precipitation of calcium phosphates

The aqueous phase of milk is supersaturated with respect to most of the calcium phosphate salts (Table 4) and is thermodynamically unstable.[14] However, although precipitation of calcium phosphates must eventually take place, the equilibrium is usually obtained very slowly if at all.[15]

Table 4. Various forms of precipitated calcium salts. pKs from Lyster[16] and Holt.[11] S (saturation) is the ion activity product (measured in the solution) divided by K_S.

Formula	Name	Code	Solubility constant $pK_S = - \lg K_S$	S
$CaHPO_4$	Calcium hydrogen phosphate (monetite)	CHP	6.9	6
$CaHPO_4 \cdot 2\ H_2O$	Calcium hydrogen phosphate dihydrate (brushite)	CHPD	6.6	3
$Ca_8H_2(PO_4)_6 \cdot 5\ H_2O$	Octacalcium dihydrogen hexa-phosphate pentahydrate	OCP	47	60
β-$Ca_3\ (PO_4)_2$	Tricalcium diphosphate (whitlockite)	TCP	28.9	200
Amorphous $Ca_3(PO_4)_2 \cdot xH_2O$	Tricalcium diphosphate (hydrated)	ACP	28.6 estim.	90
$Ca5(PO4)_3(OH)$	Hydroxyapatite	HAP	57.7	8.10^8
	Micellar calcium phosphate	MCP	/	
$Ca_3(Cit)_2 \cdot 4\ H_2O$	Tricalcium dicitrate tetrahydrate	TCCT	17.6	0.4

Inorganic precipitation is a complex phenomenon for the milk system in which MCP may be considered as hydrated calcium-phosphate clusters surrounded by casein phosphate centres: $[Ca(HPO_4)_{0,7}\ (PO_4)_{0,2}\ (H_2O)_x]_{50}.[Ca_{2.5}Mg_{0.5}Ser\ P_4$-casein$]_5$. These types of microgranules (solid state) are stabilized by their size as well as by the effect of casein polypeptide chains slowing down transport to and from the surface (inhibition effect). Thermodynamically, they are expected to lead to HAP, which in fact is not the case in milk, even at high temperature, as caseins and Mg^{2+} ions act as powerful growth inhibitors.[17] Nucleation of a calcium phosphate phase can occur on casein P(V) centres as well as on the wall surface of an exchanger. The first phase is amorphous (even by treating 45 min. at 120°C). Only afterwards will a succession of more stable compounds be observed. They are never purely inorganic species and the initial Ca/PO_4 ratio lies near 1.5.[18]

Completely different is the situation with less complex systems. In purely inorganic solutions in the 40-60°C temperature range, a CHPD-CHP mixture can be observed, richer in CHP when temperature or supersaturation increases, and the first precipitating phase is amorphous ACP above a supersaturation which decreases rapidly as temperature increases. For slightly higher pH values (pH > 6),

OCP is the final product, rapidly formed by introducing OCP nuclei in the solution.[19] In the presence of various additives, the same phases are built-up,[14,20] but strong inhibiting effects may be observed for the nucleation time, e.g. with cheese whey, acid whey or whole casein. CHPD is also formed in preconcentrated industrial wheys (from 6 to 30 % dry matter), together with small amounts of citrate compounds more difficult to characterize.[21]

Daufin et al.[22] has shown, by fouling a laboratory exchanger at 72°C with various model fluids, that inorganic solutions indeed give an OCP deposit, whereas poorly crystallized HAP is observed with a synthetic ultrafiltrate (containing urea, lactose, citrate ions: pH 6.6). Sweet whey is the first complex system giving rise (together with denatured proteins) to amorphous hydrated calcium phosphates, also observed with milk. Such amorphous phases with varying Ca/PO_4 ratios (1.4-2.1 at 50°C) are also to be obtained on CHPD as seeds on alkalinizing cheese whey permeates.[23] It can be concluded that HAP, the most likely phase from a thermodynamical point of view,[19] is but seldom observed in real systems. Calculating that - from purely ionic equilibria - milk solutions are supersaturated with respect to HAP above pH 6 at all temperatures[24] makes us understand that such a precipitation must be inhibited (nucleation and/or growth) for other phases to be observed. In the presence of complex organic matter it is not possible, at the present time, to predict in all cases which calcium salt will precipitate, or when. Performing the experiment is also the only means to know how much time will be necessary to initiate a recrystallization and obtain a given crystallinity rate.

Changes in milk on heating

The changes in calcium and phosphate partition which occur when milk is cooled or gently heated in the range 80-100°C for a short period are almost completely reversible. The free calcium ion activity very nearly reverts to its original value if given enough time. No drastic change is to be found in the structure of Colloidal Calcium Phosphate under normal milk processing conditions. Simply, an exchange of calcium and phosphate ions occurs between milk serum and micelle. As already quoted, casein is an effective inhibitor of what might also be considered as ectopic calcification in a dairy processing plant. Milk is a very stable system and a considerable difference with a purely inorganic solution is the equilibrium of calcium and phosphate ions with MCP. It is wrong to extend the phase precipitation observed in a milk ultrafiltrate to the situation in milk where such an additional source of stability exists. Suggestions that "calcium phosphate precipitates on the casein micelles on heating" is, in fact, a mistaken view in which the milk serum behaves much as it would in the absence of caseins. Only under aggressive conditions does an eggshell-type deposit gradually accumulate on heat exchanger surfaces as milk flows past, containing ACP with small amounts of poorly crystallized β-TCP. Conversely, an irreversible process in the case of an ultrafiltrate where precipitation is easy, establishes an equilibrium with at least one new phase.

Another type of irreversible process is the breakdown, at temperatures above 80°C, of lactose to lactic, formic and other acids, producing a pH decrease on cooling.

The action of heat on milk proteins

Casein micelles are remarkably stable to processing environments, that is they do not aggregate for some appreciable time at temperatures above 100 °C. Nonetheless, both serum proteins and caseins are affected by heating. Heat provides a means of denaturation, principally of serum proteins, and also provides the energy to enable a number of chemical reactions to occur. In addition to direct effects upon the protein themselves, heating also increases the amount of Ca^{2+} bound to caseins, and may also initiate the precipitation of solid calcium phosphates, in addition to MCP.

At temperatures about 80 °C, β-lactoglobulin gets denatured. Unfolding of the molecule results in a thiol group being free to react by an exchange reaction with disulphide groups. That is, a single structure is replaced by multiple structures with new disulphide bridges also being intermolecular instead of intramolecular. The result is aggregation and gel building-up. Other new disulphide bridges may emerge from thiol groups of κ-casein and possibly also with α_{s2}-casein. Such complexes are formed with denatured β-lactoglobulin and also α-lactalbumin bound to the casein micelles.

At temperatures in excess of 100 °C, the serum proteins may be extensively bound to casein micelles. This in itself will lead to changes in the properties of the latter, which may explain certain aspects of the heat stability of milk, especially the effect of forewarming. When milk is held for some time at 90 °C prior to extensive heating at 120 or 140 °C, aggregation at the higher temperature is modified. It is to be assumed that forewarming allows time for some of these denaturation reactions and for binding β-lactoglobulin to κ-casein, which have insufficient time to occur when the milk is heated directly at the higher temperatures.

The causes of heat-instability are not uniquely defined, but it is possible to list several known reactions, which help defining the course of the process : (i) the alteration of micelle surface properties by binding denatured serum proteins. (ii) the decrease in pH which accompanies heating (down to as low as 5.5). (iii) alteration of lysyl residues by Maillard reaction, or by reaction with cyanate. (iv) deposition of calcium phosphates. Heated milk serum rapidly precipitates additional poorly soluble calcium phosphates. The deposition in this way may well be a contributory cause of casein precipitation.

Heat effects in whey and whey products

At pH close to neutrality (e.g., pH 6.8), heating of whey will produce increased turbidity but little sedimentation. In contrast, whey proteins become extremely heat stable at pH values below 3.9, even after extensive heating (90 °C for 20 min).[25,26] In the intermediate pH range, typical for most whey processing operations, heating will result in formation of sedimentable aggregates containing some or all the individual heat-labile whey proteins. Many of the underlying mechanisms are still poorly understood, including the involvement of the individual whey proteins in the inorganic fraction.

As soon as sweet wheys ($5.80 < pH < 6.20$) and acid wheys (mean pH = 4.60) arrive at the factory, they contain precipitated citrate, phosphate and calcium. After preconcentrating (30 % of dry matter) and storage below 10 °C, the inorganic instability of acid wheys is more pronounced than that of sweet wheys. Brushite is detected in the precipitate but the nature of calcium citrate is not yet determined.[21]

The cheese whey ultrafiltration permeate is very sensitive to calcium phosphate precipitation initiated by pH increase or heat treatment, even at temperatures as low as 20 or 50 °C.[23]

FOULING FROM MILK AND HEAT PROCESSING OF MILK-PRODUCTS

Introduction

Heat transfer plays a key role in food engineering. Different kinds of heat exchangers have been developed for different processes. Their function is to transmit heat to a fluid or between two fluids according to a given thermal specification. While flow and heat transmission in exchangers are now well known phenomena, fouling of surfaces remains one of the major problems not yet completely solved.[2,3,18] This is especially true in the case of food liquids such as milk which contain heat sensitive compounds. Fouling in the food industry is more severe than in other industries. In the petrochemical industry it is common to clean only annually or less frequently, compared to the daily (or even more frequent) cleaning needed in food plants.

The processing of milk products represents a large portion of the food industry. The large quantities of milk and derivatives treated are generally processed in plate and tubular heat exchangers with temperatures ranging from 70 °C (pasteurization) to 130 °C (cooking of milky desserts) or 140 °C (UHT milk sterilization).

Fouling dynamics

Milk is always heat-pretreated whatever the transformation process but other products derived from milk are also of industrial interest (e.g., cream, whey, ultrafiltration permeate and retentate, milky desserts).

Milk

Lalande and Corrieu[27] and Roignant et al.[28] established the rate of growth of milk deposits to be approximately proportional to the square of time t (s) (between 1.85 and 2.05):

$$m_f = at^2 \tag{3}$$

where m_f is the mass of deposit per unit area ($kg.m^{-2}$) and a is a constant.

This implies, in terms of heterogeneous reaction, that the deposition flux φ_d is governed by an autocatalytic reaction with an order of 0.5.

$$\varphi_d = b\, m_f^{0.5} \tag{4}$$

where φ_d is the rate of deposition (in $kg\ m^{-2}\ s^{-1}$) and b is a constant.

Concerning the main parameters which influence this deposition flux, most of the works have highlighted the key role played by the bulk temperature at which milk is heated and the temperature gradient. In the 120-140°C temperature range (UHT sterilization) the mechanism of fouling seems more complex than for $T < 120°C$ since a linear law as well as a decreasing or a slightly increasing rate has been observed at pilot-plant scale where the stability of the fouling conditions was not ascertained. The absolute fluid velocity in a heat exchanger affects deposit formation.[29]

Milk derivatives

The fluids directly related to milk and most intensively investigated are whey,[22] whipping cream,[30] and chocolate flavoured desserts.[31] With the aim of understanding the mechanisms of fouling by milk components the following fluids were studied: whey and citrate, desalted whey, UF permeate (decalcified or not), phosphate and calcium solutions and whipping cream with a modified casein/whey protein ratio.[2] It can be concluded that when proteins (especially whey proteins) are present, fouling occurs at an increasing rate (as for milk); sweet whey will give more fouling than milk. Owing to the absence of casein micelles, whey protein aggregates are not hindered in transport to the surface and more calcium ions will associate with serum proteins and with inorganic phosphates, resulting in aggregates more prone to fouling. Acid whey will give more fouling than sweet whey. The pH of the former is closer to the isoelectric pH of the whey proteins: a lower electrostatic repulsion results in a higher probability of sticking onto the protein layer already deposited. Whey protein concentrates will give more fouling simply because of their higher concentration as long as the total solid content is lower than 25 % (at higher total solids a changed fouling mechanism leads to a reduction of deposit mass).

Sour cream/cultured butter milk will give strong fouling owing to the poor stability of the casein micelles at the prevailing pH; when the solutions are purely inorganic, the dynamics are linear. This indicates that the autocatalytic rate previously mentioned is mainly related to protein behaviour.

Composition of milk fouling deposits

Milk deposits do not contain any carbohydrates, they show a low fat content, their protein phase is mainly composed of serum proteins and the inorganic phase is represented by calcium salt precipitates (essentially phosphates). The heat exchanger sections containing the most troublesome deposit are those operating at temperatures around 100 °C where the deposit is composed of 50 to 60 % proteins, 30 to 35 % inorganic matter and 15 to 20 % fat. This deposit called "type A", looks spongy and has a low density, which causes a major problem in exchanger functioning, namely an increase in pressure drop. In sections where temperature ranges from 130 °C to 140 °C (sterilization) the "type B" deposit is different in appearance and in composition. It has a high density and corresponds to the well-known "milkstone". It is composed of up to 75 % inorganic salts.

In the eighties, it was demonstrated that "type A" deposit is much richer in proteins and inorganic compounds as compared to milk (referred to dry matter). But, most importantly, the casein fraction decreases in deposits while the soluble protein fraction increases greatly, mainly from an increase in β-lactoglobulin concentration. The spongy structure of "type A" deposits is due to a network mainly composed of the protein phase. The inorganic part appears to be included in this network (as aggregates) but also forms a thin (tens of μm) but dense layer between the spongy structure and the exchanger wall.[32]

Fouling by different milk components

Serum proteins

As soon as a stainless steel surface comes into contact with a serum protein solution, even at room temperature, a monolayer of protein adsorbs. Then, a prerequisite for continued deposition is the presence of some intermediate of the denaturation reaction of the serum proteins. The deposition process on top of a monolayer can be described as follows: A. formation of some "fouling intermediate" in the bulk of the solution; B. transport of this intermediate to the surface; C. deposition.[18]

Formation of "fouling intermediate"

The partly unfolded and so-called activated β-lactoglobulin (β-LG) molecules act as the "fouling intermediate" in the model for denaturation and aggregation of β-LG. The activation energy for protein denaturation (300 $kJ.mol^{-1}$)[33] is significantly higher than the activation energy for milk fouling: 87 $kJ.mol^{-1}$ by Fryer[33] and 90 $kJ.mol^{-1}$ by Lalande and Corrieu.[27] It is likely that different controlling reactions have different rate-limiting steps and therefore different activation energies. Depending on the authors, the governing reaction in deposit formation may be β-lactoglobulin denaturation or aggregation in the bulk of the fluid and, furthermore, aggregation on the surface above 90 °C.

Whatever the model, the aggregation of β-LG is described by analogy with a radical chain reaction; upon rising the temperature, a thiol group of β-LG is exposed and becomes accessible to disulphide - thiol exchange reactions. This activated molecule acts like a "radical". The rate of deposition is proportional to the concentration of these activated β-LG molecules.

Transport to the surface and deposition
Transport to the surface is determined by the flow rate and the geometry of the apparatus. During this transport activated molecules of β-LG can be inactivated (reaction with another activated β-LG molecule or with κ-casein at the surface of the casein micelles). This may explain, when comparing the amount of denatured β-LG in the milk to the corresponding amount of protein deposited, why only 0.14% of the denatured β-LG finally ends up in the deposit layer.[35] The deposition and binding of an activated protein molecule most probably occurs through the formation of a disulphide bond with a deposited molecule.

Not only β-LG but also α-lactalbumin, bovine serum albumin and immunoglobulins contribute to the fouling process. They are likely to take part in the fouling reaction through thiol and/or cystein groups. Besides, if casein micelles and fat globules covered with β-LG molecules reach the wall while β-LG is still in its active form, they may be involved in the fouling process.

Inorganic compounds
In milk calcium may react mainly with phosphate and citrate ions. The first phase of calcium phosphate observed is usually CHPD : $CaHPO_4 \cdot 2\ H_2O$[36], which may later be converted into a more stable form of calcium phosphate. At pH 6.7 simulated milk ultrafiltrate (SMUF) becomes turbid upon heating at 45°C or higher. Precipitation of calcium phosphate may thus already take place on the stainless steel wall of a heat exchanger from 45 °C upwards. The presence of casein and citrate reduces such a precipitation.

In principle, it is possible to calculate from the solubility product variation how much salt would precipitate upon heating at a given temperature. Unfortunately, the calculation does not hold for the situation in a heat exchanger where high temperatures are applied and where no equilibrium is reached.

From the results of Pouliot et al.,[37] it can be assessed that only 0.2 % of inorganic salts transferred from the serum to the colloidal phase is deposited on the stainless steel wall. Apparently, such salts do have a much higher preference for deposition onto the stainless steel wall (0.5 m^2 available area) than for association with the proteins (4 000 m^2 available area). This could mean that inorganic deposition is partly driven by the temperature difference (ΔT) between the bulk and the surface. This is further confirmed by the observation that a larger amount of inorganic compounds is found for the deposit in the final pass of the heating section than in the subsequent holding section, where $\Delta T = 0$.

Calcium phosphate may precipitate directly on the stainless steel wall or deposit as particles formed in the bulk of the solution, as observed on heating SMUF. In milk, however, it is expected that calcium and phosphates hardly form particles, but associate with the proteins. Consequently, in milk, calcium phosphates precipitate, in addition to the direct deposition, together with the milk proteins, mainly the serum proteins.

Apart from the role of calcium phosphates in the process of fouling, calcium also influences protein deposition in its ionic form. In whey the presence of calcium is essential for the growth of a protein deposit layer. The presence of positively charged calcium ions means a reduction in the electrostatic repulsion and hence the formation of larger aggregates upon heating serum proteins.

Caseins
Upon heating to 80-90 °C both the size and the mutual interactions of the micelles increase through association with b-LG. Since b-LG is reactive towards casein micelles and stainless steel, this molecule may act as a sticking agent between the micelles and the stainless steel surface. In the same conditions, part of k-casein dissociates from the casein micellar surface, and may participate in the deposit. If the colloidal stability of casein micelles in milk is decreased (proteolytic action, change in calcium ion activity, lowering of pH) the denaturation process may no longer be controlled by denaturing serum proteins but directly by coagulation of casein micelles starting where the temperature is higher, i.e., at the stainless steel wall. Then, the mechanisms for deposition of casein micelles onto heat exchanger wall may be similar to that of the heat coagulation of milk.

Fat
Although fat is present in deposit layers it does not play an important role in the process of fouling. In spite of the high fat content of whipping cream (85 % of total solids) the fouling behaviour of this cream resembles that of whole milk.

Lactose
Lactose will contribute to a deposit only at high temperatures (> 100 °C) when caramelization or Maillard reaction can take place.

Dissolved air
On heating milk, air bubbles arise as a result of the presence of dissolved air. If these bubbles are formed at the heating surface they may act as nuclei for deposit initiation. The surface dries up, hence ΔT increases, causing an evaporation of water at the vapor/liquid interface. Then, milk is transported from the bulk to the surface where the air/vapor bubble is attached, and milk proteins accumulate and may coagulate and deposit on the surface. The contribution of air/vapour bubbles to the deposit is determined by the amount of air in the milk, ΔT, the operational pressure in the heat exchanger and the wall shear stress. In film evaporators for

milk products, a similar phenomenon may occur through an insufficient distribution (parts of the surface poorly wetted) or as a result of nucleation on the wall after local boiling.

Influence of various parameters on fouling

Decreasing the pH to 6.4 causes a strong increase in fouling, mainly due to additional deposition of caseins, while a pH increase to 6.9 results in much less deposit. Either action on calcium concentration in milk leads to more fouling, owing to the effect of calcium on the stability of the casein micelles. The presence of air bubbles at the surface enhances fouling. If serum proteins are absent from milk, fouling is reduced by two thirds. Increasing the serum protein concentration results in increased fouling, and so does increasing T or ΔT (Table 5), especially because of the denaturation of serum proteins and precipitation of calcium salts. In particular, the temperature at which maximum deposition occurs in a heat exchanger corresponds to the temperature of maximum protein denaturation.[38] Higher Reynolds numbers (Re) reduce fouling, likely in relation with the wall shear stress. Aged milk causes more severe fouling as compared to fresh milk, owing to proteolytic enzymes produced by psychotrophic bacteria.

Table 5. The effect of various parameters on fouling as compared to normal milk (↑ = parameter increase ; ↓ = parameter decrease ; + = more fouling ; - : less fouling)

	pH	[Ca]	Air	[b-LG]	T	ΔT	Re	Ageing
↑	-	+++	++	+++	+	+	--	++
↓	+++	++	--	--	-	-	+	

Prevention of fouling

Upon heating milk there is a region in the heat exchanger where fouling is most severe. This critical region is determined by the concentration of activated serum protein molecules. Fouling in heat exchangers can be reduced by controlling the formation of these activated molecules. Possible ways of achieving this include:

- installing a holding section with a high volume to area ratio. By using a stirred tank as a holder, the amount of deposit is reduced by more than 50 %.[39]

- applying steam injection/infusion to shorten the period during which the concentration of activated molecules is high.

- applying such high shear stresses in the critical region that deposition of foulants will be prevented.

- applying a temperature profile in such a way that fouling is evenly spread over the various passes of a plate heat-exchanger to prevent one pass from being so rapidly fouled that it blocks the whole equipment.[40] A similar effect is achieved by applying a preheating of milk before pasteurization ; in practice, the selection of preheating temperatures is a matter of compromise to reduce downstream fouling without causing too much deposition in the preheating section.[41]

Other ways to reduce fouling include :

- degassing the milk, applying high operational pressures and high shear rates ; these procedures may prevent air bubbles from sticking to the wall, resulting in less fouling ;

- preventing psychotrophic bacteria from producing extracellular proteolytic enzymes in milk by applying thermalization to raw milk, by avoiding dead spaces in the processing lines and by draining tanks and pipelines completely ;

- desalting, adding Ca^{2+} binders and adjusting pH ; for certain milk products these may be tools to diminish fouling. However, in many cases, this may be precluded because of constraining product specifications.

Conclusions

- The two major mechanisms of fouling during heat treatment of milk are the formation and the subsequent deposition of activated serum protein molecules (heat denaturation) and the precipitation of calcium phosphates (directly or together with proteins). These foulants are formed in the bulk of the fluid and are continuously transported to the surface where they can be deposited. A third fouling mechanism, driven by a high wall temperature, i.e. a large ΔT, plays a minor role ; it only may enhance the direct precipitation of calcium phosphate particles onto the stainless steel heating surface.

- Calcium not only contributes to the deposit by precipitation of calcium salts, but also by its influence on the denaturation and aggregation of serum proteins and on the stability of casein micelles.

- An overview of the fouling mechanisms during heating of whey and milk can be schematically presented. At room temperature, a monolayer of proteins is immediatly adsorbed. Upon heating, active β-LG molecules and aggregates form, their deposition being enhanced by the presence of calcium. Calcium phosphates may precipitate directly on the stainless steel wall, promoted by a large ΔT, or associate with β-LG aggregates, or with micellar calcium phosphate if casein micelles are present. In milk the active β-LG molecules may thus entrap micelles in the deposit.

- If conditions in milk are different from normal, other processes than serum protein denaturation and inorganic precipitation may take over as the rate determining step in fouling ; for example, the coagulation of casein micelles at low pH, or the formation of protein films when air bubbles, surrounded by proteins, are present at the heating surface and collapse.

FOULING IN MEMBRANE PROCESSING OF DAIRY PRODUCTS

Introduction

Four main membrane processes, ultrafiltration (UF), reverse osmosis (RO), microfiltration (MF) and nanofiltration (NF) have successively developped in the dairy industry since the beginning of the seventies. The two major fluids processed are milk and whey with the aim of concentrating proteins, removing bacteria, separating caseins from serum proteins and fractionating proteins. As in heat exchangers, proteins and inorganic compounds are involved in membrane fouling.

Protein fouling

Protein deposition
It is generally accepted that in UF, NF and RO a surface layer forms on the membrane surface and dominates the subsequent behaviour of the membrane. In MF, both pore plugging and surface layer formation have been observed, even with proteins considerably smaller than the pores. It is also clear that, once a surface layer has formed on an MF membrane, the selectivity of the membrane is ruled by the surface layer regarless of the original pore size of the membrane. Nonetheless MF fouling is far from beeing fully understood.

The proteinaceous fouling layer is made-up of both adsorbed and multilayer deposited proteins. Protein adsorption is an irreversible process and is usually entropically driven by the structural rearrangement of the adsorbed protein molecules.[42] Both hydrophobic and ion-charge interactions influence protein adsorption. As a general rule, higher permeate fluxes have been achieved with hydrophillic than with hydrophobic membranes. However, where the protein concentration at the interface is high, the effect of hydrophobicity is masked.[43] On the other hand, the charge on the protein (either negative or positive), rather than the difference in charge between the membrane and the protein, determines the degree of protein deposition. The permeate flux is controlled by protein species loosely or dynamically bound to the membrane rather than strongly bound or adsorbed on the membrane.

Influence of feed parameters on protein fouling
The permeate flux is smallest at the isoelectric point (IEP) of the whey proteins and increases as the pH is adjusted on either side of the IEP. With milk, the permeate flux decreases when the pH is decreased. This is related to the form of the casein

present, which has a tendency to favor an increasingly compact fouling structure with minimum interstitial pore space.

Variations of the ionic strength have also been shown to affect the quantity of protein deposited and the permeate flux. At the IEP in the presence of ions, anion binding leads to an increase in the size of the protein, thus increasing the porosity of the deposited layer. Away from the IEP, added salts tend to reduce the electrostatic repulsion by shielding charges, causing molecular contraction and thereby decreasing permeability.

Inorganic fouling: calcium and phosphate

As previously discussed, precipitation of calcium phosphates is a complex phenomenon, and even more so within a membrane system where the concentration and fractionation of the milk components induce severe changes in the protein and inorganic equilibria. Interesting results have been obtained by combining infrared and x-ray photoelectron spectroscopies to investigate the nature of the material deposited on zirconium oxide membranes. For example, irreversible fouling of an M-Milk Carbosep membrane (Molecular Mass Cut-off, MMCO = 70 $kg.mol^{-1}$) by milk was reported to be the result of proteins, 2 % (of the ZrO_2 mass present in the fouled membrane) compared to 0.4 % with whey. The corresponding values for calcium phosphate fouling were 0.18 % with milk, and up to 0.5 % with whey.[43] Therefore, the contribution of calcium phosphates to fouling is more severe with whey than with milk.

In UF of a defatted clarified whey protein concentrate, the permeate flux was smaller at pH 6.9 than at pH 5.6,[45] whereas a higher protein fouling was measured on the surface of the membrane at pH 5.6 with increased phosphate fouling on the surface and within the pores at pH 6.9. Moreover, a change was noticed for phosphates. A precipitation as HAP (pH 6.9) was observed supplementary to a simple phosphate adsorption at lower pH values (pH 5.6).

From an extensive study,[46,47] it can be concluded that, when calcium and phosphate ions are present in the whey with operating conditions (temperature, pH, calcium and phosphate concentration, etc.) favoring precipitation of calcium phosphates, the latter get important relative to proteins, the fouling layer being dense with poor permeability. When the content of calcium and phosphate ions is lowered by using appropriate pretreatment conditions, proteins become the main constituent of the fouling layer which is then more permeable.

Membrane processing of milk and whey

UF concentration of milk proteins

The major application of milk UF in the dairy industry is for protein concentration before cheesemaking.

During milk UF using 0.2 μm pore size Membralox MF membrane, the surface deposit (5 μm thickness) contained all the initial constituents of milk in the upper layer and a larger quantity of serum proteins close to the membrane surface. Calcium and phosphate contents of the deposit were high, suggesting possible precipitation of calcium phosphate within this layer.[48]

UF and MF processing of whey

Ultrafiltration of whey has emerged as a successful technique to concentrate whey proteins, which have high nutritional and functional properties, for the production of whey protein concentrates (WPC) of various purities (35-95 % protein/dry matter) depending on whey pretreatment and membrane process used (ultrafiltration or diafiltration).

A few investigations pinpoint the role of proteins and that of inorganic compounds with respect to their participation in flux decline with time. The following phenomena involved during UF with M4 Carbosep membranes (MMCO = 10 $kg.mol^{-1}$) were suggested: A. interaction of proteins with ZrO_2 and B. formation of a ZrO_2 - phosphate (Na-Ca) - protein complex on the membrane surface. Traces of fat also found within the fouling layer were not attributed any significant role in fouling growth.

Pretreatment of whey by MF has been suggested to remove residual fat and phospholipids which alter WPC functionality and inorganic compounds which hinder UF performance. An industrial process is based on the tendency of part of residual fat to aggregate to calcium phosphates under pH change (>7) and heat treatment (> 55°C, 8-10 min). MF is used to separate the aggregate mass from the whey. During the process, fouling is due to pore blocking, which induces a decline of both permeability and protein transmissicn in the filtrate. There is a synergistic contribution of aggregates and proteins to fouling in MF of pretreated whey. Proteins form an irreversible layer on the surface as well as deeper in the membrane itself. Aggregates made up of a calcium phosphate-protein complex form a reversible cake which is the major contributor to the overall fouling.[49]

Fouling in NF of dairy products

NF permeation flux of dairy products is affected mainly by concentration polarization. It accounts for 67-95 % of the flux decline in NF of skim milk, depending on pH and temperature.[50] Kelly et al.[51] show severe fouling due to calcium phosphate precipitation when the pH of acid casein whey is raised above the IEP of whey protein (around 5.2). Jeantet[52] reports that the precipitation of brushite is likely. The precipitate would be associated to the carboxylate moities of aspartic and glutamic acid of whey proteins. These protein - $CaHPO_4$ structures, which participate in the formation of the protein layer, would constitute nuclei for further calcium phosphate precipitation.

Fouling in RO of cheese whey

Hiddink et al.[53] state that protein and calcium phosphates are responsible for membrane fouling during reverse osmosis of cheese whey. Again, it is claimed that special attention should be paid to the interaction between calcium phosphates and proteins. Both amorphous and crystalline calcium phosphates form complexes with proteins, the former precipitating with the latter.

Okazaki and Kimura[54] suggested a model for the precipitation of calcium phosphate at the membrane surface, which can also be applied to the deposit formation during reverse osmosis of cheese whey.[55] In this model a second order reaction from the crystallization kinetics is used. The model contains several parameters which are not exactly known or totally unknown. However it can also be evaluated by using lumped parameters :

$$\frac{dR_f}{dt} = K'\left(e^{-J/K} - \frac{C'}{C}\right)^2 \qquad (5)$$

where $K' = \varepsilon K C^2/\rho$ and $R_f = 0$ at $t = 0$

with R_f = time dependent part of hydraulic resistance caused by fouling ($Pa.s.m^{-1}$)
t = time (s)
K' = constant
J = permeate flux ($m.s^{-1}$)
k = mass transfer coefficient related to calcium phosphate ($m.s^{-1}$)
C = concentration of solid calcium phosphate in the bulk ($kg.m^{-3}$)
C' = saturation concentration of calcium phosphate ($kg.m^{-3}$)
ε = resistance per unit of fouling layer thickness ($Pa.s.m^{-2}$)
K = precipitation rate constant ($m^4.kg^{-1}.s^{-1}$)
ρ = density ($kg.m^{-3}$)

The values of K', K and the ratio C' /C are to be estimated from experimental data; k depends on flow velocity and K' is related to the bulk concentration.

Prevention of fouling: pretreatment of the feed

In order to select and/or design the most appropriate pretreatments for dairy fluids, it is necessary to consider first the fluid composition with A. residues from processing (casein fines, residual fat, products from proteolysis, enzymes, microorganisms) and B. unstable or reactive physicochemical systems (proteins, calcium, phosphates, soluble casein).[56] Considering the nature and complexity of the composition factors responsible for long term fouling in membrane processing, two types of pretreatment options have emerged, namely physical and chemical pretreatments.

Physical pretreatments

Conventional prefiltration can be utilized to remove suspended material such as casein fines, which tend to produce pore blocking, especially when large pore size MF membranes (0.2 - 1.4 μm) are used. Centrifugation desludging (disc-bowl type separation) has the advantage that residual fat and even bacteria can sometimes be removed from the feed in the same unit operation.

Treatments involving heating and holding the feed at moderate temperature for an extended time (50-60 °C, 8 min to 2 h) have been suggested to minimize fouling in the case of cheese whey. As a general procedure, heat treatments can be chosen in order to precipitate calcium phosphate salts and reduce ionic calcium in the feed. If the dairy fluid contains a colloidal phase (casein micelles), heating and holding will lead to the enrichment of this phase with respect to Ca and P, while measurable amounts of Ca^{2+} and ionic phosphates will decrease. Such heat treatments prior to milk RO cause an improvement of permeate flux.[57] However, if no casein micelles are present in the fluid, calcium phosphate will precipitate and nucleating sites such as impurities or soluble phosphorylated compounds (peptides, phospholipids) will promote the growth of an amorphous precipitate. Owing to the lowering of the ionic content (Ca and inorganic phosphates) the risk of further crystallization in the microporous matrix of the membrane upon separation will diminish.

The combination heat treatment-crossflow microfiltration (0.2 - 0.1 μm) is highly performing to achieve clarification, bacteria removal, defatting, ionic calcium and phosphate lowering. Operating conditions that caused the maximum reduction in the level of calcium and phosphate in rennet whey[47,58] also result in the lowest fouling in ultrafiltration. Besides, in MF of pretreated whey aggregates of lipids and calcium phosphates as a reversible layer is the major contribution to the overall fouling. Among the measures that should be considered to reduce such fouling the first to be considered is altering the layer porosity, that is the size of the aggregates. This is best achieved by a heat treatment (55 °C, 6 min) at constant pH 7.5, which results in bigger particles (above 50 μm) and low overall ionic calcium and phosphate content.[58] Such a pretreatment, combined to microfiltration or not, is reported to be an efficient means to limit considerably the fouling in NF of acid whey.[52]

Chemical pretreatments

Long term fouling of membranes can be minimized by adjusting the physicochemical properties of the feed. The control of soluble calcium and phosphate species is often essential when demineralization of the feed is either too expensive or not feasible owing to the further utilization of the feed. The adjustment of pH at lower values will improve the solubility of whey proteins, calcium and phosphate, whereas increasing the pH will lower the level of ionic calcium. A similar control of the latter can also be achieved by addition of sequestering agents[59] which will limit or prevent calcium phosphate precipitation. Alternatively, an increase of ionic strength in the feed (addition of NaCl or an evaporative concentration) are other means of limiting further calcium phosphate precipitation.

Nonetheless, it should be considered that the use of chemicals requires a very accurate control and monitoring of the physicochemical parameters (pH, Ca^{2+}, etc.) of the feed. The extent of chemical modification is often very limited, and thus detrimental conditions can easily be reached, leading to other fouling problems associated with extreme pH values, excess of chemicals, or even destabilization of other components in the fluid being processed.

Conclusions

- The decline in permeability of solvent and solutes experienced in all pressure-driven membrane processes is due to concentration polarization, particle cake build-up and fouling.

- Concentration polarization (diffusive molecules) and cake build-up (particles with small diffusion coefficient) are unavoidable in membrane processing. Their influence on the membrane performance can be minimized by improving the hydrodynamics over the membrane surface, what is most simply achieved by increasing the wall shear stress.

- In the processing of dairy fluids, fouling is very complex because of the multiple contributions from different milk components. The general consensus is that a fouling layer forms on the membrane surface and that its properties control the membrane behaviour. Generally protein deposition and precipitation of inorganic compounds dominate fouling.

- The conditions that lead to the precipitation of calcium salts are not well characterized and it is not possible at the present time to predict accurately whether this will occur. Furthermore, problems are experienced only when calcium phosphates precipitate during the filtration process. Precipitation is more likely with high values of temperature, pH, calcium concentration, and also when casein micelles (whey/milk) and proteins (permeate/milk and whey) have been removed.

- With operating conditions resulting in calcium phosphate precipitation, fouling is severe and dominated by insoluble particles (with the participation of proteins). When this precipitation is avoided, proteins remain the main constituent of the fouling layer.

- In almost all instances, protein adsorption and deposition occur on the membrane surface. By and large, the membrane permeability is controlled by the fouling layer thickness and porosity rather than by proteins being strongly bound to the membrane. These parameters are controlled by the properties of the feed stream : pH, ionic strength, hydraulic parameters. With skim milk, the characteristics of casein micelles control the properties of the fouling layer. With whey, the permeate flux is lowest at the IEP of the proteins.

- In addition to its role in inorganic precipitation, calcium makes an important contribution to protein fouling, either through protein-protein interactions in the fouling layer or through protein-membrane interactions.

- Irreversible membrane fouling can result in major process cost increases owing to permanent membrane damage, formation of hard-to-remove deposits leading to lower process performance as well as requirements for excessive cleaning. Minimization of fouling phenomena by appropriate pretreatments is thus important, not only for a short term flux increase, but especially for a long term process reliability. Numerous pretreatments have been successfully proposed, the aim of which is to limit the effect of proteins and inorganic compounds. Physical pretreatments (heat, separation by centrifugation or by microfiltration) and chemical pretreatments (pH and ionic strength adjustment, addition of sequestering agents, etc.) are among the most appropriate means to get rid of undesirable components (fat, casein fines, micro-organisms, ionic calcium and phosphate) prior to membrane processing (MF, UF, RO). It should not be forgotten, however, that any pretreatment is associated with the corresponding cost and thus the appropriate selection criteria must be realistic in view of the overall cost effectiveness of the operation.

SUMMARY

With the same fluid to be treated, a striking similarity exists between a heat exchanger and a membrane, as both are physical surfaces on which nucleation can occur. The mechanisms are not identical in the detail, but a mixed material results from the simultaneous presence of complex polymer organic materials (proteins in general, including caseins) and ionic species capable to build-up poorly soluble compounds (calcium phosphates) interacting with the former, both by calcium and phosphate ends. Whether the first reaction with the surface offered is of hydrophillic or hydrophobic nature is just a detail of mechanism.

The main difference lies in the temperature level. When it is lower (for membranes), the prevailing factors include pH, total phosphate and calcium contents together with the exact nature of the proteins (linked to Ca^{2+}). For heat exchangers, at higher temperature levels, the additional phenomenon of protein denaturation results in a more complex fouling, beginning with type A deposit that is rich in proteins. As the temperature increases, the protein content progressively decreases to the highly inorganic "milkstone" for the higher temperatures.

In both cases, regardless of differences in the mechanism, total fouling figures or composition of the layers formed, the result is always a decrease of performance (heat or mass transfer) and therefore a necessity for frequent cleaning.

REFERENCES

1. Sandy C, Singh RK. Energy increase in operation and cleaning due to heat exchanger fouling in milk pasteurization. Food Techn., 1991;45(Dec):84-91

2. Lalande M, René F, Tissier JP. Fouling and its control in heat exchangers in the dairy industry. Biofouling, 1989;1:233-250

3. Fryer PJ, Belmar-Belny MT, Schreier PJR. Fouling and cleaning in milk processing. In *Heat-induced changes in milk* (P.F. Fox, Ed). Int. Dairy Fed., 1995.

4. Taddéi C, Daufin G, Aimar P, Sanchez V. Role of some whey components on mass transfer in ultrafiltration. Biotechnol. Bioeng., 1989;34:171-179

5. Marshall AD, Munro PA, Tragardh G. The effect of protein fouling in microfiltration and ultrafiltration on permeate flux, protein retention and selectivity: a literature review. Desalination, 1993;91:65-108

6. Int. Dairy Fed. Fouling and cleaning in pressure driven membrane processes. Special issue 9504. 1995

7. Marschall AD, Daufin G. Physico-chemical aspects of membrane fouling by dairy fluids. In Fouling and cleaning in pressure-driven membrane processes. Int. Dairy Fed., Special issue 9504, 1995: 9-35

8. Holt C. Effect of heating and cooling on the milk salts and their interaction with caseins. In *Heat-induced changes in milk* (P.F. Fox, Ed.). Int. Dairy Fed., 1995.

9. Fox PF. The milk protein system. In *Developments in Dairy Chemistry - 4*: PF Fox (Editor),. Elsevier Applied Science, London, 1989.

10. Rollema HS. Casein association and micelle formation. In *Advanced Dairy Chemistry - 1 : Proteins*: PF Fox (Editor),. Elsevier Applied Science, London, 1992.

11. Holt C. The milk salts: their secretion, concentrations and physical chemistry. In *Developments in Dairy Chemistry - 3*. PF Fox (Editor), Elsevier Applied Science, London, 1985.

12. White JCD, Davies DT. The relation between the chemical composition of milk and the stability of the caseinate complex. I. General introduction, description of samples, methods and chemical composition of samples. J. Dairy Res. 1958;25:236-255

13. Brulé G. The minerals in milk. Rev. Lait. Fr. 1981;400:61-65

14. Schmidt DG, Both P. Studies on the precipitation of calcium phosphate. I. Experiments in the pH range 5.3 to 6.8 at 25 and 50 °C in the absence of additives. Neth. Milk Dairy J. 1987;41:105-120

15. Walstra P, Jenness R. *Dairy Chemistry and Physics*. New-York :Wiley-Interscience, 1984.

16. Lyster RLJ. Calculation by computer of individual concentrations in a simulated milk salt solution. II. An extension of the previous model. J. Dairy Res. 1981;48:85-89

17. Fox PF. (Editor), *Heat-induced changes in milk.*, Belgium:International Dairy Federation, 1995.

18. Jeurnink TJM, Walstra P, de Kruif CG. Mechanisms of fouling in dairy processing. Neth. Milk and Dairy J. 1996;50,(3):407-426

19. Heughebaerts JC, Nancollas GH. Kinetics of crystallization of octacalcium phosphate. J. Phys. Chem. 1984;88:2478-2481

20. Schmidt DG, Both P, Visser S, Slangen KJ, van Rooijen PJ. Studies on the precipitation of calcium phosphate. II. Experiments in the pH range 7.3 to 5.8 at 25 and 50 °C in the presence of additives. Neth. Milk Dairy J. 1987;41:21-136

21. Saulnier F, Ferrero F, Bottero JY, Linden F. Variation of the composition and nature of the insoluble precipitate from industrial wheys. Lait 1995;75:93-100

22. Daufin G, Labbé JP, Quémerais A, Brulé G, Michel F, Roignant M, Priol M. Fouling of a heat exchange surface by whey, milk and fluid models. An analytical study. Lait 1987;67:339-364

23. Pouliot Y, Landry J, Giasson J. Induction of calcium phosphate precipitation in sweet whey permeate. Lait 1991;71:313-320

24. Lyster RJJ. The equilibria of calcium and phosphate ions with the micellar calcium phosphate in cow's milk. J. Dairy Res. 1979;46:343-346

25. Jelen P, Rattray W. Thermal denaturation of whey proteins. In *Heat-induced changes in milk* PF Fox, Ed. Int. Dairy Fed., Special issue 9501, 1995.

26. Britten M, Giroud HJ, Jean Y, Rodrifue N. Composite blends from heat denatured and undenatured whey proteins: emulsifying properties. Int. Dairy J. 1994;4(1):25-36

27. Lalande M, Corrieu G. Fouling of a plate heat exchanger by milk. In *Fundamentals and applications of surface phenomena associated with fouling and cleaning in food processing*. B Hallstrom, D Lund, C Tragardh, Eds. Food chemical Center, University of Lund (S), 1981.

28. Roignant M, Daufin G, Michel F, Kerhervé FL, Brulé G. Heat treatment of milk ultrafiltration retentate (x2) in a plate heat exchanger: thermal and hydrodynamical aspects and fouling as compared to milk and to whey, Lait 1986;66:207-232

29. Belmar-Belny MT, Gotham SM, Fryer PJ, Pritchard AM. The effect of Reynolds number and fluid temperature in whey protein fouling. J. Food Engng. 1993;19:119-139

30. Hiddink J, Lalande M, Maas AJR Streuper A. Heat treatment of whipping cream. 1: Fouling of the pasteurization equipment. Milchwissenschaft 1986;41:542-546

31. René F, Leuillet JC, Goldberg M, Lalande M. Fouling of a plate heat exchanger while UHT treating of chocolate flavoured dessert; comparison to milk. Lait 1988;68:85-102

32. Tissier JP, Lalande M. Experimental device and methods for studying milk deposit formation on heat exchange surfaces. Biotechnol. Prog. 1986;2:218-229

33. Kessler HG, Fiedler J, Hege WU. Formation of deposits from liquid foods on heated surfaces during heating and evaporation. Chem. Ing. Tech. 1986;58:475-485

34. Fryer PJ. Modelling heat exchanger fouling. Ph.D. Thesis, University of Cambridge, GB. 1986.

35. Jeurnink TJM, de Kruif CG. Calcium concentration in milk in relation to heat stability and fouling. Neth. Milk Dairy J. 1995;49:151-165

36. Van Kemenade T. Influence of casein on precipitation of calcium phosphates. Thesis University of Utrecht, The Netherlands. 1988.

37. Pouliot Y, Boulet M, Paquin P. Observations on the heat induced salt balance changes in milk. I. Effect of heating time between 4 and 90 °C. J. Dairy Res. 1989;56:185-192

38. Hegg PO, Castberg HB, Lundh G. The fouling of whey proteins on stainless steel at different temperatures. J. Dairy Res. 1985;52:213-218

39. de Jong P, Bouman S, van der Linden HJLJ. Fouling of heat treatment equipment in relation with the denaturation of ß-lactoglobulin. J. Soc. Dairy Technol. 1992;45(1):3-8

40. Delplace F, Leuliet JC, Tissier JP. Fouling experiments of a plate heat exchanger by whey protein solutions. Trans I. Chem. Eng. 1994;72 (Part C):63-169

41. Burton H. Fouling of heat exchangers. In *Ultra-high-temperature processing of milk and milk products* H. Burton Ed. Elsevier Applied Science, London (England) 1988.

42. Norde W. Adsorption of proteins from solution at the solid liquid interface. Adv. Colloid Interface Sci. 1986;25:267-340

43. Rolchigo PM, Raymond WA, Hildebrandt JR. Improved control of ultrafiltration using vorticular hydrodynamics and hydrophillic membranes. Process Biochem. 1989;24(2),Pro Bio Tech iii - vii

44. Daufin G, Mérin U, Labbé JP, Quémerais A, Kerhervé FL. Cleaning of inorganic membranes after whey and milk ultrafiltration. Biotechnol. Bioengng. 1991;38:82-89

45. Labbé JP, Quémerais A, Michel F, Daufin G. Fouling of inorganic membranes during whey ultrafiltration : analytical methodology. J. Membr. Sci. 1990;51:293-307

46. Daufin G, Labbé JP, Quémerais A, Michel F. Fouling of an inorganic membrane during ultrafiltration of defatted whey protein concentrates. Neth. Milk Dairy J. 1991;45:259-272

47. Daufin G, Labbé JP, Quémerais A, Michel F, Mérin U. Optimizing clarified whey ultrafiltration: influence of pH. J. Dairy Res. 1994;61:355-363

48. Vetier C, Bennasar M, Tarodo de la Fuente B. Study of the fouling of a mineral microfiltration membrane using scanning electron microscopy and physicochemical analyses in the processing of milk. J. Dairy Res. 1988;55:381-400

49. Gésan G, Daufin G, Mérin U, Labbé JP, Quémerais A. Fouling during constant flux cross-flow microfiltration of pretreated whey. Influence of transmembrane pressure gradient. J. Membr. Sci. 1993;80:131-145

50. Eckner KF, Zottola EA. Effects of temperature and pH during membrane concentration of skim milk on fouling efficiency. Milchwissenschaft 1993;48:187-191

51. Kelly PM, Horton BS, Burling H. Partial demineralization of whey by nanofiltration. In *New Applications ofMembrane Processes*. Int. Dairy Fed. Special Issue, 9201, 1992.

52. Jeantet R, Nanofiltration of dairy fluids. Ph D. Thesis. National School of Agriculture ENSA of Rennes. 1995.

53. Hiddink J, de Boer R, Nooy PFC. Reverse osmosis of dairy products. J. Dairy Sci. 1980;63:204-214

54. Okazaki M, Kimura S. Scale formation on reverse osmosis membranes. J. Chem. Eng. Jpn. 1984;17 (2):145-151.

55. Van Boxtel AJB, Otten ZEH, van der Linden HJLJ. Evaluation of process models for fouling control of reverse osmosis of cheese whey. J. Membr. Sci. 1991;58:89-111

56. Pouliot Y, Jelen P. Pretreatments of dairy fluids to minimize long-term membrane fouling. In *Fouling and cleaning in pressure driven membrane processes*. Int. Dairy Fed. Special issue 9504, 1995.

57. Glover FA. Ultrafiltration and reverse osmosis for the dairy industry. Technical Bulletin 5. The National Institute for Research in Dairying, Reading , UK. 1985

58. Gésan G, Daufin G, Mérin U, Labbé JP, Quémerais A. Microfiltration performance: physicochemical aspects of whey pretreatment. J. Dairy Res. 1995;62:269-279

59. Patocka J, Jelen P, Calcium chelation or other pretreatments for flux improvements in ultrafiltration of cottage cheese whey. J. Food Sci. 1987;52:1241-1244

20

PHOSPHATE REMOVAL FROM WASTE WATER

M.M. Seckler

IPT Institute for Technological Research of the State of São Paulo
São Paulo, S.P., BRAZIL

O.S.L. Bruinsma and G.M. van Rosmalen

Delft University of Technology
Delft, The NETHERLANDS

INTRODUCTION

Eutrophication is commonly known as the state of a waterbody which is manifested by an intense proliferation of algae and higher aquatic plants and their accumulation in excessive quantities.[1] Eutrophication can be most efficiently fought by reducing the input of aquatic plant nutrients, mainly nitrogen and phosphorus, to surface waters. Therefore, policies to reduce the phosphate (P) input from several sources have been implemented worldwide.

Over the last decade emphasis has been placed on phosphate removal from point sources of nutrients such as municipal and industrial sewage. However, current practice has not been sufficient to reduce eutrophication to acceptable levels, so it is likely that stricter policies will be adopted in the future. In recent years, growing attention has been paid to diffuse nutrient sources such as surface runoff from urban and agricultural areas, atmospheric precipitation and regeneration of sediments. Methods have been developed that lead to recovery in a concentrated form, suitable for re-use, thus reducing the need for disposal of sludge as well as the phosphorus input to the environment.

Since virtually all methods for P-removal are implemented as a third step in conventional wastewater treatment processes, these are described first. Then the different phosphate removal methods are presented from a technological point of view. Finally, the phosphate removal by precipitation of calcium phosphate in a fluidized bed as a modern method for the recovery of phosphate from waste streams is discussed in more detail.

PHOSPHATE REMOVAL IN WASTEWATER TREATMENT PLANTS

Primary And Secondary Waste Water Treatment

Conventional treatment of domestic wastewater[2,3,4] usually consists of the so-called primary and secondary treatment, as illustrated in Figure 1. In the primary treatment a

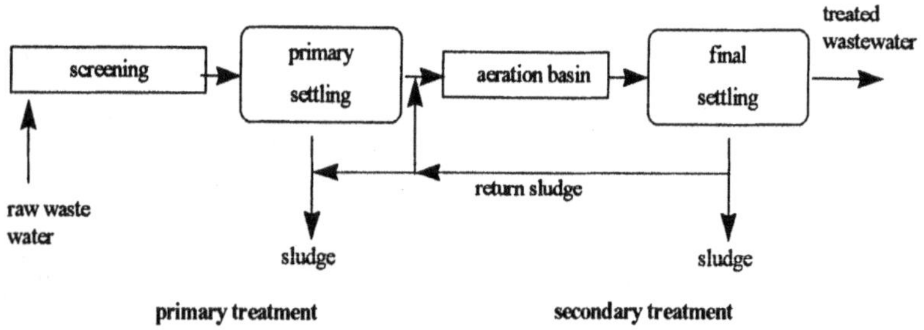

Figure 1. Conventional waste water treatment plant with primary and secondary treatment.

portion of the suspended solids and organic matter is removed from the wastewater through physical unit operations such as screening and sedimentation. Secondary treatment is directed mainly towards the removal of organic material and additional suspended solids through a combination of physical unit operations (e.g., sedimentation) and biological processes (e.g., activated sludge and fixed-media filter processes). The biological solids from the primary as well as from the secondary treatment are further treated for reuse, disposal or incineration. The purified water may be disinfected with chlorine prior to discharge to a receiving watercourse. Tertiary treatment is sometimes applied to remove nutrients, toxic compounds, organic material or suspended solids.

Industrial wastewater is often treated by the same methods as domestic wastewater, e.g., for the food industry (e.g., dairy industry, potato processing) and for the manure processing industry. Sometimes the domestic and industrial wastes are treated together in a single municipal treatment plant. In these cases, the P-load to the water is often substantially increased. So the treatment process must be designed accordingly.

Typically, domestic wastewater contains approximately 10 mg P/ℓ, mainly as orthophosphate (5 mg P/ℓ) and condensed phosphates (4 mg P/ℓ) as well as smaller amounts of organic phosphates.[5,6] However, during the aerobic biological process, much of the condensed phosphates are converted into orthophosphate. About 10 to 30% of the phosphate entering conventional treatment plants are removed naturally as part of the biological step. Effluent limitations are usually 1-2 mg total P/ℓ or lower, so additional measures for removal are required.[7]

Methods For Phosphate Removal

Since phosphate has been considered as a source of eutrophication a variety of techniques have been developed. In Table 1 a short overview is given of these methods together with some typical details of operation.

Table 1. Overview of phosphate removal methods.

Method	Details	References
	Calcium phosphate methods	
Lime addition	- pH = 8-10: $P_{effluent}$ = 1-2 mg/ℓ - pH = 10-11.5: $P_{effluent}$ = 0.3-1 mg/ℓ - lime is added to secondary clarifier (Figure 1) - lime consumption depends on water alkalinity not on P-content - carbonate is co-precipitated as $CaCO_3$	8, 5, 2, 9
Hydroxy apatite precipitation in a fixed bed	- $P_{influent}$ = 1-3 mg/ℓ; $P_{effluent} \approx 0.1$ mg/ℓ - effluent of secondary clarifier is treated (Figure 2) - fixed bed is phosphate ore or magnesia clinker - decarbonation and fines filtration necessary	10, 11
Calcium phosphate precipitation in a fluidized bed	- pH = 8-9 - single pass efficiency≈50%; efficiency is improved by recycle - water line: $P_{influent}$ = 5-10 mg P/ℓ; $P_{efflent}$ = 1-2 mg P/ℓ - sludge line: $P_{influent}$ = 20-100 mg P/ℓ; $P_{efflent}$ =1-5 mg P/ℓ - decarbonation unit and fines filtration required	12, 13, 14
Magnetic phosphate removal	- magnetite and flocculation agent are required - $P_{effluent} \approx 0.5$ mg/ℓ - phosphate is recovered as a slurry	16, 17
	Other methods	
Chemical addition of aluminum and iron salts	- primary, secondary or tertiary treatment - biological or phosphate sludge - stoichiometric consumption of chemicals - $P_{effluent}$ = 0.5-1 mg/ℓ	7, 18, 19
Ion exchange and struvite precipitation	- $P_{influent} \approx 4$ mg/ℓ; $P_{effluent} \approx 0.5$ mg/ℓ - NH_4^+ and $H_2PO_4^-$ are concentrated and exchanged with NaCl - $NH_4MgPO_4 \bullet 6H_2O$ is precipitated with a Mg-solution	20
Biological phosphate removal	- anaerobic/aerobic sequence - integrated in waste water treatment - $P_{efflent}$ = 1-2 mg P/ℓ	8, 7

Combined biological and chemical phosphate removal	- preconcentration in biological section - removal from side stream to chemical section -$P_{effluent} \approx 1$ mg P/ℓ	21, 7

Calcium Phosphate Precipitation

The first four methods in Table 1 are chemical precipitation methods where some form of calcium phosphate is formed. The simplest way is to add lime, as in the first method, resulting in a sludge which is hard to separate from the water. For this reason the three other methods have been developed in order to obtain a calcium phosphate that can be recovered and used in fertilizer industry: precipitation of hydroxy apatite in a fixed bed, precipitation in a fluidized bed and magnetic phosphate removal.

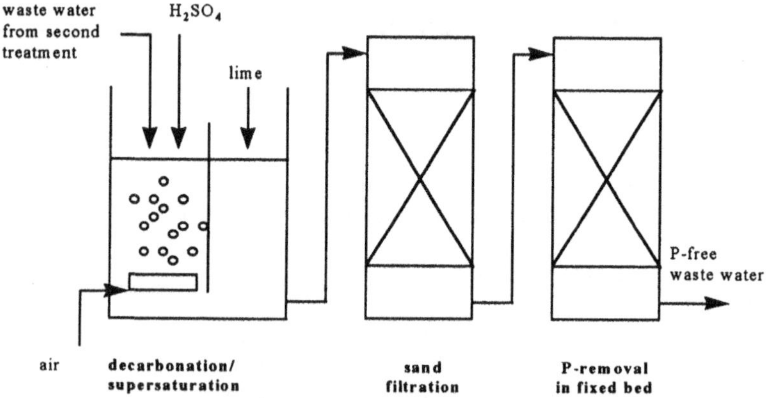

Figure 2. Fixed bed process for phosphate removal.

A method based on the crystallization of hydroxyapatite in fixed beds has been developed in Japan[10,11] (see Figure 2). The effluent of the secondary wastewater treatment is mixed with a base and eventually with a calcium source to create a supersaturated solution. Process conditions are chosen such that spontaneous precipitation of calcium phosphate is minimized. The supersaturated solution then passes through a fixed bed containing large apatite grains. P-removal is claimed to take place mainly by crystal growth on the apatite grains. A decarbonation unit is placed upstream the fixed bed to minimize calcium carbonate co-crystallization. A sand filter is also used to remove the spontaneously formed calcium phosphate.

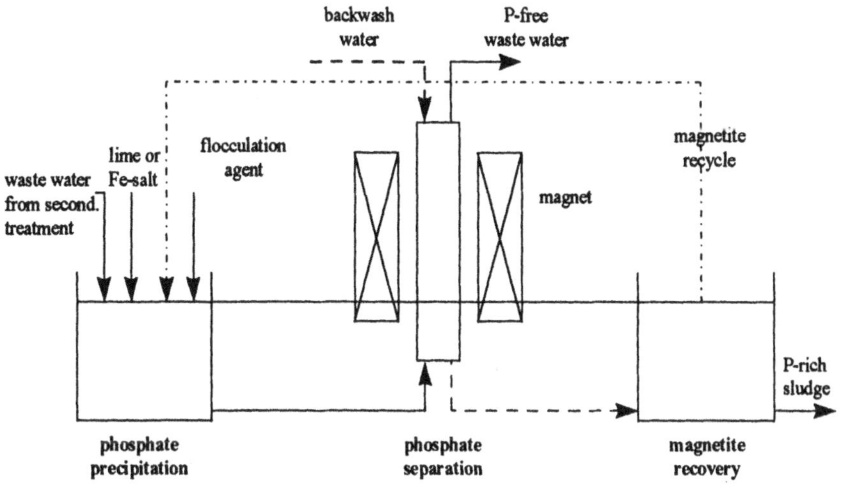

Figure 3. Magnetic phosphate removal.
——— : waste water; ········:magnetite slurry; – · –: backwash water

In the magnetic method for P-removal (Figure 3), first phosphate precipitates by chemical addition of lime or an iron salt. Magnetic grains (magnetite) and a flocculating agent are then added to the suspension to promote the attachment of the phosphate precipitate to the magnetic grains. The suspension is then passed through a magnet where the phosphorus containing magnetic grains are captured. These grains are removed intermittently by backwashing and submitted to intense mixing and ultrasonic treatment to separate the magnetite, which is recycled, from the phosphate precipitate. The phosphate is recovered as a slurry which can be reused in the agriculture or as a raw material for the fertilizer industry. The method has been developed to full scale.

Since in most cases, except for the fixed bed method, the phosphate is precipitated as an amorphous calcium phosphate (ACP) some characteristics of the types of calcium phosphates that can be formed are given in Table 2.

Several calcium phosphate modifications can be formed by precipitation from solution.[22] Normally a metastable phase is formed first and undergoes one or more solution mediated recrystallization steps until the thermodynamically stable modification precipitates. For low supersaturations the sequence in which each phase is formed is in accordance with the Ostwald rule of stages. For higher supersaturations and pH >7 one or more amorphous compounds are formed first. These latter conditions are normally encountered in processes for phosphorus removal. Some types of amorphous calcium phosphates (ACP) are shown in Table 2. For low concentrations of the ions calcium and phosphate (< 120 mM^2) and pH values between 7 and 10, ACP is a well defined compound, which is amorphous to both X-ray and electron diffraction and has a Ca/P molar ratio between 1.45 and 1.51. For pH values higher than 10 ACP has variable solubility and Ca/P molar ratio, suggesting a breakdown in the short-range order of the amorphous phase. At higher concentrations (>100 mM^2) a spherular form of ACP, called ACP1 is formed first and within 2 to 60 min transforms into a floccular amorphous phase (ACP2). The Ca/P molar ratios for both ACP1 and ACP2 are within the range 1.33 and 1.37. Other phases have also been observed as precursors of HAP crystallization within the range of concentrations and pH values just mentioned: Abbona and Franchini-Angela[23] found poorly crystalline HAP whereas Francis and Webb[24] precipitated an amorphous DCPD with a Ca/P ratio of 1.22.

Domestic waste waters contain ~10 mg total P/ℓ (~ 1 mM^2) whereas industrial waste waters may contain phosphate concentrations one or two orders of magnitude higher. In both cases, amorphous calcium phosphate formed during phosphorus removal processes are likely to be the ACP form just mentioned. For industrial waste waters, if pH values below 7 are applied, dicalcium phosphate dihydrate may precipitate as supermicrometric platelike particles.

Table 2 Summary of literature data on the characteristics of amorphous calcium phosphate (ACP).

solution pH (-)	solution Ca x P (mM^2)	crystal Ca/P (mol/mol)	note	references
7-10	7-120	1.48±0.03		24-30
9.25-12.5	24	variable	ACP structure varies with pH	27
7-8	25	-	HAP, poorly crystalline	23
7.4-9.4	300	1.35±0.02 1.35±0.02	ACP1, spherular ACP2, floccular	31
7.4	16-100	1.22	amorphous DCPD	32
9.8	37500	1.42 1.52	ions Cl^- ions NO_3^-	33,34

Other Methods

A widely applied method to remove phosphate is co-precipitation by adding aluminum or iron salts or lime, again producing a sludge that is hard to separate.[7] The chemicals may be added prior to the primary treatment, at the secondary treatment (see Figure 1) or as an independent tertiary treatment. Separation of the phosphate from the wastewater occurs respectively in the primary settler, secondary settler, or in an additional filtration step either incorporated in the biological sludge or as a separate phosphate slurry. The amount of chemicals needed to reach a given phosphorus concentration in the treated water is directly related to the stoichiometry of the precipitation reaction (except for lime). Therefore, the consumption of chemicals is proportional to the P-concentration in the raw water, making these methods attractive for waste waters with low P-concentration levels. Implementation of these methods require hardly any changes in existing sewage treatment plants, except in the case of chemical addition as a tertiary treatment. The processes have been applied on an industrial scale in many countries for several years. The concentration of the treated water is about 1 mg P/ℓ.[7,18] Chemical addition in two steps, at the secondary treatment and downstream of it, leads to effluents with a concentration of about 0.5 mg P/ℓ.[19]

Pre-concentration by ion exchange followed by precipitation of struvite or by biological methods are successful techniques to reduce the consumption of chemicals.

In Figure 4 the ion exchange/struvite precipitation process is shown as has been developed in Italy.[20] The process is based on the selective ion exchange of NH_4^+ and HPO_4^{2-}, both present in the waste water. The resins are regenerated with a 0.6 M NaCl solution. The regeneration eluates from both columns are mixed with a magnesium salt and NaOH and struvite is precipitated. The struvite is obtained by settling and filtration and the NaCl solution is returned to the storage vessel.

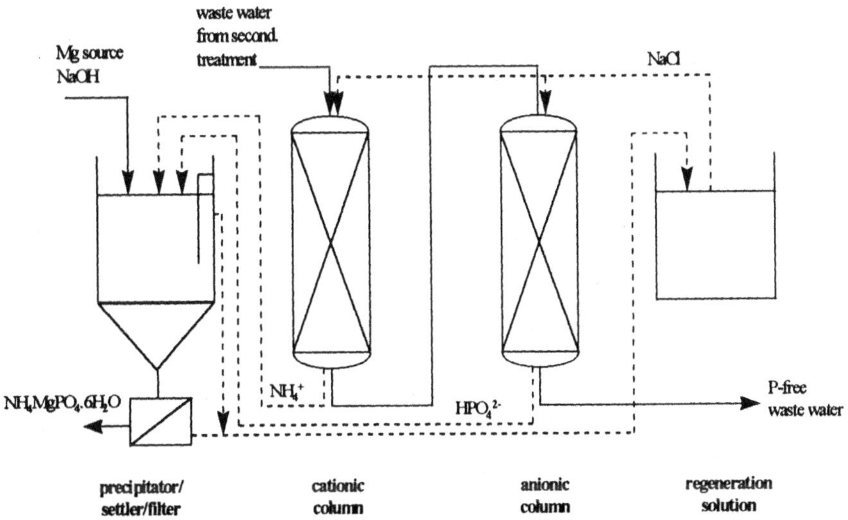

Figure 4. Flow diagram for the preconcentration of NH_4^+ and phosphate containing waste water followed by struvite ($NH_4MgPO_4.6H_2O$) precipitation. Solid lines: waste water, dotted lines: regeneration solution.

Biological phosphate removal[7,8] is based on the so-called 'luxury uptake' by certain microorganisms. Such uptake is stimulated by subjecting the microorganisms first to anaerobic conditions, where they release phosphate to the liquid phase, and next to an aerobic environment, where P-uptake occurs. The phosphate incorporated into the cell mass is removed in a so-called P-stripper (Figure 5, above dotted line). No additional sludge is produced in these processes compared to conventional activated sludge systems, but care must be taken to avoid phosphate dissolution and return to the system during the sludge handling. The method has been developed to full scale. The P-concentration in the treated wastewater depends on the wastewater characteristics (mainly the ratio between the biological oxygen demand and the phosphate concentration), but in general varies within the range of 1 to 2 mg P/ℓ. For low

P-concentrations in the raw wastewater, a higher phosphate removal efficiency is expected.[35]

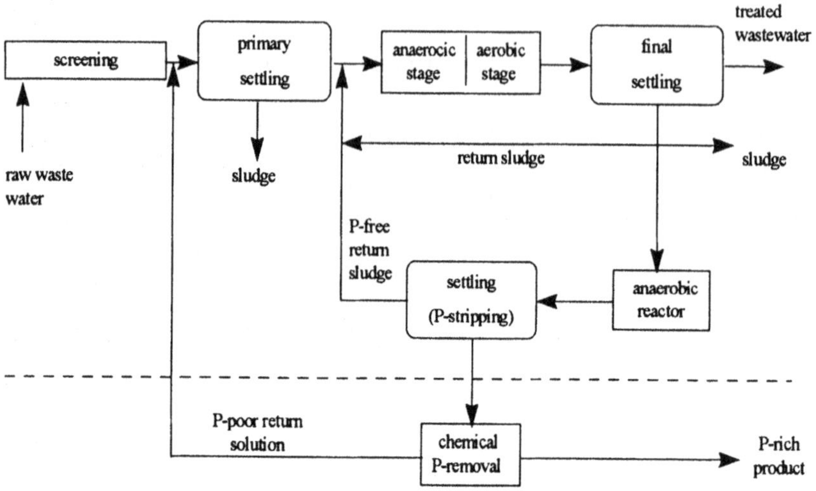

Figure 5. Flow diagram for the biological/chemical phosphate removal process. Section above the dotted line is the conventional biological method.

In the combined biological and chemical addition method,[7,21] Figure 5, the enriched P-stream from the stripper is treated by chemical methods for P-removal. Since this stream represents only 20 to 30% of the total plant flow, its treatment is economically more feasible. The method has been developed to full scale. This process, unlike other biological methods, is not very sensitive to the wastewater composition (biological oxygen demand to phosphorus concentration ratio) and leads to effluent concentrations of about 1 mg P/ℓ.

PHOSPHATE REMOVAL IN A FLUIDIZED BED

The Process

The fluidized bed process, shown schematically in Figure 6, is based on the precipitation of calcium phosphate by mixing a phosphate solution with calcium ions and a base. The waste water stream is first decarbonated and together with the recycle stream fed to the bottom of the fluidized bed. Calcium phosphate is precipitated on sand grains by adding a base and, if not yet present in the waste water, calcium either as lime or in solution. The P-rich grains are removed at the bottom and fresh seed grains are fed at the top.

The effluent of the bed contains dissolved P (at equilibrium level) and a certain amount of fines and is passed over a sand filter, where the fines are removed. The filtrate is the effluent but in order to improve the P-removal efficiency the filtrate is partly recycled over the fluidized bed. Intermittently the sand filter is cleaned with backwash water that is returned to the primary treatment. Optimization of the pH level at the end of the fluidized bed is focused on a high P-recovery on the grains, a low amount of fines and a small recycle stream. The fluidized bed can be operated in the water line, where all waste water is treated, or in the so-called sludge line (Figure 5) where the concentrated effluent of the P-stripper is the feed to the fluidized bed. The latter option considerably reduces the use of chemicals and the size of the installed equipment.

Chemistry And Efficiency

The addition of base to the phosphate stream results in a shift to the left in the following phosphate equilibria:

$$PO_4^{3-} \overset{H^+}{\Leftrightarrow} HPO_4^{2-} \overset{H^+}{\Leftrightarrow} H_2PO_4^- \qquad (1)$$

resulting in an increase in the supersaturation for amorphous calcium phosphate (ACP), and thus creating the driving force for ACP precipitation:

$$\beta = \frac{1}{5}\ln\frac{(Ca^{2+})^3.(PO_4^{3-})^2}{K_{s,ACP}} \qquad (2)$$

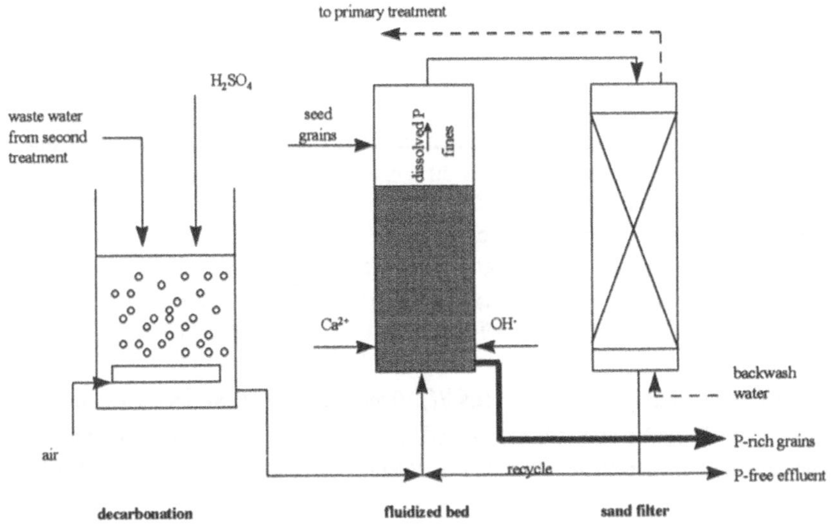

Figure 6. Fluidized bed process for phosphate removal as P-rich grains with decarbonation unit and sand filter.

The subsequent processes of nucleation, molecular growth and agglomeration are all kinetically controlled by the supersaturation levels in the fluidized bed. The conversion of dissolved phosphate into ACP is the result of nucleation and molecular growth, but in order to have an efficient P-removal this has to be achieved by agglomeration of the ACP onto the grains and by cementing of the agglomerates to a particle that is mechanically stable against attrition in the fluidized bed.

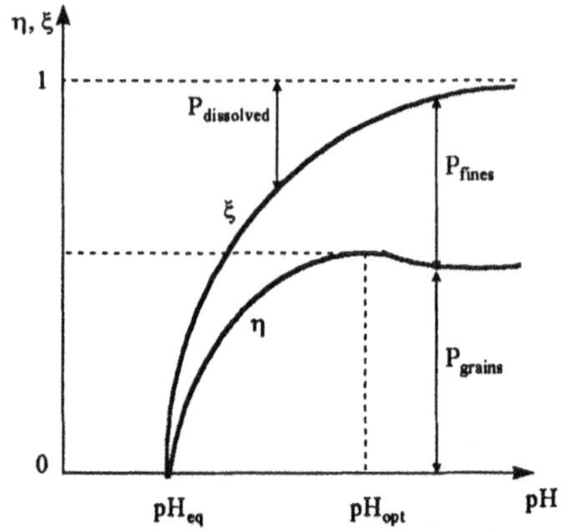

Figure 7. The conversion, ξ, and the efficiency, η, in the fluidized bed.

These complex phenomena result in conversion, ξ, and efficiency, η, curves as shown in Figure 7. The pH level has to be above the saturation level, pH_{eq}, in order to obtain any precipitation. While the conversion curve steadily approaches full conversion with rising pH, the efficiency passes through a maximum at pH_{opt}. Beyond this maximum the phosphate is converted at increasing amounts into fines that will leave the fluidized bed and eventually block the sand filter. The fluidized bed should be operated close to this optimum pH level.

The shape of the η - pH curve depends on the composition of the waste water and a maximum efficiency (without recycle) of 50% can be achieved. In general, the calcium ions supply should be enough to bring the inlet Ca/P molar ratio to the value of 3. In the absence of magnesium and carbonate the base supply should be enough to reach a conversion of 50-60%. In their presence a conversion of 80-95% leads an optimum efficiency. The outlet pH values corresponding with these conditions are in the range 7.5-9. Generally, a low inlet phosphate concentration favors high efficiencies. The ACP that is formed can suitably be represented by the chemical formula $Ca_3(PO_4)_2$, having a pH independent solubility of 3×10^{-27} $(mol/\ell)^5$ (based on activities). Decarbonation to a level of 1.8×10^{-3} mol/ℓ or lower is required in order not to disturb the P-removal and magnesium ions in the waste water can be tolerated at inlet M/P molar ratios of 2 or less.

Physical Processes And Grain Characteristics

The most important mechanism for phosphate removal in a fluidized bed is the agglomeration of primarily formed calcium phosphate particles with the grains constituting the bed. It accounts for ~ 60% of the total phosphate removed. The remaining 40% is removed by molecular growth. The supersaturation in the fluidized bed decays with a time constant of ~ 7 s, a short time compared to the mean residence time of the liquid phase of ~ 100s. This implies that the upper part of the bed is practically saturated and both agglomeration and molecular growth are therefore restricted to only a small section at the bottom of the bed.

Typical SEM-pictures of grains covered with a phosphate layer are shown in Figure 8. Crystalline structures were identified by X-ray diffraction. Figure 8a shows a grain covered with an ACP-layer, smoothed by abrasion at the bottom of the fluidized and produced at a pH above 7 in the absence of magnesium and carbonate. Figure 8b is obtained from the middle of the bed and exhibits an irregular shape that is attributed to fines that are agglomerated and cemented to the grain. These grains will move down the bed by removal of P-covered grains at the bottom and enter the zone of increased energy dissipation where the ACP layer will transform towards that in Figure 8a. At higher P-concentrations in the influent (~ 20 mg P/ℓ) calcium phosphate can be precipitated at pH 6-7 platelike dicalcium phosphate dihydrate (DCPD = $CaHPO_4.H_2O$) crystals are formed (Figure 8c).In the presence of magnesium ions hexagonal platelike $Mg_3(PO_4)_2{\cdot}22H_2O$ crystals are formed together with ACP (Figure 8d).

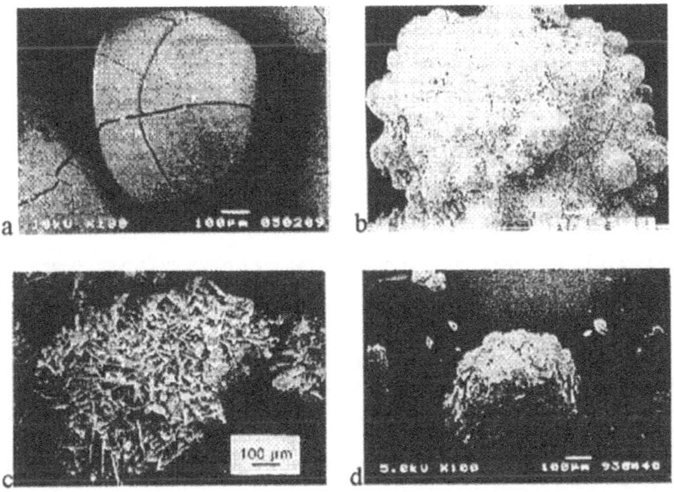

Figure 8. SEM views of grains covered with (a) ACP as formed at the bottom of the fluiduzed bed, (b) ACP from the middle of the bed, (c) $CaHPO_4.H_2O$ crystals and (d) $Mg_3(PO_4)_2.22H_2O$ crystals.

Hydrodynamic Regime

The fluidized bed for phosphate removal is filled with grains in the range 0.2-0.6 mm and is operated at a superficial velocity of about 40 m/hr, corresponding with a residence of about 100 s. The waste water inlet is responsible for mixing at the bottom. For the precipitation of compounds other than ACP (like $CaCO_3$ in softening of drinking water) the inlet jet is usually operated at a velocity in the order of 1 m/s in order to reduce the local supersaturation by mixing and to limit primary nucleation. However, in the case of calcium phosphate this is not successful and it is better to lower the jet velocity in order to avoid abrasion (see Figure 8a) and to spread the chemical dosing in order to enhance the agglomeration over the full bed height (Figure 8b). Seckler et al.[36] showed that with two dosing points for the chemicals and the use of a low superficial velocity in combination with small grains increased the optimum P-removal efficiency to ~ 80%.

REFERENCES

1. Ryding SO and Rast W, *The control of eutrophication of lakes and reservoirs, Man and the biosphere series, Volume I*, U.N.E.S.C.O., Paris and The Parthenon Pub. Group, Carnforth, U.K., 1989

2. Fresenius W, Schneider W, Böhnke B and Pöppinghaus, *Waste water technology: origin, collection, treatment and analysis of waste water*, Springer-Verlag, Berlin, p.1136, 1989

3. Hammer M.J. *Water and wastewater technology*, John Wiley & Sons, New York, 1986

4. Tchobanoglous G and Burton FL. *Wastewater engineering: treatment, disposal and reuse*, McGraw-Hill Pub. Co., New York, 1991

5. Eilbeck WJ and Mattock G. *Chemical processes in waste water treatment*, Ellis Horwood Ltd., Chichester, England, 1987

6. Corbitt RA. Wastewater disposal, in: *Standard handbook of environmental engineering*, Corbitt, R.E. (Ed.), McGraw-Hill Inc., New York, N.Y., p.6.1-6.274, 1990

7. Bowker RPG and Stensel H.D. *Phosphorus removal from wastewater*, Noyes Data Corp., New Jersey, p. 114, 1990

8. Henze M, Harremoës J les CJ and Arvin E. *Wastewater treatment, biological and chemical processes*, Springer-Verlag, Berlin, 1995

9. Fergusson JF, Jenkins D and Eastman J. Calcium phosphate precipitation at slightly alkaline pH values, *J.W.P.C.F.* 45 (4), p.621-631, 1973

10. Hirasawa I and Toya Y. Fluidized bed process for phosphate removal by calcium phosphate crystallization, ACS Symp. Series 438, p.355-363, 1990

11. Joko I., Sawada,S, Goto C and Toyokura K. Phosphorus removal from wastewater by crystallization method: preparation and performance of artificial seed materials, Industrial Crystallization 84, Jancic, S.J. and de Jong, E.J. (Eds.), Elsevier Sci. Pub., Amsterdam, p.431-434, 1984

12. Seckler MM, Bruinsma OSL, van Rosmalen GM, van Dijk JC, Delgorge F and Eggers E. Phosphate removal by means of a full scale pellet reactor, 11th Int. Symp. Indust. Cryst., 18-20 Sept, Garmisch-Partenkirchen, Germany, 1990

13. van Dijk, J.C. and Eggers, E., Defosfatering met een korrelreactor: een Nederlandse ontwikkeling met toekomst, H_2O 20(3), p.63-68, 1987

14. van Dijk JC and Braakensiek H. Phosphate removal by crystallization in a fluidized bed, Wat. Sci. Tech. 17(2-3), p.133-142, 1985

15. Hirasawa I., Shimada K and Osanai M. Phosphorous removal process from wastewater by contact crystallization of calcium apatite, Proc. Third Pacific Chem. Eng. Congr., Seoul, Korea, Volume IV, p. 259-264, 1983

16. van Velsen AFM, de Reuver JL and van der Vlugt AJ. Ontwikkeling van magnetische defosfatering, H_2O 23(1) p.2-8, 1990

17. van Velsen AFM, van der Vos G and de Reuver JL. Defosfatering door magnetische separatie, H_2O 21(15), p.402-411, 1988

18. de Jong P and van Starkenburg W. Defosfateren van huishoudelijk afvalwater, H_2O 22(4), p.122-123, 1989

19. Pelloni L. Removal of residual phosphorus and suspended solids by contact filtration, Prog. Wat. Tech. 10(5/6), p.255-271, 1978

20. Liberti L, Limoni N, Longobardi C, Lopez A, Passino R, Kang SJ and Horvat INPJ. Field demonstrations of the rim-nut process for nutrients recovery from municipal wastewater, Nucl. and Chem. Waste Management, 8, p.83-86, 1988

21. Rensink JH. High biological nutrient removal from domestic wastewater in combination with phosphorus recycling, Wat. Sci. Tech. 23,1991

22. Nancollas GH, Lore M, Perez L, Richardson C and Zawacki SJ. Mineral phases of calcium phosphate, Anat. Rec. 224(2), p.234-241, 1989

23. Abbona F and Franchini-Angela M. Crystallization of calcium and magnesium phosphates from solutions of low concentration, J. Cryst. Growth 104, p.661-671, 1990

24. Francis MD and Webb NC. Hydroxyapatite formation from a hydrated calcium monohydrogen phosphate precursor, Calc. Tiss. Res. 6, p.335-342, 1971

25. Füredi-Milhofer H, Brecevic L and Purgaric B. Crystal growth and phase transformation in the precipitation of calcium phosphates, Faraday Discuss. Chem. Soc. 61(Precipitation), p.184-193, 1976

26. Lundager Madsen HE, Lopez VI., Lopez AV and Boistelle R. The formation product of amorphous tricalcium phosphate at 37°, J. Crystal Growth 75(3), p.429-434, 1986

27. Meyer JL and Weatherall CC. Amorphous to crystalline calcium phosphate phase transformation at elevated pH, J. Colloid and Interface Sci. 89(1), p.257-267, 1982

28. Termine JD and Posner AS. Calcium phosphate formation in vitro. I. Factors affecting initial phase separation, Arch. Biochem. Biophys. 140, p.307-317, 1970

29. Tropp J, Blumenthal NC and Waugh JS. Phosphorus NMR study of solid amorphous calcium phosphate, J. Am. Chem. Soc. 105, p.22-26, 1983

30. Walton AG, Bodin WJ, Furedi H and Schwartz A. Nucleation of calcium phosphate from solution, Canadian J. Chem., 45, p.2696-2701, 1967

31. Christoffersen J, Christoffersen MR, Kibalczyc W and Andersen FA. A contribution to the understanding of the formation of calcium phosphates, J. Cryst. Growth 94, p.767-777, 1989

32. Boskey AL and Posner AS. Conversion of amorphous calcium phosphate to microcrystalline hydroxyapatite. A pH dependent, solution mediated, solid-solid conversion, J. Phys. Chem. 77(19), p.2313-2317, 1973

33. Eanes ED and Posner AS. Kinetics and mechanism of conversion of noncrystalline calcium phosphate to crystalline hydroxyapatite, Trans. New York Academy of Sciences, Meeting of the Div. of Biophysics, Nov. 18, 1965

34. Eanes ED, Gillessen IH and Posner AS. Intermediate states in the precipitation of hydroxyapatite, Nature 208, p.365-367, 1965

35. Janssen PMJ, Smoes E and Rensink JM. Drastische fosfaatverlaging in afvalwater en de gevolgen voor biologische defosfatering, H_2O 23(1), p.6-8, 1990

36. Seckler MM, van Leeuwen MLJ, Bruinsma OSL and van Rosmalen GM. Wat. Res. 30 (7), p. 1589-1596, 1996

21

PHOSPHOROUS SPECIATION IN SEDIMENTS AND ITS RELEASE CONTROL

Kazuaki Ito and Mitsumasa Okada

Department of Environmental Science, Faculty of Engineering
Hiroshima University, Higashi-Hiroshima 739 JAPAN

INTRODUCTION

For the prevention of eutrophication and algal bloom, it is necessary to effectively reduce the concentration of essential nutrients such as phosphorous (P) and nitrogen (N) in the water bodies to growth-limiting levels. The reduction of P input by pretreatment and reduction of influent water by using bypass are promising methods. If water flows rapidly and the nutrient recycling from sediments is small, the eutrophic state of water bodies can be significantly improved. At Lake Washington, for example, the treated wastewater which was discharged into the lake was redirected such that it is now discharged directly to sea.[1] This approach resulted in a significant decrease in the concentration of dissolved phosphorous to one-fourth its initial value, improving the water quality.

However, the improvement of the water quality is often more difficult to accomplish than simply reducing the amount of nutrients which flow into water bodies. One reason for this is due to release of nutrients from sediments. These sediments act as a constant source of nutrients. It was estimated that an increase (65-100%) in phosphorous content in two eutrophicated Ohio lakes during the summer season was from such internal loading.[2]

The general techniques commonly employed to suppress the P release from sediments are dredging, aeration, scattering of chemical reagents, and capping the sediments with relatively inert or active materials including solid wastes. Of these, sediment removal is the most effective technique for the removal of P but is often prohibitively expensive. Thus, alternative techniques have been examined.

This chapter summarizes the speciation of sediment phosphorous (particulate P) and its bioavailability. Secondly, phosphorous release from sediments is presented under the various environmental conditions, in relation to chemical background of phosphorous retention and release. Finally, the techniques for control of P release from sediments are described with respect to P retention mechanisms, cost of each procedure, and their effects on living organisms.

PHOSPHOROUS IN SEDIMENTS

Phosphorous fractionation in sediments

The sediment phosphorous is generally divided into inorganic and organic phosphorous. Fractionation of the sediments was carried out to characterize the overall phosphorous dynamics including availability for plants and planktons.

The initial work involving the sequential chemical extraction and separation of phosphorous compounds from sediments was reported by Chang and Jackson.[3] They divided the sediment phosphorous into labile P, aluminum-bound P, iron-bound P, calcium-bound P, reductant-soluble P, and refractory P. However, the procedure used by these investigators has several shortcomings, including the extraction of Fe-bounded P in the extraction procedure of aluminum-bound P with 0.5 M ammonium fluoride and the resorption of P with calcium fluoride precipitates.[4,5] Subsequently, some sequential chemical extraction and other techniques concerning P exchangeability across sediment - water interface were developed and modified by other researchers for better characterization of sediments.[6-8] Despite this continued research, the procedure of P fractionation has still not been established because of inherent uncertainty in the structure of fractionated P compounds. Therefore, instead of identifying the discrete phosphorous compounds, sediment phosphorous constituents have been mainly divided into non-apatitic inorganic P (NAI-P), apatite P, and organic P.

The chemical extraction scheme of phosphorous in sediments by Hieltjes and Lijklema,[6] has been popularly recognized and is shown in Figure 1. Sediment phosphorous is divided into labile (loosely bounded) P (NH_4Cl-RP), iron- and aluminum-bound P (NaOH-RP, NaOH-nRP), calcium-bound P (HCl-RP), and residual P. Residual P consists mainly of organic phosphorous and is given by the difference between total P and the sum of reactive (inorganic) fractions.

The ratio of each P fraction in sediments with phosphorous content 0.6-6.5 mg-P/g (dry weight) has been reported by Petterson et al[8-10] as follows: A. inorganic P content is roughly comparable to organic P content with large deviations. B. of the inorganic P forms, iron- and aluminum-bound P or non-apatite inorganic phosphorous (NaOH-RP), are major fractions in most lake sediments with up to 62% of total phosphorous. C. the apatite phosphorous (HCl-RP), with some exceptions, is a larger fraction for some lake sediments. D. labile phosphorous (NH_4Cl-RP) is generally small (in the range of 0 - 7.6%).

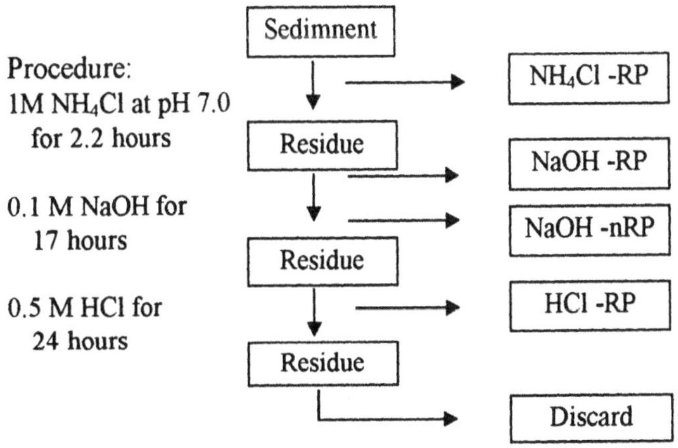

Figure 1. The phosphorous fractionation scheme (RP, soluble reactive P; nRP, the difference between total dissolved P and soluble reactive P).[6]

The bioavailability data of each P fraction by using two types of bioassay techniques, batch and two-chambered systems with algae, were summarized and reviewed by Boström et al.[11] According to Boström et al. the bioavailability of P fractions can be defined into two categories. The first category consists of those P fractions which are immediately available for planktonic algae and bacteria. Orthophosphate ($H_2PO_4^-$, HPO_4^{2-}, or PO_4^{3-}) seems to be the only directly available P source except for direct uptake of some organic forms of phosphorous. In the second category are those P fractions which can be transformed to a bioavailable form by naturally occurring processes such as desorption, dissolution, enzymatic degradation, etc. Thus, if the P fractions in Figure 1 can be transformed to orthophosphate under natural conditions, the fraction can be considered to be potentially bioavailable P fraction.

The general conclusions reached by Boström et al.[11] were as follows: A. NaOH-extractable P (Al-, Fe-bound P) was the most bioavailable P fraction in most studies, although the rates for bioavailability were different between samples. B. HCl-extractable P (apatite P) was inversely correlated to algal-extractable P. C. organic P was usually less available by algae but present to some extent in sediment of eutrophicated area.

Vertical distribution

The vertical distribution of P in sediment can be considered to be the history of sedimentation and diagenetic processes within sediments. Figure 2 shows the depth profile of idealized distribution of P in sediments with the assumption of a constant sedimentation.[12] Similar profiles have been observed in lake sediments.[13] Particulate phosphorous in mineral (inorganic) and organic tends to be deposited in the sediment. However, the distribution of various P compounds will vary with depth. For example, Fe- and Al-bound P (NAI-P) fractions decrease strongly with sediment depth due to decreasing redox potential. Apatite P, which is an inorganic P form, retains its structure at most depths. Additionally, apatite can be formed in the sediment by diagenetic processes. Organic P forms are also transformed to mineral P. Organic P decreases with sediment depth, suggesting slower mineralization compared to sedimentation. This occurs despite the observation that metabolic degradability is usually higher at the surface of sediment due to higher oxygen concentration. Of course this ideal P profile is naturally affected by environmental factors including soil erosion, oxygen levels and pH which affect the sorption capacity of the surface sediment, physical and biological activity of the benthic organisms, and hydrodynamics.

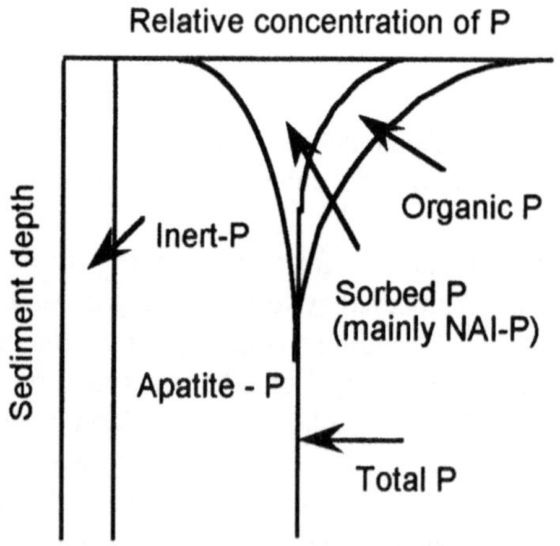

Figure 2. Idealized distribution of sediment phosphorous assuming a constant accumulation.[12]

Figure 2 also indicates that the increase in sediment depth results mostly in the mineralization, especially apatite formation by diagenetic processes, and compaction of sediment leading to reduced water content. Also, the larger P content of surface sediment on a dry weight basis is explained by P loading from the water body, a delay in mineralization, and accumulation of precipitated phosphorous by diffusion from the deeper sediments layers.

Chemical background on sorption

The sorption of phosphorous on the compounds of iron, aluminum, and calcium has been reported.[14] Aluminum (Al) compounds such as aluminum sulfate or other aluminum salts have been used for P precipitation in processes involving wastewater treatment, drinking water, and P inactivation due to retarding P release from lake sediments. When aluminum- and iron-based compounds, for example, are added to lake water, P in water is removed by sorption onto the surface of colloidal/amorphous $Al(OH)_3$ flocs which are then removed by sedimentation. The sorption of phosphate ions on aluminum can occur by ligand exchange with aqua- and hydroxo-groups as shown in Figure 3.[14] The removal of P is most effective in the pH range of 6-8.[15]

Similarly, phosphate also sorbs onto iron hydroxide, ($Fe(OH)_3$), particles forming large flocs. The P removal due to sorption by $Fe(OH)_3$ is maximum at pH 5-7.[15] However, under reducing condition, i.e., lower dissolved oxygen levels, ferric ion (Fe^{3+}) becomes an alternative electron acceptor leading to the formation of more soluble ferrous iron (Fe^{2+}) and releasing Fe-bound P to the water body. Mawson et al. showed that the P flux rate from sediment increases with decreasing dissolved oxygen level.[16] It was found on a diurnal basis that P was sorbed to ferric complexes during the day and was released during the night in littoral areas of some lakes.[17] According to Caraco et al.,[18] sulfate reduction and the formation of FeS would prevent ferrous phosphate and iron oxide formation, which is based on the fact that low (60 mM) sulfate concentrations have low P release from sediment under both aerobic and anaerobic conditions, while high P results in high sulfate release.

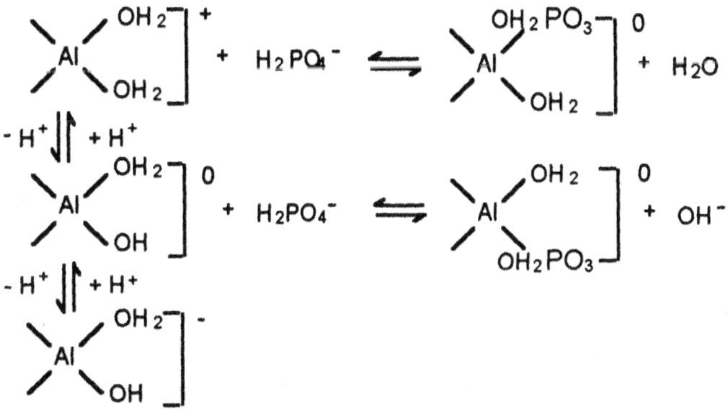

Figure 3. Aluminum hydrous oxides and P sorption.

Calcium ions in aqueous system precipitate as $CaCO_3$ (calcite) at higher pH and high concentration of carbonate ions. Calcite sorbs P and is slowly transformed to hydroxyapatite form.

$$Ca^{2+} + CO_3^{2-} \leftrightharpoons CaCO_3(s)$$

$$10\,CaCO_3 + 6\,HPO_4^{2-} + 2H_2O \leftrightharpoons Ca_{10}(PO_4)_6(OH)_2(s) + 10\,HCO_3^-$$

Hydroxyapatite is less solubile than $Fe(OH)_3$ or $Al(OH)_3$, and is thermodynamically the most stable state among various calcium phosphate phases. However, the solubility of calcite and hydroxyapatite increases with an increase in CO_2 concentration or a pH decrease which might occur, for example, due to the intense respiration in a hipolimnion or dark littoral zone.

Exchange of phosphorous between sediment and water

Figure 4 shows the factors regulating the P release from sediments[12,19] These are based on the ideas by Boström et al.[12] that two different mechanisms occur simultaneously or within a short period: A. mobilization of particulate P to the pore water and B. its transport to water body.

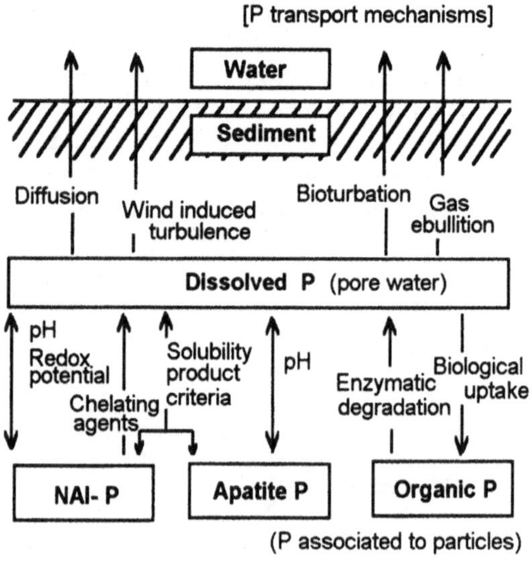

Figure 4. Factors regulating the P release from sediment.[12,19]

The pore (interstitial) water is the mobile water in sediment except for small amount of water attached to sediment and is highly important transition medium for the transport of solute species between sediment and water body. The P concentration in the pore water is roughly 3-20 times[9,12] higher than the P concentration in the overlying waters, although there are large deviations because of some lake conditions such as trophic state of lake ecosystem and temporal and spatial variation.[20] The P in pore water can be transported from the sediment to water body via diffusion, wind-induced sediment resuspension, bioturbation and gas ebullition. Particulate P (NAI-P, apatite P, organic P) in sediment is mobilized in pore water via desorption, dissolution due to the pH, redox potential changes, ligand exchange and enzymatic degradation. Additionally, direct release of particulate P in top-sediment, mainly Fe-bound P into water body, may occur under reductive condition and pH change.[20]

CONTROL OF PHOSPHOROUS RELEASE

Sediment Removal

Sediment removal has several advantages including lake deepening, limiting nutrient recycling, reducing macrophyte nuisances, and removing toxic substances. A sediment removal project can be very effective if it can be performed without resuspension of sediment and with proper disposal planning of dredged materials.

Dredging depth depends on the purpose of the project and characteristics of the lake sediments. For example, in Lake Trummen, Sweden, dredging depth was limited to top-50 cm because the P and N content in that sediment was very high compared to that under 50 cm. The P concentration in water body after dredging was dramatically decreased to about one-tenth, suggesting that P release from sediments had a major effect.[21,22] Despite the dramatic results obtainable with dredging, this method suffers from the disadvantage that it is more expensive than chemical treatment involving the use of such chemicals as aluminum sulfate.[23]

Another problem which must be addressed with dredging is resuspension of sediments which might result in liberation of nutrients. These nutrients, with higher concentrations, are contained in pore waters of the eutrophic area and agitation by dredging tends to move nutrients in sediments into the eutrophic zone of the lake, creating the potential for algal blooms.

Aeration or Circulation

Two techniques, namely, artificial circulation and hypolimnetic aeration are currently recommended as convenient and inexpensive lake management techniques.[24] One of the main objectives of aeration or circulation is to recover oxygen depletion, viz., to prevent the persistence of anaerobic bottom water and subsequently the release of nutrients such as phosphorous and nitrogen.

Regarding artificial circulation, better results have been obtained in terms of improved levels of DO (dissolved oxygen), Fe and Mn content, ammonium, and pH. The control of water flow rate and also the use of compressed air or a pump, are critical for complete mixing and preventing micro-stratification and lower DO in the lower water. This technique was effective for most lakes and reservoirs.

However, some problems have been encountered with this technique. According to the investigation of American Water Works Association, aeration caused an increase in turbidity and the appearance of water bloom in 46% of treated reservoirs.[25]

If the water column is sufficiently deep, hypolimnetic aeration will provide increased DO, decreased iron, manganese and ammonium, and often moderately decreased P concentration in the hypolimnion. In this case, the technique increases DO only in the lower water and does not break temperature stratification. Soltero et al. examined the effect of partial lift and full lift aerators for hypolimnetic aeration and they indicated that partial lift aerator was effective for reduction of P release in spite of no measurable increase in dissolved oxygen.[26] In another study the addition of iron along with aeration showed better results in terms of decreasing P release from sediment.[27]

Use of Chemical Reagents

The P precipitation by the addition of aluminum salts is effective between pH 6 to 8 provided aluminum is present as $Al(OH)_3$. The P precipitates complexed with aluminum are stable and do not dissolve into water body under anaerobic condition. $Al(OH)_3$ floc is also effective for entrapment of P from sediments. However, there are some toxicity issues related to the use of aluminum salts Kennedy et al. suggested that the maximum Al dose should be decided from the concentration of dissolved aluminum ions after scattering.[28] The appropriate concentration they determined was less than 50 μg Al/ℓ, based on studies of rainbow trout the result that rainbow trout did not show any obvious changes in the behavior or physiological activity with chronic exposure of 52 μgAl/ℓ.

A different technique using chemical reagents was developed by Ripl in order to reduce internal P loading from anaerobic sediment.[29] The purpose was not direct removal of phosphorous by instead to oxidize the top 15-20 cm of anaerobic lake sediments and subsequently to suppress P release from sediments. It is based on the observation that iron redox reaction is a controlling factor for P interchange between sediment and underlying water. The technique is as follows: a solution of $Ca(NO_3)_2$ is injected into the sediment to stimulate denitrification, which depletes the organic matter in sediments, and thereby restores an oxidized state. Nitrate is preferred as an electron acceptor as it penetrates into the sediment more readily and to a greater degree than added oxygen. In this treatment process, ferric chloride ($FeCl_3$) and lime ($Ca(OH)_2$) are added initially for removal of hydrogen sulfide as FeS, binding of P with ferric hydroxide ($Fe(OH)_3$) in interstitial water, and pH control for microbial denitrification. The addition of the reagents may not be necessary in some cases, depending on the sediment conditions. Since the redox potential of nitrate reduction is higher than that of iron reduction, the latter would be inhibited and therefore the release of Fe-bound P would be suppressed.[30] Anderson showed that the higher concentration of NO_x-N in underlying water correlated well with the suppression of the P release from sediments based on the data of 31 lakes in Denmark.[31]

Use of Solid Reagents and Wastes

The use of solid reagents and wastes has been examined for suppression of eutrophication which is caused by nutrients, especially phosphate. The removal of phosphate has been carried out by surface fixation on the solids and has been applied to synthesized wastewaters, secondary effluents, and natural waters under various conditions. The solids used include granular activated magnesia clinker (95% MgO, CaO),[32] half-burned dolomite (MgO, $CaCO_3$),[33] red mud activated (Al_2O_3, $F_{e2}O_3$, SiO_2, $N_{a2}O$; waste product of aluminum industry),[34] magnesium silicate,[35] aluminum oxide,[36] fly ash,[37-39] and slag.[40-44] In the following section, the application of solid wastes to lake water and seawater is described.

As an alternative of aluminum salts to remove phosphorous in lake or reservoir waters, the use of fly ash, an economical absorbent, was examined by Theis et al.[37] Fly ash (20g/ℓ) was added for controlling P release from the sediment followed by the addition of CaO (4g/ℓ) to provide buffer action to the system. Because fly ash is an effective absorbent and having a large surface area (4,500 to 5,500 cm^2 cm^{-3}) its addition reduces the P concentration from 255 to 150 mg/ℓ. Additionally, laboratory experiments also indicated that the P release rate of treated sediment was half of that observed for untreated sediments. Unfortunately, concentration of heavy metals, Fe, Mn, Zn, and Ni increased, due to slight solubility of fly ash in water. Guthrie et al. reported bioaccumulation of heavy metals in organisms of several trophic levels.[38] Theis et al pointed out that use of fly ash may pose some potential hazards to lakes and reservoirs in terms of high pH of treated waters, dissolved oxygen depletion, biological reduction of sulfate to sulfide, release of heavy metals, and clogging of biota or biotic activities etc.[39] These results suggest that a moritorium on the use of fly ashes to lakes and reservoirs in order to control P release from sediments should be put in place until laboratory and field studies are completed on these problems.

Slag is a by-product in iron and steel production. Approximately 300 kg and 130 kg of furnace slag and iron slag (converter slag and electric furnace slag) are generated in the production of 1 ton pig iron and crude iron, respectively. Both slag materials contain CaO and SiO_2 as its main components as well as Al_2O_3, MgO (furnace slag) and FeO, MgO, and MnO (iron slag). Slag output in Japan is about 37 million ton per year. Yamada et al. examined the possibility of phosphorous removal in seawater by using furnace slags with different crevice rate (particle size range, 63-2830 mm).[42,43] Figure 5 shows the suppression effect for P release by furnace slag under anaerobic conditions. For control without capping materials, the liberation of phosphorous increased rapidly with time. On the other hand, the capping by river-sand and sea-sands resulted in the decrease in P release and the use of slags almost completely suppressed the P release. The difference between the various slags was attributed to the surface area. The suppression by slags was thus attributed to capping effect (50%), adsorption (40%), and others (10%). From x-ray microanalysis, the site of P adsorption was the region where elements such as Ca, Mg, Al, and Si, but not Fe, predominated. Although there is no data on the influence to the ecosystem by scattering slags, the results indicated that the concentration of Pb, Cu, Cd, Zn, and total Cr in slag was lower than that in the earth's crust or in the sediment in the coastal area like Tokyo Bay. This is due to the fact that most of heavy metals can be removed in the formation process of slag.

The P adsorption of $1 cm^3$ converter slag was examined on the bottom sediments in eutrophicated inland sea for 2 years.[44] The P_2O_5 content (3.0%) in the region containing CaO and SiO_2 of the slag increased to 6.0% (surface) and 5.2% (5 mm deep), suggesting the possibility of the use of larger solids.

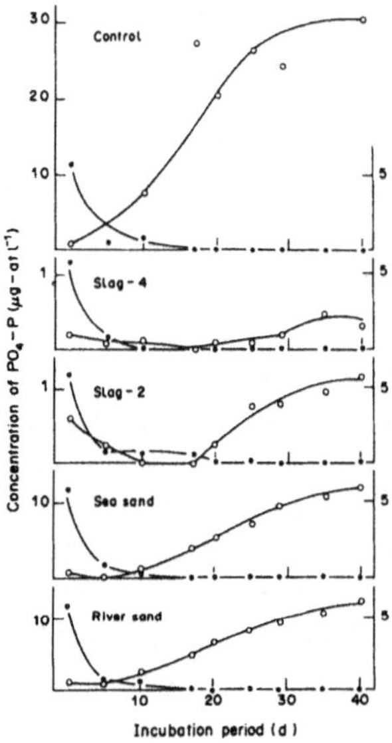

Figure 5. The change in dissolved oxygen and phosphate-P concentration in sea water under aerobic conditions (250 g material, 2.5 kg sediment, 3.2 ℓ sea water.[43]

CONCLUSION

The phosphorous species in sediments do not have discrete chemical structures. The P release from sediments, especially top sediment is mainly from non-apatite inorganic P under anaerobic condition. The sediment removal (i.e., dredging) is the most effective technique, but is expensive. Thus, it is necessary to increase the dissolved oxygen in underlying water by using artificial circulation (by compressed air or a pump) or hypolimnetic aeration. A more inexpensive procedure is to use chemical reagents, especially solid wastes. However, the influence on ecosystem of water bodies must be studied for problems such as dissolution of toxic heavy metals and the bioaccumulation of toxic metals in fishes and benthic organisms before using solid wastes.

REFERENCES

1. Edmondson WT, Lehman JT. The effects of changes in the nutrient income on the condition of Lake Washington. Limnol Oceanogr 1981; 26:1-29.

2. Cooke GD, McComas MR, Walter DW, Kennedy RH. The occurrence of internal phosphorous loading in two small, eutrophic glacial lakes in Northeastern Ohio. Hydrobiologia 1977;56:129-135.

3. Chang SC, Jackson ML. Fractionation of soil phosphorous, Soil Sci 1957;

4. Williams JDH, Syers JK, Armstrong DE, Harris F. Fractionation of inorganic phosphate in calcareous lake sediments. Soil Sci Soc Amer Proc

5. Williams JDH, Jaquet J-M, Tohmas RL. Forms of Phosphorous in surficialsediments of Lake Erie. J Fish Res Bd Can 1976;33:413-429.

6. Hieltjes AHM, Lijklema L. Fractionation of iorganic phosphates in calcareous sediments. J Environ Qual 1980;9:405-407.

7. Psenner R, Pucsko R, Sager M. Fraktionierung organischer und anorganisher phisphorverbindungen von sedimenten. Versuch einer definition okologisch wichtiger fraktionen. Arch Hydrobiol/Suppl 1985;70:111-115.

8. Petterson K, Boström B, Jacobsen O-S. Phosphorous in sediment and analysis. Hydrobiologia 1988;170:91-101.

9. Istvanovics V. Fractional composition, adsorption and release of sediment phosphorous in the Kis-Balaton reservir. Wat Res 1994;28:717-726.

10. Furumai H, Ohgaki S. Adsorption-desorption of phosphorous by lake sedimentsunder anaerobic conditions. Wat Res 1989;23:677-683.

11. Boström B, Persson G, Broberg B. Bioavailability of different phosphorous forms in freshwater systems. Hydrobiol 1988;170:133-155.

12. Boström B, Jansson M, Forsberg C. Phopphorus release from sediments. Arch Hydrobiol Beih Ergebn Limnol 1982;18:5-59.

13. Moore Jr. PA, Reddy KR, Graetz DA. Phosphorous release from sediments. Arch Hydrobiol Beih Ergebn Limnol 1982;18:5-59.

14. Berkheiser VE, Street JJ, Rao PSC, Yuan TL. Patitioning of inorganic phosphate in soil-water systems. CRC Crit Rev Envir Control 1980;10:179-

15. Anderson JM. Influence of pH on release of phosphorous from lake sediments. Arch Hydrobiol 1975;76:411-419.

16. Mawson SJ, Gibbons Jr HL, Funk WH, Hartz KE. Phosphorous flux rates in lake sediments. J Wat Pollut Control Fed 1983;55:1105-1110.

17. Carton RG, Wetzel RG. Phosphorous flux from lake sediment: effect of epipelic algal oxygen production. Limnol Oceanogr 1980;33:562-570.

18. Caraco NF, Cole TT, Likens GE. Evidence for sulfate-controlled phosphorous release from sediments of aquatic systems. Nature 1989;341:316-318.

19. Håkansson L, Jansson M. *Princeple of Lake Sedimentology* p.316. Berlin: Springer-Verlag, 1983.

20. Enell M, Löfgren S. Phosphorous in interstitial water: methods and dynamics. Hydrobiologia 1988;170:103-132.

21. Bengtsson L, Gelin C. Articitial aeration and suction dredging methods for controlling water quality. Proc, Symp. Effects of storage on water quality. Water Research Center, Medmenham, 1975;314-342.

22. Cronberg G, Gelin C, Larsson K. The lake Trummen restration project.II.: bacteria, phytoplankton and phytoplankton-producbility. Verh Int Ver Limnol 1975;19:1088-1096.

23. Cooke GD, Welch EB, Peterson SA, Newyork PR. "Sediment removal" p.522-525 In Resoration and Management of Lakesand Reservoirs, 2nd ed. Lewis Publishers, 1993.

24. Cooke GD, Welch EB, Peterson SA, Newyork PR. "Hypolimnetic aeration" and "Artificial circulation" In Resoration and Management of Lakesand Reservoirs, 2nd ed. Levis Publishers, 1993.

25. American Water Works Association. Artificial destratification in reservoirs, Committee Report, J Amer Wat Works Assoc 1971;63:597-604.

26. Soltero RA, Sexton LM, Ashley KI, McKee KO. Partial and full lift hypolimnetic aeration of Medial Lake, WA to improve water quality. Wat Res 1994;28:2297-2308.

27. Björk. *Scandinavian lake restoration activities* in Lake Pollution and Recovery. Int Comp European Water Pollut Cont Assoc Found 1985,293-301.

28. Kennedy RH, Cooke GD. Control of lake phosphorous with alminum sulfate: dose determination and application techniques. Wat Res Bull 1982;18:389- 395.

29. Ripl W. Biochemical oxidation of polluted lake sediment with nitrate: a new restoration†method. Ambio 1976;5:132.

30. Foy RH. Suppressions of Phophorus release from lake sediments by the addition of nitrate. Wat Res 1986;20:1345-1351.

31. Anderson JM. Effect of nitrate concentration in lake water on phophorus release from the sediment. Wat Res 1982;16:1119-1126.

32. Kaneko S, Kakajima K. Phosphorous removal by crystallization using a granular activated magnesia clinker. J Wat Pollut Control Fed 1988;60:1239-1244.

33. Roques H, Nugroho-Jeudy L, Lebugle A. Phosphorous removal from wastewater by half-burned dolomite. Wat Res 1991;25:959-965.

34. Shiao SJ, Akashi.K. Phosphate removal from aqueous solution from activated fly ash. J Wat Pollut Control Fed 1977;49:280-285.

35. Smith RW, Hwang M-Y. Phosphate adsorption of magensium silicates. J Wat Pollut Control Fed 1978;50:2189-2197.

36. Huang CP. Removal of phosphate by powdered aluminum oxide adsorption. J Wat Pollut Control Fed 1977;49:1811-1817.

37. Theis TL, Greene RW, Sturm TW, Spencer DF, McCabe PJ, Higgins BP, Yeung H, Irvine RL. Treatment of Lake Charles East, Indiana, sediments with fly ash. 1979, EPA-600/3-79-060.

38. Guthrie RK, Cherry DS. Trophic level accumulation of heavy metals in a coal ash drainage basin. Wat Res Bull 1979;15:244-248.

39. Theis TL, Depinto JV. Studies on the reclamation of stone lake, Michigan, 1976, EPA-600/3-76-106.

40. Sunahara H, Xie W, Kayama M. Phosphate removal by column packed blast furnace†slag.I: fundamental research by synthetic wastewater. Environ Technol 1987;8:589-598.

41. Xie W, Zhang X, Kitaide T, Sunahara H. Phosphate removal by column packed blast furnace slag.II: practical application of secondary effluent. Environ Technol 1987;8:589-598.

42. Yamada H, Kayama M, Saito K, Hara M. A fundamental research on phosphate removal by using slag. Wat Res 1986;20:547-557.

43. Yamada H, Kayama M, Saito K, Hara M. Suppression of phosphate liberation from sediment by using iron slag. Wat Res 1987;21:325 333.

44. Ito K, Nishijima W, Shoto E, Okada M. Control of phosphorous release from coastal bottom sediments by converter slag: laboratory and field experiments. J Jap Soc Water Environ 1996;19:501-507.

22

MECHANISTIC ASPECTS OF CALCIUM PHOSPHONATES PRECIPITATION

Lynn M. Frostman, Amy T. Kan, and Mason B. Tomson

Department of Environmental Science and Engineering
Rice University, Houston, Texas 77005 USA

INTRODUCTION

Organic phosphonates play a role in a wide range of water treatment processes. In particular, phosphonate inhibitors are commonly used to prevent the formation of mineral scale in brine during production of gas and oil. The phosphonate inhibitors work by adsorbing on extremely small crystallites of scale, thus preventing them from growing large enough to become stable in solution and causing the crystallites to re-dissolve. These inhibitors are often referred to as threshold inhibitors because they are effective at very low concentrations, typically less than a few ppm. Phosphonate inhibitors are generally "squeezed" into a reservoir by pumping the inhibitors into the well and shutting-in the well for one to two days. During this time, the inhibitor precipitates and / or adsorbs on the mineral surfaces. When the well is returned to production, the inhibitor slowly dissolves and / or desorbs into the produced brine. As long as the inhibitor concentration in the brine is sufficiently high, scaling in prevented.

The overall objective of this work is to predict inhibitor retention and release in reservoirs and improve squeeze lifetimes for NORM (Naturally-Occurring Radioactive Material) and calcite scale control. In particular, an understanding of the important mechanisms for inhibitor fate and transport in reservoirs is needed. While adsorption and mass transfer are most certainly important, this work focuses on precipitation as the primary mechanism controlling inhibitor retention and release. In fact, since materials which would strongly adsorb would also tend to precipitate, and it is very difficult to distinguish between the two mechanisms, it is most likely that either mechanism could be used to describe the system, regardless of which mechanism was actually controlling the system.[1] Precipitation could then be combined with the appropriate mass transfer model to account for the reservoir system.

Previous studies have suggested that the Ca-inhibitor phases are much less soluble than the salts of other divalent cations.[2,3] Hence the Ca-inhibitor precipitates are the focus of this study. However, Fe(II)-inhibitor salts may also be important is gas and oil production systems, as some evidence suggests these salts are slightly less soluble than the corresponding Ca salts.[3] Experiments are currently in progress to study Fe(II)-inhibitor salts; results will be reported at a later date.

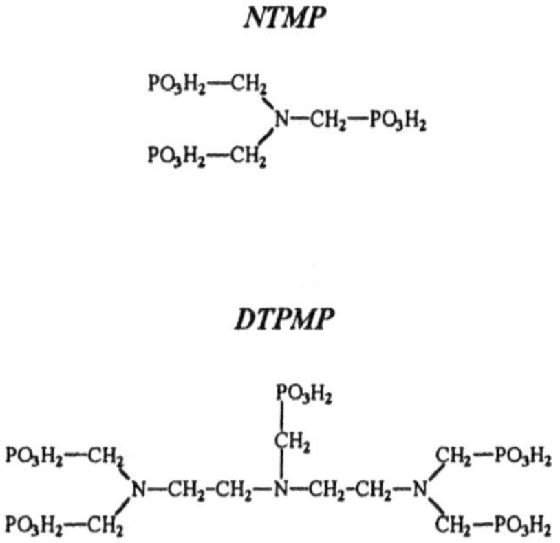

Figure 1. Structures of inhibitors used in this study.

The two phosphonate inhibitors investigated in this study were NTMP [nitrilotri(methylene phosphonic acid)] and DTPMP [diethylenetriamine penta(methylene phosphonic acid)] (see Figure 1). Some data for DTPMP was reported in previous publications[4,5,6] and is included here for comparison with NTMP. These inhibitors were chosen based on their common use in gas and oil wells, and their general interest to prominent gas, oil, and service companies. Each of these inhibitors forms an unstable initial precipitate which is either amorphous or microcrystalline and exhibits a fairly high solubility in brine. However, after extended diafiltration with fresh brine, these solids transform to more crystalline materials which have significantly lower solubilities in brine. The precipitates prior to diafiltration will be referred to as the initial phases while those after diafiltration will be referred to as the crystalline phases. The solubilities of both types of precipitates are presented, along with some data concerning the transformation between phases. These results are then compared against a variety of field observations.

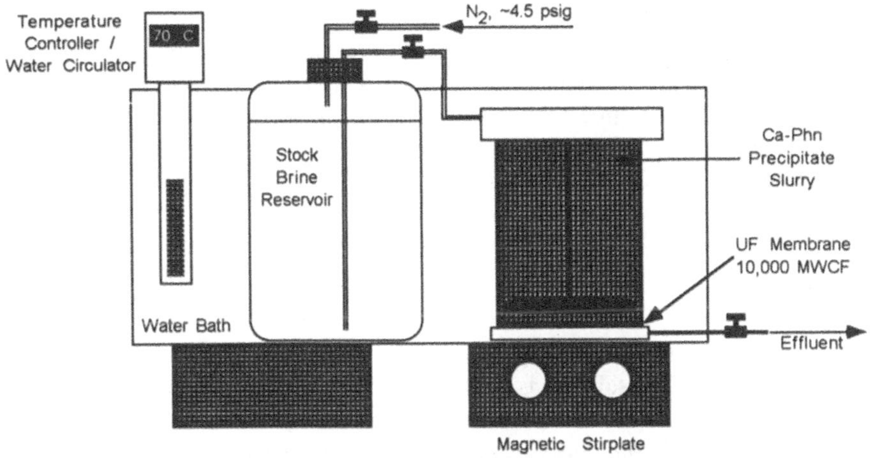

Figure 2. Diafiltration apparatus used to "mature" calcium phosphonate precipitates. As shown, nitrogen pressure is used to force fresh brine past the precipitate slurry. Alternatively, a piston pump can be used to produce the appropriate flow rate

EXPERIMENTAL

Materials

Commercial grade phosphonate inhibitors were used as received from the manufacturers: NTMP = 50 wt % solution from Monsanto (Dequest 2000), DTPMP = 50 wt % solution from Monsanto (Dequest 2060). Water was deionized using a U.S. Filter purification system. All salts were reagent grade or better.

Apparatus

The precipitations are performed in a 1000 mℓ jacketed beaker fitted with a Teflon lid and situated on an insulated magnetic stirplate. The beaker is connected to a temperature controlled circulating bath (VWR 1135) that allows the temperature of the beaker to be controlled to ±0.5°C. The pH is monitored with a Ross combination pH electrode (Orion #815600) and a Fisher Scientific Accumet 950 pH meter. Maturation experiments are performed using a diafiltration apparatus, comprising an Amicon diafiltration cell (either 50 mℓ or 350 mℓ volume) fitted with an Amicon YM 10 membrane (molecular weight cut-off = 10,000 MW), a 2 ℓ reagent bottle with several inlets and outlets, a heated water bath, and a magnetic stirrer (see Figure 2). All connectors, tubing, and valves are Teflon. Either ultra-high purity nitrogen (~4.5 psig) or a Pharmacia LKB piston pump (model P-500) are used to produce a constant flow rate of stock brine, typically 70-90 mℓ/hr. Solubilities are determined in Nalgene bottles held at a constant temperature.

Solution Preparation

Stock brines are prepared with deionized water to be 0.8 molal NaCl, 0.1 molal $CaCl_2 \cdot 2H_2O$, and 0.001 molal $NaC_2H_3O_2 \cdot 3H_2O$ (buffer). The pH is adjusted to 5.5 with 1N HCl. The final ionic strength is 1.1 molal. Stock inhibitor solution is prepared by diluting commercial grade inhibitor with deionized water and neutralizing to pH 6.8 with NaOH. The final concentration is ~200 mg phosphonate/ℓ.

Precipitation

At the beginning of each experiment, 750 mℓ of stock brine are added to the sample beaker. The solution is heated to 70 °C and degassed with nitrogen for at least 15 minutes with constant stirring. The pH electrode is then calibrated with phosphate (pH~7) and phthalate (pH~4) buffers that have also been preheated and degassed. Phosphonate and calcium (1 molal) stock solutions are alternately added in small (0.25-0.50 mℓ) increments using 2 mℓ Gilmont syringes. The addition of calcium serves to help maintain a constant calcium concentration during the precipitation. The calcium phosphonate salt immediately precipitates in the brine. The pH of the solution is periodically monitored and the pH is adjusted back to 5.5 if necessary with 1 N HCl or 1.6 N NaOH. By the end of the experiment, a total of 7.500 mℓ of phosphonate stock solution and 3.750 mℓ of calcium stock solution have been added. The solution is then allowed to stir continuously for an hour. A portion of the resulting slurry is transferred to the diafiltration cell while the remainder is used to determine amorphous solubilities.

Diafiltration

The water bath and stock brine are preheated to 70 °C. The diafiltration cell is filled with the initial precipitate slurry, sealed, connected to the stock brine reagent bottle, and immersed in the constant temperature bath. Flow is initiated either using a piston pump or pressure from an ultra-high purity nitrogen supply. The effluent from the diafiltration cell is periodically sampled to track the change in phosphonate concentration.

Solubility and Stoichiometry

Solubilities are determined by diluting some of the precipitate slurry in a Nalgene bottle, adjusting the ionic strength to the desired level, and holding the solution at the desired temperature for 24 hours. The final phosphonate and calcium concentrations, along with the pH, are then measured. Samples of each solid are washed alternately with borax buffer and deionized water to remove all excess calcium from solution. The solid is then dissolved in dilute HCl and the resulting phosphonate and calcium concentrations are measured to determine the solid stoichiometry.

Analytical Techniques

Phosphonate concentrations were determined by first oxidizing the phosphonate using potassium persulfate and UV light, then using the Hach "yellow" (molybdovanadate) or "blue" (ascorbic acid) method, diluting the samples as necessary.[7] Calcium concentrations were determined by EDTA titration or ICP analysis.[8] X-ray powder patterns were collected on a Philips diffractometer and SEM micrographs were obtained using a Cameca SX-50 Electron Microprobe.

DATA ANALYSIS

The solubility of a calcium phosphonate precipitate is given by the dissolution reaction:

$$Ca_xH_{n-2x}Phn \rightarrow xCa^{2+} + (n-2x)H^+ + Phn^{n-} \qquad (1)$$

where *Phn* is the phosphonate ligand (either DTPMP or NTMP in this study), *n* is the number of ionizable protons on the phosphonate group ($n = 10$ for DTPMP and 6 for NTMP), and *x* is the stoichiometric coefficient. The corresponding solubility product is then:

$$K_{sp} = (Ca^{2+})^x \{H^+\}^{(n-2x)} (Phn^{n-}) \qquad (2)$$

where parentheses represent concentrations in molality (m), $\{H^+\}$ refers to the proton activity in solution and is given by $\{H^+\} = 10^{-pH}$, the superscripts inside the parentheses or brackets refer to the charges on the ions, and the superscripts outside the parenthesis refer to the power to which that concentration is raised. Thus, in order to calculate the solubility products for the calcium phosphonate precipitates, it is necessary first to determine the concentrations of free calcium ions (Ca^{2+}) and free phosphonate ions (Phn^{n-}) in solution. These can be calculated from the total solution phase calcium and phosphonate concentrations using a speciation model, such as the one presented in earlier papers.[5,9] It should be noted, however, that the choice of speciation models will affect the resulting Ksp values and so when using the Ksp values presented herein, one must also use the speciation model presented herein.

Detailed descriptions of the speciation model used in this study are presented elsewhere.[5,9] Only a brief summary is included here. It has been previously shown that the first proton of NTMP and the first two protons of DTPMP associate with the basic nitrogens.[10,11] The corresponding association constants have also been

reported: $\log K_1^N$ (NTMP) = 12.3,[10] $\log K_1^N$ (DTPMP) = 12.58, and $\log K_2^N$ (DTPMP) = 11.18.[9]
The model chosen considers only electrostatic effects and assumes that the protonation constants are a linear function of charge. For an arbitrary phosphonic acid with n ionizable protons, H_nPhn, the general equilibrium equation is given by:

$$H_{i-1}Phn^{(n-i+1)-} + H^+ \leftrightarrow H_iPhn^{(n-i)-} \tag{3}$$

$$K_i = \frac{(H_iPhn^{(n-i)-})}{\{H^+\}(H_{i-1}Phn^{(n-i+1)-})} \tag{4}$$

and the equilibrium constant is modeled as:

$$\log K_i = a_{H^+} + b_{H^+}|q_{i-1}| \quad \text{for } i \geq nN + 1 \tag{5}$$

where parentheses represent concentrations in molality (m), the superscripts refer to the charge on the molecule or ion, n is the number of ionizable protons on the phosphonate, nN is the number of protons which associate with nitrogens ($nN = 1$ for NTMP and 2 for DTPMP), a_{H+} and b_{H+} are speciation constants determined from the experimental titration data, and $|q_{i-1}|$ is the absolute value of the charge on the (i-1)th species ($|q_{i-1}| = n-i+1$).

A similar procedure is used to calculate the stability constants for calcium-phosphonate complexes. It is assumed that the protons associated with nitrogens are not displaced by calcium. The speciation constants are then given by:

$$Ca_{j-1}H_iPhn^{(n-i-2j+2)-} + Ca^{2+} \leftrightarrow Ca_jH_iPhn^{(n-i-2j)-} \tag{6}$$

$$K_{ij} = \frac{(Ca_jH_iPhn^{(n-i-2j)-})}{(Ca^{2+})(Ca_{j-1}H_iPhn^{(n-i-2j+2)-})} \tag{7}$$

and the equilibrium constant is modeled as:

$$\log K_{ij} = a_{Ca^{2+}} + b_{Ca^{2+}}|q_{2j-2+i}| \quad \text{for } i \geq nN + 1, j \geq 1 \tag{8}$$

Analysis of the titration data showed that including the a_{Ca2+} term did not significantly improve the model fit to the data for either NTMP or DTPMP, so only the b_{Ca2+} term was used. Similarly, the a_{H+} term for DTPMP was not statistically different from zero and so was not included. The parameters a_{H+}, and b_{H+}, and b_{Ca2+} are given in Table 1 for DTPMP and NTMP. The speciation constants K_i and K_{ij} can then be determined using the above equations.

Table 1. Speciation constants for DTPMP and NTMP.

Parameter	DTPMP[5]	NTMP
a_{H+}	-	$2.296-.567\sqrt{IS}+.184IS-314/T$
b_{H+}	$1.040-0.0047\sqrt{IS}-5.97/T$	$1.439-.160\sqrt{IS}+.0255IS-54.3/T$
b_{Ca2+}	$1.109+0.021\sqrt{IS}-176/T$	$1.569-.606\sqrt{IS}+.201IS-206/T$

where: IS = ionic strength (molal) and T = temperature (K)

RESULTS

Batch precipitation was used to produce the initial Ca-inhibitor salts. The solubility of the resulting salts was then determined as a function of ionic strength and temperature using batch experiments. A sample of the initial precipitate was also transferred to a diafiltration cell where it was continually flushed with fresh brine for extended periods of time. The resulting matured solid was also studied by batch experiments to determine solubility as a function of ionic strength and temperature. The stoichiometry and solubility for both phases of each inhibitor are listed in Table 2. Figure 3 shows the pKsp of Ca-NTMP precipitates as a function of ionic strength and temperature, before and after diafiltration. Curve fits to the pKsp data are also given in Table 2.

Analyses of the resulting solids by X-ray diffraction and SEM (data not shown) were also determined. The X-ray patterns of the initial Ca-inhibitor phases lack coherent peaks, suggesting that the materials are amorphous. This is also consistent with the SEM picture for the initial phase of Ca-DTPMP; however, the SEM of the initial phase of Ca-NTMP suggests it may be microcrystalline with large aggregates of ~1 micron particles. The X-ray patterns suggest that after diafiltration, the Ca-NTMP and Ca-DTPMP precipitate phases are highly crystalline. This is confirmed by the SEM photographs, which show the Ca-NTMP to be ~30 x 70 micron rice-shaped particles composed of a multitude of plate shaped crystals, and Ca-DTPMP to be aggregates of rod-shaped crystals.

Table 2. Stoichiometry and solubility of ca-inhibitor precipitates.

Precipitate Phase	Stoichiometry	Solubility: 70 °C, pH ~ 5.5, 1.1 m IS, 0.1 m Ca^{2+}	pKsp
NTMP			
Initial Precipitate	$Ca_{2.5}HNTMP$	285 mg/ℓ	$32.96 - 6.88\sqrt{IS} + 2.17IS - 2380/T$
After Diafiltration	Ca_3NTMP	0.6 mg/ℓ	$26.67 - 4.46\sqrt{IS} + 1.48IS - 1850/T$
DTPMP[4]			
Initial Precipitate	Ca_3H_4DTPMP	193 mg/ℓ	~ 51
After Diafiltration	Ca_3H_4DTPMP	2.3 mg/ℓ	$58.95 + 0.048\sqrt{IS} - 2084.5/T$

where: IS = ionic strength (m) and T = temperature (K)

The maturation process was monitored by measuring the effluent phosphonate concentration during the flow-through experiment (Figure 4). It is interesting to note that Ca-NTMP matures significantly faster than Ca-DTPMP. However, these curves should not be interpreted as representing the thermodynamic solubility as a function of time, as subsequent batch dissolution experiments indicate the actual solubility is slightly higher than is indicated in the flow-through experiments.

Since these systems are inherently metastable, the reported solubilities must be interpreted cautiously. The solubilities of the initial precipitates are very sensitive to preparation and measurement protocols. Also, while the diafiltration process produces phases which are more stable than the initial precipitates, the matured phases may still not be the thermodynamically most stable phases. In fact, the solubility measurements of the matured phases did depend somewhat on the measurement technique, most notably the time allowed for the solubility measurement. For instance, the solubility of the matured Ca_3NTMP at 70°C, 1.1 molal ionic strength, 0.1 molal Ca^{2+}, and 5.5 pH varied between 0.6 and 3 mg/ℓ, depending on the time allowed for dissolution. This could be due to several factors. The precipitate may still be undergoing a phase transformation to a more stable and less-soluble form. Since the solubility experiments are performed in closed containers, this would cause the phosphonate concentration in solution, and thus the apparent solubility, to rise with time. Similarly, it is possible that occluded impurities are slowly diffusing out of the precipitate and dissolving, again leading to a rise in solution phosphonate concentration. Finally, the precipitates may be slow to dissolve and may not have time to reach equilibrium with the solution. The data reported here reflects a 24 hour equilibration time and is believed to be the most consistent and accurate. It should also be noted that the pH of the solutions must be carefully monitored. The Ca-inhibitor salts are known to form phases with different stoichiometries depending on the pH of the solution.[4] Consequently, if the solution pH changes during an experiment, a surface coating of a different phase from the bulk phase could form. Such surface coatings are known to persist for extended periods of time on Ca-phosphate precipitates,[12] and so would be expected to affect solubility measurements of Ca-phosphonate precipitates.

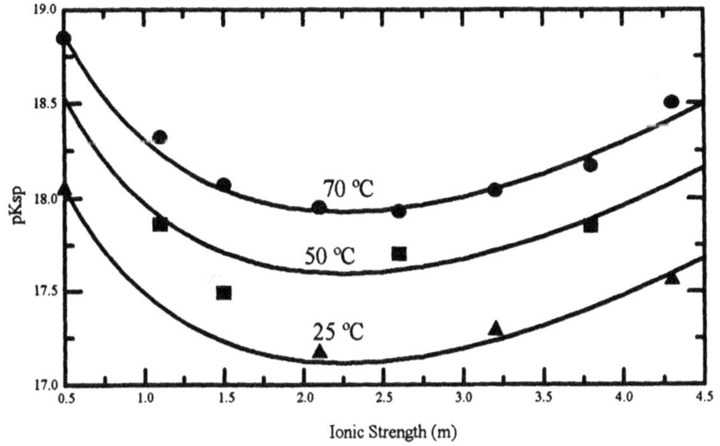

Figure 3. Solubility of Ca-NTMP at 5.5 pH. Top: initial precipitate, $Ca_{2.5}HNTMP$; Bottom: matured crystalline phase, Ca_3NTMP. Points represent actual data, while the lines represent the model fit. Note: even though the pKsp values in the top graph are larger than those in the lower plot, they represent higher solubilities due to the different stoichiometries.

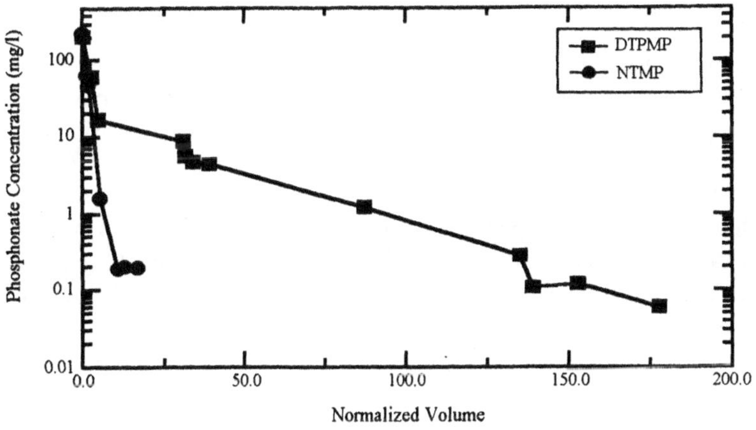

Figure 4. Maturation of Ca-inhibitor precipitates during diafiltration experiments, where the normalized volume is the volume of liquid passed through the diafiltration cell divided by the diafiltration cell volume (350 mℓ for NTMP and 50 mℓ for DTPMP).

In general, the initial precipitates are 1-3 orders of magnitude more soluble than the corresponding matured crystalline phases. As is typical for many calcium salts, the solubility increases as the temperature decreases. The solubilities of Ca-NTMP and Ca-DTPMP precipitates show a slight dependence on ionic strength. Both Ca-NTMP precipitates show a maximum in solubility around an ionic strength of 2-2.5 molal, while the maximum solubility of the crystalline Ca-DTPMP occurs between 0.5-1 molal. However, in the region of interest for most gas and oil wells, ~0.7-2.5 molal ionic strength, the measured solubilities exhibit virtually no dependence on ionic strength. Conversely, the temperature dependence is expected to be more significant. In the typical range of downhole temperatures in gas and oil wells, ~50-200 °C, the pKsp of a Ca-inhibitor precipitate can vary on the order of 1-2 log units.

DISCUSSION

This study employed commercial grade inhibitors, rather than the purified inhibitors typically used in previous studies. The commercial formulations contain numerous impurities –including phosphoric acid, phosphorous acid, and various other starting materials and intermediates – which will affect and may actually dictate (see below) the measured solubilities of the Ca-inhibitor salts. However, it was decided not to purify the commercially available formulations since it is these impure formulations which are actually used in the field. The results thus represent "effective" solubilities, but in practice are more applicable to field work than the corresponding data for the pure components.

Calcium phosphates exhibit complicated precipitation and dissolution chemistries, including the formation of multiple phases, persistence of thermodynamically unstable phases for extended times, development of non-stoichiometric phases due to incorporation of impurities and lattice defects, and formation of surface phases which differ in composition from the bulk phase. As a consequence, the measured solubilities of calcium phosphates often represent the kinetics of the system rather than a thermodynamic equilibrium.[12] Though the data is far less extensive for calcium phosphonates, it appears these materials exhibit a similar richness in solid-state chemistry.

Table 3. Inhibitor precipitation - survey of solubilities reported in the literature.

Source	Precipitate	Conditions	Cphn
Oddo & Tomson[6]	Ca-DTPMP	2 M IS, 70 °C, pH 5	660 mg/ℓ
Samakaev, *et al.*[2]	Ca-NTMP	Water, 75 °C	53 mg/ℓ
Browning & Fogler[13]	Ca-HEDP[a] (prepared at pH 1.5)	DI Water, 25 °C, pH 6.5	1380 ppm
	Ca-HEDP[a] (prepared at pH 6)	DI Water, 25 °C, pH 6.5	12 ppm
Ebrahimpour, *et al.*[14]	Ca-Risedronate[b]	0.15 M IS, 37 °C, pH 7	4.95 mg/ℓ
	Ca-3-pic AMBP[c]	0.15 M IS, 37 °C, pH 7	9.87 mg/ℓ
	Ca-Alendronate[d]	0.15 M IS, 37 °C, pH 7	0.84 mg/ℓ
	Ca-3-pic AMPMP[e]	0.15 M IS, 37 °C, pH 7	7.39 mg/ℓ

Notes: a) HEDP = 1-hydroxyethylidene-1,1-diphosphonic acid
b) Risedronate = 2-(3-pyridyl)ethylidene-1,1-hydroxy bisphosphonic acid
c) 3-pic AMBP is a bisphosphonic acid (see reference for structure)
d) Alendronate = (4-amino-1-hydroxybutylidene) bisphosphonic acid
e) 3-pic AMPMP is a phosphonoalkylphosphinate (see reference for structure)

Previous studies had produced much inconsistent data concerning the solubility of Ca-inhibitor precipitates. Some previously reported solubilities are given in Table 3. In addition, the resulting inhibitor solubilities were too high to account for inhibitor return concentrations obtained in the field, which are usually less than 1 mg/l. The work by Kan *et al.*[4] served to elucidate the inherent reason for these difficulties: the initial precipitates of Ca-DTPMP salts are metastable in brines and slowly transform to lower-solubility phases. Hence the solubility and stoichiometry are expected to depend strongly on the method of preparation and history of each precipitate. This current study serves to confirm that additional Ca-inhibitor salts exhibit multiple phases of varying stability.

Table 4. Estimated return concentrations based on solubility data.

Well[a]	Ca (mg/l)	T (F)	TDS[b] (mg/l)	pH	Calculated Final Concentration (mg/ℓ) based on NTMP	Calculated Final Concentration (mg/ℓ) based on DTPMP	Observed (mg/ℓ)
Gladys McCall	4130	298	96340	5.71	0.31	0.13	0.10-0.14
Pleasant Bayou	9600	305	130000	6.03	0.04	0.18	0.07-0.15
Aramco Well A	4000	178	51833	5.72	0.47	1.15	0.20-3.0
NRSmith	480	160	50899	6.69	0.59	3.1	0.35-0.50
AEGuerra	6500	340	70908	6.18	0.08	0.16	0.41-1.1
Huff A	448	210	42831	6.33	0.91	0.91	0.60-1.6
O'Daniel	760	285	43624	5.39	23	0.36	0.80-1.2
Aramco Well B	10960	178	128830	5.61	0.15	1.46	1.1-3.0
N. Riley #302	3800	118	54365	5.34	2.8	4.4	~13
High Island	1230	282	68050	5.27	32	0.38	~19

Notes: a) all wells were squeezed with NTMP except A.E. Guerra and High Island, which were squeezed with BHPMP = bis-hexamethylenetriamine penta(methylene phosphonic acid); the calculated return concentrations are adjusted to account for the actual inhibitor used
b) TDS = total dissolved solids

Ongoing work suggests that the calcium salt of bis-hexamethylenetriamine penta(methylene phosphonic acid) behaves similarly.[15] In addition, preliminary evidence suggests the same occurs for the polymeric inhibitors.[16]

While it is clear that a transformation between phases takes place during diafiltration, the mechanism of this transformation remains uncertain. The observations are consistent with the Ostwald theory of reaction steps, which suggests that the first phase to form should be the least stable, most disordered, and most soluble phase.[17] Subsequent transformation to more stable, more crystalline, and less soluble phases may take place through a series of intermediate phases. These transformations may occur in the form of solid-state phase transitions or by dissolution of the initial phases and nucleation of new phases. The kinetics of these transformations can be exceedingly slow. Another hypothesis which has been suggested is that the initial precipitate includes all of the impurities in the commercial inhibitor cocktail, and the subsequent maturation process is simply the dissolution of the impurities followed by the recrystallization of the pure Ca-inhibitor precipitate.[18]

Since the diafiltration process is similar to the flow of brine past precipitated inhibitor in a reservoir, the results of these experiments are expected to be applicable to field data. In particular, the solubilities of crystalline Ca_3NTMP and Ca_3H_4DTPMP appear to be comparable to inhibitor return concentrations observed in the field. To test the applicability of these results, the correlations in Table 2 were used to predict the inhibitor return concentrations in ten gas and oil wells with which the authors have been involved. The predictions are calculated by assuming the reservoir brine at downhole conditions is in equilibrium with the crystalline inhibitor precipitate. The resulting concentrations are then converted to reflect the inhibitor actually used in the well, if necessary. Table 4 shows the conditions of each well, along with the observed return concentration and the predicted concentrations based on the Ca_3NTMP and Ca_3H_4DTPMP data. These wells represent a range of conditions typically observed in gas and oil wells. In general, predictions based on either NTMP or DTPMP give good estimates for most of the observed return concentrations, especially when the analytical uncertainties in the field data are considered. The good agreement between observed and predicted concentrations suggests that the solubilities of matured crystalline Ca-inhibitor precipitates can be used to model the return of inhibitors in real gas and oil wells.

CONCLUSIONS

The solubilities of the calcium precipitates of two commonly used scale inhibitors (NTMP [nitrilotri(methylene phosphonic acid)] and DTPMP [diethylenetriamine penta(methylene phosphonic acid)]) were determined. These salts exhibit multiple phases and interesting kinetics of phase transformation, dissolution, and mass transfer. Each of these materials forms an unstable initial precipitate, which upon undergoing extended diafiltration with fresh brine, transforms into a more stable and more crystalline phase. The solubilities of both the initial precipitate and the matured material are presented as a function of ionic strength and temperature. The results suggest that the return of inhibitors from real gas and oil well can be predicted based on the solubility of matured crystalline Ca-inhibitor precipitates.

ACKNOWLEDGMENTS

This work was supported by members of the Rice University Brine Chemistry Consortium – Amoco, Aramco, Champion Technologies, Chevron, Conoco, Petrolite, Texaco, Unichem, and UNOCAL. Portions of this work were also supported by the Gas Research Institute under contract number 5095-250-3352, but in no way does this constitute an endorsement by GRI of any products or views contained herein. The authors also acknowledge Milton Pierson, Department of Geology, Rice University, for his help with the x-ray and SEM analyses.

REFERENCES

1. Sposito G. *The Surface Chemistry of Soils*; Oxford University Press: New York, 1984.

2. Samakaev RK, Dyatlova NM, Dytyuk LT. Solubility of group II element nitrilomethylene phosphonates in water. Russ. J. Inorg. Chem. 1984,29,819-1820

3. Salimi MH, Tomson MB. *unpublished results*

4. Kan AT, Oddo JE, Tomson MB. Formation of 2 calcium diethylenetriaminepentakis (methylene phosphomnic acid) orecipitates and their physical properties. Langmuir 1994,10,1450-1455

5. Tomson M, Kan AT, Oddo JE, Gerbino AJ. "Solution and Precipitation Chemistry of Phosphonate Scale Inhibitors"; in *Mineral Scale Formation and Inhibition,* Z. Amjad, Ed.; Plenum Press: New York, 1995, pp 307-322.

6. Oddo JE, Tomson MB. The solubility and stoichiometry of calcium diethylenetriaminepenta (methylene phosphonate) at 70 degrees C in brine solutions at 4.7 and 5.0 pH. Appl. Geochem. 1990 5, 527-532

7. *Water Analysis Handbook*; Hach Company: Loveland, CO, 1989.

8. *Standard Methods for the Examination of Water and Wastewater*; 18th ed.; Greenberg, A. E.; Clesceri, L. S.; Eaton, A. D., Ed.; American Public Health Association: Washingtion, D. C., 1992.

9. Tomson MB, Kan AT, Oddo JE. Acid-base and metal-complex solution shemistry of the phosphonate DTPMP versus temperature and ionic strength. Langmuir 1994, 10, 1442-1449

10. Carter RP, Carroll RL, Irani RR. Nitrilotri (methylene phosphonic acid), ethyliminodi (methylene phosphonic acid), and diethylaminomethyl phosphonic acid: acidity and calcium (II) and magnesium (II) complexing. Inorg. Chem. 1967, 6, 939-942

11. Tikhonova LI. Complex formation of diethylenetriamine-N,N,N',-N',N''-pentakis (methyl phosphonic acid) with some alkaling earth and rare earth elements. Russ. J. Inorg. Chem. 1968, 13, 1384-1388

12. Nancollas GH. "The Nucleation and Growth of Phosphate Minerals"; in *Phosphate Minerals,* J.O. Nriagu and P. B. Moore, Ed.; Springer-Verlag: Berlin, 1984, pp 137-154.

13. Browning FH, Fogler HS. Effect of precipitation conditions on the formation of calcium-HEDP precipitates. Langmuir 1995, 11, 4143-4152

14. Ebrahimpour A, Ebetino FH, Sethuraman G, Nancollas GH. "Determination of Solubility and Calcium Ion Stability Constants of a Phosphonoalkylphosphinate (PAP) and Bisphosphonates (BPs) such as EHDP, Risedronate, Alendronate, 3-pic AMBP, and 3-pic AMPMP" ; in *Mineral Scale Formation and Inhibition,* Z. Amjad, Ed.; Plenum Press: New York, 1995, pp 295-305.

15. Frostman LM, Kan AT, Tomson MB, *in preparation*

16. Xiao J, Kan AT, Tomson MB, *in preparation*

17. Morse JW, Casey WH. Ostwald processes and mineral paragenesis in sediments. Amer. J. Sc. 1988, 288, 537-560

18. Gerbino AJ. *Quantifying the Retention and Release of Polyphosphonates in Oil and Gas Producing Formations using Surface Complexation and Precipitation Theory*; Ph.D. Thesis; Rice University: Houston, 1996.

INDEX

- C -

- G -

- H -

- I -

- K -

- L -

- M -

- N -

- O -

- T -

GPSR Compliance

The European Union's (EU) General Product Safety Regulation (GPSR) is a set of rules that requires consumer products to be safe and our obligations to ensure this.

If you have any concerns about our products, you can contact us on ProductSafety@springernature.com

In case Publisher is established outside the EU, the EU authorized representative is:

Springer Nature Customer Service Center GmbH
Europaplatz 3
69115 Heidelberg, Germany

Batch number: 10761599

Printed by Printforce, the Netherlands